THE ILLUSTRATED

PETROLEUM
REFERENCE
DICTIONARY

SECOND EDITION

THE ILLUSTRATED
PETROLEUM REFERENCE DICTIONARY

SECOND EDITION

EDITED BY

ROBERT D. LANGENKAMP

PennWell Books
PennWell Publishing Company
Tulsa, Oklahoma

Copyright © 1980, 1982 by
PennWell Publishing Company
1421 South Sheridan Road/P.O. Box 1260
Tulsa, Oklahoma 74101

Library of Congress Cataloging in Publication Data
Main entry under title:

The Illustrated petroleum reference dictionary.

   1. Petroleum—Dictionaries.   I. Langenkamp, R. D.
TN865.I43   1982   553.2'8'0321   82-412
ISBN 0-87814-160-X   AACR2

Library of Congress Catalog Card Number: 78-71326
International Standard Book Number: 0-87814-160-X
Printed in the United States of America

2 3 4 5   86 85 84 83

CONTENTS

FOREWORD

A petroleum reference dictionary with thousands of entries and hundreds of illustrations can be more than a collection of words and pictures. Rightly, it should be a rich and informative distillation of the history of the industry. From the expressions and the descriptions of tools, jobs, and processes, one should be able to trace the growth of oil from a "cottage industry" to one of the giants of the industrial world whose activities affect nearly everyone.

It all began 120 years ago, in 1859, when Colonel Edwin Drake drilled his now historic well near Titusville, Pennsylvania. Since then every person who ever drilled for oil must have felt some mystical kinship with the world's first and most famous "wildcatter" who "kicked down" his well to a depth of 69½ ft and found oil.

Following Drake's discovery there was a frantic rush to get in on the "play." The new "oil men" were hastily converted farmers, tradesmen, blacksmiths, and draymen plus a liberal sprinkling of gamblers and adventurers looking for quick fortunes.

From this mixed bag of "diggers" working along the creeks and in the hollows of western Pennsylvania and West Virginia came an expressive and colorful vocabulary describing the still-unfamiliar business of prospecting for oil with improvised tools.

By the turn of the century, the industry had gathered some momentum, and a measure of know-how and had spread into the Southwest, Texas, Oklahoma, and then California. Cable tool rigs, steam boilers and engines, wooden derricks, and horses and mules were still doing the work and providing the bulk of the transportation. Among those whose vernacular and idiomatic phrases few non-oil people understood were the drillers, "roughnecks," tool dressers, mule skinners, well shooters, pipeliners, and tankies. These and others of the hardy breed, in greasy overalls and with a quid of tobacco in the jaw, made up the rough and profane world of the boom town.

During the ensuing decades, the search for "black gold" intensified and grew progressively more sophisticated. Operators relied more on scientific methods than "creekology" as more advanced tools and instruments were developed. The rotary rig was gradually displacing the lower and less efficient cable tool rigs; steam power was giving way to the diesel and gas engine.

The petroleum geologist, the geophysicist, the well logger took a lot of the guesswork out of exploration, drilling, and well completion. But even today there are many questions still unanswered concerning where and how.

Beginning in the late 1930s the search for more and cheaper oil to keep pace with the nation's ever-increasing demands sent the oil men overseas where they were rewarded by the discovery of huge new reserves of oil. As a result we have words relating to oil in a global context, reflecting the industry's multi-national character. Terms such as *consortium, concession, fixer's fee, participation, buy-back oil,* and the portentous acronym *OPEC* are a few of the new additions to oil's ever-expanding vocabulary.

Yes, the petroleum industry has assumed an international character—its exploration activities are global in scope, its advanced technology has an influence worldwide. As a prime energy supplier, it is a dynamic force in the Free World. Today, oil men and women are at work on many frontiers, each as demanding in energy and innovation as any encountered in the industry's 120-year history. The physical frontiers of deep water offshore, the Arctic, the deserts, and remote, uncharted jungles—environments as varied as the world's geography affords—are awesome and call for good measures of courage and fortitude.

Equally demanding are the scientific and technological frontiers, where efforts are concentrated in developing synthetic fuels, recovering more oil from known reservoirs, and furthering the development of exploration techniques for new discoveries deeper in the earth. As even the casual observer is aware, these physical and technological challenges are being met head-on as notable advances are being made in all segments of the industry. And with these advances come words to describe new concepts, new processes, new tools, and new procedures.

To make note of and record what is new in the growing lexicon of oil, the second edition of the *Illustrated Petroleum Reference Dictionary* has been revised and considerably enlarged. There are now more than 3,000 entries, a sizable increase in the number of definitions and descriptions, plus hundreds of additional illustrations to aid in understanding and appreciating the colorful and expressive language of the oil patch. Also included are Steven Gerolde's *Universal Conversion Factors* and the Desk & Derrick Club's Abbreviator, bringing together all that is venerable and historic plus the newest in oil terminology and measurement.

The author, with more than 30 years in the business, hopes that this book will prove useful as well as interesting to those in the oil business, to writers, students, lawyers, and investors, as well as members of the public who would like to learn more about an industry that, along with automobile, has affected their lives profoundly.

Robert D. Langenkamp

A

AAODC
American Association of Oilwell Drilling Contractors.

AAPG
American Association of Petroleum Geologists.

ABANDONED OIL
Oil permitted to escape from storage tanks or pipeline by an operator. If the operator makes no effort to recover the oil, the land owner on whose property the oil has run may trap the oil for his own use.

ABANDONED WELL
A well no longer in use; a dry hole that, in most states, must be properly plugged.

ABSOLUTE ALCOHOL
One hundred percent ethyl alcohol.

ABSORPTION OIL
An oil with a high affinity for light hydrocarbons but containing few if any of the light compounds composing gasoline or natural gas. The oil used in an absorption plant (q.v.).

ABSORPTION PLANT
An oil field facility that removes liquid hydrocarbons from natural gas, especially casinghead gas. The gas is run through oil of a proper character which absorbs the liquid components of the gas. The liquids are then recovered from the oil by distillation.

ABSORPTION TOWER
A bubble tower; a tall cylindrical column in which absorption of a gas by a liquid occurs. A liquid hydrocarbon fraction is piped to the top of the tower and moves to the bottom over baffles or horizontal trays through which a gas is rising from the bottom. As the liquid percolates downward making contact with the gas, a quantity of the gas is absorbed by the liquid.

ACCELERATED AGING TEST
A procedure whereby an oil product is subjected to intensified but controlled conditions of heat, pressure, radiation, or other variables to produce, in a short time, the effects of long-time storage or use under normal conditions.

ACCUMULATOR
A small tank or vessel to hold air or liquid under pressure for use in a hydraulic or air-actuated system. Accumulators, in effect, store a source of pressure for use at a regulated rate in mechanisms or equipment in a plant or in drilling or production operations.

ACCUMULATOR SYSTEM
A hydraulic system designed to provide power to all closure elements of the rig's blowout preventer stack. Hydraulic oil is forced into one or more vessels by a high-pressure, small-volume pump and its charge of inert

Absorption tower

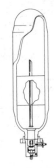
Accumulator bottle

Accumulator system
(Courtesy Hydril)

1

gas, usually nitrogen. The gas is compressed and stores potential energy. When the system is actuated, the oil under high pressure is released and opens or closes the valves on the BOP stack.

ACETONE
A flammable, liquid compound used widely in industry as a solvent for many organic substances.

ACETYLENE
A colorless, highly flammable gas with a sweetish odor; used with oxygen in oxyacetylene welding. It is produced synthetically by incomplete combustion of coal gas and also by the action of water on calcium carbide (CaC_2). Also can be made from natural gas.

ACID BOTTLE INCLINATOR
A device used in a well to determine the degree of deviation from the vertical of the well bore. The acid is used to etch a horizontal line on the container, and from the angle the line makes with the wall of the container, the angle of the well's course can be arrived at.

ACIDIZING A WELL
A technique for increasing the flow of oil from a well by the use of a quantity of acid pumped downhole and into the rock formation. Hydrochloric acid is pumped or forced under high pressure into a limestone formation which dissolves the limestone, enlarging the cavity and increasing the surface area of the hole opposite the producing formation. The high pressure of the treatment also forces the acid into cracks and fissures enlarging them and resulting in an increased flow of oil into the well bore.

ACID-RECOVERY PLANT
An auxiliary facility at some refineries where sludge acid is separated into acid oil, tar, and weak sulfuric acid. The sulfuric acid is then reconcentrated.

ACID SLUDGE
The residue left after treating petroleum with sulfuric acid for the removal of impurities. The sludge is a black, viscous substance containing the spent acid and the impurities which the acid has removed from the oil.

ACID TREATMENT
A refining process in which unfinished petroleum products such as gasoline, kerosene, diesel fuels, and lubricating stocks are treated with sulfuric acid to improve color, odor, and other properties.

ACOUSTIC PLENUM
A sound-proof room; an office or "sanctuary" aboard an offshore drilling platform protected from the noise of drilling engines and pipe handling.

ACOUSTIC REENTRY
A method used in offshore operations, particularly in deep water, for repositioning a drillship or semisubmersible drilling platform over a hole previously drilled and cased. The technique employs acoustic signals emitted by equipment onboard the ship which are "bounced off" the submerged wellhead indicating to receivers the location of the hole. In hostile environments such as a severe storm or encroaching ice, a drillship may be forced

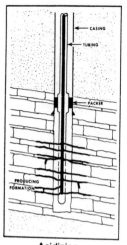

Acidizing

to move off location before the well is completed. And although the operators of the drilling vessel know the approximate location of the temporarily abandoned well, to be able to reenter the hole without diver assistance, which often is not possible, requires the pinpointing of the location by electronic means.

ACRE-FOOT OF SAND
A unit of measurement applied to petroleum reserves; an acre of producing formation one foot thick.

ACS
American Chemical Society.

ACT SYSTEM
Automatic Custody Transfer System (q.v.).

ACTUATOR
See Operator.

ACV
Air cushion vehicle. *See* Air Cushion Transport.

ADA MUD
A material which may be added to drilling mud to condition it in order to obtain satisfactory core samples.

ADAPTER
A device to provide a connection between two dissimilar parts or between similar parts of different sizes. *See* Swage.

ADDITIVE
A chemical added to oil, gasoline, or other products to enhance certain characteristics or to give them other desirable properties.

ADSORPTION
The attraction exhibited by the surface of a solid for a liquid or a gas when they are in contact.

ADVANCE PAYMENT AGREEMENT
A transaction in which one operator advances a sum of money or credit to another operator to assist in developing an oil or gas field. The agreement provides an option to the "lender" to buy a portion or all of the production resulting from the development work.

AEC
Atomic Energy Commission.

AERIFY
To change into a gaseous form; to infuse with or force air into; gasify.

AFRA
Average freight rate assessment (for tankers).

A-FRAME
A two-legged, metal or wooden support in the form of the letter "A" for hoisting or exerting a vertical pull with block and tackle or block and winch line, the block fastened to the apex of the A-frame. Such a frame fixed to

Actuator *(Courtesy Fisher Controls Co.)*

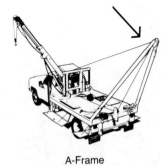
A-Frame

the end of a truck bed is a "gin pole" (q.v.). A truck-mounted A-frame is a useful item of equipment around pipeline or plant construction sites. A pump, engine, or other equipment can be lifted off the ground by the truck's winch line and gin pole and transported, dangling at the end of the line, to another location.

AGA
American Gas Association.

AGENCY CONTRACT
A type of agreement which in many cases has replaced the concession (q.v.) as the form of petroleum development agreement in the Middle East and with OPEC countries elsewhere. Under an agency contract, title to oil installations and oil produced are held by the host government, but the government bears none of the costs of initial exploration. Also, the foreign company does not have a long-term, exclusive right to exploit the minerals as is the case under a concession agreement.

AIChE
American Institute of Chemical Engineers.

AIMME
American Institute of Mining and Metallurgical Engineers.

Air-balanced pumping
unit

AIR-BALANCED BEAM PUMPING UNIT
An oil well pumping jack equipped with a piston and rod that works in an air chamber to balance the weight of the string of sucker rods. The device is attached to the well end of the walking beam and, acting as a shock absorber, does away with the need for counterweights on the rear end of the walking beam.

AIR BOTTLE
A steel cylinder of oxygen for oxyacetylene welding; an air chamber (q.v.); a "bottle" or welded steel tank with air under pressure for use in starting certain types of gas engines on leases or at pumping stations. The compressed air at a pressure of 200 pounds or so per square inch is piped from the tank to the engine's cylinderhead. To start the engine a hand-operated, quick-opening and closing valve is instantly opened admitting the high-pressure air into the firing chamber. This pushes the piston as on a power stroke. After a burst of air, the valve is shut off until the piston moves back on another compression stroke. Then the valve is opened for another shot of air. By this time the engine is rolling so the fuel (gas) valve is opened, the engine fires and begins running.

Underwater seismic
explosion, now replaced
by air bursts

AIR BURSTS
A geophysical technique used in marine seismic work in which bursts of compressed air from an air gun towed by the seismographic vessel are used to produce sound waves. Air bursts do not destroy marine life as did explosive charges.

AIR CHAMBER
A small tank or "bottle" connected to a reciprocating pump's discharge chamber or line to absorb and dampen the surges in pressure from the rhythmic pumping action. Air chambers are charged with sufficient air

pressure to provide an air cushion that minimizes the pounding and vibration associated with the pumping of fluids with plunger pumps.

AIR-COOLED ENGINE
An engine in which heat from the combustion chamber and friction is dissipated to the atmosphere through metal fins integral to the engine's cylinder head and block assemblies. The heat generated flows through the engine head and cylinder walls and into the fins by conductance and is given off by the fins acting as radiators. A small, two-cycle engine without water jacketing, water pump, or conventional radiator.

Air cushion transport
(Courtesy Bell
Aerospace Co.)

AIR CUSHION TRANSPORT
A vehicle employing the hovercraft principle of down-thrusting airstream support, developed to transport equipment and supplies in the arctic regions. The air cushion protects the tundra from being cut by the wheels or treads of conventional vehicles.

AIR DRILLING
The use of air as a drilling fluid. In certain types of formations, air drilling is considered a better medium than conventional drilling mud. It is more economical (mud is expensive and the preparation of the slurry and maintaining its condition is time consuming), drilling rates are higher, penetration is faster, and bit life is longer. Although air does a good job of cooling the bit and bringing out the pulverized rock, it has severe limitations. With air drilling, water in the subsurface formations and downhole gas pressure cannot be controlled. When drilling in an area where these two types of intrusions may occur, a mud system must be on standby to avert possible trouble.

Air drilling rig

AIRED UP
Refers to a condition in a plunger pump when the suction chamber is full of air or gas blocking the intake of oil into the chamber. Before the pump will operate efficiently, the air must be bled off and vented to the atmosphere through a bleeder line or by loosening the suction valve covers to permit the escape of the air.

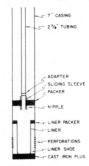

Air hoist (Courtesy
Ingersoll-Rand)

AIR HOIST
A hoist, a mechanism for lifting operated by a compressed air motor; pneumatic hoist.

AIR-INJECTION METHOD
A type of secondary recovery to increase production by forcing the oil from the reservoir into the well bore. Because of the dangers inherent in the use of air, this method is not a common practice except in areas where there is insufficient gas for repressuring.

AIR WRENCH
See Impact Wrench.

Air-injection well
(after Franco)

ALIPHATICS
One of the two classes of organic petrochemicals; the other is the aromatics (q.v.). The most important aliphatics are the gases, ethylene, butylene, acetylene, and propylene.

ALKYLATION
A refining process that, simply stated, is the reverse of cracking. The alkylation process starts with small molecules and ends up with larger ones. To a refining engineer, alkylation is the reaction of butylene or propylene with isobutane to form an isoparaffin, alkylate, a superior gasoline blending component.

ALLOWABLE
The amount of oil or gas a well or leasehold is permitted to produce under proration orders of a state regulatory body. Under a lease allowable, the lease is considered a producing unit. In some instances if a lease has eight wells, for example, and one of the eight is unable to make its production for some reason, the other seven wells can increase their flow to make up the loss of the ailing well.

ALL-THREAD NIPPLE
A short piece of small-diameter pipe with threads over its entire length; a close nipple.

ALLUVIAL FAN
Pertains to the silt, clay, sand, and other sediment deposited by a stream as it spreads out on a plain or overflows its banks and then recedes. Also the silt laid down by a tributary stream as it joins the main stream.

ALTERNATE FUELS
Fuels—gas, gasoline, heating oil—made from coal, oil shales, or tar sands by various methods. Alternate fuels may also include steam from geothermal wells where super-heated water deep in the earth is used to generate steam for electric power generation.

ALUMINUM CHLORIDE
A chemical used as a catalytic agent in oil refining and for the removal of odor and color from cracked gasoline.

AMERIPOL
The trade name for products made from a type of synthetic rubber.

AMINE
Organic bases used in refining operations to absorb acidic gases (H_2S, COS, CO_2) occurring in process streams. Two common amines are monoethanolamine (MEA) and diethanolamine (DEA).

AMINE UNIT
A natural gas treatment unit for removing contaminants—H_2S, COS, CO_2—by the use of amines (q.v.). Amine units are often skid-mounted so they can be moved to the site of new gas production. Gas containing H_2S and other impurities must be cleaned up before it is acceptable to gas transmission pipelines.

AMMONIUM SULFATE
A salt having commercial value which is obtained in the distillation of shale oils.

AMYL HYDRIDE
This fraction in the distillation of petroleum was used as an anesthetic by J. Bigelow and B. Richardson in the year 1865.

The floating drilling vessel *Discoverer*

ANCHOR BOLT
A stud bolt; a large bolt for securing an engine or other item of equipment to its foundation.

ANEMOMETER
An instrument for measuring and indicating the force or speed of the wind.

ANNULAR BLOWOUT PREVENTER
See Spherical Blowout Preventer.

ANNULAR SPACE
The space between the well's casing and the wall of the borehole. The ring of space surrounding the well's casing, defined by the wall of the borehole. *See* Annulus of a Well.

ANNULUS OF A WELL
The space between the surface casing or conductor casing and the producing or well-bore casing. Annular and annulus are often used interchangeably as both derive from the Latin stem word *annularis* which means "relating to or forming a ring."

ANODE
A block of nonferrous metal buried near a pipeline, storage tank, or other facility and connected to the structure to be protected. The anode sets up a weak electric current that flows to the structure thus reversing the flow of current associated with the corrosion of iron and steel. *See* Rectifier Bed.

ANODE, BUOYANT
A source of electric current (DC) for protecting offshore platforms and other steel structures resting on the sea floor against corrosion. The anode is anchored to the sea floor a few hundred feet away from the structure but is held off bottom by its buoyancy. The anode is connected to a source of DC current on the platform by an insulated cable. The weak current is supplied by a transformer-rectifier, the negative terminal of which is grounded to the steel structure. Thus the completion of the circuit from rectifier to anode to structure is through the sea water. The weak current moving from anode to the structure reverses the flow of current associated with the corrosion of metal. *See also* Rectifier Bed.

ANSI
American National Standards Institute.

ANTICLINAL FOLD
A subsurface formation resembling an anticline.

ANTICLINE
A subsurface geological structure in the form of a sine curve or an elongated dome. Historically this type of formation has been found favorable to the accumulation of oil and gas. The ideal location to find production is on the crest of the dome; less favorable are the flanks where one may encounter less oil and quantities of salt water which usually underlie the oil in the reservoir.

ANTIKNOCK COMPOUNDS
Certain chemicals which are added to automotive gasolines to improve

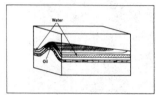

Anticline

their performance, to reduce "ping" or knock in high-compression, internal combustion engines. Tetraethyl lead is one well-known compound. The addition of small amounts of these chemicals has the effect of slowing down the explosion or burning of the air-gasoline vapor mixture in the combustion chamber, thus giving the piston a "push" instead of a sudden explosive blow. The compounds contribute also to smoother performance of the engine and more miles per gallon of fuel.

API
American Petroleum Institute; the oil industry's trade association. The API, through its research and engineering work, establishes operating and safety standards for all segments of the petroleum industry; issues specifications for manufacture of pipe, pressure vessels, and other equipment; and furnishes statistical and other information to government agencies having to do with the industry.

API logo

API BID SHEET AND WELL SPECIFICATIONS
A form many operators use in soliciting bids on a well to be drilled and completed. The form is submitted to drilling contractors in the area of the proposed well. The operator asking for bids fills out the part of the form giving the name and location of the well, commencement date, depth or formation to be drilled into, and other information. When a drilling contractor submits a bid, he lists the rig and equipment to be furnished by him: draw works, slush pumps, derrick or mast size, make, capacity, drillpipe, tool joints, etc. The bid sheet brings operator and contractor together, as it were; then they arrive at rates and other matters.

API GRAVITY
Gravity (weight per unit of volume) of crude oil or other liquid hydrocarbons as measured by a system recommended by the API. API gravity bears a relationship to true specific gravity but is more convenient to work with than the decimal fractions which would result if petroleum were expressed in specific gravity.

APPRAISAL DRILLING
Wells drilled in the vicinity of a discovery or wildcat well in order to evaluate the extent and the importance of the find.

APRON RING
The bottommost ring of steel plates in the wall of an upright, cylindrical tank.

AQUAGEL
A specially prepared bentonite (clay) widely used as a conditioning material in drilling mud.

AQUIFER
A water-bearing rock strata. In a water-drive field the aquifer is the water zone of a reservoir, underlying the oil zone. When wells are drilled in the reservoir, the water pushes the oil toward the wells' boreholes. If the wells are produced at an excessive rate the water may bypass the oil and break into the well bore, leaving much of the oil behind. *See* Channeling, *also* Water Coning.

ARBITRAGE, PRODUCT (PETROLEUM)
The buying, selling, or trading of petroleum or products in various markets to make a profit from short-term differences in prices in one market as compared to those in another. A sophisticated method of trading in world petroleum markets.

ARC WELDER
(1) An electric welding unit consisting of a gasoline engine direct connected to a DC generator. In the field the welding unit is usually skid mounted. (2) A person who uses such a machine in making electric welds. The term arc welding derives from the arc of electricity which spans the gap between the tip of the welding rod (the electrode) and the piece of metal being welded. The flow of electricity jumping the gap produces the heat to melt the rod onto the molten area of the metal.

Arc welding

AREAL GEOLOGY
The branch of geology that pertains to the distribution, position, and form of the areas of the earth's surface occupied by different types of rocks or geologic formations; also, the making of maps of such areas.

AREOMETER
An instrument for measuring the specific gravity of liquids; a hydrometer (q.v.).

ARGON
An inert, colorless, odorless gaseous element sometimes and in some locations produced with natural gas.

AROMATICS
A group of hydrocarbon fractions that forms the basis of most organic chemicals so far synthesized. The name aromatics derives from their rather pleasant odor. The unique ring structure of their carbon atoms makes it possible to transform aromatics into an almost endless number of chemicals. Benzene, toluene, and xylene are the principal aromatics and are commonly referred to as the BTX group (q.v.).

ARTIFICIAL DRIVES
Methods of producing oil from a reservoir when natural drives—gas-cap, solution gas, water drive, etc.—are not present or have been depleted. Waterflood, repressuring or recycling, and in situ combustion are examples of artificial drives.

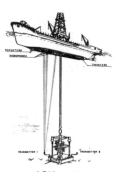

ASK system

ASK SYSTEM
Automatic station-keeping system; the name applied to a sophisticated drillship positioning technique consisting of subsurface acoustical equipment linked to shipboard computers that control ship's thrusters. The thrusters fore and aft reposition the ship, compensating for drift, wind drag, current, and wave action. *See* Dynamic Stationing.

ASME
American Society of Mechanical Engineers.

ASPHALT
A solid hydrocarbon found as a natural deposit. Crude of high asphaltic content when subjected to distillation to remove the lighter fractions such

as naphtha and kerosene leave asphalt as a residue. Asphalt is dark brown or black in color and at normal temperatures is a solid. *See* Brea.

ASPHALT-BASE CRUDE
Crude oil containing very little paraffin wax and a residue primarily asphaltic. Sulfur, oxygen, and nitrogen are often relatively high. This type crude is particularly suitable for making high-quality gasoline, lubricating oil, and asphalt. *See* Paraffin-base Crude.

ASPHALTENES
At the very bottom of the crude oil barrel are the asphaltenes, composed of complex molecules. Asphaltenes are polyaromatic compounds with high carbon—hydrogen ratios in their molecules from which asphalt is made.

ASPHALTIC PETROLEUM
Petroleum which contains sufficient amounts of asphalt in solution to make recovery commercially practical by merely distilling off the solvent oils.

ASPHALTIC SANDS
Natural mixtures of asphalts with varying proportions of loose sand. The quantity of bituminous cementing material extracted from the sand may run as high as 12 percent. This bitumen is composed of soft asphalt.

ASSEMBLY
A term to describe a number of special pieces of equipment fitted together to perform a particular function; e.g., a drill assembly may include other pieces of downhole equipment besides the drill bit, such as drill collars, damping subs, stabilizers, etc.

ASSET, WASTING
See Wasting Asset.

ASSIGNEE
A recipient of an interest in property or a contract; in oil and gas usage, the recipient of an oil or gas lease; a transferee.

ASSOCIATED GAS
Gas that occurs with oil, either as free gas or in solution. Gas occurring alone in a reservoir is unassociated gas.

ASTM
American Society for Testing and Materials.

ASTM DISTILLATION
A test of an oil's distillation properties standardized by the American Society for Testing and Materials. A sample of oil is heated in a flask; the vapors pass through a tube where they are cooled and condensed; the liquid is collected in a graduate. When the first drop of "distillate" is obtained the temperature at which this occurs is the "initial boiling point" of the oil. The test is continued until all distillable fractions have distilled over and have been measured and their properties examined.

ATMOSPHERE, ONE
The pressure of the ambient air at sea level; 14.69 pounds per square inch. Air at sea level, 29.92 inches of mercury or 33.90 feet of water.

ATMOSPHERIC STILL
A refining vessel in which crude oil is heated and product is distilled off at the pressure of one atmosphere.

ATOMIZER, FUEL OIL
A nozzle or spraying device used to break up fuel oil into a fine spray so the oil may be brought into more intimate contact with the air in the combustion chamber. *See* Ultrasonic Atomizer.

AUSTRALIAN OFFSET
A humorous reference to a well drilled miles away from proven production.

AUTOMATIC CUSTODY TRANSFER
A system of oil handling on a lease; receiving into tankage, measuring, testing, and turning into a pipeline the crude produced on a lease. Such automatic handling of oil is usually confined to leases with settled production.

AUTOMATIC TANK BATTERY
A lease tank battery (two or more tanks) equipped with automatic measuring, switching (full tank to empty and full tank into the pipeline), and recording devices. *See* Automatic Custody Transfer.

AUTOMATIC WELDING MACHINE
After two joints of pipe are joined by tack welds, automatic wire-welding machines are used to put on the filler beads.

AXLE GREASE
A cold-setting grease made of rosin oil, hydrated lime, and petroleum oils. *See* Grease.

Automatic welding machine

B

BABBITT
A soft, silver-colored metal alloy of relatively low melting point used for engine and pump bearing; an alloy containing tin, copper and antimony.

BACKFILL
To replace the earth dug from a ditch or trench; also, the earth removed from an excavation.

BACKHOE
A self-propelled ditching machine with a hydraulically operated arm equipped with a toothed shovel that scoops earth as the shovel is pulled back toward the machine.

BACK-IN FARM-OUT
A farm-out agreement (q.v.) in which a retained nonoperating interest of the lessor may be converted, at a later date, into a specified individual working interest (q.v.).

BACK-IN PROVISION
A term used to describe a provision in a farm-out agreement whereby the person granting the farm-out (the farmor) retains an option to exchange a retained override for a share of the working interest.

Backhoe *(Courtesy Robinson-Gerrard, Inc.)*

BACK OFF
To raise the drill bit off the bottom of the hole; to slack off on a cable or winch line; to unscrew.

BACK-OFF WHEEL
See Stripper Wheel.

BACK PRESSURE
The pressure against the face of the reservoir rock caused by the control valves at the wellhead, hydrostatic head of the fluid in the hole, chokes, and piping. Maintenance of back pressure reduces the pressure differential between the formation and the borehole so that oil moves into the well with a smaller pressure loss. This results in the expenditure of smaller volumes of gas from the reservoir, improves the gas-oil ratio, and ensures the recovery of more oil.

BACK-PRESSURE VALVE
A check valve (q.v.).

BACKSIDE PUMPING
See Pumping, Backside.

BACK-UP MAN
The person who holds (with a wrench) one length of pipe while another length is being screwed into or out of it.

BACKWASHING
Reversing the fluid flow through a filter to clean out sediment that has clogged the filter or reduced its efficiency. Backwashing is done on closed-system filters and on open-bed, gravity filters.

BAD OIL
Cut oil (q.v.).

BAFFLES
Plates or obstructions built into a tank or other vessel that change the direction of the flow of fluids or gases.

BAIL
To evacuate the liquid contents of a drill hole with the use of a long, cylindrical bucket (bailer).

BAIL DOWN
To reduce the level of liquid in a well bore by bailing.

BAILER
A cylindrical, bucket-like piece of equipment used in cable-tool drilling to remove mud and rock cuttings from the borehole.

BAILER DART
The protruding "tongue" of the valve on the bottom of a bailer. When the dart reaches the bottom of the hole, it is thrust upward opening the valve to admit the mud-water slurry.

BAIT BOX
A pipeliner's lunch pail.

Ball joint

Ball valve

Reel barge

BALL AND SEAT
A type of valve used in a plunger pump.

BALLING OF THE BIT
The fouling of a rotary drilling bit in sticky, gumbo-like shale which causes a serious drag on the bit and loss of circulation.

BALL JOINT
A connector in a subsea, marine riser assembly whose ball and socket design permits an angular deflection of the riser pipe caused by horizontal movement of the drillship or floating platform of 10° or so in all directions.

BALL VALVE
A type of quick-opening valve with a spherical core, a ball with a full-bore port, that fits and turns in a mating cavity in the valve body. Like plug valves, ball valves open or close by a quarter turn, 90 degrees, of the valve handle attached to the spherical core.

BAND WHEEL
In a cable tool rig, the large vertical wheel that transmits power from the drilling engine to the crank and pitman assembly that actuates the walking beam. Used in former years in drilling with cable tools. Old pumping wells still use a band wheel.

BAREFOOT CHARTER
A contract or charter agreement between the owner of a drilling rig, semi-submersible, or drillship and a second party in which the owner rents or leases his equipment (usually short-term) barefoot, i.e., without the owner or his representative taking any part in the operation or maintenance of the equipment. The lessee agrees to man or staff the equipment and operate it without assistance from or responsibility by the owner. Also bareboat charter for boats or ships.

BAREFOOT COMPLETION
Wells completed in firm sandstone or limestone that show no indication of caving or disintegrating may be finished "barefoot," i.e., without casing through the producing interval.

BARGE, REEL
See Reel Barge.

BARITE
A mineral used as weighting material in drilling mud; a material to increase the density or weight per gallon or cubic foot of mud.

BARKER
A whistle-like device attached to the exhaust pipe of a one-cylindered oil field engine so that the lease pumper can tell from a distance whether or not the engine is running. The noise the device makes resembles the bark of a hoarse fox.

BARNSDALL, WILLIAM
William Barnsdall and W. H. Abbott built the first refinery in Pennsylvania in 1860, shortly after Col. Edwin Drake discovered oil near Titusville in 1859. By the end of the Civil War there were more than 100 plants using 6,000 barrel a day. Kerosene was the main product.

BAROID
A specially processed barite (barium sulfate) to which Aquagel has been added, used as a conditioning material in drilling mud in order to obtain satisfactory cores and formation samples.

BARREL
(1) Petroleum barrel; a unit of measure for crude oil and oil products equal to 42 U.S. gallons. (2) Pump barrel; cylindrical body of an oil well pump.

BARREL HOUSE
A building on the refinery grounds where barrels are filled with various grades of lubricating and other oils, sealed, and made ready for shipment; oil house. *See* Drum.

BARREL MILE
The cost to move a barrel of oil or an equivalent amount of product one mile.

BASEMENT ROCK
Igneous or metamorphic rock lying below the sedimentary formations in the earth's crust. Basement rock does not contain petroleum deposits.

BASIC SEDIMENT
Impurities and foreign matter contained in a tank of crude oil, e.g., water-sand, oil-water emulsions. When produced, crude oil may contain one or more of these impurities. In the lease tank the impurities settle to the bottom of the tank, with the relatively clean oil on top. After repeated filling and emptying of the tank, the sediment builds up to the pipeline connection and must be removed. This is done by unbolting the plate from the cleanout box on the back of the tank and shoveling out the heavy, accumulated sediment.

BASIN
A synclinal structure in the subsurface, once the bed of a prehistoric sea. Basins, composed of sedimentary rock, are regarded as good prospects for oil exploration.

BASTARD
(1) Any nonstandard piece of equipment. (2) A kind of file. (3) A word used in grudging admiration, or as a term of opprobrium.

BATCH
A measured amount of oil or refined product in a pipeline or a tank; a shipment of oil or product by pipeline.

BATCHING SPHERE
An inflated, hard rubber sphere used in product pipelines to separate "incompatible" batches of product being pumped one behind the other. Fungible (q.v.) products are not physically separated, but gasoline is separated from diesel fuel and heating oils by batching spheres.

BATCH INTERFACE
See Interface.

Tank battery *(Courtesy Cities Services)*

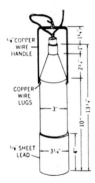

Beaker sampler

Ball bearings

BATHOLITH

A great mass of intruded igneous or metamorphosed rock found at or near the surface of the earth. The presence of a batholith, often referred to as a shield, usually precludes drilling for oil or gas as there are no sedimentary formations above it. The largest batholith in the U.S. is in Idaho, underlying nearly two-thirds of the state.

BATTERY

Two or more lease or stock tanks connected together to receive oil production from a well or a producing lease; a tank battery. The lease tank battery is the starting point for crude oil on its way to the refinery—via gathering line, pump station, and trunk line to the refinery tank farm. It is at the tank battery where the pipeline gauger measures and tests the oil, and after making out a run ticket (q.v.) opens the tank valve, turning the oil into the pipeline system. A battery is two or more units of equipment linked together in the performance of a function, e.g., separator battery; heater battery; filter battery; and tank battery.

BATTERY, TRICKLE-CHARGED

See Trickle-charged Battery.

BAUME, ANTOINE

The French chemist who devised a simple method to measure the relative weights of liquids using the hydrometer (q.v.).

BCD

Barrels per calendar day. *See* Stream Day.

BEAD

A course of molten metal laid down by a welder (electric or oxyacetylene) in joining two pieces of metal. *See* Pipeline Welding.

BEAKER SAMPLER

A metal or glass container with a small opening fitted with a stopper that is lowered into a tank of oil to obtain a sample of oil.

BEAM WELL

A well whose fluid is being lifted by rods and pump actuated by a walking beam (q.v.).

BEAN

A choke used to regulate the flow of fluid from a well.

BEAN JOINT

In early pipeline parlance, the joint of line pipe laid just before the break for lunch. When the bean joint was bucked in (q.v.) the pipeliners grabbed lunch buckets from the gang truck and found a comfortable place to eat.

BEARING, BALL

A type of revolving bearing. The other type is the roller bearing (q.v.).

BEARING, INSERT

Thin, bimetal, half-round bearings that fit in the journal box around a shaft to provide a smooth, hard surface. One-half of the insert (in cross section, a semicircle) fits into the journal box, the other half into the journal box cap. Insert bearings are designated bimetal because although the bearing sur-

face is made of babbitt (q.v.) it is backed with a layer of bronze, brass, or steel. There are also trimetal insert bearings. They are made with steel backing, a "soft" alloy middle layer, and a babbitt outer layer. Babbitt metal, an alloy of tin, copper, and antimony, was invented by Isaac Babbitt in 1862.

BEARING, OUTBOARD
A shaft-supporting bearing outside the body or frame of a pump's gear box or engine's crankcase; a bearing on a pump's pinion shaft outside the gear box; a line-shaft bearing.

BEARING, ROLLER
A type of cylindrical, revolving bearing. The other type is the ball bearing (q.v.).

BEARING, SADDLE
See Saddle Bearing.

BEARING, STIRRUP
A bearing and its frame in the shape of a saddle stirrup; e.g. the bearing connecting the pitman and the walking beam on an early-day cable tool drilling or pumping well.

BEARING, THRUST
A bearing to support the endwise or downward thrust or weight of a machinery part against another. Thrust bearings can be constructed of ball bearings or cylindrical roller bearings held in a circular frame or housing that fits over a shaft.

Roller bearing

BEHIND THE PIPE
Refers to oil and gas reservoirs penetrated or passed through by wells but never tapped or produced. Behind the pipe refers usually to tight formations of low permeability which, although recognized, were passed through because they were uneconomical to produce at the time. Today, however, with the growing scarcity of oil and high prices, many of these passed-through formations are getting a second look by producers.

BELL AND SPIGOT JOINT
A threaded pipe joint; the spigot or male end is threaded and screwed into the bell or female coupling. The female end of a coupling has threads on the inside circumference.

BELL HOLE
An excavation dug beneath a pipeline to provide room for the use of tools by workers; a hole larger in diameter at the bottom than at the top.

BELL-HOLE WELDER
A welder who can do oxyacetylene or electric welding in a bell hole. This requires a great deal of skill as the molten metal from the welding rod is being laid on upside down and tends to fall away from the weld; a skilled welder.

BELL NIPPLE
A large swage nipple for attaching casinghead fittings to the well's casing above the ground or at the surface. The bell nipple is threaded on the casing end and has a plain or weld end to take the casinghead valves.

BELT HALL
A wooden shed built to protect the wide belt that runs from the engine to the bandwheel on a cable tool rig or an old beam pumping well. The belt hall extends from the engine house to the derrick.

BENCH-SCALE TEST
Testing of methods or materials on so small a scale that it can be carried out on a laboratory table or specially constructed bench.

BENZINE
An old term for light petroleum distillates in the gasoline and naphtha range.

BENZOL
The general term which refers to commercial benzene which may contain other aromatic hydrocarbons.

BEVELING MACHINE
An oxyacetylene pipe-cutting machine. A device that holds an acetylene cutting torch so that the ends of joints of pipe may be trimmed off at an angle to the pipe's long axis. Line pipe is beveled in preparation for welding joints together.

B.H.P.
Brake horsepower (q.v.).

BIG INCH PIPELINE
A 24-inch pipeline from Longview, Texas, to Norris City, Illinois, built during World War II to meet the problem caused by tanker losses at sea as a result of submarine attacks. Later during the war the pipeline was extended to Pennsylvania. Following the war the line was sold to a private company and converted to a gas line.

BIG SPROCKET, ON THE
Said of a person who is moving in influential circles or has suddenly gone from a small job to one of considerably larger responsibility; a big operator, often used pejoratively.

BIOCHEMICAL CONVERSION
The use of bacteria to separate kerogen from oil shale. Certain bacteria will biodegrade the minerals in oil shale, releasing the kerogen from the shale in liquid or semiliquid form. (From studies made by Dr. Ten Fu Yen and Dr. Milo D. Appleman, University of Southern California, Professors of Bacteriology.)

BIOMASS
Wood and other plant materials used to make methanol as a supplement to petroleum.

BIRD CAGE
(1) To flatten and spread the strands of a cable or wire rope. (2) The slatted or mesh-enclosed cage used to hoist workmen from crew boats to offshore platforms.

Bird cage

BIRD DOG
To pay close attention to a job or to follow a person closely with the intent to learn or to help; to follow up on a job until finished.

BIT
The cutting or pulverizing tool or head attached to the drillpipe in boring a hole in underground formations. *See* Drill Bit.

BITUMEN, OIL SANDS
A heavy petroleum-like substance found in certain consolidated sand formations at the surface of the earth or at relatively shallow depths where it can be surface mined after the removal of a few feet of overburden. The extraction process is complicated, but basically it involves the heating of the oil sands to separate the oil. The oil is floated off and undergoes treatment before it is piped to a refinery.

BITUMINOUS SAND
Tar sand; a mixture of asphalt and loose sand which, when processed, may yield as much as 12 percent asphalt.

BLACK OILS
(1) A term denoting residual oil; oil used in ships' boilers or in large heating or generating plants; bunkers. (2) Black-colored oil used for lubricating heavy, slow-moving machinery where the use of higher-grade lubes would be impractical.

BLEED
To draw off a liquid or gas slowly. To reduce pressure by allowing fluid or gas to escape slowly; to vent the air from a pump.

BLEEDER VALVE
A small valve on a pipeline, pump, or tank from which samples are drawn or to vent air or oil; sample valve.

BLEEDING
The tendency of a liquid component to separate from a lubricant, as oil from a grease; to seep out.

BLEEDING CORE
A core sample of rock highly saturated and of such good permeability that oil drips from the core.

BLEED LINE
A line on the wellhead or blowout preventer stack through which gas pressure can be bled to prevent a threatened blowout.

BLENDING
The process of mixing two or more oils having different properties to obtain a lubricating oil of intermediate or desired properties. Certain classes of lube oils are blended to a specified viscosity. Other products, notably gasolines, are also blended to obtain desired properties.

BLIND FLANGE
A companion flange with a disc bolted to one end to seal off a section of pipe.

A drill bit *(Courtesy Reed Tool Co.)*

BLM
Bureau of Land Management.

BLOCK
(1) A pulley or sheave in a rigid frame. (2) To prevent the flow of liquid or gas in a line. (3) A chock.

BLOCK AND BLEED VALVE
A heavy-duty mainline valve made to hold "bubbletight" against high pressure. The valve is made with a small bleeder line and valve which are tapped into the block valve's bonnet. When the block valve is closed, its effectiveness may be checked by opening the bleeder valve for evidence of any leakage from the upstream or high-pressure side.

BLOCK AND TACKLE
An arrangement of ropes and blocks (pulleys) used to hoist or pull.

BLOCK GREASE
A grease of high melting point that can be handled in block or stick form. Block grease is used on large, slow-moving machinery, on axles and crude bearings. In contact with a hot journal bearing, the grease melts slowly lubricating the bearing.

BLOCKING
Pumping measured amounts of crude oil or product through a pipeline in batches or blocks. For example when moving different grades or types of crude, e.g., sweet or sour, the two are not mixed but are blocked through the line, one behind the other. And when each arrives at destination, it is switched into separate tankage. In a packed or full line pumping under high pressure, very little mixing of the two types of crude occurs at the interface, the point in the line where the two are in contact.

Block valve

BLOCK LEASE
A lease executed by owners of separate tracts or, sometimes, separate leases executed by owners of individual tracts which provide that drilling of one or more test wells within the combined area or block will satisfy the conditions of the lease as to each of the tracts in the block.

BLOCK TREE
A type of well-completion Christmas tree in which a number of control and production valves are made as a unit in one block of steel. Valve pockets for the special valve assembly are bored in the steel forging which makes the valve assembly a strong, rigid unit integral to the forging. Block trees are often used on multiwell offshore platforms to conserve space.

BLOCK VALVE
A large, heavy-duty valve on a crude oil or products trunk line placed on each side of a pipeline river crossing to isolate possible leaks at the crossing.

Blooie pipe

BLOOIE PIPE
A horizontal vent pipe extending from the wellhead a couple of hundred feet from the rig floor to a burn pit. The blooie pipe, named for the noise it makes, vents the returns during air or gas drilling. In air drilling no mud is used; the pulverized rock from the action of the bit is brought up from the

bottom of the hole by compressed air and blown through the blooie pipe into the burn pit. Should gas or oil be encountered, it too is vented to the burn pit. Should the well need to be controlled because of oil or gas in quantity and under high pressure, the well must be mudded up. Drilling mud is pumped into the hole and circulated as in conventional rotary drilling.

BLOOM
The irridescent cast of color in lubricating oil.

BLOWBY
The escape of combustion or unburned fuel past the engine's piston and piston rings into the crankcase. Blowby occurs during the power stroke but unburned fuel can also escape during the compression stroke on spark-ignition engines.

BLOWDOWN
The venting of pressure in a vessel or pipeline; the emptying of a refinery vessel by relieving pressure at a discharge valve to direct the contents into another vessel or to the atmosphere.

BLOWDOWN STACK
A vent or stack into which the contents of a processing unit are emptied when an emergency arises. Steam is injected into the stack to prevent ignition of volatile material or a water quench is sometimes used.

BLOWING A WELL
Opening a well to let it blow for a short period to free the well tubing or casing of accumulations of water, sand, or other deposits.

BLOWING THE DRIP
To open the valve on a drip (q.v.) to drain off the "drip gasoline" and to allow the natural gas to "blow" for a moment to clear the line and drip of all liquid.

BLOWOUT
Out of control gas and/or oil pressure erupting from a well being drilled; a dangerous, uncontrolled eruption of gas and oil from a well; a wild well.

BLOWOUT PREVENTER
A stack or an assembly of heavy-duty valves attached to the top of the casing to control well pressure; a Christmas tree (q.v.).

BLOWOUT PREVENTER, SPHERICAL
See Spherical Blowout Preventer.

BLOWPIPE (WELDING AND CUTTING)
See Welding Torch.

BLUE SKY LAW
A statute which regulates the issuance and sale of securities. The term usually is restricted to state statutes. The corresponding federal statutes and regulations are the Federal Securities Act and the Securities and Exchange (SEC) regulations. States differ in subjecting the sale of property interests in oil and gas to Blue Sky regulations.

Blowout

Blowout preventer

Spherical blowout preventer *(Courtesy Hydril)*

BNOC
British National Oil Corporation (q.v.).

BOBTAIL
A short-bodied truck.

BOBTAIL PLANT
A gas plant which extracts liquid hydrocarbons from natural gas but does not break down the liquid product into its separate components. The liquid instead is pumped to a fractionator plant where it is separated into various components or fractions in fractionator towers.

BODY
Colloquial term for the viscosity of an oil.

BOGIES
Colloquial term for small transport dollies. A low, sturdy frame or small platform with multiple wheels (4 to 8) for moving heavy objects short distances.

BOILER HOUSE
(1) A lightly constructed building to house steam boilers. (2) To make a report without doing the work; to fake a report.

BOILING POINT, INITIAL
The temperature at which a product being distilled comes to a boiling point; the beginning of the distillation of a particular product; the temperature at which this occurs.

BOIL OFF
The vaporization or gasification of liquefied natural gas (LNG) or other gases liquefied by applying high pressure and severe cooling. Boiloff occurs when the holding vessel's insulation fails to maintain the low temperature required to keep the gas in liquid form. Boiloff is a problem for shippers of LNG in the specially built ocean carriers.

BOLL WEEVIL
An inexperienced worker or "green hand" on a drilling crew.

BOLSTER
A support on a truck bed used for hauling pipe. The heavy wooden or metal beam rests on a pin that allows the forward end of the load to pivot as the truck turns a corner.

BONNET
The upper part of a gate valve that encloses the packing gland and supports the valve stem.

BONUS
Usually, the bonus is the money paid by the lessee for the execution of an oil and gas lease by the landowner. Another form is called an oil or royalty bonus. This may be in the form of an overriding royalty reserved to the landowner in addition to the usual one-eighth royalty.

BONUS BIDDING
Competitive bidding for oil and gas leases in which the lease providing for a

fixed royalty is offered to the prospective lessee offering to pay the largest bonus to the lessor. *See* Royalty Bidding.

BOOK VALUE
The worth of an oil company; properties and all facilities, less depreciation.

BOOM
A beam extending out from a fixed foundation or structure for lifting or hoisting; a movable arm with a pulley and cable at the outer end for hoisting or exerting tension on an object. *See* Boomcats.

BOOMCATS
Caterpillar tractors equipped with side booms and winches; used in pipeline construction to lift joints of pipe and to lower sections of the line into the ditch.

A boomcat at work

BOOMER
(1) A link and lever mechanism used to tighten a chain or cable holding a load of pipe or other material. (2) A worker who moves from one job to another.

BOOSTER STATION
A pipeline pumping station usually on a main line or trunk line; an intermediate station; a field station that pumps into a tank farm or main station.

BOOT
A tall section of 12 or 14-inch pipe used as a surge column at a lease tank battery, downstream of the oil/gas separator. The column, 15 to 25 feet high, provides a hydrostatic head (q.v.) and permits the escape of gas which prevents gas or vapor locks in the pipeline gathering system.

BOP STACK
Blowout preventer stack (q.v.).

BORE AND STROKE
See Pump Specifications.

BOP stack *(Courtesy Atlantic-Richfield Co.)*

BOREHOLE
The hole in the earth made by the drill; the uncased drill hole from the surface to the bottom of the well.

BORING MACHINE
A power-driven, large-diameter auger used to bore under roads, railroads, and canals for the purpose of installing casing or steel conduits to hold a pipeline.

BOTTLENECKING
The deformation of the ends of the casing or tubing in the hanger resulting from excessive weight of the string of pipe and the squeezing action of the slips.

BOTTOM FRACTION
The last cut; the "bottom of the barrel" in petroleum distillation.

BOTTOM-HOLE ASSEMBLY (BHA)
A drilling string comprised of a drillbit and several drill collars is a simple

Boring a road crossing

bottom-hole assembly. Such an assembly may also include a bottom-hole reamer above the bit or above the first drill collar. When in addition to drill collars and reamers there are two or three stabilizers in the string, it is referred to as packed-hole assembly (q.v.). The main purpose of a packed-hole assembly is to keep the bit drilling as straight down as possible.

BOTTOM-HOLE CHOKE
A device placed at the bottom of the tubing to restrict the flow of oil or regulate the gas-oil ratio.

BOTTOM-HOLE HEATER
Equipment used in the bottom of the well bore to increase bottom-hole temperature in an effort to increase the recovery of low gravity or heavy oil.

BOTTOM-HOLE LETTER
An agreement by which an operator, planning to drill a well on his own land, secures the promise from another to contribute to the cost of the well. In contrast to a dry-hole letter, the former requires payment upon completion of the well whether it produces or not. A bottom-hole letter is often used by the operator as security for obtaining a loan to finance the drilling of the well.

BOTTOM-HOLE PRESSURE
The reservoir or formation pressure at the bottom of the hole. If measured under flowing conditions, readings are usually taken at different rates of flow in order to arrive at a maximum productivity rate. A decline in pressure indicates the amount of depletion from the reservoir.

BOTTOM-HOLE PUMP
A pump located in the bottom of the well and not operated by sucker rods and surface power units. Bottom-hole pumps are compact, high-volume units driven by an electric motor or hydraulically operated.

BOTTOM OUT
To reach total depth, to drill to a specified depth.

BOTTOM WATER
Free water in a permeable reservoir rock beneath the space in the reservoir trap that contains oil and gas. If the water zone underlies the entire reservoir it is called bottom water; if it occurs at the sides of the reservoir only it is referred to as edge water.

BOURDON TUBE
A small, crescent-shaped tube closed at one end, connected to a source of gas pressure at the other, used in pressure recording devices or in pilot-operated control mechanisms. With increases in gas pressure the Bourdon tube flexes (attempts to straighten) and this movement, through proper linkage, actuates recording instruments.

Bourdon tube

BOWL
A device that fits in the rotary table and holds the wedges or slips that support a string of tubing or casing.

BOWLINE
A knot used to form a loop in a rope which will neither slip nor jam.

BOX AND PIN JOINT
A type of screw coupling used to connect sucker rods and drill pipe. The box is a thick-walled collar with threads on the inside; the pin is threaded on the outer circumference and is screwed into the box.

BOYLE'S LAW
"The volume of any weight of gas is inversely proportional to the absolute pressure, provided the temperature remains constant."

BRADENHEAD GAS
Casinghead gas. Bradenhead was an early-day name for the wellhead or casinghead.

BRAKE HORSEPOWER (B.H.P.)
The power developed by an engine as measured at the drive shaft; the actual or delivered horsepower as contrasted to "indicated horsepower" (q.v.).

BRASS POUNDER
A telegrapher, especially one who uses a telegraph key (q.v.). *See* Telegrapher's Bug. Until the 1940s or so, much of the communication from oil patch to division and head offices was by telegraph. (The editor was once an oilfield telegrapher.)

BREA
A viscous, asphaltic material formed at oil seepages when the lighter fractions of the oil have evaporated, leaving the black, tar-like substance.

BREAK CIRCULATION
To resume the movement of drilling fluid down the drillpipe, through the "eyes" of the bit, and upward through the annulus to the surface.

BREAK TOUR
To begin operating 24 hours a day on three eight-hour shifts after rigging up on a new well. Until the derrick is in place and rigged up, mud pits dug, pipe racked, and other preparatory work done, the drill crew works a regular eight-hour day. When drilling commences, the crews break tour and begin working the three, eight-hour tours.

BREAKING DOWN THE PIPE
Unscrewing stands of drillpipe in one-joint lengths usually in preparation for stacking and moving to another well location.

BREAK OUT
(1) To isolate pertinent figures from a mass of data; to retrieve relevant information from a comprehensive report. (2) To loosen a threaded pipe joint.

BREAK-OUT TANKAGE
Tankage at a take-off point or delivery point on a large crude oil or products pipeline.

BREATHING
The movement of oil vapors and air in and out of a storage tank owing to the alternate heating by day and cooling by night of the vapors above the oil in the tank.

Bridle

Bright spots *(Courtesy Shell Ecolibrium)*

Bristle or foam pig

BRIDGE OVER

The collapse of the walls of the borehole around the drill column.

BRIDGE PLUG

An expandable plug used in a well's casing to isolate producing zones or to plug back to produce from a shallower formation; also to isolate a section of the borehole to be filled with cement when a well is plugged.

BRIDLE

A sling made of steel cable fitted over the "horsehead" on a pumping jack and connected to the pump rod; the cable link between "horsehead" and pump rod on a pumping well.

BRIGHT SPOTS

White areas on seismographic recording strips which may signal to the geologist or trained observer the presence of hydrocarbons.

BRIGHT STOCKS

High-viscosity, fully refined, and dewaxed lubricating oils; used for blending with lower viscosity oils. The name originated from the clear, bright appearance of the dewaxed lubes.

BRING BOTTOMS UP

To wash rock cuttings from the bottom of the hole to the surface by maintaining circulation after halting the drilling operation. This allows time for closer inspection of the cuttings and for a decision as to how to proceed when encountering a certain formation.

BRISTLE PIG

A type of pipeline pig or scraper made of tough plastic covered with flame-hardened steel bristles. Bristle or foam pigs are easy to run, do not get hung up in the line, and are easy to "catch." They are usually run in newly constructed lines to remove rust and mill scale (q.v.).

BRITISH NATIONAL OIL CORPORATION

The United Kingdom government agency that "participates" in drilling and production activities in the British sectors of the North Sea with U.S. oil companies and others; the "corporation" through which Britain assumes ownership of the U.K.'s share of the North Sea oil.

BROKE OUT

To be promoted; "He broke out as a driller at Midland"; to begin a new job after being promoted.

BRONC

A new driller promoted from helper; a new tool pusher up from driller; any newly promoted oil field worker whose performance is still untried.

BRUCKER SURVIVAL CAPSULE

A patented, sef-contained survival vessel that can be lowered from an offshore drilling platform or semisubmersible in the event of a fire or other emergency. The vessel, of spheroid shape, is self-propelled and is equipped with first-aid and life-support systems. Some models can accommodate 26 persons. *See* Whitaker System.

BSD

Barrels per stream day. *See* Stream Day.

B S & W
Short for basic sediment and water often found in crude oil.

BTU
British thermal unit; the amount of heat required to raise one pound of water one degree Fahrenheit.

BTX
Benzene—toluene—xylene; basic aromatics used in the manufacture of paints, synthetic ruber, agricultural chemicals, and chemical intermediates. The initials are used by refinery men in designating a unit of the refinery.

BUBBLE-CAP TRAYS
Shelves or horizontal baffles inside a fractionating tower or column that are perforated to allow the fluid charge to run down to the bottom of the column and the vapors to rise through the trays to the top where they are drawn off. The perforations in the trays are made with small umbrella-like caps called bubble caps whose purpose is to force the rising vapors to bubble through the several inches of liquid standing on each tray before the vapors move upward to the next tray. The hot vapors bubbling through the liquid keep the liquid charge heated.

BUBBLE POINT
The pressure at which gas, held in solution in crude oil, breaks out of solution as free gas; saturation pressure.

BUBBLE POINT PUMP
A type of downhole oil pump very sensitive to gas. When the saturation pressure is reached, gas is released which gas-locks the pump until pressure is again built up by the oil flowing into the well bore. This type of pump regulates, in effect, oil production from a reservoir with a gas drive.

BUBBLE TOWER
Any of the tall cylindrical towers at an oil refinery. *See* Fractionator.

Bubble tower

BUCK UP
To tighten pipe joints with a wrench.

BUCKING THE TONGS
Working in a pipeline gang laying screw pipe; hitting the hooks (q.v.).

BUG BLOWERS
Large fans used on or near the floor of the drilling rig to keep mosquitoes and other flying insects off the rig crew.

BULK PLANT
A distribution point for petroleum products. A bulk plant usually has tank car unloading facilities and warehousing for products sold in packages or in barrels.

BULLDOGGED
Said of a fishing tool lowered into the well bore that has latched onto lost pipe or another object being fished out and won't unlatch or cannot be disengaged, owing to a malfunction of the tool.

BULLET TANKS
Colloquial term for horizontal pressure tanks made in the shape of a very fat bullet. Bullet tanks are for storing gasoline or butane under pressure. Other liquefied petroleum gases (LPG) with higher vapor pressures (q.v.) are stored in Hortonspheres or spheroids that can withstand higher pressures per square inch.

BULL GANG
Common laborers who do the ditching and other heavy work on a pipeline construction job.

BULL NOSE
A screw-end pipeline plug; a pipeline fitting one end of which is closed and tapered to resemble a bull's nose; a nipple-like fitting, one end threaded, the other end closed.

BULL PLUG
A short, tapered pipe fitting used to plug the open end of a pipe or throat of a valve.

BULL WAGON
A casing wagon (q.v.).

BULL WHEEL
On a cable-tool rig, the large wheels and axle located on one side of the derrick floor used to hold the drilling line. *See* Calf Wheel.

BUMPER SUB
A slip joint that is part of the string of drillpipe used in drilling from a drillship to absorb the vertical motion of the ship caused by wave action. The slip joint is inserted above the heavy drill collars in order to maintain the weight of the collars on the drill bit as the drillpipe above the slip joint moves up and down with the motion of the ship.

BUMPER SUB (FISHING)
A hydraulically actuated tool installed in the fishing string above the fishing tool to produce a jarring action. When the fishing tool has a firm hold on the lost drillpipe or tubing, which may also be stuck fast in the hole, the bumper sub imparts a jarring action to help free the "fish."

BUMP OFF A WELL
To disconnect a rod-line well from a central power unit.

BUNKER "C" FUEL OIL
A heavy, residual fuel oil used in ships' boilers and large heating and generating plants.

BUNKERING
To supply fuel to vessels for use in the ships' boilers; the loading of bunker fuel on board ship for use by the ship's boilers.

BUNKHOUSE
Crew quarters, usually a portable building used on remote well locations to house the drilling crew and for supplies; quarters for single, oil field workers in the days when transportation to a nearby town was primitive or unavailable.

BURNER
A device for the efficient combustion of a mixture of fuel and air. *See* Ultrasonic Atomizer.

BURNING POINT
The lowest temperature at which a volatile oil in an open vessel will continue to burn when ignited by a flame. This temperature determines the degree of safety with which kerosene and other illuminants may be used.

BURN PIT
An excavation in which waste oil and other material are burned.

BURTON, WILLIAM M.
The petroleum chemist who developed the first profitable means of cracking low-value middle distillates into lighter fractions (gasolines) by the use of heat and pressure.

BUTANE
A hydrocarbon fraction; at ordinary atmospheric conditions, butane is a gas but it is easily liquefied; one of the most useful LP-gases, widely used household fuel.

BUTTERFLY VALVE
A type of quick-opening valve whose orifice is opened and closed by a disk that pivots on a shaft in the throat of the valve.

BUTT-WELDED PIPE
Pipe made from a rectangular sheet of steel which is formed on mandrels. The two edges of the sheet are butted together and welded automatically.

BUY-BACK CRUDE OIL
In foreign countries, buy-back oil is the host government's share of "participation crude" it permits the company holding the concession (the producer) to buy back. This occurs when the host government has no market for its share of oil received under the joint-interest or participation agreement.

BYPASS VALVE
A valve by which the flow of liquid or gas in a system may be shunted past a part of the system through which it normally flows; a valve that controls an alternate route for liquid or gas.

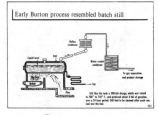

Burton process

Butterfly valve with
manual actuator
Courtesy Fisher)

C

CABLE, DRILLING
A heavy cable, 1 to 2 inches in diameter, made of strands of steel wire.

CABLE TOOLS
The equipment necessary for cable-tool drilling of a well. A heavy metal bar, sharpened to a chisel-like point on the lower end is attached to a drilling rope or wire line (cable) which is fastened to a walking beam above the rig floor that provides an up and down motion to the line and the metal drilling tools. The drilling tool, called a bit, comes in a variety of cutting-edge configurations.

Drilling cable

6"   16"

Caliper log

Canning line

CALCINING (OF COKE)
Calcining is the heating of a substance to drive off moisture and other gaseous impurities or to make it more friable or crushable. Petroleum coke is calcined, crushed, and heated to drive off any remaining liquid hydrocarbons and water.

CALF WHEEL
The spool or winch located across the derrick floor from the bull wheels (q.v.) on a cable tool rig. The casing is usually run with the use of the calf wheels which are powered by the bandwheel (q.v.). A line from the calf wheels runs to the crown block and down to the rig floor.

CALIPER LOG
A tool for checking casing downhole for any bending or flattening or other deformation prior to running and setting a packer or other casing hardware. *See* Drift Mandrel.

CAMP, COMPANY
A small community of oil field workers; a settlement of oil company employees living on a lease in company housing. In the early days, oil companies furnished housing, lights, gas, and water free or at a nominal charge to employees working on the lease and at nearby company installations—pumping stations, gasoline plants, tank farms, loading racks, etc. Camps were known by company lease or simply the lease name, e.g. Gulf Wolf Camp, Carter Camp, and Tom Butler.

CANNING LINE
A facility at refinery where cans are filled with lubricating oil, sealed, and put in cases. Modern canning lines are fully automated.

CAP BEAD
The final bead or course of metal laid on a pipeline weld. The cap bead goes on top of the hot passes or filler beads to finish the weld.

CAPILLARY ATTRACTION
The attraction of the surface of a liquid to the surface of a solid. Capillary attraction or capillarity adversely affects the recovery of crude oil from a porous formation because a portion of the oil clings to the surface of each pore in the rock. Flooding the formation with certain chemicals reduces the capillary attraction, the surface tension, permitting the oil to drain out of the pores of the rock. *See* Tertiary Recovery.

CAPITAL ASSETS
Assets acquired for investment and not for sale and requiring no personal services or management duties. In Federal income tax law, oil and gas leases are, ordinarily, property used in the taxpayers' trade or business and are not capital assets. Royalty, if held for investment, is usually considered a capital asset.

CAPITAL EXPENDITURES
Nondeductible expenditures which must be recovered through depletion or depreciation. In the oil industry, these items illustrate expenditures that must be capitalized: geophysical and geologic costs, well equipment, and lease bonuses paid by lessee.

CAPITAL-GAP DILEMMA
The growing disproportion of capital investment to oil reserves discovered; the increasing need for investment capital coupled with diminishing results in terms of oil and gas discovered; spending more to find less oil.

CAPITAL-INTENSIVE INDUSTRY
Said of the oil industry because of the great amounts of investment capital required to search for and establish petroleum reserves.

CAPITAL STRING
Another name for the production string (q.v.).

CAPPING
Closing in a well to prevent the escape of gas or oil.

CAP SCREW
A bolt made with an integral, hexagonal head; cap screws are commonly used to fasten water jackets and other auxiliary pieces to an engine or pump, and have slightly pointed ends, below the threads, to aid in getting the "screw" into the tapped hole and started straight.

CAPTURED BOLT
A bolt held in place by a fixed nut or threaded piece. The bolt can be tightened or loosened but cannot be removed completely because of a shoulder at the end of the bolt. Captured bolts are in reality a part of an adjustable piece and are so made to preclude the chance of being removed and dropped or because of limited space and accessibility in an item of equipment.

CARBON BLACK
A fine, bulky carbon obtained as soot by burning natural gas in large horizontal "ovens" with insufficient air.

CARBON PLANT
A plant for the production of carbon black by burning natural gas in the absence of sufficient air. Carbon plants are located close to a source of gas and in more or less isolated sections of the country because of the heavy emission of smoke.

CARRIED INTEREST
A fractional interest in an oil or gas property, most often a lease, the holder of which has no obligation for operating costs. These are paid by the owner or owners of the remaining fraction who reimburse themselves out of profits from production. The person paying the costs is the carrying party; the other person is the carried party.

CARRIED WORKING INTEREST
A fractional interest in an oil and gas property conveyed or assigned to another party by the operator or owner of the working interest. In its simplest form a carried working interest is exempt from all costs of development and operation of the property. However, the carried interest may specify "to casing point," "to setting of tanks," or "through well completion." If the arrangement specifies through well completion, then the carried interest may assume the equivalent fractional interest of operating costs upon completion of the well. There are many different types of carried

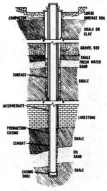

Capital string

interests, the details varying considerably from arrangement to arrangement. One authority has observed, "The numerous forms this interest is given from time to time make it apparent the term carried interest does not define any specific form of agreement but serves only as a guide in preparing and interpreting instruments."

CARVED-OUT INTEREST
An interest; an oil payment or overriding royalty (q.v.) conveyed to another party by the owner of a larger interest, i.e., a working interest. The owner of the working interest in a producing property may grant an oil payment to a bank to pay off a loan. For other considerations, the owner of the larger interest may convey an overriding royalty, one-sixteenth, for example, which he has "sliced off" or carved out of his interest.

CASH BONUS
See Bonus.

CASING
Steel pipe used in oil wells to seal off fluids from the borehole and to prevent the walls of the hole from sloughing off or caving. There may be several strings of casing in a well, one inside the other. The first casing put in a well is called surface pipe which is cemented into place and serves to shut out shallow water formations and also as a foundation or anchor for all subsequent drilling activity. *See* Production String.

CASINGHEAD
The top of the casing set in a well; the part of the casing that protrudes above the surface and to which the control valves and flow pipes are attached.

CASINGHEAD GAS
Gas produced with oil from an oil well as distinguished from gas from a gas well. The casinghead gas is taken off at the top of the well or at the separator.

CASINGHEAD GASOLINE
Liquid hydrocarbons separated from casinghead gas by the reduction of pressure at the wellhead or by a separator or absorption plant. Casinghead gasoline or natural gasoline is a highly volatile, water-white liquid.

CASING PACKER, EXTERNAL
See External Casing Packer.

CASING POINT
A term that designates a time when a decision must be made whether casing is to be run and set or the well abandoned and plugged. In a joint operating agreement, casing point refers to the time when a well has been drilled to objective depth, tests made, and the operator notifies the drilling parties of his recommendation with respect to setting casing, a production string, and completing the well. On a marginal well, the decision to set pipe is often difficult. To case a well often costs as much as the drilling. On a very good well there is no hesitation; the operators are glad to run casing and complete the well.

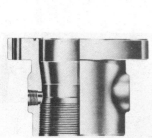

Casing

Casinghead

CASING PRESSURE
Pressure between the casing and the well's tubing.

CASING, SHALLOW-WELL
Small-diameter casing of lighter weight than conventional casing used in deep wells. The lighter-weight casing is less costly, easier to handle, and adequate for certain kinds of shallow, low-pressure wells.

CASING SHOE
A reinforcing collar of steel screwed onto the bottom joint of casing to prevent abrasion or distortion of the casing as it forces its way past obstructions on the wall of the borehole. Casing shoes are about an inch thick and 10 to 16 inches long and are an inch or so larger in diameter in order to clear a path for the casing.

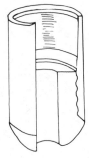

Casing shoe

CASING, SPIRALED-WALL
Well casing made with spiral grooves on the outer circumference of the pipe. The purpose of the patented axial grooves is to aid in running casing or a liner (q.v.) in deviated or crooked holes. The pipe with its grooves, like screw threads, is said to be less susceptible to wall or differential sticking (q.v.).

CASING WAGON
A small, low cart for moving casing from the pipe rack to the derrick floor. Two wagons are used. The forward wagon holds the pipe in a V-shaped cradle; the rear wagon is in reality a lever on wheels which raises the end of the casing so it is free to be pulled.

CAT
Short for Caterpillar tractor, a crawler-type tractor that moves on metal tracks made in segments and connected with pins to form an "endless" tread.

CATALYSIS
The increase or speeding up of a chemical reaction caused by a substance that remains chemically unchanged at the end of the reaction. Any reaction brought about by a separate agent.

Cat cracker (*Courtesy Phillips Petroleum Co.*)

CATALYST
A substance which hastens or retards a chemical reaction without undergoing a chemical change itself during the process.

CAT CRACKER
A large refinery vessel for processing reduced crude oil, naphthas, or other intermediates in the presence of a catalyst. *See* Fluid Catalytic Cracking Unit.

CATENARY
The sag or curve of a cable or chain stretched between two supports.

CAT HEAD
A spool-shaped hub on a winch shaft around which a rope is wound for pulling and hoisting; a power take-off spool used by the driller as he operates the cat line(q.v.).

Cat head

CATHODIC PROTECTION

An anticorrosion technique for metal installations—pipelines, tanks, buildings—in which weak electric currents are set up to offset the current associated with metal corrosion. Carbon or nonferrous anodes (q.v.) buried near the pipeline are connected to the pipe. Current flowing from the corroding anode to the metal installation control the corrosion of the installation.

CAT LINE

A hoisting or pulling rope operated from a cat head (q.v.). On a drilling rig, the rope used by the driller to exert a pull on pipe tongs in tightening (making up) or loosening (breaking out) joints of pipe.

CAT SHAFT

The shaft on the draw works on which the cat heads are mounted. One cat head is a drum, and by using a large rope wrapped around it a few turns the drilling crew can do such jobs as make-up and break-out and light hoisting. The other end of the cat shaft has a manual or air-actuated quick-release friction clutch and drum to which the tong jerk line or spinning chain is attached.

CATTLE GUARD

A ground-level, trestle-like crossing placed at an opening in a pasture fence to prevent cattle from getting out while permitting vehicles to cross over the metal or wooden open framework.

CATWALK

A raised, narrow walkway between tanks or other installations.

CAT WORKS

A part of the rig's draw works; the gear and chain linkage that powers the cat heads. The cat works are used by the driller to make up and break out tool joint connections; spin in the casing and apply torque to the pipe tongs. *See* Cat Shaft.

CAVEY FORMATION

A formation that trends to cave or slough into the well's borehole. In the parlance of cable-tool drillers, "the hole doesn't stand up."

CAVITATION

The creation of a partial vacuum or a cavity by a high-speed impeller blade or boat propeller moving in or through a liquid. Cavitation is also caused by a suction pump drawing in liquid where there is an insufficient suction or hydrostatic head to keep the line supplied.

CD

Contract depth; the depth of a well called for or specified in the drilling contract.

CELLAR

An excavation dug at the drill site before erecting the derrick to provide working space for the casinghead equipment beneath the derrick floor. Blowout preventer valves (BOP stack) are also located beneath the derrick floor in the cellar.

Catwalk

CELLAR DECK
Lower deck on a large, double-decked, semisubmersible drilling platform.

CELLAR, WELLHEAD
See Wellhead Cellar.

CEMENT
To fix the casing firmly in the hole with cement, which is pumped through the drillpipe to the bottom of the casing and up into the annular space between the casing and the walls of the well bore. After the cement sets (hardens) it is drilled out of the casing. The casing is then perforated to allow oil and gas to enter the well.

A wellhead cellar
(bottom)

CEMENTATION
The filling in of the pore spaces of reservoir rock by the natural concretion of limestone.

CEMENT SLURRY
A soupy mixture of water (or other liquid) and cement. Slurries are thin so they can be pumped to enable the cement to penetrate cracks and crevices and to fill all voids.

CEMENT SQUEEZE
A method whereby perforations, large cracks, and fissures in the wall of the bore hole are forced full of cement and sealed off.

Cement slurry sampling

CENTRALIZERS, CEMENTING
Cyclindrical, cage-like devices fitted to a well's casing as it is run to keep the pipe centered in the borehole. Cementing centralizers are made with two bands that fit the pipe tightly with spring steel ribs that arch out to press against the wall of the borehole. By keeping the pipe centered, a more uniform cementing job is assured. Centralizers are especially useful in deep or deviated holes.

CENTRAL POWER
A well-pumping installation consisting of an engine powering a large-diameter, horizontal band wheel with shackle-rod lines attached to its circumference. The band wheel is an excentric and as it revolves on a vertical axle a reciprocating motion is imparted to the shackle rods. A central power may pump from 10 to 25 wells on a lease.

Old central power or
jack-well plant

CENTRIFUGAL PUMP
A pump made with blades or impellers in a close-fitting case. The liquid is pushed forward by the impellers as they rotate at high speed. Centrifugal pumps, because of their high speed, are able to handle large volumes of liquid.

CENTRIFUGE
A motor-driven machine in which samples of oil or other liquids are rotated at high speed causing suspended material to be forced to the bottom of a graduated sample tube so that the percent of impurities or foreign matter may be observed. Some centrifuges are hand-operated. *See* Shake Out.

CENTRIFUGE, DECANTING
A large centrifuge machine for separating or removing pulverized rock and

Single-stage centrifugal
pump

Decanting centrifuge
(Courtesy Pioneer)

Chain wheels

fines from drilling mud returning from downhole. A decanting centrifuge located between the rig and mud pits removes the fine particles of rock from the mud by centrifugal action and discharges the clean mud to the working pits.

CESSATION OF PRODUCTION
The termination of production from a well. It may be owing to mechanical breakdown, reworking operations, governmental orders, or depletion of oil or gas. Temporary cessation usually does not affect the lease, but a permanent shutdown terminates the ordinary oil and gas lease.

CETANE NUMBER
A measure of the ignition quality of diesel fuel. The cetane number of diesel fuel corresponds to the percent of cetane ($C_{16}H_{34}$) in a mixture of cetane and alpha-methyl naphthalene. When this mixture has the same ignition characteristic in a test engine as the diesel fuel, the diesel fuel has a cetane number equal to the percent of cetane in the mixture. Regular diesel is 40−45 cetane; premium is 45−50.

CFM
Cubic feet per minute.

CHAIN TONGS
A pipe wrench with a flexible chain to hold the toothed wrench head in contact with the pipe. The jointed chain can be looped around pipes of different diameters and made fast in dogs on the wrench head.

CHAIN WHEELS
Some gate valves are operated from a distance either for safety or convenience. Such valves have a gate wheel made to accept a chain in the wheel's outer circumference. The chain is reeved or passed over a drum or windlass which the operator turns to open or close the valve from a distance.

CHANNEL
A "vacation" or void in a cement squeeze job allowing salt water or other fluid into the production zone or another interval in the annular space. Also, in waterflooding, a natural void or "path" in a formation permitting the injection fluid to break through to a producing well from the injection well subverting the waterflooding project. *See* Squeeze a Well.

CHANNELING
A condition which arises in oil production when water bypasses the oil in the formation and enters the well bore through fissures or fractures. There are two general types of channeling: (1) coning off in which a small amount of oil "rides" on top of the encroaching water; (2) bypassing, water breaks through to the well bore through fractures or more permeable streaks or sections of the formation leaving the oil behind.

CHARCOAL TEST
A test to determine the gasoline content of natural gas.

CHARGING STOCK
Oil that is to be "charged" or treated in a particular refinery unit.

CHASE THREADS
To straighten and clean threads of any type.

CHATTER
A noisy indication that a mechanical part is behaving erratically and de-structively. In the case of a spring-loaded relief valve, chatter is caused by the valve disk opening and closing rapidly and repetitively, striking against the seat sharply many times a second. Chatter in a bearing is caused by an improperly fitted bearing or from excessive wear that permits lateral motion. As the shaft rotates at high speed, the journal (q.v.) strikes the bearing surface repetitively and rapidly.

CHEATER
A length of pipe used to increase the leverage of a wrench; anything used to lengthen a handle to increase the applied leverage.

CHECKERBOARD LEASING
The acquisition of mineral rights (oil and gas) in a checkerboard pattern. A company may be forced to lease land over a wide area before it has completed geological and geophysical studies. Leases then may be taken on one-quarter section (160 acres) in each section of land.

CHECK VALVE
A valve with a free-swinging tongue or clapper that permits fluid in a pipeline to flow in one direction only; back-pressure valve.

CHECK VALVE, TILTING DISC
A type of check valve, usually for large-diameter pipelines, with the disc mounted on trunnions instead of a hinge as in more conventional check valves. One advantage of the tilting disc is its quiet operation, the absence of "slam" as in other types of check valves.

CHEESE BOX
An early-day, square, box-like refining vessel; a still to heat crude oil for distilling the products in those days—kerosene, gas oil, and lubricating oil.

CHEMICAL FEEDER PUMP
A small-volume pump used on oil leases to inject chemicals into flow lines. The pump may be located at the wellhead and be actuated by the motion of the pumping jack. The chemical is used to break down water/oil emulsions that may be contained in the crude oil stream.

CHEMICAL INJECTION PUMP
A small-volume, high-pressure pump for injecting chemicals into produc-ing wells or pipelines. Chemicals are injected into oil streams to reduce any emulsified oil to free oil and water. When the droplets of water are freed of their film of oil, the water will drop out, settle out of the oil stream, and can be drawn off.

CHILLERS
Refinery apparatus in which the temperature of paraffin distillates is lowered preparatory to filtering out the solid wax.

CHOCK
A wedge or block to prevent a vehicle or other movable object from shifting position; a chunk.

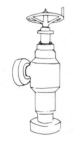

Manual production
choke

CHOKE
A heavy steel nipple inserted into the production tubing that closes off the flow of oil except through an orifice in the nipple. Chokes are of various sizes. It is customary to refer to the production of a well as so many barrels through (or on) a (e.g.) 22/64th-inch choke.

CHRISTMAS TREE
An assembly of valves mounted on the casinghead through which a well is produced. The Christmas tree also contains valves for testing the well and for shutting it in if necessary.

CHRISTMAS TREE (MARINE)
A subsea production system similar to a conventional land tree except it is assembled complete for remote installation on the sea floor with or without diver assistance. The marine tree is installed from the drilling platform; it is lowered into position on guide cables anchored to foundation legs implanted in the ocean floor. The tree is then latched mechanically or hydraulically to the casinghead by remote control.

Christmas tree

CHURN DRILLING
Another name for cable tool drilling because of the up and down, churning motion of the drill bit.

CID
Cubic inch displacement; the volume "swept out" or evacuated by the pistons of an engine in one working stroke; used to describe the size (and by implication, the power) of an automobile engine.

C.I. PLUG
A cast-iron plug; a flat plug used to close the end of a pipe or a valve.

CIRCLE JACK (CABLE-TOOL RIG)
A device used on the floor of a cable-tool rig to make up and break out (tighten and loosen) joints of drilling tools, casing or tubing; a jacking device operated on a toothed or notched metal, circular track placed around the pipe joint protruding from the borehole, above the floor. The jack is operated manually with a handle, and is connected to a wrench which tightens the pipe joint as the jack is advanced, notch by notch.

CIRCULATE
To pump drilling fluid into the borehole through the drillpipe and back up the annulus between the pipe and the wall of the hole; to cease drilling but to maintain circulation for any reason. When closer inspection of the formation rock just encountered is desired, drilling is halted as circulation is continued to "bring bottoms up" (q.v.).

CIRCULATION
The round trip made by drilling mud; down through the drillpipe and up on the outside of the drillpipe, between the pipe and the walls of the borehole. If circulation is "lost," the flow out of the well is less than the flow into the well; the mud may be escaping into some porous formation or a cavity downhole. *See* Lose Returns.

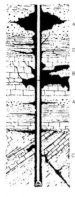

Lost circulation

CITY GATE
The measuring point at which a gas distributing utility receives gas from a gas transmission company.

CLADDING
Coating of one material with another; to cover one metal with another by bonding the two. In the oil patch cladding or "trimming" steel pipe and valves with corrosion-resistant metal alloy is necessary when H_2S (hydrogen sulfide) and other corrosive gases must be handled. Cladding of pipe, valves, and fittings is less costly than making them of expensive, anticorrosion alloys. *See* Sour Service Trim.

CLAMPS, PIPELINE LINE-UP
See Line-up Clamps.

CLAMPS, RIVER
See River Clamps.

CLAMSHELL BUCKET
A hinged, jaw-like digging implement suspended at the end of a cable running down from the boom of an excavating machine. A drag-line bucket.

CLAPPER
The internal moving part, the "tongue" of a check valve that permits a liquid or gas to flow in one direction only in a pipeline. Like a trap door, the check-valve clapper works on a hinge attached to the body of the valve. When at rest the clapper is a few degrees off the vertical or, as in certain valves, completely horizontal.

CLAUS PROCESS
A process for the conversion of hydrogen sulfide (H_2S) to plain sulfur developed in 1885 by the German chemist Claus.

CLAY
The filtering medium, especially Fuller's earth, used in refining; a substance which tends to adsorb the coloring materials present in oil which passes through it.

CLAY PERCOLATER
Refinery filtering equipment employing a type of clay to remove impurities or to change the color of lubricating oils.

CLEAN CARGO
Refined products—distillates, kerosene, gasoline, jet fuel—carried by tankers, barges, and tank cars; all refined products except bunker fuels and residuals (q.v.).

CLEAN CIRCULATION
The circulation of drilling mud free of rock cuttings from the bottom of the borehole. This condition may be caused by a worn bit; circulating to clean the hole or by a broken or parted drillstring.

CLEAN OIL
Crude oil containing less than one percent sediment and water (BS&W); pipeline oil; oil clean enough to be accepted by a pipeline for transmission.

Clevis or shackle
(Courtesy Crosby)

CLEAN-OUT BOX
A square or rectangular opening on the side of a tank or other vessel through which the sediment that has accumulated can be removed. The opening is closed with a sheet of metal (a door) bolted in place.

CLEAN-UP TRIP
Running the drillpipe into the hole for circulation of mud only; to clean the borehole of cuttings.

CLEVIS
A U-shaped metal link or shackle with the ends of the U drilled to hold a pin or bolt; used as a connecting link for a chain or cable.

CLOSED IN
Refers to a well, capable of producing, that is shut in (q.v.).

CLOSE NIPPLE
A very short piece of pipe having threads over its entire length; an all-thread nipple.

CLOUD POINT
The temperature at which paraffin wax begins to crystallize or separate from the solution, imparting a cloudy appearance to the oil as it is chilled under prescribed conditions.

CO_2 INJECTION
A secondary recovery technique in which carbon dioxide is injected into service wells in a field as part of a miscible recovery program. CO_2 is used in conjunction with waterflooding.

CO_2-SHIELDED WELDING
See Welding, CO_2-Shielded.

COAL GAS
Also referred to as town gas. An artificial gas produced by pyrolysis (heating in the absence of air) of coal. Coal gas has a BTU content of 450 per cubic foot; natural gas, on average, has 1,030 BTU per cubic foot, more than twice the thermal value.

COAL GASIFICATION
A process for producing "natural gas" from coal. Coal is heated and brought in contact with steam. Hydrogen atoms in the vapor combine with coal's carbon atoms to produce a hydrocarbon product similar to natural gas.

COAL OIL
Kerosene made from distilling crude oil in early-day pot stills; illuminating and heating oil obtained from the destructive distillation of bituminous coal.

COAL-SEAM GAS
Methane found in certain coal fields in higher concentrations than is common. In some coal operations in areas where the beds are tilted up 30 to 45° from horizontal, holes are drilled along the seams to permit the methane to escape and be brought to the surface. In other tilt-bed fields, the coal is burned in situ and the resulting gases are collected and piped away.

COATING & WRAPPING

A field operation in preparing a pipeline to be put in the ditch (lowered in). The line is coated with a tar-like substance and then spiral-wrapped with tough, chemically impregnated paper. Machines that ride the pipe coat and wrap in one continuous operation. Coating and wrapping protects the pipeline from corrosion. For large pipeline construction jobs, the pipe may be coated and wrapped at the mill or in yards set up at central points along the right of way.

Coating and wrapping a section of pipeline

COGENERATION PLANT

A coal or gas-fired plant that generates both process (commercial) steam and electricity for in-plant use or for sale.

COIL CAR (OR TRUCK)

A tank car or transport truck equipped with heating coils in order to handle viscous liquids that will not flow at ordinary temperatures.

COKE DRUMS

Large vertical, cylindrical vessels which receive their charge of residue at very high temperatures (1000°F.). Any cracked lighter products rise to the top of the drum and are drawn off. The remaining heavier product remains and, because it is still very hot, cracks or is converted to petroleum coke, a solid coal-like substance. In a large refinery that makes a lot of coke, the drums are in batteries of four to eight drums.

COKE, NEEDLE

A form of petroleum coke that gets its name from its microscopic elongated crystalline structure. Needle coke is of a higher quality than the more ordinary sponge coke (q.v.). The manufacture of needle coke requires special feeds to the coker and more severe operating conditions. Severe conditions in refining parlance usually means higher temperatures and pressures in a process.

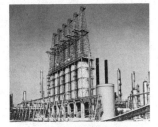

Coking drums

COKE, PETROLEUM

Solid or fixed-carbon that remains in refining processes after distillation of all volatile hydrocarbons; the hard, black substance remaining after oils and tars have been driven off by distillation.

COKE, SPONGE

Petroleum coke that looks like a sponge, hence the name. Sponge coke is used for electrodes and anodes. The weak physical structure of sponge coke makes it unfit for use in blast furnaces and foundry work. *See* Needle Coke.

COKING

(1) The process of distilling a petroleum product to dry residue. With the production of lighter distillable hydrocarbons, an appreciable deposit of carbon or coke is formed and settles to the bottom of the still. (2) The undesirable building up of carbon deposits on refinery vessels.

COLD PINCH

To flatten the end of a pipe with a hydraulically powered set of pinchers. Pinching the pipe end is done to make a quick, temporary closure in the event a loaded pipeline is accidentally ruptured.

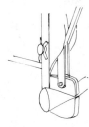

A cold-pinch tool

Drill collar with bit
(*Courtesy Union Oil*)

Collet connector

COLLAR
A coupling for two lengths of pipe; a pipe fitting with threads on the inside for joining two pieces of threaded pipe of the same size.

COLLAR CLAMP
A device fitted with rubber gaskets bolted around a leaking pipe collar. The clamp is effective in stopping small leaks but is used only as a temporary measure until permanent repairs can be made.

COLLAR POUNDER OR PECKER
A pipeline worker who beats time with a hammer on the coupling into which a joint of pipe is being screwed by a tong gang. The purpose is twofold: to keep the tong men pulling in unison and to warm up the collar so that a tighter screw joint can be made.

COLLET CONNECTOR
A component of a subsea drilling system; a mechanically or hydraulically operated latching collar connecting the marine riser (q.v.) to the blowout preventer stack.

COMBINATION DRIVE
A condition in an oil reservoir where there are two or more natural-drive mechanisms present to force the oil to the surface, e.g., water drive and gas-cap drive.

COMBINATION TRAP
A subsurface formation that exhibits characteristics of both a structural and stratigraphic trap. For example, a monocline that loses porosity and permeability up dip is a combination trap. The monocline is its structural character; the change in the reservoir rock gives it the characteristic of a stratigraphic trap.

COME-ALONG
A lever and short lengths of chain with hooks attached to the ends of the chains; used for tightening or pulling a chain. The hooks are alternately moved forward on the chain being tightened.

COMMERCIAL WELL
A well of sufficient net production that it could be expected to pay out in a reasonable time and yield a profit for the operator. A shallow, 50-barrel-a-day well in a readily accessible location on shore could be a commercial well whereas such a well in the North Sea or in the Arctic Islands would not be considered commercial.

COMMINGLING
(1) The intentional mixing of petroleum products having similar specifications. In some instances products of like specifications are commingled in a product pipeline for efficient and convenient handling. (2) Producing two pay zones in the same wellbore. Very often the two or more pay zones have different royalty interests. When this is true the well's zones are produced separately, each through its own tubing and into separate tankage on the lease. In effect the single well with two pay zones is treated as two separate wells. *See* Dual Completion.

COMMON CARRIER
A person or company having state or Federal authority to perform public transportation for hire; an organization engaged in the movement of petroleum products—oil, gas, refined products—as a public utility and common carrier.

COMMUNITY LEASE
A single lease covering two or more separately owned parcels of land. A community lease may result from the execution of a single lease by the owners of separate tracts or by the execution of separate but identical leases by the owners of separate tracts when each lease purports to cover the entire consolidated acreage. Usually the result of the execution of a community lease is the apportionment of royalties in proportion to the interests owned in the entire leased acreage.

COMPANION FLANGE
A two-part connector or coupling: one part convex, the other concave. The two halves are held together by nuts and bolts. This type flange or "union" is used on small-diameter piping.

COMPLETE A WELL
To finish a well so that it is ready to produce oil or gas. After reaching total depth (TD) casing is run and cemented; casing is perforated opposite the producing zone, tubing is run, and control and flow valves are installed at the wellhead. Well completions vary according to the kind of well, depth, and the formation from which it is to produce.

COMPRESSION CUP
A grease cup; a container for grease made either with a screw cap or spring-loaded cap for forcing the grease onto a shaft bearing.

COMPRESSION-IGNITION ENGINE
A diesel engine (q.v.); a four-cycle engine whose fuel charge is ignited by the heat of compression as the engine's piston comes up on the compression stroke. *See* Hot-plug Engine.

COMPRESSION RATIO
The ratio of the volume of an engine's cylinder at the beginning of the compression stroke to the volume at the end or the top of the stroke. High compression engines are generally more efficient in fuel utilization than those with lower compression ratios. A cylinder of 10 cubic inch volume at the beginning of the compression stroke and one cubic inch at the top of the stroke indicates a 10:1 compression ratio.

COMPRESSOR, AXIAL
A gas compressor that takes in gas at the inlet and moves the charge axially over the compressor's long axis to the discharge port. This is accomplished by the action of a central impeller shaft studded with hundreds of short, fixed blades. The impeller and its paddle-like blades rotate at speeds of 3–6,000 rpm. Large compressors move up to 300,000 cubic feet per minute.

Compressor (*Courtesy Union Pump Co.*)

COMPRESSOR PLANT
A pipeline installation to pump natural gas under pressure from one loca-

tion to another through a pipeline. On large interstate gas transmission lines repressuring stations are located every 100 miles, more or less, depending on terrain and other factors, to boost the gas along to its destination. The gas stream arrives at a compressor station at a few hundred pounds per square inch and is discharged from the station's multistage compressors at 1,000 to 1,500 pounds pressure to begin the next leg of its journey to the consumers' gas stoves and furnaces.

COMPRESSOR, SKID-MOUNTED
A "portable" gas compressor and engine module for use in repressuring (q.v.) or to inject gas into a high-pressure gas trunk line.

CONCESSION
An agreement (usually with a foreign governemnt) to permit an oil company to prospect for and produce oil in the area covered by the agreement.

CONDEMNATION
The taking of land by purchase, at fair market value, for public use and benefit by state or federal government, as well as by certain other agencies and utility companies having power of eminent domain (q.v.).

CONDENSATE
Liquid hydrocarbons produced with natural gas which are separated from the gas by cooling and various other means. Condensate generally has an A.P.I. gravity of 50° to 120° and is water-white, straw or bluish in color.

CONDENSATE, LEASE
See Lease Condensate.

CONDENSATE, RETROGRADE GAS
See Retrograde Gas Condensate.

CONDENSATE WATER
Water vapor in solution with natural gas in the formation. When the gas is produced the water vapor condenses into liquid as both pressure and temperature are reduced. See Retrograde Gas Condensate.

CONDENSATION
The transformation of a vapor or gas to a liquid by cooling or an increase in pressure or both simultaneously.

CONDENSER
A water-cooled heat exchanger used for cooling and liquefying vapors.

CONDUCTOR CASING
A well's surface pipe used to seal off near-surface water, prevent the caving or sloughing of the walls of the hole, and as a conductor of the drilling mud through loose, unconsolidated shallow layers of sand, clays, and shales. See Casing.

CONE ROOF
A type of tank roof built in the form of a flat, inverted cone; an old-style roof for large crude storage tanks, but still employed on tanks storing less volatile products. See Floating Roof.

Skid-mounted compressor (Courtesy Fry Assoc. Inc.)

CONFIRMATION WELL
A well drilled to "prove" the formation or producing zone encountered by an exploratory or wildcat well. *See* Step-out Well.

CONGLOMERATE
A type of sedimentary rock compounded of pebbles and rock fragments of various sizes held together by a cementing material, the same type material that holds sandstone together. Conglomerates are a common form of reservoir rock.

CONICAL-TOWER PLATFORM
A type of offshore drilling platform made of reinforced concrete for use in Arctic waters where pack ice prevents the use of conventional platform construction. The structure is a truncated cone supporting a platform from which the wells are drilled.

Conglomerate *(Courtesy Core Lab)*

CONNATE WATER
The water present in a petroleum reservoir in the same zone occupied by oil and gas. Connate water is not to be confused with bottom or edge water. Connate water occurs as a film of water around each grain of sand in granular reservoir rock and is held in place by capillary attraction.

CONNECTION FOREMAN
The supervisor, the boss of a pipeline connection gang (q.v.).

CONNECTION GANG
A pipeline crew that lays field gathering lines, connects stock tanks to gathering lines, and repairs pipelines and field pumping units in their district. Connection gangs also install manifolds and do pipe work in and around pumping stations. A typical gang of 8 or 10 men has a welder and a helper, a gang-truck driver and swamper (helper), 3 or 4 pipeliners, and a connection foreman.

CONSORTIUM
An international business association organized to pursue a common objective, e.g., to explore, drill, and produce oil.

CONSUMER GAS
Gas sold by an interstate gas pipeline company to a utility company for resale to consumers.

CONTINENTAL SHELF
See Outer Continental Shelf (OCS).

CONTOUR LINE
A line (as on a map) connecting points on a land surface that have the same elevation above or below sea level.

CONTOUR MAP
A map showing land surface elevations by the use of contour lines (q.v.). Structure contour maps are used by geologists and geophysicists to depict subsurface conditions or formations. *See* Isopachous Map.

Control panel *(Courtesy Gulf Oil Canada Ltd.)*

CONTROL PANEL
An assembly of indicators and recording instruments—pressure gauges,

Electrically operated control valve *(Courtesy Magnatrol)*

Cooling tower

Cord road

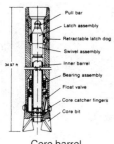

Core barrel

warning lamps, and other digital or audio signals—for monitoring and controlling a system or process.

CONTROL VALVE, ELECTRICALLY OPERATED
A small-diameter valve used in process piping that is opened or closed by a quick-acting solenoid.

COOLING TOWER
A large louvered structure (usually made of wood) over which water flows to cool and aerate it. Although most cooling towers are square or rectangular in shape, some are cylindrical, open at the bottom and top, which produces strong air currents through the center of the structure for more rapid cooling.

CORD ROAD
A passable road made through a swampy, boggy area by laying logs or heavy timbers side by side to make a bumpy but firm surface; a log road.

CORE BARREL
A device with which core samples of rock formations are cut and brought to the surface; a tube with cutting edges on the bottom circumference, lowered into the well bore on the drill pipe and rotated to cut the plug-like sample.

CORE BIT
A special drill bit for cutting and removing a plug-shaped rock sample from the bottom of the well bore. (See illustration, p. 47)

CORE BOAT
A seagoing vessel for drilling core holes in offshore areas.

CORE RECORD
A record showing the depth, character, and fluid content of cores taken from a well.

CORE SAMPLE
A solid column of rock, usually from two to four inches in diameter, taken from the bottom of a well bore as a sample of an underground formation. Cores are also taken in geological studies of an area to determine its oil and gas prospects. (See illustration, p. 47)

CORRELATIVE RIGHTS, DOCTRINE OF
The inherent right of an owner of oil or gas in a field to his share of the "reservoir energy" and his right to be protected from wasteful practices by others in the field.

CORROSION
The eating away of metal by chemical action or an electrochemical action. The rusting and pitting of pipelines, steel tanks, and other metal structures is caused by a complex electrochemical action. *See* Anode.

COST CRUDE OIL
Crude oil produced from an operator's own wells; oil produced at "cost" on a lease or concession acreage as compared to purchased crude. Also, in another context, that portion of oil produced which is applied to paying off

the cost of the well, to the recovery of the costs of drilling, completing, and equipping the well for production. In some production-sharing contracts particularly with governments, foreign as well as our own, on offshore wells, 40 percent of the oil produced is cost oil; the remaining 60 percent is "profit oil" and is divided according to royalty interests. As the costs are paid off the percent of cost oil is reduced and the percentage of profit oil increases accordingly.

COUPLING
A collar; a short pipe fitting with both ends threaded on the inside circumference; used for joining two lengths of line pipe or casing or tubing.

COUPLING POLE
The connecting member between the front and rear axles of a wagon or four-wheel trailer. To lengthen the frame of the vehicle, a pin in the pole can be removed and the rear-axle yoke (which is fastened to the pole by the pin) moved back to another hole. On pipe-carrying oil field trailers, the coupling pole is a telescoping length of steel tubing. The trailer can be made as long as necessary for the load.

CRACK A VALVE
To open a valve so slightly as to permit a small amount of fluid or gas to escape.

CRACKING
The refining process of breaking down the larger, heavier, and more complex hydrocarbon molecules into simpler and lighter molecules. Cracking is accomplished by the application of heat and pressure, and in certain advanced techniques, by the use of a catalytic agent. Cracking is an effective process for increasing the yield of gasoline from crude oil.

CRACKING A VALVE
Opening a valve very slightly.

CRANE BARGE
A derrick barge (q.v.).

CRANK
An arm attached at right angles to the end of a shaft or axle for transmitting power to or from a connecting rod or pitman (q.v.).

CRANKCASE VENTILATION SYSTEM, POSITIVE
See Positive Crankcase Ventilation System.

CRATER
(1) A bowl-shaped depression around a blowout well caused by the caving in and collapse of the surrounding earth structure. (2) To fail or fall apart (colloquial).

CROSSHEAD
A sliding support for a pump or compressor's connecting rod. The rod is attached to a heavy "head" which moves to and fro on a lubricated slide in the pump's frame. Screwed into the other end of the crosshead is the pump's piston rod or plunger rod. A crosshead moves back and forth in a

Core bit

Core samples

Coupling

Crane barge

Crown block

Crown platform

horizontal plane or up and down in a vertical plane transmitting the power from the connecting rod to the pump's piston rod.

CROSSOVER
A stile; a step-and-platform unit to provide access to a work platform or an elevated crossing. *See* Stile.

CROWBAR CONNECTION
A humorous reference to an assembly of pipe fittings so far out of alignment that a crowbar is required to force them to fit.

CROWN BLOCK
A stationary pulley system located at the top of the derrick used for raising and lowering the string of drilling tools; the sheaves and supporting members to which the lines of the traveling block (q.v.) and hook are attached.

CROWN PLATFORM
A platform at the very top of the derrick that permits access to the sheaves of the crown block and provides a safe area for work involving the gin pole (q.v.).

CRUDE OIL
Oil as it comes from the well; unrefined petroleum.

CRUDE OIL, BUY-BACK
See Buy-back Crude Oil.

CRUDE OIL, REDUCED
See Reduced Crude Oil.

CRUDE OIL, VOLATILES-LADEN
A crude oil stream carrying condensate, natural gasoline, and butane. Sometimes it is convenient and economical to move certain natural gas liquids to refineries by injecting them into crude oil pipelines to be pumped with the crude.

CRUDE STILL
A primary refinery unit; a large vessel in which crude oil is heated and various components are taken off by distillation.

CRUMB BOSS
A person responsible for cleaning and keeping an oil field bunkhouse supplied with towels, bed linen, and soap; a construction camp housekeeper.

CRUMB OUT
To shovel out the loose earth in the bottom of a ditch; also, to square up the floor and side of the ditch in preparation for laying of pipe.

CRYOGENICS
A branch of physics that relates to the production and effects of very low temperatures. The process of reducing natural gas to a liquid by pressure and cooling to very low temperatures employs the principles of cryogenics.

CUBES
Short for cubic inch displacement; CID (q.v.).

CULTIVATOR WRENCH
Any square-jawed, adjustable wrench that is of poor quality or worn out. *See* Knuckle Buster.

CUP
Disc with edges turned at right angles to the body used on plungers in certain kinds of pumps; discs of durable plastic or other tough, pliable material used on pipeline pigs or scrapers to sweep the line.

CUP GREASE
Originally, a grease used in compression cups (q.v.) but today the term refers to grease having a calcium fatty-acid soap base. *See* Grease.

CUT
(1) A petroleum fraction, a product such as gasoline or naphtha distilled from crude oil. (2) Crude oil contaminated with water so as to make an oil-water emulsion. (3) To dilute or dissolve.

CUT OIL
Crude oil partially emulsified with water; oil and water mixed in such a way as to produce an emulsion in which minute droplets of water are encased in a film of oil. In such case the water, although heavier, cannot separate and settle to the bottom of a tank until the mixture is heated or treated with a chemical. *See* Roll a Tank.

CUT POINTS
The temperatures at which various distilling products are separated out of the charge stock. One cut point is the temperature at which the product begins to boil or vaporize, the initial boiling point. The other cut point is the temperature at which the product is completely vaporized; this is the end point.

CUTTING OILS
Special oils used to lubricate and cool metal-cutting tools.

CUTTINGS
Chips and small fragments of rock as the result of drilling that are brought to the surface by the flow of the drilling mud as it is circulated. Cuttings are important to the geologist who examines them for information concerning the type of rock being drilled. *See* Sample.

CUTTING TORCH
A piece of oxyacetylene welding and cutting equipment; a hand-held burner to which the oxygen and acetylene hoses are attached. The gases when ignited by the welder's lighter produce a small, intense flame that "cuts" metal by melting it. *See* Welding Torch.

CYCLING (OF GAS)
Return to a gas reservoir of gas remaining after extraction of liquid hydrocarbons for the purpose of maintaining pressure in the reservoir, and thus increasing the ultimate recovery of liquids from the reservoir.

CYCLING PLANT
An oil field installation that processes natural gas from a field, strips out the gas liquids, and returns the dry gas to the producing formation to maintain reservoir pressure.

CYLINDER OIL
Oils used to lubricate the cylinders and valves of steam engines.

CYLINDER STOCK
A class of highly viscous oils so called because originally their main use was in preparation of products to be used for steam cylinder lubrication.

D

DAMPING SUB
Essentially, a downhole "shock absorber" for a string of drilling tools; a 6 to 8-foot-long device, a part of the drill assembly, that acts to dampen bit vibration and impact loads during drill operations. Damping subs are of the same diameter as the drillpipe into which they are screwed to form a part of the drillstring.

D & P PLATFORM
A drilling and production platform. Such an offshore platform is a large structure with room to drill and complete a number of wells; as many as 60 have been drilled from a large platform by the use of directional drilling techniques. When many wells are producing from a single platform, the oil can be treated (put through oil-water separators), measured and pumped ashore, as though the platform were simply a land lease. The production from single-well platforms is usually piped, along with other wells on the offshore lease, to a production platform for treatment and pumping ashore. *See* Directional Drilling.

D & P Platform

DARCY
A unit of permeability of rock. A rock of one darcy permeability is one in which fluid of one centipoise viscosity will flow at a velocity of one centimeter per second under a pressure gradient of one atmosphere per centimeter. As a darcy is too large a unit for most oil producing rocks, permeabilities used in the oil industry are expressed in units one-thousandths as large, i.e., millidarcies (0.001 darcy). Commercial oil and gas sands exhibit permeabilities ranging from a few millidarcies to several thousand.

D'ARCY'S LAW
During experimental studies on the flow of water through consolidated sand filter beds, Henry D'Arcy, in 1856, formulated a law which bears his name. D'Arcy's Law states that the velocity of a homogenous fluid in a porous medium is proportional to the pressure gradient and inversely proportional to the fluid's viscosity. This law has been extended to describe, with certain limitations, the movement of other fluids including miscible fluids in consolidated rocks and other porous substances.

DAY-WORK BASIS
Refers to a drilling contract in which the work of drilling and completing a well is paid for by the days required for the job instead of by the feet drilled. *See also* Turnkey Contract.

DC-DC RIG
See Drill Rig, Electric.

DEAD LINE

The anchored end of the drilling line that comes down from the crown block through a fixed sheave at ground level, called a dead-line anchor, and onto a storage drum. When stringing up the drilling line, the big traveling block is set on the rig floor and the free end of the line is threaded over the crown block and through the traveling block a sufficient number of times to lift the anticipated load with a good margin of safety. The free end is then attached to the draw works drum which is rotated until one layer of the line is spooled on. The traveling block is then hoisted into the derrick. The other end of the threaded or reeved line is the dead line or, one might say, the anchored line. *See* Fast Line.

DEADMAN

(1) A substantial timber or plug of concrete buried in the earth to which a guy wire or line is attached for bracing a mast or tower; a buried anchoring piece capable of withstanding a heavy pull. (2) A land-side mooring device used with lines and cables when docking a vessel.

DEADMAN CONTROL

A device for shutting down an operation should the attendant become incapacitated. The attendant using such a device must consciously exert pressure on a hold-down handle or lever to work the job. When pressure is relaxed owing to some emergency, the operation will automatically come to a halt.

DEAD OIL

Crude oil containing no dissolved gas when it is produced.

Decanting centrifuge

DEAD WELL

A well that will not flow, and in order to produce must be put on the pump.

DEADWOOD

Material inside a tank or other vessel such as pipes, supports, and construction members that reduce the true volume of the tank by displacing some of the liquid contents.

DECANTING CENTRIFUGE

See Centrifuge, Decanting.

DECK BLOCK

A pulley or sheave mounted in a steel frame which is securely fixed to the metal deck of a ship or barge. Deck blocks which lie horizontal to the vessel's deck are for horizontal pulls with hawser or cable.

DEEP RIG

A specially designed drilling derrick built to withstand the extreme hook loads of ultra-deep (20,000 to 30,000-foot) wells. Deep rigs, in addition to extra-strong structural members, have massive substructures 25 to 35 feet high to accommodate the large and tall blowout preventer stacks flanged to the wellhead. Hook loads on deep rigs often exceed 800 tons, 1,600,000 pounds.

Deep rig

DEFICIENCY GAS

The difference between a quantity of gas a purchaser is obligated by contract either to take or pay for if not taken and the amount actually taken.

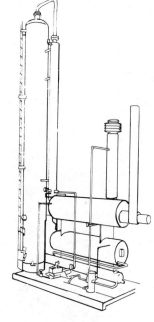

Natural gas dehydrator

DEGASSING DRILLING MUD

An important part of well drilling operations is keep the drilling mud free of entrained gas, bubbles that enter the mud as it circulates downhole through gaseous formations. One of the three functions of mud is to provide sufficient hydrostatic head to control a kick when high-pressure oil or gas is encountered. When mud of a certain density is circulated, it can become infused with gas to an extent that although the volume of mud may increase the density is severely reduced. To guard against this dangerous situation occurring, the mud is degassed at the surface. Several kinds of equipment get the gas out, but all have one aim in common: to make it possible for the gas bubbles to free themselves. One method flows the mud over wide sheets so the slurry is no more than one-eighth to three-eighths thick so the bubbles may come to the surface and escape. Another method sprays the mud against a baffle in a spray tank which squeezes out the gas. A third method directs the mud through a vacuum tank where, under reduced atmospheric pressure, the bubbles of gas expand and break out of the slurry.

DEHYDRATOR, NATURAL GAS

A tank or tower through which gas is run to remove entrained water. A common method of gas dehydration is through the use of various glycols — diethylene, triethylene, and tetraethylene. Dehydration is accomplished by contact of the wet gas with a pure or "lean" glycol solution. Gas is fed in to the bottom of a trayed or packed column in the presence of the glycol solution. As the gas percolates upward through the solution, the lean glycol absorbs the entrained water, and dry gas is taken off at the top of the tower. Gas must be extremely dry to meet pipeline specifications; it may not contain more than 7 pounds of water per million standard cubic feet.

DELAY RENTAL

Money payable to the lessor by the lessee for the privilege of deferring drilling operations or commencement of production during the primary term of the lease. Under an unless lease, the nonpayment of delay rental by the due date is the cause for automatic termination of the lease unless the lease is being held by drilling operations, by production, or by some special provision of the lease.

DELINEATION WELLS

Wells drilled outward from a successful wildcat well to determine the extent of the oil find, the boundaries of the productive formation. See Development Wells.

DEMULSIFIER

A chemical used to "break down" crude oil/water emulsions. The chemical reduces the surface tension of the film of oil surrounding the droplets of water. Thus freed, the water settles to the bottom of the tank.

DEMURRAGE

The charge incurred by the shipper for detaining a vessel, freight car, or truck. High loading rates for oil tankers are of utmost importance in order to speed turnaround and minimize demurrage charges.

DENSMORE, AMOS

The man who first devised a method of shipping crude oil by rail. In 1865 he mounted two "iron banded" wooden tanks on a railway flatcar. The tanks or tubs held a total of 90 barrels. Densmore's innovation was the forerunner of the "unit train" for hauling oil and products, and the latest development, Tank Train (q.v.).

DEPLETION ALLOWANCE

See Percentage Depletion.

DEPOSIT

An accumulation of oil or gas capable of being produced commercially.

DEPROPANIZER

A unit of a processing plant where propane, a liquid hydrocarbon, is separated from natural gas.

DERRICK

A wooden or steel structure built over a well site to provide support for drilling equipment and a tall mast for raising and lowering drillpipe and casing; a drilling rig.

Derrick

DERRICK BARGE

A type of work boat on which a large crane is mounted for use offshore or other over-water work. The larger derrick or crane barges are self-propelled and are, in effect, a boat or ship with full-revolving crane, a helicopter pad, and tools and equipment for various tender work. A crane barge.

DERRICK FLOOR

The platform (usually 10 feet or more above the ground) of a derrick on which drilling operations are carried on; rig floor.

DERRICKMAN

A member of the drilling crew who works up in derrick on the tubing board, racking tubing or drillpipe as it is pulled from the well and unscrewed by other crew members on the derrick floor.

Derrick barge

DERRICK, PUMPING

In the early days, before the widespread use of portable units for pulling and reconditioning a well, the original derrick, used for drilling, was often replaced by a smaller, shorter derrick called a pumping derrick or pumping rig. Well workovers could be done with these rigs; the well also could be pumped by pumping jack or by a walking beam.

DESALTING PLANT

An installation that removes salt water and crystalline salt from crude oil streams. Some plants use electrostatic precipitation; others employ chemical processes to remove the salt.

DESICCANT DRYING

The use of a drying agent to remove moisture from a stream of air or gas. In certain product pipelines great effort is made to remove all water vapor before putting the line in service. To accomplish this, desiccant-dried air or an inert gas is pumped through the line to absorb the moisture that may be present even in the ambient air in the line.

Derrick floor

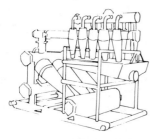

Desilter

Deviated hole

DESILTER - DESANDER

A filtering device on a drilling well's mud system that removes harmful abrasive material from the mud stream.

DETERGENT OILS

Lubricating oils containing additives that retard the formation of gums, varnishes, and other harmful engine deposits. The detergents act to keep all products of oxidation and other foreign matter in suspension which permits it to be removed by the engine's filtering system.

DETRIUS

Fragments of disintegrated rock; an accumulation of material as the result of the wearing away or erosion of rock. *See* Talus.

DETROIT IRON

A humorous reference to a large, old car or truck.

DEVELOPMENT

The drilling and bringing into production of wells in addition to the discovery well on a lease. The drilling of development wells may be required by the express or implied covenants of a lease.

DEVELOPMENT CLAUSE

The drilling and delay-rental clause of a lease; also, express clauses specifying the number of development wells to be drilled.

DEVELOPMENT WELLS

Wells drilled in an area already proved to be productive.

DEVIATED HOLE

A well bore which is off the vertical either by design or accident. All wells are off the vertical, some considerably farther off than others. In drilling a 15,000-foot hole, for example, through many types of hard and soft formations, it is not unusual to have such a deviation that the actual bottom of the hole is 2,000 feet or so away from the well location, and on someone else's lease. This can be litigious business if the other lease owner gets picky about it. Deviated or crooked holes as they are commonly referred to are more expensive to drill than a straight hole for several reasons. For example, to reach a particular formation at 10,000 feet, a crooked hole, on a severe angle from the vertical, may have to be dug several hundred to 1,000 feet farther. In addition to more drilling time it would also mean 1,000 feet more casing and an equal amount of tubing to complete the well. Every effort is made to drill a straight hole. But there are times when drilling at an angle from the vertical serves a useful purpose. And the technique for digging such a hole is quite advanced. *See* Slant-hole Technique.

DEW POINT

The temperature at which water vapor condenses out of a gas at 14.7 psia, (pounds per square inch absolute) or at sea level.

DIAMOND DRILL BIT

A drill bit with many small industrial (man-made) diamonds set in the nose or cutting surface of the bit. Diamonds are many times harder than the hardest steel, so a diamond bit makes possible longer bit runs before a round trip is necessary to change bits.

DIE
A replaceable, hardened steel piece; an insert for a wrench or set of tongs that bites into the pipe as the tool is closed on the pipe; a tong key. Also, in the plural, dies are cutters for making threads on a bolt or pipe.

DIESEL ENGINE
A four-stroke cycle internal combustion engine that operates by igniting a mixture of fuel and air by the heat of compression, and without the use of an electrical ignition system.

DIESEL FUEL
A fuel made of the light gas-oil range of refinery products. Diesel fuel and furnace oil are virtually the same product. Self-ignition is an important property of diesel fuel, as the diesel engine has no spark plugs; the fuel is ignited by the heat of compression within the engine's cylinders. *See* Diesel Engine; *also* Cetane Number.

DIESELING
The tendency of some gasoline engines to continue running after the ignition has been shut off. This is often caused by improper fuel or carbon deposits in the combustion chamber hot enough to ignite the gasoline sucked into the engine as it makes a few revolutions after being turned off.

DIESEL, RUDOLPH
The German mechanical engineer who invented the internal combustion engine that bears his name.

DIFFERENTIAL-PRESSURE STICKING
Another name for wall sticking (q.v.), a condition downhole when a section of drillpipe becomes stuck or hung up in the deposit of filter cake on the wall of the borehole.

DIGGER
One who digs or drills a well; a driller.

DIGGING TOOLS
Hand tools used in digging a ditch, i.e., shovels, picks, mattocks, spades.

DIP
The angle that a geological stratum makes with a horizontal plane (the horizon); the inclination downward or upward of a stratum or bed.

DIRECTIONAL DRILLING
The technique of drilling at an angle from the vertical by deflecting the drill bit. Directional wells are drilled for a number of reasons: to develop an offshore lease from one drilling platform; to reach a pay zone beneath land where drilling cannot be done, e.g., beneath a railroad, cemetery, a lake; and to reach the production zone of a burning well to flood the formation. *See* Killer Well.

DIRTY CARGO
Bunker fuel and other black residual oils.

DISCOVERY WELL
An exploratory well that encounters a new and previously untapped petro-

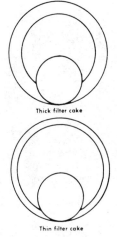

Thick filter cake

Thin filter cake

Pipe stuck in wall cake

Directional drilling

leum deposit; a successful wildcat well. A discovery well may also open a new horizon in an established field.

DISCOVERY WELL ALLOWABLE
An allowable above that of wells in a settled field. Some states allow the operators of a discovery well to produce at the maximum efficiency rate (MER) until the costs of the well have been recovered in oil or gas.

DISPATCHER
One who directs the movement of crude oil or product in a pipeline system. He receives reports of pumping rates, and line pressures and monitors the movement of batches of oil; he may also operate remote, unmanned stations.

DISPOSAL WELL
A well used for the disposal of salt water. The water is pumped into a subsurface formation sealed off from other formations by impervious strata of rock; a service well.

DISSOLVED GAS
Gas contained in solution with the crude oil in the reservoir. *See* Solution Gas.

Distilling columns
(Courtesy Shell)

DISSOLVED-GAS DRIVE
The force of expanding gas dissolved in the crude oil in the formation that drives the oil to the well bore and up to the surface through the production string.

DISTILLATE
Liquid hydrocarbons, usually water-white or pale straw color, and of high API gravity (above 60°) recovered from wet gas by a separator that condenses the liquid out of the gas stream. *See* Condensate. (Distillate is an older term for the liquid; today, it is called condensate or natural gasoline.)

DISTILLATE FUEL OILS
A term denoting products of refinery distillation sometimes referred to as middle distillates, i.e., kerosene, diesel fuel, home heating oil.

DISTILLATION
The refining process of separating crude oil components by heating and subsequent condensing of the fractions by cooling. The basic principle in refining is that of heating crude oil until various vapors or gases "boil off" and then condensing them to form a condensate or distillate. First to boil off or come over would be the very light, dry gases, then highly volatile natural gasoline, gasoline, kerosene, gas oil, light lube oil stock, and so on until a heavy residue is left. This would be broken up by other, more sophisticated processing.

Distillation system

DISTILLATION SYSTEM
A small, temporary "refinery" (200 to 1,000 b/d) set up at a remote drilling site to make diesel fuel and low-grade gasoline from available crude oil for the drilling engines and auxiliary equipment.

DITCHING MACHINE
See Trencher.

Ditching machine

DIVERTER SYSTEM
An assembly of nipples and air-actuated valves welded to a well's surface or conductor casing for venting a gas kick (q.v.) encountered in relatively shallow offshore wells. In shallow wells there is often insufficient overburden pressure around the base of the conductor casing to prevent the gas from a substantial kick from blowing out around the casing. When a kick occurs, the blowout preventer is closed and the valves of the diverter system open to vent the gas harmlessly to the atmosphere.

DIVESTITURE
Specifically as it relates to the industry, to break up, to fragment an integrated oil company into individual, separate companies, each to be permitted to operate only within a single phase of the oil business: exploration, production, transportation, refining or marketing.

DIVISION ORDER
A contract of sale to the buyer of crude oil or gas directing the buyer to pay for the product in the proportions set forth in the contract. Certain amounts of payment go to the operator of the producing property, the royalty owners, and others having an interest in the production. The purchaser prepares the division order after determining the basis of ownership and then requires that the several owners of the oil being purchased execute the division order before payment for the oil commences.

DMWD or MWD
Downhole measurement while drilling (q.v.).

DOCTOR SWEET
A term used to describe certain petroleum products that have been treated to remove sulfur compounds and mercaptans that are the sources of unpleasant odors. A product that has been so treated is said to be "sweet to the doctor test."

DOCTOR TEST
A qualitative method of testing light fuel oils for the presence of sulfur compounds and mercaptans, substances that are potentially corrosive and impart an objectionable odor to the fuel when burned.

Doghouse

DOGHOUSE
A portable, one-room shelter (usually made of light tank iron) at a well site for the convenience and protection of the drilling crew, geologist, and others. The doghouse serves as lunchroom, change house, dormitory, and storage room for small supplies and records.

DOG IT
To do less than one's share of work; to hang back; to drag one's feet.

DOGLEG
A deviation in the direction of a ditch or the bore hole of a well; a sharp bend in a joint of pipe. *See* Key Seat.

DOG ROBBER
A loyal aid or underling who does disagreeable or slightly unorthodox (shady) jobs for his boss; a master of the "midnight requisition."

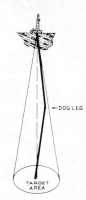

Doglegged hole

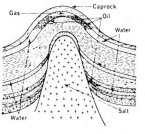

Dome

Dope chopper *(Courtesy Evans)*

DOLLY
Metal rollers fixed in a frame and used to support large diameter pipe as it is being turned for welding; a small, low platform with rollers or casters used for moving heavy objects.

DOME
An incursion of one underground formation into the one above it; an upthrust, as a salt dome, that penetrates overlying strata.

DONKEY PUMP
Any small pumping device used on construction jobs or other temporary operations.

DOODLE BUG
A witching device; a twig or branch of a small tree (peach is favored by some witchers) which when held by an "expert" practitioner as he walks over a plot of land is supposed to bend down locating a favorable place to drill a well; a popular term for any of the various geophysical prospecting equipment.

DOOR MATS
Colloquial term used in the early days to describe small tracts, 1/20 of an acre, just large enough to accommodate an oil derrick. The concept of pooling (q.v.) had not yet been accepted.

DOPE
Any of various viscous materials used on pipe or tubing threads as a lubricant and to prevent corrosion; a tar-base coating for pipelines to prevent corrosion.

DOPE CHOPPER
A machine for removing tar and asphalt coatings from line pipe. The pipe is placed in the chopper where guillotine-like blades cut through the dope but do not damage the pipe. The chunks of coating fall onto a conveyor belt and are carried away from the job.

DOPE GANG
Workers who clean and apply a coat of enamel primer to a pipeline in preparation for coating with a tar-based anticorrosion material and wrapping with tough paper bandage. On large pipeline projects, a machine rides the pipe, cleaning it with rotating metal brushes and then spraying on a primer. A second machine, also riding the pipe, coats and wraps the line in one operation.

DOUBLE-ACTING PUMP
A reciprocating pump (plunger pump) with two sets of suction and discharge valves permitting it to pump fluid during the forward and backward movement of each plunger. (Single-action pumps discharge on the forward stroke and draw in fluid on the return stroke.)

DOUBLES
Drillpipe and tubing pulled from the well two joints at a time. The two joints make a stand (q.v.) of pipe that is set back and racked in the derrick. Three-joint stands are called "thribbles," fours are "fourbles."

DOUBLING YARD
An area convenient to a large pipeline construction project where line pipe is welded together in two-joint lengths preparatory to being transported to the job and strung along the right of way.

DOWNCOMER
A pipe in which the flow of liquid or gas is downward.

DOWNHOLE
A term to describe tools, equipment, and instruments used in the well bore; also, conditions or techniques applying to the well bore.

DOWNHOLE MEASUREMENT WHILE DRILLING
A downhole "real time" data gathering and transmitting system that sends information from the drill bit to the surface by one of several means. The data transmitted by some form of telemetry—hardwire or electronics or hydraulic impulse—includes drilling angle and rate, temperature, type formation, and condition of the bit. The MWD system is the most advanced yet developed to keep the driller and geologist informed on conditions several thousand feet downhole.

DOWN IN THE BIG HOLE
A slang expression meaning to shift down into the lowest low gear.

DOWNSTREAM
Refers to facilities or operations performed after those at the point of reference. For example, refining is downstream from production operations; marketing is downstream from refining.

DOWSING RODS
See Doodle Bug.

DOZER
Bulldozer, a crawler-type tractor equipped with a hydraulically operated blade for excavating and grading.

DRAG LINE
A type of large excavating machine made with a long boom over which a line runs down to a clamshell bucket. The bucket at the end of the line is swung into position and is then dragged into the material to be moved or dug out.

DRAG UP
To draw the wages one has coming and quit the job; an expression used in the oil fields by pipeline construction workers and temporary or day laborers.

DRAINAGE
Migration of oil or gas in a reservoir owing to a pressure reduction caused by production from wells drilled into the reservoir. Local drainage is the movement of oil and gas toward the well bore of a producing well.

DRAINAGE TRACT
A lease or tract of land, usually offshore, immediately adjacent to a tract with proven production; an offshore Federal lease contiguous to producing property whose subsurface geologic structure is a continuation of the pro-

Doubling yard *(Courtesy Reading & Bates)*

Drag line

A modern draw works
(Courtesy National)

Spindletop's draw works

Dresser sleeves

Insert drill bit

ducing acreage and therefore more or less valuable as a source of additional oil or gas.

DRAINAGE UNIT
The maximum area in an oil pool or field that may be drained efficiently by one well so as to produce the maximum amount of recoverable oil or gas in such an area.

DRAKE, COL. EDWIN L.
The man who drilled the country's first oil well near Titusville, Pennsylvania, in 1859 to a depth of 69½ feet using crude cable-tool equipment.

DRAWING THE FIRES
Shutting down a refinery unit in preparation for a turnaround (q.v.).

DRAW WORKS
The collective name for the hoisting drum, cable, shaft, clutches, power take off, brakes and other machinery used on a drilling rig. Draw works are located on one side of the derrick floor, and serve as a power-control center for the hoisting gear and rotary elements of the drill column.

DRESSER SLEEVE
A slip-type collar that connects two lengths of plain-end (threadless) pipe. This type sleeve connection is used on small-diameter, low-pressure lines.

DRESS-UP CREW
A right-of-way gang that cleans up after the construction crews have completed their work. The dress-up crew smooths the land, plants trees, grass, and builds fences and gates.

DRIFTING THE PIPE
Testing casing or tubing for roundness; making certain there are no kinks, bends, or flat places in the pipe by use of a drift mandrel (q.v.) or jack rabbit. Pipe must be of proper diameter throughout to be able to run downhole tools such as packers, plugs, etc.

DRIFT MANDREL
A device used to check the size of casing and tubing before it is run. The drift mandrel (jack rabbit) is put through each joint of casing and tubing to make certain the inside diameters are sizes specified for the particular job.

DRILL BIT, DIAMOND
See Diamond Drill Bit.

DRILL BIT, DRAG
A type of old-style drilling tool in which the cutting tooth or teeth were the shape of a fish tail. Drilling was accomplished by the tearing and gouging action of the bit, and was efficient in soft formations; the forerunner of the modern, three-cone roller bit.

DRILL BIT, FISH-TAIL
A drag bit. See Drill Bit, Drag.

DRILL BIT, INSERT
A bit with super-hard metal lugs or cutting points inserted in the bit's cutting cones; a rock bit with cutting elements added that are harder and more durable than the teeth of a mill-tooth bit (q.v.).

DRILL BIT, MILL-TOOTH
A bit with cutting teeth integral to the metal of the cones of the bit; a non-insert bit. Mill-tooth bits are used in relatively soft formations found at shallow depths.

DRILL BIT, ROTARY
The tool attached to the lower end of the drillpipe; a heavy steel "head" equipped with various types of cutting or grinding teeth, some are fixed, some turn on bearings. A hole in the bottom of the drill permits the flow of drilling mud being pumped down through the drillpipe to wash the cuttings to the surface and also cool and lubricate the bit.

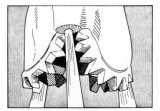

Mill-tooth drill bit

DRILL COLLAR
A heavy, tubular connector between drillpipe and bit. Originally, the drill collar was a means of attaching the drill bit to the drillpipe and to strengthen the lower end of the drill column which is subject to extreme compression, torsion, and bending stresses. Now the drill collar is used to concentrate a heavy mass of metal near the lower end of the drill column. Drill collars were once a few feet long and weighed 400 or 500 pounds. Today, because of increased bit pressure and rapid rotation, collars are made up in 1,000-foot lengths and weigh 50-100 tons.

DRILL COLLAR, SQUARE
A type of drill collar whose cross section is square instead of circular as in a more conventional collar. Square drill collars are used to prevent or minimize the chances of becoming hung up or stuck in a dogleg downhole. The square corners on the collar, which is located just above the drill bit in the string, act as a reamer and tend to keep the hole passable for the drillpipe.

DRILLED SOLIDS
Rock particles broken and pulverized by the bit and picked up by the drilling mud as it circulates. If the minute rock particles do not drop out in the mud pits or are not removed by surface equipment, they add to the mud's density. This condition can cause serious drilling and circulation problems. *See* Drilling Mud Density.

DRILLER
One who operates a drilling rig; the person in charge of drilling operations and who supervises the drilling crew.

DRILLER'S LOG
A record kept by the driller showing the following: when the well was spudded in, the size of the hole, the bits used, when and at what depth various tools were used, the feet drilled each day, the point at which each string of casing was set, and any unusual drilling condition encountered. In present-day wells, the driller's log is supplemented by electrical well logs.

DRILLING AND BELLING TOOL
A long, large-diameter, cylindrical drill with articulating cutting blades folded into the body of the drill for digging holes for piling-in offshore installations: drilling, producing or production platforms. The tool, two to four feet in diameter, is so constructed that when it reaches the required depth of a few hundred feet the hinged cutting blades are extended to cut out a bell-shaped cavity at the bottom of the borehole. Piles then can be

Drilling and belling tool
(Courtesy Calweld)

inserted and cemented. This operation is "drilling in" the piling instead of the more common method of driving the piling.

DRILLING BLOCK
An area composed of separate contiguous leaseholds large enough to drill an exploratory well. Before drilling such a well, particularly a deep well, the operator will usually try to acquire a sizeable block of leases surrounding the site of the proposed exploratory well.

DRILLING CONTRACT
An agreement setting forth the items of major concern to both the operator and drilling contractor in the drilling of a designated well at a given location and at a specified time. One standard drilling contract form is the American Petroleum Institute's (API); another is the American Association of Oilwell Drilling Contractors' (AAODC). *See* API Bid Sheet and Well Specifications.

DRILLING CONTRACTOR
A person or company whose business is drilling wells. Wells are drilled on a per foot basis, others are contracted on a day rate. *See* Turnkey Contract.

DRILLING COSTS, INTANGIBLE
See Intangible Drilling Costs.

DRILLING FLOOR
Derrick floor; the area where the driller and his crew work.

DRILLING FLUID SPECIALIST
A mud engineer (q.v.).

DRILLING FUND
A general term describing a variety of organizations established to attract venture capital to oil and gas exploration and development. Usually the fund is established as a joint venture or limited partnership with minimum investments of $5–$10,000. Such funds attract "high-bracket" persons who will receive certain tax benefits.

Drilling floor

DRILLING HEAD, ROTATING
A heavy casting bolted to the top of the blowout preventers on the casing through which air or gas drilling is done. The kelly joint fits in the rotating element of the drilling head. Compressed air, as the drilling fluid, enters the drillstring through a flexible hose attached to the kelly. As the bit pulverizes the rock, the chips are brought back up the annulus by the force of the high-pressure air and are vented through the blooie pipe to the burn pit.

DRILLING, INFILL
See Infill Drilling.

DRILLING ISLAND
A man-made island constructed in water 10 to 50 feet deep by dredging up the lake or bay bottom to make a foundation from which to drill wells. This procedure is used for development drilling, rarely in wildcatting.

DRILLING JARS
A jointed section in a string of cable tools made with slack or play between the joints. If the bit becomes lodged in the hole, the sudden jar or impact developed by taking up of the slack in the jars aids in freeing the bit.

Drilling and production
island

DRILLING JARS, HYDRAULIC
A tool used in the drillstring for imparting an upward or downward jar or jolt to the drillpipe should it get stuck in the hole while drilling or making a trip. The jars' jolting action is initiated either by the weight or tension of the drillstring which the driller can apply.

DRILLING LOG
See Driller's Log.

DRILLING MAST
A type of derrick consisting of two parallel legs, in contrast to the conventional four-legged derrick in the form of a pyramid. The mast is held upright by guy wires. This type mast is generally used on shallow wells or for reconditioning work. An advanced type of deep-drilling rig employs a mast-like derrick of two principal members with a base as an integral part of the mast.

Drilling mast

DRILLING MUD
A special mixture of clay, water, and chemical additives pumped downhole through the drillpipe and drill bit. The mud cools the rapidly rotating bit; lubricates the drillpipe as it turns in the well bore; carries rock cuttings to the surface; and serves as a plaster to prevent the wall of the borehole from crumbling or collapsing. Drilling mud also provides the weight or hydrostatic head to prevent extraneous fluids from entering the well bore and to control downhole pressures that may be encountered.

DRILLING MUD DENSITY
The weight of drilling mud expressed in pounds per U.S. gallon or in pounds per cubic foot. Density of mud is important because it determines the hydrostatic pressure the mud will exert at any particular depth in the well. In the industry, mud weight is synonymous with mud density. To "heavy up on the mud" is to increase its density.

DRILLING PERMIT
In states that regulate well spacing, a drilling permit is the authorization to drill at a specified location; a well permit.

Drilling platform

DRILLING PLATFORM
An offshore structure with legs anchored to the sea bottom. The platform, built on a large-diameter pipe frame, supports the drilling of a number of wells from the location. As many as 60 wells have been drilled from one large offshore platform.

DRILLING PLATFORM, MAT-SUPPORTED
See Mat-supported Drilling Platform.

DRILLING PLATFORM, MONOPOD
See Monopod Drilling Platform.

DRILL RIG, DC-DC
See Drilling Rig, Electric.

DRILLING RIG, ELECTRIC
A drilling rig that receives its power from a system comprized of diesel engine–DC generator–DC motor. A typical engine–generator–motor

Split-level drilling rig

Flanged drilling spool

Jack-up rig (left) and drilling tender (right)

Drillship

rig-up would include four such sets: two for the mud pumps, one for the draw works and rotary table, and one somewhat smaller set for lighting and auxiliary loads. Another type of electric rig uses the same power-flow system but the generators are AC, whose current is converted to DC current to drive DC motors for variable-speed drilling operations.

DRILLING RIG, MECHANICAL
The most common type of drilling rig is the mechanical compound rig. Mechanical rigs use diesel engines coupled directly to the equipment or through compound shafts to drive the rotary, draw works, and mud pumps. Separate engine-AC generator sets provide lighting and power for auxiliary functions. *See* Drilling Rig, Electric.

DRILLING RIG, SPLIT-LEVEL
A land rig design in which the diesel engines, gear compound, and draw works are at or near ground level, 12 to 15 feet below and behind the rig floor. On the rig floor are the cat works and the rotary table as on a conventional rig. The power from the high-speed (1,800 rpm) diesel engines is transmitted through clutches and compound to the rotary table through a torque tube, rising at about a 45° angle to the gear and chain drive on the rig floor.

DRILLING SPOOL
(1) The part of the draw works that holds the drilling line; the drum of drilling cable on which is spooled the wire line that is threaded over the crown block sheaves and attached to the traveling block. (2) A flanged connector installed within the blowout preventer stack to which the mud access lines and choke and kill lines are attached.

DRILLING TENDER
A barge-like vessel that acts as a supply ship for a small, offshore drilling platform. The tender carries pipe, mud, cement, spare parts, and in some instances provides crew quarters.

DRILLPIPE
Heavy, thick-walled steel pipe used in rotary drilling to turn the drill bit and to provide a conduit for the drilling mud. Joints of drillpipe are about 30 feet long.

DRILL, ROTARY-PERCUSSION
A drill bit that rotates in a conventional manner but at the same time acts as a high-frequency pneumatic hammer, producing both a boring and a fracturing action simultaneously. The hammer-like mechanism is located just above the bit and is actuated by air, liquid, or high-frequency sound waves.

DRILLSHIP
A self-propelled vessel, a ship equipped with a derrick amidships for drilling wells in deep water. A drillship is self-contained, carrying all of the supplies and equipment needed to drill and complete a well.

DRILLSTEM
The drillpipe. In rotary drilling, the bit is attached to the drillstem or drill column which rotates to "dig" the hole.

DRILLSTEM TEST
A method of obtaining a sample of fluid from a formation using a "formation-tester tool" attached to the drillstem. The tool consists of a packer (q.v.) to isolate the section to be tested and a chamber to collect a sample of fluid. If the formation pressure is sufficient, fluid flows into the tester and up the drillpipe to the surface.

DRILLWELL (DRILLSHIP)
See Moonpool.

DRIP
A small in-line tank or condensing chamber in a pipeline to collect the liquids that condense out of the gas stream. Drips are installed in low places in the line and must be "blown" or emptied periodically.

DRIP GASOLINE
Natural gasoline recovered at the surface of a well as the result of the separation of certain of the liquid hydrocarbons dissolved in the gas in the formation; gasoline recovered from a drip (q.v.) in a field gas line; casing-head gasoline.

DRIP OILER
See Wick Oiler.

DRIVE
The energy or force present in an oil reservoir which causes the fluid to move toward the well's borehole and up to the surface when the reservoir is penetrated by the drill. A reservoir is very much like a pressure vessel; when a well is drilled into the reservoir it is as if a valve were opened to vent the pressure. There are several kinds of reservoir drives: gas cap, solution gas, water, and artificial drives (q.v.).

DRIVE PIPE
A metal casing driven into the borehole of a well to prevent caving of the walls and to shut off surface water. The drive pipe, first used in an oil well by Colonel Drake, was the forerunner of the modern conductor or surface pipe. *See* Casing.

DRIVE THE HOOPS
To tighten the staves of a wooden stock tank by driving the metal bands or hoops down evenly around the circumference of the tank. Early-day lease tanks were made of redwood in the shape of a truncated cone (nearly cylindrical). Metal bands like those on a wooden barrel held the staves together. Once a year or so, the hoops had to be driven to tighten the seams between the staves to prevent leaks. Today wooden tanks are used on leases to handle salt water and other corrosive liquids. Their staves are held together with steel rods equipped with turnbuckles for keeping the tank watertight.

DROWNING
A colloquial term for the encroachment of water at the well bore into a formation that once produced oil but now produces more and more water.

DRUM
A 55-gallon metal barrel; a standard container used for shipping lubricating oil and other petroleum products.

Drums

DRY GAS
Natural gas from the well free of liquid hydrocarbons; gas that has been treated to remove all liquids; pipeline gas.

DRY HOLE
An unsuccessful well; a well drilled to a certain depth without finding oil; a "duster" (q.v.).

DRY-HOLE MONEY
Money paid by one or more interested parties (those owning land or a lease nearby) to an operator who drills a well that is a dry hole. The well whether successful or dry serves to "prove their land," providing useful information. Before the well is drilled, the operator solicits dry hole "contributions" and in return for financial assistance agrees to furnish certain information to the contributors.

DRY-HOLE PLUG
A plug inserted in a well that is dry to seal off the formations that were penetrated by the borehole. This treatment prevents salt water, often encountered in "dry holes," from contaminating other formations. *See* Plugging a Well.

DRY TREE
A Christmas tree installed on land or above water as distinguished from a "wet tree," one installed on the sea bed or under water.

DST
Drillstem test (q.v.).

DUAL COMPLETION
The completion of a well in two separate producing formations, each at different depths. Wells sometimes are completed in three or even four separate formations with four strings of tubing inserted in the casing. This is accomplished with packers (q.v.) that seal off all formations except the one to be produced by a particular string of tubing.

DUAL DISCOVERY
A well drilled into two commercial pay zones, two separate producing formations, each at a different depth.

DUAL-FUEL ENGINES
Engines equipped to run on liquid as well as gaseous fuel. Stationary engines in the field have modifications made to their carburetors that permit them to operate either on gasoline or natural gas. In some installations, when the gasoline supply is used up, the engine is switched to natural gas automatically.

DUBAI STORAGE TANKS
A specially designed, underwater storage tank the shape of an inverted funnel, built by Chicago Bridge & Iron for Dubai Petroleum Company. The tanks have no bottoms and rest on the sea floor supported on their rims. Oil from fields on shore is pumped into the top of the tanks under pressure forcing the sea water out the bottom. The offshore tanks which are more

Christmas tree of a triple
completion well

than 100 feet tall also serve as single-point moorings for tankers taking on crude.

DUBBS, CARBON PETROLEUM
Mr. Dubbs, a petroleum chemist, developed a cracking process that found wide acceptance in the 1920s, and was almost as popular as the Burton still which was developed earlier for Standard Oil Company of Indiana.

DUCK'S NEST
Colloquial term for a standby drilling mud tank or pit used to hold extra mud, or as an overflow in the event of a gas "kick" (q.v.).

DULLS
Drill bits badly worn or have lost inserts or have broken teeth.

DUMP BOX
A heavy wooden or metal box where the contents of a cable-tool well's bailer are emptied. The end of the bailer is lowered into the box which pushes the dart upwards, unseating the ball valve and permitting the water, mud, and rock cuttings to empty into the box and slush-pit launder (q.v.).

Worn inset bit

DUMP FLOODING
An unusual secondary recovery technique that uses water from a shallow water bed above the producing pay to flood the oil-producing interval. The water from the aquifer (q.v.) enters the injection string by its own pressure. The weight of the hydrostatic column (water column) produces the necessary force for it to penetrate the oil formation, pushing the oil ahead of it to the producing wells in the field.

DUMP GAS
Gas delivered under a dump gas contract, i.e., a gas purchase and delivery contract that does not call for the delivery of a specified amount of gas but which does call for delivery of surplus gas after meeting the terms of a firm gas contract.

DUSTER
A dry hole; a well that encounters neither gas nor liquid at total depth.

DUTCHMAN
The threaded portion of a length of pipe or nipple twisted or broken off inside a collar or other threaded fitting. Threads thus "lost" in a fitting have to be cut out with a chisel or cutting torch.

DWT
Deadweight ton; a designation for the size or displacement of a ship, e.g., 100,000-dwt crude oil tanker.

DYNAMIC STATIONING
A method of keeping a drill ship or semisubmersible drilling platform on target, over the hole during drilling operations where the water is too deep for the use of anchors. This is accomplished by the use of thrusters (q.v.) activated by underwater sensing devices that signal when the vessel has moved a few degrees off its drilling station.

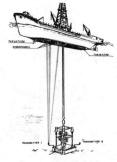
Dynamic stationing

E

EARNEST MONEY
A sum of money paid to bind a financial transaction prior to the signing of a contract; hand money.

EASY DIGGING
A soft job; an assignment of work that can be handled without much exertion.

ECONOMIC DEPLETION
The reduction in the value of a wasting asset (q.v.) by removing or producing the minerals.

EFFLUENT
The discharge or outflow from a manufacturing or processing plant; outfall; drainage.

ELASTOMER
Any of various elastic materials or substances resembling rubber. The petrochemical industry has produced many types of elastomers that are used for gasket material, guides, swab cups, valve seats, machinery vibration-absorber mounts, etc. Elastomers are highly resistant to chemical decomposition (hydrolysis) in the presence of hydrocarbons which make them desirable for use in the petroleum industry, much more so than natural rubber.

ELECTRIC TRACING
See Heat Tracing.

ELECTRIC (WELL) LOG
An electrical survey of a well's borehole before it is cased, which reflects the degree of resistance of the rock strata to electric current. From the results of the survey, geologists are able to determine the nature of the rock penetrated by the drill and some indication of its permeability.

ELECTRIC WELL LOGGING
The procedure of lowering electrical instruments into the well bore to test the density and other characteristics of rock formations penetrated.

Zip-lift elevator

ELEVATORS
A heavy, hinged clamp attached to the hook and traveling block by bail-like arms, and used for lifting drillpipe, casing, and tubing and lowering them into the hole. In hoisting a joint of drillpipe, the elevators are latched on to the pipe just below the tool joint (coupling) which prevents the pipe from slipping through the elevators.

ELK HILLS
Located in the lower end of the San Joaquin Valley of California, Elk Hills is a major part of the U.S. Naval Petroleum Reserves, and is classed as one of the world's giant petroleum fields. Teapot Dome in Wyoming and the North Slope in Alaska are two other large Naval Petroleum Reserves. Elk Hills was set aside as a U.S. oil reserve in 1909 by President Taft upon recommendation of the U.S. Geological Survey (USGS).

EMBAYMENT
A large identation of a coastline; a bay. An embayed coastline.

EMINENT DOMAIN
The right of a government body or public utility (common carrier) to take private property for public use by condemnation proceedings (q.v.).

EMULSION, CRUDE OIL-WATER
Very small droplets of water suspended in a volume of crude oil, each droplet surrounded or encased in a film of oil. The water, although heavier than oil, cannot settle to the bottom of the tank until, through the application of heat or mixing with a chemical, the surface tension of the film of oil is reduced sufficiently to free the water droplets. When this occurs, the small droplets join others to form larger ones which have enough mass or weight to settle to the bottom.

EMULSION TREATER
A tall cylindrical vessel, a type of oil heater for "breaking down" oil-water emulsions with heat and the addition of certain chemicals. Emulsion treaters have a gas-fired furnace at the bottom of the vessel to heat the stream of oil piped through from the well to the stock tanks; a heater-treater.

Emulsion treater

END-O
The command given by one worker to another or to a group to lift together and move an object forward; a signal to "put out" in a big lift.

ENDOTHERMIC
Refers to a process or chemical reaction that requires the addition of heat to keep it going. Exothermic is the reverse; a process or reaction that once begun gives off heat.

END POINT
The point indicating the end of an operation or the point at which a certain definite change is observed. In the analysis of liquids such as gasoline, the end point is the temperature at which the liquid ceases to distill over. End points are of value in predicting certain performance characteristics of gasoline.

END PRODUCTS
Material, substances, goods for consumer use; finished products.

END USE
Ultimate use; consumption of a product by a commercial or industrial customer.

ENERGY SOURCES
Petroleum, coal, hydropower, nuclear, geothermal (q.v.), synthetic fuels (q.v.), tides, solar, wind.

ENERGY VALUE OF PETROLEUM AND PRODUCTS
Million BTU per barrel: crude oil, 5.6; distillate fuel oil, 5.8; residual fuel oil, 6.3; gasoline, 5.3; kerosene, 5.7; petroleum coke, 6.0; and asphalt, 6.6. BTU per cubic foot: dry natural gas, 1,031; wet gas, 1,110.

ENGINE, HOT-PLUG
See Hot-plug Engine.

ENGINE HOUSE
On a cable tool rig the engine house held the steam-powered drilling engine. Attached to the engine house was the belt hall which housed the wide, fabric belt which transmitted power from the engine to the bandwheel.

ENHANCED OIL RECOVERY
Sophisticated recovery methods for crude oil which go beyond the more conventional secondary recovery techniques of pressure maintenance and waterflooding. Enhanced recovery methods now being used include micellar surfactant (q.v.), steam drive, polymer, miscible hydrocarbon, CO_2, and steam soak. EOR methods are not restricted to secondary or even tertiary projects. Some fields require the application of one of the above methods even for initial recovery of crude oil.

ENRICHED-GAS INJECTION
A secondary recovery method involving the injection of gas rich in intermediate hydrocarbons or enriched by addition of propane, butane, or pentane on the surface or in the well bore as the gas is injected.

ENTITLEMENTS PROGRAM
A program instituted in 1974 by the federal government to equalize the access to domestic crude by all U.S. refiners—crude oil which was price controlled substantially below world price. The reasoning was that disproportionate access to inexpensive domestic crude would give an unfair advantage to some refiners, those with a large supply of price-controlled oil. The program's aim was to make available to each refiner the same fraction of low-priced oil. Refiners with more price-controlled oil than a calculated national average were required to buy entitlements. Refiners with a lower than average amount of price-controlled oil could sell entitlements. The buying and selling of entitlements was between traditional suppliers and purchasers. For example, if Gulf Oil were the traditional supplier of crude oil to Bradford Refining Co., and Gulf had available a larger percent of price-controlled oil than the national average and Bradford had less than the average, Bradford could sell its entitlements for crude-cost equalization to Gulf and Gulf would be required to buy them. In effect, Gulf, the traditional supplier, would pay to Bradford Refining a certain amount of money for each barrel of uncontrolled crude oil it had to buy in the world market.

ENTRAINED OIL
Oil occurring as part of the gas steam, but as a relatively small percentage of total flow. Special separators are used to remove the liquid from the gas stream.

ENTRY POSITION
A starting job with a company usually sought by a young man or woman just out of school who wishes to get into the business at whatever level— with the expectation of becoming president in due time.

EOR
Enhanced oil recovery (q.v.).

EPA
Environmental Protection Agency.

E.P. LUBRICANTS
Extreme pressure lubricating oils and greases which contain substances added to prevent metal to metal contact in highly loaded gears and turn-tables.

EQUITY CRUDE
In cases where a concession is owned jointly by a host government and an oil company, the crude produced which belongs to the oil company is known as equity crude, as opposed to buy-back (participation) crude. The cost of equity crude is calculated according to the posted price. *See* Buy-back Crude Oil.

ESCAPE BOOMS
Devices used on offshore drilling or production platforms for emergency escape of personnel in the event of a fire or explosion. They consist of counterweighted arms supporting a buoyant head. When the arms are snapped loose from the platform, they fall outward, the head descending to the water. The workers then slide down a lifeline to the floating head.

ESCROW MONEY
See Suspense Money.

ET ALS
And others; unnamed participants or interest holders in a deal or a con-tract; et al. made plural and used colloquially by oil men.

ETHANE
A simple hydrocarbon associated with petroleum. Ethane is a gas at ordi-nary atmospheric conditions.

ETHANOL
Alcohol; one component of gasohol (q.v.).

EVAPORATION PIT
An excavation dug to contain oil field salt water or brine which is disposed of by evaporation. Great amounts of salt water are produced with crude oil in some oil fields, particularly older ones.

EXOTHERMIC
Refers to a process or chemical reaction that gives off heat. Endothermic is the reverse; a process or reaction that requires the addition of heat to keep it going.

EXPANSION FIT
See Shrink Fit.

EXPANSION JOINT
A section of piping constructed in such a way as to allow for expansion and contraction of the pipe connections without damaging the joints. Specially fabricated, accordion-like fittings are used as expansion joints in certain in-plant hookups where there are severe temperature changes.

Deethanizer vessel
(Courtesy Motherwell Bridge)

Expansion loop

EXPANSION LOOP

A circular loop (360° bend) put in a pipeline to absorb expansion and contraction caused by heating and cooling, without exerting a strain on pipe or valve connections.

EXPANSION ROOF TANK

A storage or working tank (q.v.) with a roof made like a slip joint. As the vapor above the crude oil or volatile product expands with the heat of the day, the roof-and-apron section of the tank moves upward permitting the gas to expand without any loss to the atmosphere. The telescoping roof, as it moves up and down, maintains a gastight seal with the inner wall of the tank.

EXPLOITATION WELL

A development well; a well drilled in an area proven to be productive. *See* Infill Drilling.

EXPLORATION ACTIVITIES

The search for oil and gas. Exploration activities include aerial surveying, geological studies, geophysical surveying, coring, and drilling of wildcat wells.

EXPLORATION VESSEL

A seagoing, sophisticated research ship equipped with seismic, gravity, and magnetic systems for gathering data on undersea geologic structures. On the more advanced vessels of this type there are onboard processing and interpretation capabilities for the information gathered as the vessel cruises on the waters of the Outer Continental Shelf around the world.

EXPLOSION-PROOF MOTORS

A totally enclosed electric motor with no outside air in contact with the motor windings; an enclosed brushless motor. Cooling is by conduction through the frame and housing.

EXPLOSIVE FRACTURING

Using an explosive charge in the bottom of the well to fracture the formation to increase the flow of oil or gas. *See* Well Shooter.

EXPLOSIVE WELDING

A method of welding is which a shaped explosive charge is used to "fast-expand" the end of a section of pipe into the bore of a special steel sleeve to form a solid bond. The shaped charge is inserted into the end of the pipe over which the sleeve is placed. When the charge is detonated the force expands the pipe's outer circumference forcibly to the sleeve's inner circumference making a secure, pressure-tight bond. This welding technique creates little heat which, for certain jobs, is more desirable than fusion welding in which both pieces of metal must be heated to a high temperature.

EXTENSION TEST

See Outpost Well.

EXTERNAL CASING PACKER

A device used on the outside of a well's casing to seal off formations or to protect certain zones. Often used downhole in conjunction with cementing.

The packer is run on the casing and when at the proper depth, it may be expanded against the wall of the borehole hydraulically or by fluid pressure from the well.

EYEBALL
To align pipe connections or a temporary construction with the eye only; to inspect carefully.

F

FABRICATED VALVE
A type of valve or other fitting that is built and welded together from wrought iron and forged steel pieces to make a particularly strong high-pressure valve. Most valves are steel castings with bodies, bonnets, and packing glands cast separately and assembled.

FACING MACHINE, PIPE
A device for beveling or putting a machined face on the ends of large-diameter line pipe. The facing machine essentially is a revolving disc-chuck holding a number of cutting tools. The chuck is held in alignment against the pipe end by a hydraulically actuated mandrel inserted into the pipe similar to internal line-up clamps used to align pipe for welding. The facing machine is transported and brought into position by a modified boomcat.

Pipe facing machine

FAIL-SAFE
Said of equipment or a system so constructed that, in the event of failure or malfunction of any part of the system, devices are automatically activated to stabilize or secure the safety of the operation.

FAIRLEAD
A guide for ropes or lines on a ship to prevent chafing; a sheave supported by a bracket protruding from the cellar deck of a semisubmersible drilling platform over which an anchor cable runs. Some large floating platforms have anchor lines made up of lengths of chain and cable.

FAIRWAY
A shipping lane established by the U.S. Coast Guard in Federal offshore waters. Permanent structures such as drilling and production platforms are prohibited in a fairway which significantly curtails oil activity in some off-shore areas.

Fairlead

FANNING THE BOTTOM (OF THE BOREHOLE)
Drilling with very little weight on the drill bit in the hope of preventing the bit from drifting from the vertical and drilling a crooked hole. Fanning the bottom, however, is considered detrimental to the drillstring, by some authorities, as reduced weight on the bit causes more tension on the drill-pipe, resulting in pipe and collar fatigue.

FARM BOSS
A foreman who supervises the operations of one or more oil-producing leases.

FARMER'S OIL

An expression meaning the landowner's share of the oil from a well drilled on his property; royalty, usually one-eighth of the produced oil free of any expense to the landowner.

FARMER'S SAND

A colloquial term for "the elusive oil-bearing stratum which many landowners believe lies beneath their land, regardless of the results of exploratory wells."

FARM-IN

An arrangement whereby one oil operator "buys in" or acquires an interest in a lease or concession owned by another operator on which oil or gas has been discovered or is being produced. Often farm-ins are negotiated to assist the original owner with development costs and to secure for the buyer a source of crude or natural gas. See Farm-out Agreement.

FARM-OUT

The name applied to a leasehold held under a farm-out agreement (q.v.).

FARM-OUT AGREEMENT

A form of agreement between oil operators whereby the owner of a lease who is not interested in drilling at the time agrees to assign the lease or a portion of it to another operator who wishes to drill the acreage. The assignor may or may not retain an interest (royalty or production payment) in the production.

FAST LINE

On a drilling rig, the fast line is the cable spooled off or on the hoisting drum of the draw works; the line from the hoisting drum that runs up through the derrick to the first sheave in the crown block.

FAT OIL

The absorbent oil enriched by gasoline fractions in an absorption plant. After absorbing the gasoline fractions, the gasoline is removed by distillation, leaving the oil "lean" and ready for further use to absorb more gasoline fractions from the natural gas stream.

FAULT

A fracture in the earth's crust accompanied by a shifting of one side of the fracture with respect to the other side; the point at which a geological strata "breaks off" or is sheared off by the dropping of a section of the strata by settling.

FEA

Federal Energy Agency.

FEDERAL LEASE

An oil or gas lease on Federal land issued under the Act of February 25, 1920, and subsequent legislation.

FEE

The title or ownership of land; short for "owned in fee." The owner of the fee holds title to the land.

Fast line on the drum

NORMAL   REVERSE

THRUST   LATERAL

Four common faults

FEEDER LINE
A pipeline; a gathering line tied into a trunk line.

FEED OR FEEDSTOCK
Crude oil or other hydrocarbons that are the basic materials for a refining or manufacturing process.

FEE ROYALTY
The lessor's share of oil and gas production; landowner's royalty.

FEE SIMPLE
Land or an estate held by a person in his own right without restrictions.

FEMALE CONNECTION
A pipe, rod, or tubing coupling with the threads on the inside.

FERC
Federal Energy Regulatory Commission.

FIELD
The area encompassing a group of producing oil and gas wells; a pool. An oil field may include one or more pools, and have wells producing from several different formations at different depths. A roughly contiguous grouping of wells in an identified area. Some of the early prolific fields were: East Texas, Seminole, Cushing, Oklahoma City, and West Texas. Large areas that used to be designated as fields are now identified as districts, e.g. Appalachian, Mid-Continent, Gulf Coast, Rocky Mountain, and Permian Basin.

Female connection

FIELD BUTANES
A raw mix of natural gas liquids; the product of gas processing plants in the field. Raw mix streams are sent to fractionating plants where the various components—butane, propane, hexane, and others—are separated. Some refineries are capable of using field butanes as 10 to 15 percent of charge stock.

FIELD COMPRESSION TEST
A test to determine the gasoline content of casinghead or wet gas.

FIELD POTENTIAL
The producing capacity of a field during a 24-hour period. In order to establish a field allowable for prorationing purposes by a state regulatory commission, a field potential was necessary to set equitable production levels.

Field tank

FIELD TANKS
Stock tanks (q.v.).

FILTER CAKE
A plaster-like coating on the borehole resulting from the solids in the drilling fluid adhering and building up on the wall of the hole. The buildup of "cake" can cause serious drilling problems including the sticking of the drillpipe. *See* Differential-pressure Sticking.

FILTRATE
The solid material in drilling mud. When filtrate is deposited on the wall of

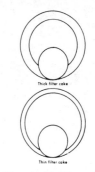

Pipe stuck in filter cake

the borehole of the well forming a thick, restrictive layer, it is referred to as filter cake.

FILTRATION-LOSS QUALITY OF MUD
A drilling-mud quality measured by putting a water-base mud through a filter cell. The mud solids deposited on the filter is filter cake (q.v.) and is a measure of the water-loss quality of the drilling mud. Mud mixtures with low water loss are desirable for most drilling operations.

FINES
Minute particles of a solid substance—rock, coal, or catalytic material—too small to be used or handled efficiently, and that must be removed by screening.

Finger board

FINGER BOARD
A rack high in the derrick made to support, in orderly fashion, the upper ends of the tubing stands that are pulled from the well and set back (q.v.) in the derrick.

FINGERING
Rivulet-like infiltration of water or gas into an oil-bearing formation as a result of failure to maintain reservoir pressure, or as the result of taking oil in excess of maximum efficiency rates (MER) (q.v.). See Channel.

FINGER PIER
A jetty or bridge-type structure extending from the shore out into a body of water to permit access to tankers and other vessels where water depth is not sufficient to allow docking at the shore.

FIRE FLOODING
See In Situ Combustion.

FIRE-TRIMMED (VALVES AND FLANGES)
A designation for valves, flanges, and other fittings made to withstand an accidental fire in a plant or process unit. Fire-trimmed valves, when subjected to fire from whatever cause, will not be damaged to the extent that they will leak and thereby add to the emergency. Such valves and flanges have metal gaskets and stuffing boxes with specially formulated fire-resistant packing or are fitted with metal-to-metal seals.

Tanker loading at finger pier

FIRE WALL
An earthen dike built around an oil tank to contain the oil in the event the tank ruptures or catches fire. Also a fireproof panel, bulkhead, or wall to block the spread of a fire or give protection for a period of time while emergency action is taken.

FIRE-WATER SUPPLY
A pond or tank containing water used exclusively for firefighting.

FIRING LINE
In pipeline construction, the part of the project where the welding is being done. Ahead of the firing line or the "front end," there may be some preparatory or alignment welding. This is true for extra-large diameter pipe. At the front end the pipe is lined up end-to-end by sideboom tractors, brought into perfect alignment by internal line-up clamps, tack welded in position, and

then a root run or first pass or course of metal is laid on by a welder. The firing line is where two or more filler passes and the final or cap pass are made.

FIRM GAS
Gas required to be delivered and taken under the terms of a firm gas purchase contract. Firm gas is priced higher than dump gas (q.v.).

FISH
Anything lost down the drill hole; the object being sought downhole by the fishing tools.

FISHING JOB
The effort to recover tools, cable, pipe, or other objects from the well bore which may have become detached while in the well or been dropped accidentally into the hole. Many special and ingeniously designed fishing tools are used to recover objects lost downhole.

FISHING TOOLS
Special instruments used in recovering objects lost in a well. Although there are scores of standard tools used in fishing jobs, some are specially designed to retrieve particular objects. There are a variety of spears, harpoon-like tools which are forced into the end of a section of tubing, casing or drillpipe; there are overshots with internal lugs or teeth which are forced over the end of lost pipe; and junk baskets for retrieving smaller objects such as lost drill bit cones, small slips, and other steel, nondrillable objects.

FISHPROOF
Describes an item of equipment used in or over the well's borehole without parts—screws, lugs, wedges or dogs—that can come loose, fall into the well, and have to be fished out. *See* Captured Bolt

FITTINGS
Small pipes (nipples), couplings, elbows, unions, tees, swages used to make up a system of piping.

FITTINGS, TRANSITION
See Transition Fittings.

FIVE-SPOT WATERFLOOD PROGRAM
A secondary recovery operation where four input or injection wells are located in a square pattern with the production well in the center, a layout similar to a five-of-spades playing card. The water from the four injection wells moves through the formation flooding the oil toward the production well.

FIXED-BED CATALYST
A catalyst in a reactor vessel through which the liquid being treated drips or percolates through the bed of catalyst material. In other methods, the catalyst is mixed thoroughly with the feedstock as it is pumped into the reactor vessel.

FLAG THE LINE
To tie pieces of cloth on the swab line at measured intervals to be able to tell how much line is in the hole when coming out with the line.

Fittings (bushing, union, ell, tee plug)

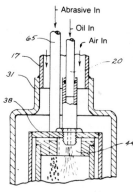

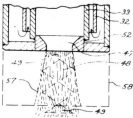

Abrasive In

Oil In

Air In

Flame jet drill *(after Browning & Fitzgerald)*

A flare

FLAMBEAU LIGHT

A torch used in the field for the disposal of casinghead gas produced with oil when the gas is without a market or is of such small quantity as to make it impractical to gather for commercial use. The use of flambeau lights is now regulated under state conservation laws.

FLAME ARRESTOR

A safety device installed on a vent line of a storage or stock tank that, in the event of lightning or other ignition of the venting vapor, will prevent the flame from flashing to the vapor inside the tank. Flame arrestors have a "membrane" of fine-mesh screen across the vent opening. The vapor venting from the tank may be ignited but the flame will not flash through the fine screen into the tank. Just as the old-fashioned miners' lamps, with their open flames encased in a fine-mesh screen, could be used in a gassy mine, the tank flame arrestor is effective for the same reason: the fine screen "breaks up" and cools the flame at the surface of the screen preventing the ignition of the vapor behind the screen.

FLAME-JET DRILLING

A drilling technique that uses rocket fuel to burn a hole through rock strata. This leaves a ceramic-like sheath on the walls of the borehole, eliminating the need for casing.

FLAME SNUFFER

An attachment to a tank's vent line that can be manually operated to snuff out a flame at the mouth of the vent line; a metal clapper-like valve that may be closed by pulling on an attached line.

FLAMMABLE (INFLAMMABLE)

Term describing material which can easily be ignited. Petroleum products with a flash point (q.v.) of 80°F. or lower are classed as flammable. The use of the word "inflammable" which means flammable but thought by some people to mean nonflammable has fallen into disuse because of the confusion and the potential hazard of someone mistaking the prefix "in" for "non."

FLANGE

(1) A type of pipe coupling made in two halves. Each half is screwed or welded to a length of pipe and the two halves are then bolted together joining the two lengths of pipe. (2) A rim extending out from an object to provide strength or for attaching another object.

FLANGE UP

To finish a job; to bring to completion.

FLARE

(1) To burn unwanted gas through a pipe or stack. (Under conservation laws, the flaring of natural gas is illegal.) (2) The flame from a flare; the pipe or the stack itself.

FLARE GAS

Gaseous hydrocarbons discharged from safety relief valves on process units in a refinery or chemical plant. Should a unit go down from an electrical or cooling water failure, making it necessary to dump a batch of liquid

feed or product, the flare stack is equipped to handle such an emergency. If it were impossible to dump both gases and liquids in an emergency, the plant personnel and the operating units would be in danger. With recovery equipment which larger plants are installing, the flare gases as well as the dumped process fluid are recovered. The gases used as fuel; the liquids reprocessed.

FLARE, SMOKELESS
See Smokeless Flare.

FLASH CHAMBER
A refinery vessel into which a process stream is charged or pumped and where lighter products flash off or vaporize and are drawn off at the top. The remaining heavier fractions are drawn off at the bottom of the vessel.

FLASH OFF
To vaporize from heated charge stock; to distill.

FLASH POINT
The temperature at which a given substance will ignite.

FLEXIBLE COUPLING
A connecting link between two shafts that allows for a certain amount of misalignment between the driving and driven shaft without damage to bearings. Flexible couplings dampen vibration and provide a way to make quick hook ups of engines and pumps which is useful in field operations.

FLOAT
(1) A long, flat-bed trailer the front end of which rests on a truck, the rear end on two dual-wheel axles. Floats are used in the oil fields for transporting long, heavy equipment. (2) The buoyant element of a fluid-level shut off or control apparatus. An airtight canister or sphere that floats on liquids and is attached to an arm that moves up and down, actuating other devices as the liquid level rises and falls.

Floater

FLOATER
(1) A barge-like drilling platform used in relatively shallow offshore work. (2) Any offshore drilling platform without a fixed base, e.g. semisubmersibles—drill ships—drill barges.

FLOATING ROOF TANK
A storage tank with a flat roof that floats on the surface of the oil thus reducing evaporation to a minimum. The roof rests on a series of pontoons whose buoyancy supports the roof proper; a floater.

FLOATING THE CASING
A method of lowering casing into very deep boreholes when there exists the danger of the casing joints separating because of the extreme weight or tension on the upper joints. In floating, the hole is filled with fluid and the casing is plugged before being lowered into the hole. The buoyant effect of the hollow column of casing displacing the fluid reduces the weight and the tension on the upper joints. When the casing is in place, the plug is drilled out.

Floating roof tank

Float-actuated valve
(Courtesy Fisher)

Floor men

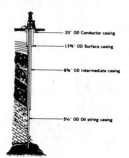

Flow string

Fluid catalytic cracking
unit

FLOAT VALVE
A valve whose stem is attached by an arm to a float; an automatic valve operated, through linkage to a float mechanism, by the change in liquid level in a tank or other vessel.

FLOODING
The use of water injected into a production formation or resevoir to increase oil recovery. *See* Secondary Recovery; Micellar-surfactant Flooding; Tertiary Recovery.

FLOODING, DUMP
See Dump Flooding.

FLOOR MEN
Members of the drilling crew (usually two) who work on the derrick floor.

FLOW BEAN
A drilled plug in the flow line at the wellhead that controls the oil flow to desired rate. Flow beans are drilled with different-sized holes for different flow rates.

FLOW CHART
A replaceable, paper chart on which flow rates are recorded by an actuated arm and pen of a flowmeter.

FLOW NIPPLE
A choke; a heavy steel nipple put in the production string of tubing that restricts the flow of oil to the size of the orifice in the nipple. It is usual to report a new well's production as a flow of a certain number of barrels per day through a choke of a certain size, e.g., 16/64 in., 9/64 in. etc.; a flow plug.

FLOW, PLASTIC
See Plastic Flow; *also* Turbulent Flow.

FLOW SHEET
A diagrammatic drawing showing the sequence of refining or manufacturing operations in a plant.

FLOW STRING
The string of casing or tubing through which oil from a well flows to the surface. *Also* Oil string; pay string; capital string; production string.

FLOW TANK
A single unit that acts as an oil and gas separator, an oil heater, and an oil and water treater.

FLUID CATALYTIC CRACKING UNIT
A large refinery vessel for processing reduced crude, naphthas, or other intermediates in the presence of a catalyst. Catalytic cracking is regarded as the successor to thermal cracking as it produces less gas and highly volatile material; it provides a motor spirit of 10-15 octane numbers higher than that of the thermally cracked product. The process is also more effective in producing isoparaffins and aromatics which are of high antiknock value.

FLUID END (OF A PUMP)

The end of the pump body where the valves (suction and discharge) and the pump cylinders are located. The fluid end of a reciprocating pump is accessible by removing the cylinder heads which exposes the pistons or pump plungers. The cylinders or liners in most pumps are removable and can be changed for others with larger or smaller internal diameters. Inserting smaller liners and pistons permits pumping at higher pressure but at a reduced volume.

FLUIDICS

Pertains to the use of fluids (and air) in instrumentation. Fluidics is defined as "engineering science pertaining to the use of fluid-dynamic phenomena to sense, control, process information, and actuate." Fluidics provide a reliable system far less expensive than explosion-proof installations required with electrical instrumentation on offshore rigs.

FLUID LEVEL

The distance between the wellhead and the point to which the fluid rises in the well.

FLUID LOSS

A condition downhole in which a water-base drilling mud loses water in a highly permeable zone causing the solids in the drilling fluid to build up on the wall of the borehole. This buildup of mud solids can result in stuck pipe, which often arises when the hydrostatic head or mud pressure is considerably higher than the formation pressure.

FLUSHING OILS

Oils or compounds formulated for the purpose of removing used oil, decomposed matter, metal cuttings, and sludge from lubricating passages and engine parts.

FLUSH PRODUCTION

The high rate of flow of a good well immediately after it is brought in.

FLUXING

To soften a substance with heat so that it will flow; to lower a substance's fusing point.

FOLDING

Buckling or other deformation of rock strata caused by movement of the earth's crust.

FOOTAGE CONTRACT

A contract for the drilling of a well in which the drilling contractor is paid on a footage basis as the well is taken down. Sometimes the price per foot changes as the well progresses and different formations are encountered.

FOOT-POUND

A unit of energy or work equal to the work done in raising one pound the height of one foot against the force of gravity.

FOOT VALVE

A type of check valve (q.v.) used on the "foot" or lower end of a suction-pipe riser to maintain the column of liquid in the riser when the liquid is being drawn upward by a pump.

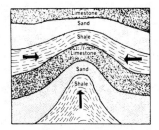

Folding *(Courtesy Petex)*

Forced-draft burner

FORCED-DRAFT BURNER
Crude oil disposal equipment on offshore platforms. The burner, mounted on a boom or an extension of the deck, burns crude oil during testing operations. Gas, air and water manifolded with the test-crude stream result in complete combustion of the oil.

FORCE MAJEURE CLAUSE
A lease clause providing that cessation or failure of production shall not cause automatic termination of the leasehold, and that the performance of lessee's covenants shall be excused when the failure of production or performance of covenants is owing to causes set forth in the clause. Such clauses usually list acts of God; adverse weather; compliance with federal, state, or municipal laws; wars; strikes; and other contingencies over which the lessee has no control.

FORCE PUMP
A barrel pump; a portable, hand-operated, one-cylinder pump for moving limited amounts of liquid, pumping out sump pits, or transferring oil or water from one small tank to another. The pump has one horizontal barrel and a plunger attached to a vertical handle. When moved back and forth, the handle, attached to a fulcrum at the base of the pump, actuates the plunger.

FOREIGN TAX CREDIT
Taxes paid a foreign government by a U.S. company on its overseas oil operations that are creditable against taxes owed the U.S. government. Production sharing by a U.S. company with a foreign government or one of its agencies represents oil royalty payments, not taxes creditable in the U.S. according to the Internal Revenue Service.

FOREIGN TRADE ZONE
An area in the United States where imported oil, reduced crude, or intermediates are processed.

FORMATION
A sedimentary bed or series of beds sufficiently alike or distinctive to form an identifiable geological unit.

FORMULATION
The product of a formula, i.e., a plastic, blended oils, gasolines; any material with two or more components or ingredients.

FOSSIL ENERGY
Energy derived from crude oil, natural gas, and coal; also shale oil and oil recovered from tar sands. Fossil energy by implication is the energy derived from sedimentary beds containing the fossilized remains of marine plants and animals; and thus oil, gas, and coal were derived from organisms living in eons past.

FOSSIL FUEL
See Fossil Energy.

FOURBLE
A stand (q.v.) of drillpipe or tubing consisting of four joints. In pulling pipe from the well, every fourth joint is unscrewed and the four-joint stand is set

back and racked in the derrick. This is not a common practice; the usual stands are of two and three joints.

FOUR-CYCLE ENGINE
An internal combustion engine in which the piston completes four strokes—intake, compression, power, and exhaust—for each complete cycle. The Otto-cycle engine; four-stroke cycle engine.

FPC
Federal Power Commission, an agency of the Federal government; a regulatory body having to do with oil and gas matters such as pricing and trade practices. As of 1977 the FPC has been superceded by the FERC, the Federal Energy Regulatory Commission which has taken over the duties and responsibilities of Federal Power Commission along with other agencies in the field of energy.

FRAC JOB
See Hydraulic Fracturing.

FRACTION
A separate, identifiable part of crude oil; the product of a refining or distillation process; a cut. Through the highly efficient and sophisticated science of petroleum refining, crude oil is divided literally into hundreds of fractions, valuable products of a spectrum ranging from highly volatile, ethereal gases through fuels and oils, and road building material to coke, hard, black, and valuable.

FRACTIONATOR
A tall, cylindrical refining vessel where liquid feedstocks are separated into various components or fractions.

FRASCH, HERMAN
A Canadian chemist who developed a process for the use of sour crude for making kerosene. The Frasch process opened the market for sour crude from Ohio and Canada just when it was thought the production from Pennsylvania and West Virginia fields had peaked and the country was running low on sweet crude for kerosene and gasoline.

FREE MARKET PRICE
Oil prices not subject to controls by the government; world prices. *See* Posted Price.

FREE RIDE
An interest in a well's oil and gas production free of any expense of that production; a royalty interest.

FREEZE BOX
An enclosure for a water-pipe riser that is exposed to the weather. The freeze box or frost box surrounding the pipe is filled with sawdust, manure or other insulating material.

FREON
A trademark applied to a group of halogenated hydrocarbons having one or more florine atoms in the molecule; a refrigerant.

Fractionator

FRESNO SLIP
A type of horse-drawn, earth-moving or cutting scoop with curved runners or supports on the sides and a single long handle used to guide the scoop blade into the earth or material being moved.

FROST UP
Icing of pipes and flow equipment at the wellhead of a high-pressure gas well. The cooling effect of the expanding gas, as pressure is reduced, causes moisture in the atmosphere to condense and freeze on the pipes.

FUEL-AIR RATIO
The ratio of fuel to air by weight in an explosive mixture which is controlled by the carburetor in an internal combustion engine.

FUEL OIL
Any liquid or liquefiable petroleum product burned for the generation of heat in a furnace; or for the generation of power in an engine, exclusive of oils with a flash point below 100°F.

FULL BORE
Designation for a valve, ram or other fitting whose opening is as large in cross section as the pipe, casing or tubing it is mounted on.

FULLER'S EARTH
A fine, clay-like substance used in certain types of oil filters.

FULLY INTEGRATED
Said of a company engaged in all phases of the oil business, i.e., production, transportation, refining, marketing. *See* Integrated Oil Company.

FUNGIBLE
Products that are or can be commingled for the purpose of being moved by product pipeline. Interchangeable.

FUNNEL VISCOSITY
The number of seconds required for a quart of drilling mud to run through a mud funnel. The funnel viscosity has only one function: to spot a change in the density of the drilling fluid. A change in density, higher or lower, tells the experienced driller and mud men a lot about drilling conditions downhole.

FURFURAL
An extractive solvent of extremely pungent odor, used extensively in refining a wide range of lubricating oils and diesel fuels; a liquid aldehyde.

FURNACE OIL
No. 2 heating oil; light gas oil that can be used as diesel fuel and for residential heating; Two oil; distillate fuel.

FUSIBLE PLUG
A fail-safe device; a plug in a service line equipped with a seal that will melt at a predetermined temperature releasing pressure that actuates shutdown devices; a meltable plug.

G

GALL
To damage or destroy a finished metal surface, as a shaft journal, by moving contact with a bearing without sufficient lubrication. To chafe by friction and heat as two pieces of metal are forcibly rubbed together in the absence of lubrication.

GAMMA RAY
Minute quantities of radiation emitted by substances that are radioactive. Subsurface rock formations emit radiation quantum that can be detected by well-logging devices, and which indicate the relative densities of the surrounding rock.

GAMMA RAY — GAMMA RAY LOGGING
A well-logging technique wherein a well's borehole is bombarded with gamma rays from a gamma ray emitting device to induce output signals that are then recorded and transmitted to the surface. The gamma ray signals thus picked up indicate to the geologist the relative density of the rock formation penetrated by the well bore at different levels.

GAMMA RAY LOGGING
See Natural Gamma Ray Logging.

GANG PUSHER
A pipeline foreman; the man who runs a pipeline or a connection gang; a pusher.

GANG TRUCK
A light or medium-sized flat-bed truck carrying a portable doghouse or man rack where the pipeline repair crew rides to and from the job. The pipeliners' tools are carried in compartments beneath the bed of the truck.

GAS
"Any fluid, combustible or noncombustible, which is produced in a natural state from the earth and which maintains a gaseous or rarified state at ordinary temperature and pressure conditions." Code of Federal Regulations, Title 30, Mineral Resources, Chap. II, Geological Survey. 221.2.

GAS ANCHOR
A device for the bottom-hole separation of oil and gas in a pumping well. The gas anchor (a length of tubing about 5 feet long) is inside a larger pipe with perforations at the upper end. Oil in the annulus between the well's casing and tubing enters through the perforations and is picked up by the pump; the gas goes out through the casing to the wellhead.

GAS BOTTLES
The cylindrical containers of oxygen and acetylene used in oxyacetylene welding. Oxygen bottles are tall and slender with a tapered top; acetylene bottles are shorter and somewhat larger in diameter.

GAS BOX
Colloquial term for a mud-gas separator at a drilling well. *See* Degassing Drilling Mud.

Hydraulic fracturing a gas well

GAS BUSTER
A drilling-mud/gas separator; a surge chamber on the mud-flow line where entrained gas breaks out and is vented to a flare line; the gas-free mud is returned to the mud tanks or mud pits.

GAS CAP
The portion of an oil-producing reservoir occupied by free gas; gas in a free state above an oil zone.

GAS-CAP ALLOWABLE
A production allowable granted an operator who shuts in a well producing from a gas cap of an oil-producing reservoir. The allowable is transferable to another lease in the same field. The shutting in of the gas-cap producer preserves the reservoir pressure which is essential to good production practice.

GAS-CAP DRIVE
The energy derived from the expansion of gas in a free state above the oil zone which is used in the production of oil. Wells drilled into the oil zone cause a release of pressure which allows the compressed gas in the cap to expand and move downward forcing the oil into the well bores of the producing wells.

GAS-CAP FIELD
A gas-expansion reservoir in which some of the gas occurs as free gas rather than in solution. The free gas will occupy the highest portion of the reservoir. When wells are drilled to lower points on the structure, the gas will expand forcing the oil down-dip and into the well bores.

GAS CONDENSATE
Liquid hydrocarbons present in casinghead gas that condense upon being brought to the surface; formerly distillate, now condensate. Also casinghead gasoline; white oil.

GAS CONDENSATE, RETROGRADE
See Retrograde Gas Condensate.

GAS-CUT MUD
Drilling mud aerated or charged with gas from formations downhole. The gas forms bubbles in the drilling fluid seriously affecting drilling operations, sometimes causing loss of circulation.

GAS DISTILLATE
See Distillate.

GAS DRILLING
The use of gas as a drilling fluid. See Air Drilling.

GAS ENGINE
A two or four-cycle internal combustion engine that runs on natural gas; a stationary field engine. Before the electrification of oil fields, all pumping wells and small pipeline booster stations were powered by stationary gas engines. Hundreds of thousands of stripper wells are pumped by small gas engines, attended by lease pumpers who not only are practical mechanics but are experienced production men as well.

GASIFICATION

Converting a solid or a liquid to a gas; converting a solid hydrocarbon such as coal or oil shale to commercial gas; the manufacture of synthetic gas from other hydrocarbons. *See* Synthetic Gas.

GAS INJECTION

Natural gas injected under high pressure into a producing reservoir through an input or injection well as part of a pressure maintenance, secondary recovery, or recycling operation.

GAS INJECTION WELL

A well through which gas under high pressure is injected into a producing formation to maintain reservoir pressure.

GAS, INTERRUPTIBLE

See Interruptible Gas.

GASKET

Thin, fibrous material used to make the union of two metal parts pressure-tight. Ready-made gaskets are often sheathed in very thin, soft metal, or they may be made exclusively of metal, or of specially formulated rubber.

GAS KICK

See Kick.

GAS LIFT

A method of lifting oil from the bottom of a well to the surface by the use of compressed gas. The gas is pumped into the hole and at the lower end of the tubing it becomes a part of the fluid in the well. As the gas expands, it lifts the oil to the surface.

GAS-LIFT GAS

Natural gas used in a gas-lift program of oil production. Lift gas is usually first stripped of liquid hydrocarbons before it is injected into the well. And because it is a "working gas" as opposed to commercial gas, its cost per thousand cubic feet (MCF) is considerably less. Gas lift and commercial gas commingle when produced, so when the combined gas is stripped of petroleum liquids only the formation gas is credited with the recovered liquids. This is necessary for oil and gas royalty purposes.

GAS, LIQUEFIED PETROLEUM

See Liquefied Petroleum Gas.

GAS LIQUIDS

See LPG.

GAS LOCK

A condition that can exist in an oil pipeline when elevated sections of the line are filled with gas. The gas, because of its compressibility and penchant for collecting in high places in the line, effectively blocks the gravity flow of oil. Gas lock can also occur in suction chambers of reciprocating pumps. The gas prevents the oil from flowing into chambers and must be vented or bled off.

GAS MEASUREMENT, STANDARD

A method of measuring volumes of natural gas by the use of conversion

factors of standard pressure and temperature. The standard pressure is 14.65 pounds per square inch; the standard temperature is 60°F. One standard cubic foot of gas is the amount of gas contained in one cubic foot of space at a pressure of 14.65 psia at a temperature of 60°F. Using the conversion table, natural gas at any temperature and pressure can be converted to standard cubic feet, the measurement by which gas is usually bought, sold, and transported.

GAS METER, MASS-FLOW
See Mass-flow Gas Meter.

GASOHOL
A mixture of 90 percent gasoline and 10 percent alcohol; a motor fuel. Gasohol was first marketed in the late 1970s as a way to stretch available gasoline stocks by using surplus agricultural products to make ethanol or grain alcohol.

GAS OIL
A refined fraction of crude oil somewhat heavier than kerosene, often used as diesel fuel.

GAS-OIL RATIO
The number of cubic feet of natural gas produced with a barrel of oil. The ratio is expressed 500:1 or 1000:1, whatever the volume of gas measured at the well that is produced per barrel of oil. A high gas-to-oil ratio is extremely undesirable because the pressure in a reservoir, the propulsive force to move the oil in the formation to the boreholes, is being depleted. And with the reservoir pressure gone, a great percentage of the oil may not be recoverable, except by a costly secondary recovery program.

GASOLINE, MARINE WHITE
See Marine White Gasoline.

GASOLINE PLANT
A compressor plant where natural gas is stripped of the liquid hydrocarbons usually present in wellhead gas.

Gasoline plant

GASOLINE, RAW
The untreated gasoline cut from the distillation of crude oil; natural gasoline; a gasoline similar to motor fuel but lower in octane and highly unstable.

GASOLINE, STRAIGHT-RUN
The gasoline-range fraction distilled from crude oil. Virgin naphtha.

GAS REGULATOR
A pressure-reducing device on gas take-off piping that can be set to deliver a supply of gas at a predetermined pressure. For example, a regulator can be adjusted to permit a flow of gas at a pressure of 8 or 10 ounces per square inch from a gas main carrying 100 pounds per square inch.

GASSER
A commercial, natural-gas well.

GAS SNIFFER
A colloquial term for a sensitive electronic device that detects the presence

Gas regulator *(Courtesy Fisher Controls Co.)*

of gas or other hydrocarbons in the stream of drilling mud returning from downhole.

Gas turbine

GAS TURBINE

A rotary engine whose power is derived from the thrust of expanding gases on blades or vanes on a spindle within the body of the engine. As natural gas enters the combustion chamber it is ignited, and as it instantaneously expands, it creates a powerful thrust that is directed against the vanes of the turbine, causing rotation of the spindle furnishing power to an attached pump or compressor.

GAS, UNASSOCIATED

See Unassociated Gas; *also* Associated Gas.

GAS, UNCONVENTIONAL NATURAL

See Unconventional Natural Gas.

GAS WELDING

Welding with oxygen and acetylene or with oxygen and another gas. *See* Oxyacetylene Welding.

GATE

Short for gate valve, common term for all pipeline valves.

Early gas welding rig

GATE, BACKFLOW

A type of swing-check valve made so the clapper's position may be changed from open to closed by an externally mounted handle. The handle is attached to the clapper's fulcrum shaft which protrudes through the side of the valve body. When the clapper is closed (resting on its seat in a normal position), fluid can flow in one direction only; when open (raised from its seat by the handle), fluid can flow in the opposite direction.

GATE VALVE

A pipeline valve made with a wedge-shaped disk or "tongue" which is moved from open to closed (up to down) by a threaded valve stem. Some valves have stems which remain in the valve bonnet (do not rise) and when they are rotated by the valve wheel screw into the disk, raising the disk and opening the valve. Other valves have a rising stem (q.v.) which is firmly attached to the wedge. The valve stem is threaded and when the threaded valve wheel is turned the stem rises through the wheel, raising the disk to open position.

GATHERING FACILITIES

Pipelines and pumping units used to bring oil or gas from production leases by separate lines to a central point, i.e., a tank farm, or a trunk pipeline.

GATHERING SYSTEM

See Gathering Facilities.

Gate valve

GAUGE HATCH

An opening in the roof of a stock or storage tank, fitted with a hinged lid, through which the tank may be gauged and oil samples taken. *See* Hatch.

GAUGE HOLE

A gauge hatch (q.v.).

GAUGE LINE

A reel of steel measuring tape, with a bob attached, held in a frame equipped with handle and winding crank used in gauging the liquid level in tanks. To prevent striking sparks, the bob is made of brass or other non-sparking material or sheathed in a durable plastic. The tip of the bob is point zero on the gauge column.

GAUGER, FIELD

A person who measures the oil in a stock or lease tank, records the temperature, checks the sediment content, makes out a run ticket, and turns the oil into the pipeline. A gauger is the pipeline company's agent and, in effect, "buys" the tank of oil for his company.

GAUGE TANK

A tank in which the production from a well or a lease is measured.

GAUGE TAPE

Gauge line (q.v.).

GAUGE, TICKET

A run ticket (q.v.).

Gauging a tank

GEAR BOX

The enclosure or case containing a gear train or assembly of reduction gears; the case containing a pump's pinion and ring gears.

GEAR PUMP

See Pump, Gear.

GEL

A viscous substance, a jelly-like material used in well stimulation and formation fracturing to suspend sand or other proppants in the fracturing medium. Gelling agents are mixed with water or light oil to form an emulsion that will carry a quantity of sand for various well workover procedures.

GEOCHEMICAL PROSPECTING

Exploratory methods that involve the chemical analysis of rocks and subsurface water for the presence of organic matter associated with oil and gas deposits.

GEOCHEMISTRY

The science of chemistry applied to oil and gas exploration. By analyzing the contents of subsurface water for presence of organic matter associated with oil deposits, geochemistry has proved to be an important adjunct to geology and geophysics in exploratory work.

GEODESY

The branch of science concerned with the determination of the size and shape of the earth and the precise location of points on the earth's surface.

GEOLOGICAL STRUCTURE

Layers of sedimentary rocks which have been displaced from their normal horizontal position by the forces of nature. Folding, fracturing, faulting (the place where the strata have fractured and slipped by one another) are geological structures that often form structural traps that are logical places to find accumulations of oil and gas — and water.

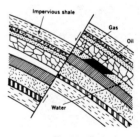

Faulting

GEOLOGIC COLUMN
The vertical range of sedimentary rock from the basement rock (q.v.) to the surface.

GEOLOGIC ERAS
See Geologic Time Scale.

GEOLOGIC PERIODS
See Geologic Time Scale.

GEOLOGIC TIME SCALE
According to authorities in the study of the earth, the Paleozoic Era represents the oldest rocks whose ages are 225 to 600 million years old; the next era is the Mesozoic with rocks 75 to 225 million years old; the most recent era, the Cenozoic, has rocks from the present to 75 million years old. Dividing these eras are periods. Beginning with the oldest, they are Cambrian, Ordovician, Silurian, Devonian, Mississippian, Pennsylvanian, and Permian, all in the Paleozoic Era. In the Mesozoic Era are the Triassic, Jurassic, and Cretaceous Periods. In the youngest era, Cenozoic, are the Tertiary and Quaternary. Geologists have divided the Quarternary Period, identifying the earlier epoch of the Quarternary Period as Pleistocene.

GEOLOGIST
A person trained in the study of the earth's crust. A petroleum geologist, in contrast to a hard-rock geologist, is primarily concerned with sedimentary rocks where most of the world's oil has been found. In general, the work of a petroleum geologist consists in searching for structural traps favorable to the accumulation of oil and gas. In addition to deciding on location to drill, he may supervise the drilling, particularly with regard to coring, logging, and running tests.

GEOLOGIST, HYDRODYNAMICS
A geologist specializing in the study of the mechanics of fluids in underground formations. His work involves analysis of the test data, the interpretation of fluid pressures from drilling wells and well logs, and applying his findings to the solution of problems associated with oil and gas well exploration and development.

GEOLOGRAPH
A device on a drilling rig to record the drilling rate or rate of penetration during each 8-hour tour.

GEOLOGY
The science that deals with the history of the earth and its life as recorded in the rocks.

GEOMETRY OF A RESERVOIR
A phrase used by petroleum and reservoir engineers meaning the shape of a reservoir of oil or gas.

GEOPHONES
Sensitive sound detecting instruments used in conducting seismic surveys. A series of geophones are placed on the ground at intervals to detect and transmit, to an amplifier-recording system, the reflected sound waves created by explosions set off in the course of seismic exploration work.

Geophone

GEOPHYSICAL CAMP
Temporary headquarters established in the field for geophysical teams working the area. In addition to providing living quarters and a store of supplies, the camp has facilities for processing geophysical data gathered on the field trips.

GEOPHYSICAL TEAM
A group of specialists working together to gather geophysical data. Their work consists of drilling shot holes, placing explosive charges, setting out or stringing geophones, detonating shot charges, and reading and interpreting the results of the seismic shocks set off by the explosive charges.

GEOPHYSICS
The application of certain familiar physical principles—magnetic attraction, gravitational pull, speed of sound waves, the behavior of electric currents—to the science of geology.

GEOTHERMAL GRADIENT
The increase in temperature of the earth the deeper a hole is drilled. The rate of increase in geothermal temperature is approximately one degree Fahrenheit for each 55 feet of depth or more than 100°F. per mile of hole. In very deep wells, the bottom-hole temperature is so hot (400 to 500°F) that special drilling mud formulations must be used. Plain water cannot be used because the water in the returning drilling mud would vaporize, turn to steam at the surface.

GEOTHERMAL POWER GENERATION
The use of underground, natural heat sources, i.e., superheated water from deep in the earth, to generate steam to power turboelectric generators.

GILSONITE
A solid hydrocarbon with the general appearance of coal; uintaite; a black, lustrous form of asphalt that, when treated and refined, yields gasoline, fuel oil, and coke. Found in deposits in Utah.

GIN POLE
(1) An A-frame made of sections of pipe mounted on the rear of a truck bed that is used as a support or fixed point for the truck's winch line when lifting or hoisting. (2) A vertical frame on the top of the derrick, spanning the crown block, providing a support for hoisting. (3) A mast (q.v.).

Gin pole

GIRBITOL PROCESS
A process used to "sweeten" sour gas by removing the hydrogen sulfide (H_2S).

GIRT
One of the braces between the legs of a derrick; a supporting member.

GLOBE VALVE
A type of pipeline valve that shuts off as the stem, rotated by the hand wheel, moves a mating part downward onto a ground seat that is integral to the valve body.

Glycol dehydrator

GLYCOL DEHYDRATOR
A facility for removing minute particles of water from natural gas not removed by the separator.

G.M.P.
Gallons of gasoline per 1,000 cubic feet of natural gas produced.

GO-DEVIL
A pipeline scraper, a cylindrical, plug-like device equipped with scraper blades, rollers, and wire brushes used to clean the inside of a pipeline of accumulations of wax, sand, rust, and water. When inserted in the line, the go-devil is pushed along by the oil pressure. Also a missile dropped into the well bore to detonate an explosive charge or to jar a downhole tool into operation.

GONE TO WATER
A well in which the production of oil has decreased and the production of water has increased to the point where the well is no longer profitable to operate.

GOOSENECK
A nipple in the shape of an inverted U attached to the top of the swivel (q.v.) and to which the mud hose is attached.

Gooseneck

GOOSING GRASS
Cutting grass and weeds around the lease or tank farm, shaving the grass off the ground with a sharp hoe-like tool, leaving the ground clean.

GOR
Gas-oil ratio (q.v.).

GPM
Gallons per minute.

GRABLE OIL WELL PUMP
A patented, drum-and-cable pumping unit that can be installed in a wellhead cellar. The unit raises and lowers the pumping rods by winding and unwinding cable on a drum or spool. The low profile of the pumping unit makes it ideal for use in populated areas, and to protect the beauty of the landscape.

GRADIENTS (TEMPERATURE AND PRESSURE)
The rates of increase or decrease of temperature or pressure are defined as gradients; the rate of regular or graded ascent or descent.

GRANNY HOLE
The lowest, most powerful gear on a truck.

GRANNY KNOT
A knot tied in such a way as to defy untying; an improperly tied square knot; a hatchet knot.

A small, modern "grass roots" refinery

GRASS ROOTS
Said of a refinery or other installation built from the ground up as contrasted to a plant merely enlarged or modernized.

GRAVEL ISLAND
In some locations (in shallow water near shore) in the Arctic, gravel islands, 40 to 50 feet in diameter, are constructed to make a foundation area from which to drill exploratory wells. Gravel is dredged from the sea bottom or transported from a nearby river or delta and dumped into holes cut in the ice. *See* Drilling Island; *also* Ice Platform.

GRAVEL PACKING
Using gravel to fill the cavity created around a well bore as it passes through the producing zone to prevent caving or the incursion of sand and to facilitate the flow of oil into the well bore.

GRAVEYARD SHIFT
A tour of work beginning at midnight and ending at 8 a.m. In pipeline operations, the graveyard shift is customarily from 11 p.m. to 7 a.m. Hoot-owl shift.

GRAVIMETER
A geophysical instrument used to measure the minute variations in the earth's gravitational pull at different locations. To the geophysicist, these variations indicate certain facts about subsurface formations.

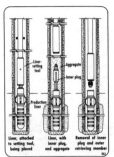

Gravel pack tool

GRAVING DOCK
A dry dock; a dock that can accept ships into an enclosure. When the water is pumped out, the ship is left high and dry for repairs.

GRAVITY
(1) The attraction of the earth's mass for bodies or objects at or near the surface. (2) Short for specific gravity; API gravity. (3) To flow through a pipeline without the aid of a pump; to be pulled by the force of gravity.

GRAVITY DRIVE
A natural drive occurring where a well is drilled at a point lower than surrounding areas of producing formations causing the oil to drain downhill into the well bore. If the reservoir rock is highly permeable and dips sharply toward the well, there is usually good oil recovery.

GRAVITY LINE
A pipeline that carries oil from a lease tank to pumping station without the use of mechanical means; a line that transports liquid from one elevation to a lower elevation by the force of gravity alone.

GRAVITY MAPS
Results of reconnaissance gravity surveys; display of gravity measurements taken in an area. *See* Gravimeter.

GRAVITY SEGREGATION
The separation of water from oil, or heavy from lighter hydrocarbons by the force of gravity, either in the producing zone or by gravity in the separators after production; the stratification of gas, oil, and water according to their densities.

GRAVITY STRUCTURE
An offshore drilling and production platform made of concrete and of such tremendous weight that it is held securely on the ocean bottom without the

Gravity structure

need for piling or anchors. One of the world's largest gravity structures is located off the Scottish coast in the North Sea. Its general configuration is that of a column mounted on a large circular base which has storage for 1 million barrels of crude. The base is 450 feet in diameter; the column is 180 feet in diameter and the overall height of the structure is about 550 feet.

GREASE
(1) A lubricating substance (solid or semisolid) made from lubricating oil and a thickening agent. The lube oils may be very light or heavy cylinder oils; the thickening agent (usually soaps) may be any material that when mixed with oil will produce a grease structure. (2) Colloquial for crude oil.

GREEN OIL
A paraffin-base crude oil. Asphalt-base crudes are sometimes referred to as black oil.

GRIEF STEM
Kelly joint; the top joint of the rotary drill string that works through the square hole in the rotary table. As the rotary table is turned by the drilling engines, the grief stem and the drillpipe are rotated. Grief stems are heavy, thick-walled tubular pieces with squared shoulders that are made to fit into the hole in the rotary table.

Grief stem

GRIND OUT
Colloquial for centrifuge; to test samples of crude oil or other liquid for suspended material—water, emulsion, sand—by use of a centrifuge machine.

GROSS PRODUCTION TAX
A severance tax (q.v.); a tax usually imposed by a state, at a certain sum per unit of mineral removed (barrels of oil, thousands or millions of cubic feet of gas, or tons of coal, sulphur, sand, and gravel).

GROUND-SEAT UNION
A pipe coupling made in two parts; one half is convex, the other half concave in shape, and both ground to fit. A threaded ring holds the halves together, pressuretight. Used on small-diameter piping.

GROUP SHOT
Geophysical exploration performed for several individuals or companies on a cost-sharing basis. The companies share the information as well as the cost. This type arrangement is usually for offshore seismic surveys in which several companies are planning to submit bids for offshore leases offered in a government lease sale.

GROUT
(1) A concrete mixture used to fill in around piling, caissons, heavy machinery beds, and foundation work. (2) To stabilize and make permanent. Grout is usually a thin mixture that can be worked into crevices and beneath and around structural forms.

GROWLER BOARD
See Lazy Board.

Growler board

GRUB STAKE AGREEMENT
An agreement whereby one person undertakes to prospect for oil and agrees to hand over to the person who furnishes the money or supplies a certain proportionate interest in the oil discovered. This type of agreement is common for solid minerals but is not often used in oil prospecting.

G.S.A.
Geological Society of America.

GUIDE SHOE
A casing shoe (q.v.).

GUMBO
A heavy, sticky mud formed downhole by certain shales when they become wet from the drilling fluid.

GUM BOOTS
Rubber boots, the kind you pull on like a cowboy boot.

GUNK
The collection of dirt, paraffin, oil, mill scale, rust, and other debris that is cleaned out of a pipeline when a scraper or a pig is put through the line.

GUN PERFORATION
A method of putting holes in a well's casing downhole in which explosive charges lowered into the hole propel steel projectiles through the casing wall. (Casing is perforated to permit the oil from the formation to enter the well.)

GUSHER
A well that comes in with such great pressure that the oil blows out of the wellhead and up into the derrick, like a geyser. With improved drilling technology, especially the use of drilling mud to control downhole pressures, gushers are rare today. *See* Blowout.

GUY WIRE
A cable or heavy wire used to hold a pole or mast upright. The end of the guy wire is attached to a stake or a deadman (q.v.).

GYP
Boiler scale; a residue or deposit from "hard water," water with high concentrations of minerals. Pipe and vessels handling hard water become gypped up as the minerals form a hard, rock-like layer on the inner surfaces. Gypsum.

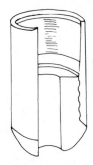
Guide shoe

An early-day gusher

H

HABENDUM CLAUSE
The clause in a lease setting forth the duration of the lessee's interest in the property. An habendum clause might read, "It is agreed that this lease shall remain in force for a term of five years from this date and for as long thereafter as oil or gas, of whatever kind, or either of them is produced or drilling operations are continued as hereinafter provided." The primary term in this case is 5 years. *See* Primary Term.

HALF SOLE

A metal patch for a corroded section of pipeline. The patch is cut from a length of pipe of the same diameter as the one to be repaired. Half soles can be from six to 12 feet in length and are placed over the pitted or corroded section of the pipe and welded in place with a bead around the entire perimeter of the half sole.

HAND MONEY

See Earnest Money.

HANDY

Hand-tight; a pipe connection or nut that can be unscrewed by hand.

HANG A WELL OFF

To disconnect the pull-rod line from a pumping jack or pumping unit being operated from a central power (q.v.). On a lease with a number of stripper wells, the pumper knows each well and how long it should be pumped. As a result he may "hang a well off" after six hours of pumping, others after 12 hours or so. Strippers which make 10 barrels or less of oil a day rarely are pumped 24 hours a day.

HANGER, ROD

See Rod Hanger.

HANGING IRON

A colloquial expression for the job of assembling a high-pressure, heavy-duty blowout preventer stack or production tree. Some of the valve assemblies weigh thousands of pounds or more so they must be hoisted into place, aligned, and bolted to their mating piece.

HANG THE RODS

To pull the pump rods out of the well and hang them in the derrick on rod hangers. On portable pulling units, the rods are hung outside the derrick. On shallow wells with short strings of rods, they may be pulled from the well by a simple pulling unit consisting only of a guyed mast. If this is the case, the rods are pulled, unscrewed and layed down, i.e., layed out horizontal on a rack.

HARDBANDING

Laying on a coating or surface of super-hard metal on a softer metal part at a point or on an area of severe wear or abrasion. Putting a hard surface on a softer metal by welding or other metallurgical process. Where it may be impractical or prohibited by structural constraints to manufacture a part from the harder metal, a coating or hard-surfacing of the part is a practical solution. Also called hardfacing.

HARDFACING

See Hardbanding.

HARDWARE

(1) Electronic and mechanical components of a computer system, e.g. storage drums, scanners, printers, computers. (2) Mechanical equipment, parts, tools.

HARDWARE CLOTH
A type of galvanized metal screen that can be bought in hole sizes, e.g. ⅛, ¼, ½-inch, etc. The holes are square.

HARDWIRE (TELEMETRY)
Describes a system of communication or information transmission using electric wire from point to point instead of electronic or wireless transmission.

HATCH
An opening in the top of a tank or other vessel through which inspections are made or samples taken; a gauge hatch.

HATCHET KNOT
A knot that defies untying and so must be cut; a granny knot.

HAT-TYPE FOUNDATION
A metal base or foundation in the shape of an inverted, rectangular cake pan. Hat-type foundations are used for small pumps and engines or other installations not requiring solid, permanent foundations.

HAUL ASS
An inelegant term meaning to leave a place with all haste; vamoose; split.

Hay rake

HAWSER
A large-diameter hemp or nylon rope for towing, mooring, or securing a ship or barge.

HAY RAKE
Another name for the finger board (q.v.) in the derrick of an oil well.

HEADACHE!
A warning cry given by a fellow worker when anything is accidentally dropped or falls from overhead toward another worker.

HEADACHE POST
A frame over a truck cab that prevents pipe or other material being hauled from falling on the cab; a timber set under the walking beam to prevent it from falling on the drilling crew when it is disconnected from the crank and pitman (q.v.).

HEADER
A large-diameter pipe into which a number of smaller pipes are perpendicularly welded or screwed; a collection point for oil or gas gathering lines. *See* Manifold.

HEAD, HYDROSTATIC
See Hydrostatic Head.

Headache post

HEADING
An intermittent or unsteady flow of oil from a well. This type of flow is often caused by a lack of gas to produce a steady flow thus allowing the well's tubing to load up with oil until enough gas accumulates to force the oil out.

HEAD WELL
A well that makes its best production by being pumped or flowed intermittently. Such a well lacks sufficient gas pressure to flow steadily and must wait for the tubing to load up with oil until enough gas accumulates to force the oil up the tubing and out to the tanks. If the well cannot accumulate enough gas pressure to overcome the hydrostatic head, represented by the column of oil in the tubing, the well must be pumped.

HEATER
(1) An installation used to heat the stream from high-pressure gas and condensate wells (especially in winter) to prevent the formation of hydrates, a residue which interferes with the operation of the separator. (2) A refinery furnace.

Electric and gas-fired
field heaters

HEATER, PIPELINE
An installation fitted with heating coils or tubes for heating certain crude oil to keep it "thin" enough to be pumped through a pipeline. Crudes with high pour points (congealing at ordinary temperatures) must be heated before they can be moved by pipeline.

HEATER TREATER
See Emulsion Treater.

HEAT EXCHANGER, FINNED TUBE
Small-diameter pipe or tubing with thin metal fins attached to the outer circumference for cooling water and other liquids or gases. Finned tube exchangers cool by giving up heat from the surface of the fins to the atmosphere in a manner similar to an automobile radiator. Heat exchangers are not only for cooling but for heat recovery systems as well. In some plants finned tube exchangers are built in ductwork through which the exhaust gas of a turbine flows at 800°F. Oil or process liquids are pumped through the exchanger tubes to use the waste heat to heat the process stream or to make steam.

Crude oil heater
(Courtesy Born, Inc.)

HEAT EXCHANGER, HAIRPIN
A type of shell and tube exchanger with tubes inside a 12 to 18-inch diameter shell which may extend 20 to 30 feet and then doubles back the same distance like a hairpin. Hairpin exchangers may have bare or finned tubes inside the shell.

HEAT EXCHANGER, PLATE
A relatively low-pressure heat exchanger that uses thin-walled plates as its heat transfer elements. Because of its thin walls, plate exchangers exhibit a much higher heat transfer coefficient than the more conventional shell and tube exchangers. However, because of their less-sturdy construction there are pressure limits to their use.

Heater treater

HEAT EXCHANGER, SHELL AND TUBE
A common type of industrial heat exchanger with a "bundle" of small-diameter pipes (tubes) inside a long, cylindrical steel shell. The tubes (50 to 100 in small units, several hundred in larger ones) run parallel to the shell and are supported, equidistant, by perforated steel end plates. The space inside the shell not filled with tubes carries the cooling water or other liquid. The liquid to be cooled is pumped through the tubes. Heat exchangers act

not only as a cooling apparatus but are often used as a waste heat recovery system. Heat normally lost to a cooling medium can be used to heat a process stream.

HEAT PIPE, GRAVITY RETURN
A type of passive heat exchanger (requiring no external energy source) that draws heat from a heat source and gives up heat to a heat sink, the atmosphere in most cases. In its basic form a heat pipe consists of a closed tube (the shell) two to six inches in diameter and as long as need be. The shell has a porous wick made of fine metal mesh in the inside circumference extending from top to bottom. The shell also contains a quantity of working liquid which may be anhydrous ammonia, liquid metals, glycerine, methanol, or acetone. Heat taken in or absorbed at the lower end of the heat pipe, the end in contact with the heat source, causes the liquid to evaporate and move up the pipe as a vapor. The dissipation of the heat at the upper end condenses the vapor which, as a liquid, moves back down the pipe in the wick by gravity or capillary action. The continuous cycle of vaporization and condenation within the closed pipe makes the heat pipe an efficient, natural-convection heat transfer loop. On the trans-Alaska pipeline, thousands of heat pipes were installed along the big line to maintain the frozen soil around the vertical support members.

HEAT TAPE
An electrical heating element made in the form of an insulated wire or tape used as a tracer line to provide heat to a pipeline or instrument piping. The heat tape is held in direct contact with the piping by a covering of insulation.

HEAT TRACING
The paralleling of instrumentation, product or heavy crude oil, lines with small-diameter steam piping or electrical heat tape to keep the lines from freezing or to warm the product or instrument fluid sufficiently to keep them flowing freely. Heat tracing lines, whether steam or electrical tape, are attached parallel to the host piping and both are covered with insulation.

Applying heat tape

HEAVE COMPENSATOR
A type of snubber-shock absorber on a floating, drilling platform or drillship that maintains the desired weight on the drill bit as the unstable platform heaves on ocean swells. Some compensators are made with massive counterweights; others have hydraulic systems to keep the proper weight on the bit constant. Without compensators, the bit would be lifted off bottom as the platform rose on each swell.

HEAVY BOTTOMS
A thick, black residue left over from the refining process after all lighter fractions are distilled off. Heavy bottoms are used for residual fuel and/or asphalt.

HEAVY CRUDE OIL
Crude oil of 20° API gravity or less. There are perhaps billions of barrels of heavy oil still in place in the U.S. which require special production techniques, notably steam injection or steam soak, to extract them from the underground formations. Because heavy crude oil is more costly to produce, it and other types of oil are eligible for free market or world prices.

HEAVY ENDS

In refinery parlance, heavy ends are the heavier fractions of refined oil—fuel oils, lubes, paraffin, and asphalt—remaining after the lighter fractions have been distilled off. *See* Light Ends.

HEAVY FUEL OIL

A residue of crude oil refining processes. The product remaining after the lighter fractions—gasoline, kerosene, lubricating oils, wax, and distillate fuels—have been extracted from the crude; residual fuel oil.

HEAVY METAL

Spent uranium or tungsten. Heavy metal is used to make drill tools to add weight to the drill assembly. Drill collars made of heavy metal weigh twice as much as those made of steel, and are used to stablize the bit and to force it to make a straighter hole, with less deviation from the vertical.

HEAVY OIL PROCESS (HOP)

A steam injection process developed by a subsidiary of Barber Oil Corporation in which steam is injected through horizontal lines into subsurface oil sands containing heavy oil, oil of 20° API gravity or less. Conventional steamflooding employs vertical holes through which steam is injected. In the horizontal method, a seven-foot diameter shaft is drilled into the relatively shallow formation. After it is cased, workmen construct a concrete cavern 25 feet in diameter and 20 feet high. From this work area, lateral holes are drilled several hundred feet in all directions. Perforated pipe is inserted in the drilled holes to carry steam. The steam, injected under pressure, soaks the formation, causing the highly viscous oil to separate from the sand and flow into the laterals after the steam injection is halted.

HEAVY-WALL DRILLPIPE

Drillpipe with thicker walls than regular drillpipe. Heavy-wall is sometimes used in the drillstring to reduce the number of larger diameter and stiffer drill collars. This is true in directional drilling and even in straight holes in certain areas of the country. Some of the advantages of heavy-wall pipe over drill collars are: Heavy-wall can be handled at the rig floor by regular drillpipe elevators and slips and can be racked in the rig like regular pipe.

HECTARE

A metric unit of land measurement equal to 10,000 square meters or 2.47 acres. Abbreviation — ha.

HEPTANE

A liquid hydrocarbon of the paraffin series. Although heptane is a liquid at ordinary atmospheric conditions, it is sometimes present in small amounts in natural gas.

HEXANE

A hydrocarbon fraction of the paraffin series. At ordinary atmospheric conditions hexane is a liquid but often occurs in small amounts in natural gas.

HHP

Hydraulic horsepower; a designation for a type of very-high-pressure plunger pump used in downhole operations such as cementing, hydrofracturing, and acidizing.

HI-BOY

A skid-mounted or wheeled tank with a hand-operated pump mounted on top used to dispense kerosene, gasoline, or lubricating oil to small shops and garages.

HIDE THE THREADS

To make up (tighten) a joint of screw pipe until all the threads on the end of the joint are screwed into the collar, hiding the threads and making a leakproof connection.

HIGH

A geological term for the uppermost part of an inclined structure where the likelihood of finding oil is considered to be the greatest. As oil and gas tend to accumulate at the top of underground structures, traps and domes, the higher up on the structure a well is drilled the better the chances of encountering oil and avoiding the underlying water.

HIGH BOTTOM

A condition in a field stock tank when BS&W (basic sediment and water) has accumulated at the bottom of the tank to a depth making it impossible to draw out the crude oil without taking some of the BS with it into the pipeline. When this condition occurs the operator (lease pumper) must have the tank cleaned before the pipeline company will run the tank of oil.

HIGH-PRESSURE GAS INJECTION

Introduction of gas into a reservoir in quantities exceeding the volumes produced in order to maintain reservoir pressure high enough to achieve mixing between the gas and reservoir oil. *See* Solution Gas.

HISTORY OF A WELL

A written account of all aspects of the drilling, completion, and operation of a well. (Well history is required in some states.) Well histories include formations encountered, depths, size and amount of casing, mud program, any difficulties, coring record, cementing and perforating, etc.

HITTING THE HOOKS

Working on a pipeline, screwing in joints of pipe using pipe tongs; an expression used by the tong crew of a pipeline gang. The tong crews on large-diameter screw pipelines (up to about 12-inch pipe) hit the hooks in perfect rhythms. With three sets of tongs on the joint being screwed in, each large tong, run by two or three men, made a stroke every third beat of the collar pecker's hammer (q.v.) until the joint was nearly screwed in. Then the three tongs, with all six or nine men, hit together to "hide the threads," to tighten the joint the final and most difficult round.

HOISTING DRUM

A powered reel holding rope or cable for hoisting and pulling; a winch. *See* Draw Works.

HOLD DOWN/HOLD UP

Oscillating anchoring devices or supports for a shackle-rod line to hold the rod line to the contour of the land it traverses. The devices are timbers or lengths of pipe hinged to a deadman or overhead support at one end, the other end attached to and supporting the moving rod line.

Hoisting drum

Hole opener *(Courtesy Grant)*

Repairing holidays

Hook or crane block
(Courtesy Crosby)

Horizontal directional
drilling *(Courtesy
Reading & Bates)*

HOLE OPENER
A type of reamer used to increase the diameter of the well bore below the casing. The special tool is equipped with cutter arms that are expanded against the wall of the hole and by rotary action reams a larger diameter hole.

HOLIDAYS
Breaks or flaws in the protective coating of a joint of line pipe. Holidays are detected by electronic testing devices as the pipe is being laid. When detected, the breaks are manually coated. *See* Jeeping.

HOOK
The hook attached to the frame of the rig's traveling block (q.v.) and which engages the bail of the swivel in drilling operations. *See* Hook-load Capacity.

HOOK BLOCK
A pulley or sheave mounted in wooden or metal frame to which a hook is attached. A hook block may have more than one sheave mounted in the frame; a traveling block (q.v.).

HOOK-LOAD CAPACITY
The maximum weight or pull a derrick and its lines, blocks, and hook are designed to handle. A rating specification for a drilling rig. (Some large rigs have a hook-load capacity of two million pounds.)

HOOKS
Pipe-laying tongs named for the shape of the pipe-gripping head of the scissors-like wrench.

HOOK UP
To make a pipeline connection to a tank, pump, or a well. The arrangement of pipes, nipples, flanges, and valves in such a connection.

HORIZON
A zone of a particular formation; that part of a formation of sufficient porosity and permeability to form a petroleum reservoir.

HORIZONTAL ASSIGNMENT
The assignment of an interest in oil or gas above or below a certain depth in a well, or an assignment can specify a particular formation.

HORIZONTAL DIRECTIONAL DRILLING
Drilling with a specially designed slant rig (q.v.) at an angle from the horizontal beneath a stream, canal, or ship channel. This type of directional drilling has been perfected and is used to make pipeline crossings where dredging a trench across a waterway is too costly or too disruptive of ship traffic and a bridge or A-frame supported line is prohibited by the authorities.

HORIZONTAL INTEGRATION
Refers to the condition in which a diversified company has resources or investments other than its principal business, and from which it makes a profit. Specifically, an oil company is said to be horizontally integrated when, besides oil and gas holdings, it owns coal deposits, is into nuclear energy, oil shale and geothermal energy. *See* Vertical Integration.

HORSE FEED
An old oil field term for unexplainable expense account items in the days of the teamster and line rider who were given an allowance for horse feed. Expenses that needed to be masked in anonymity were simply listed "horse feed."

HORSEPOWER
A unit of power equivalent to 33,000 foot-pounds a minute or 745.7 watts of electricity.

HORSEPOWER, INDICATED AND BRAKE
See Indicated and Brake Horsepower.

HORTONSPHERE
A spherical steel tank for the storage, under pressure, of volatile petroleum products, e.g. gasoline, and LP-gases; also Hortonspheroid, a flattened spherical tank, resembling somewhat a tangerine in shape. As vessels subjected to high internal pressures tend to take the shape of a sphere, the tanks designed to hold liquid or gases under pressure are made spherical or nearly so to handle internal pressures of several hundred pounds per square inch safely, without distortion and undue stress.

Hortonsphere

HOT-FLUID INJECTION
A method of thermal oil recovery, in which hot fluid (water, gas, or steam) is injected into a formation to increase the flow of low-gravity crude to production wells.

HOT FOOTING
Installing a heater at the bottom of an input well to increase the flow of heavy crude oil from the production wells. *See* Hot-fluid Injection.

HOT-HEAD ENGINE
A hot-plug engine (q.v.); a "semidiesel."

HOT OIL
Oil produced in violation of state regulations or transported interstate in violation of Federal regulations.

HOT OIL (FORIEGN)
A term applied to oil produced by a host country after the host country confiscates the assets of a foreign oil company.

HOT PASS
A term describing a "bead" or course of molten metal laid down in welding a pipeline. The hot pass is the course laid down on top of the stringer bead, which is the first course in welding a pipeline. *See* Pipeline Welding.

HOT-PLUG ENGINE
A stationary diesel-cycle engine that is started by first heating an alloy-metal plug in the cylinder head that protrudes into the firing chamber. The hot plug assists in the initial ignition of the diesel fuel until the engine reaches operating speed and temperature. Afterwards the plug remains hot, helping to provide heat for ignition; hot-tube engine; hot head. *See* Semidiesel.

Hot tapping

Hovercraft

Hydraulic workover unit
(Courtesy Otis)

HOT TAPPING
Making repairs or modifications on a tank, pipeline, or other installation without shutting down operations. *See* Tapping and Plugging Machine.

HOUDRY, EUGENE J.
A pioneer in developing the use of catalysts in cracking crude oil. Houdry, a wealthy Frenchman, was a World War I hero and auto racer. It is said his interest in cars led him to experiment with more efficient methods of refining and to work with various catalysts until he perfected the catalytic cracking process that bears his name. Although there are several cracking processes in use today, Houdry's work is credited with ushering in the era of catalytic cracking. *See* Hydrocracking.

HOUSE BRAND (GASOLINE)
An oil company's regular gasoline; a gasoline bearing the company's name.

HOVERCRAFT
See Air Cushion Transport.

H₂S
Hydrogen sulfide (q.v.).

HUMPHREYS, DR. R.E.
A petroleum chemist who worked with Dr. W.M. Burton in developing the first commercially successful petroleum cracking process using heat and pressure.

HUNDRED-YEAR STORM CONDITIONS
A specification for certain types of offshore installations—production and drilling platforms, moorings, and offshore storage facilities—is that they be built to withstand winds of 125 miles an hour and "hundred-year storm conditions"; the biggest blow on record.

HURRY-UP STICK
The name given to the length of board with a hole in one end which the cable-tool driller used to turn the T-screw at the end of the temper screw (q.v.) when the walking beam was in motion. This enabled the driller to perform the job of letting out the drilling line easily and rapidly.

HYDRAULIC FRACTURING
A method of stimulating production from a formation of low permeability by inducing fractures and fissures in the formation by applying very high fluid-pressure to the face of the formation, forcing the strata apart. Various patented techniques, using the same principle, are employed by oil field service companies.

HYDRAULIC WORKOVER UNIT
A type of workover unit that is used on high-pressure wells where it may be necessary to snub the pipe out of the hole and back in the hole when the workover is completed. *See* Snubbing.

HYDROCARBONS
Organic chemical compounds of hydrogen and carbon atoms. There are a vast number of these compounds and they form the basis of all petroleum

products. They may exist as gases, liquids, or solids. An example of each is methane, hexane, and asphalt.

HYDROCRACKATE
The main product from the hydrocracking process (q.v.); gasoline blending components.

HYDROCRACKING
A refining process for converting middle-boiling or residual material to high-octane gasoline, reformer charge stock, jet fuel and/or high-grade fuel oil. Hydrocracking is an efficient, relatively low-temperature process using hydrogen and a catalyst. The process is considered by some refiners as a supplement to the basic catalytic cracking process.

HYDRODYNAMICS
A branch of science that deals with the cause and effect of regional subsurface migration of fluids.

HYDROGEN SULFIDE (H$_2$S)
An odorous and noxious compound of sulfur found in "sour" gas. *See* Sour Gas.

HYDROMETER
An instrument designed to measure the specific gravity of liquids; a glass tube with a weighted lower tip that causes the tube to float partially submerged. The API gravity of a liquid is read on a graduated stem at the point intersected by the liquid.

HYDROPHONES
Sound-detecting instruments used in underwater seismic exploration activities. Hydrophones are attached to a cable towed by the seismic vessel. Sound waves generated by blasts from an air gun reflect from formations below the seabottom and are picked up by the hydrophones and transmitted to the mother ship.

HYDROSTATIC HEAD
The height of a column of liquid; the difference in height between two points in a body of liquid. The hydrostatic head in an oil well is the height of the column of oil in the borehole or casing. The hole full of drilling mud also represents a hydrostatic head. The pressure exerted by several thousand feet of "head" often causes the drilling mud to penetrate cracks and fissures, and even very porous formations.

HYDROSTATIC TESTING
Filling a pipeline or tank with water under pressure to test for tensile strength, its ability to hold a certain pressure without rupturing. Water is used for testing because it is noncompressible so if the pipe or tank does rupture there is no potentially dangerous expansion of the water as would be the case if a gas under very high pressure were used.

HYPERBARIC WELDING (EXCESSIVE-PRESSURE WELDING)
Welding on the sea bottom "in the dry" but under many atmospheres of pressure (compression). In hyperbaric welding of undersea pipelines, a large frame is lowered into the water and clamped to the pipeline. Then an open-bottomed, box-like enclosure is placed in the center of the frame over

the pipe. Power lines and life-support umbilicals are connected to the box. The sea water is displaced with breathing-gas mixtures for the diver-welders permitting them to do their work in the dry but high-pressure atmosphere.

I

IADC
International Association of Drilling Contractors.

I.C.C.
Interstate Commerce Commission.

ICE PLATFORM
A man-made, thick platform of ice for drilling in the high Arctic. Sea water is pumped onto the normal ocean ice, itself quite thick, where it freezes in the minus 30° to 40°C. temperatures. The platform is built up a few inches at a time with successive pumpings and freezing of the water until the ice is calculated to be thick enough to support drilling operations with a 1,000 to 1,500-ton drillrig and auxiliary equipment. Ice platform technology was pioneered by Panarctic Oil Ltd., a company with a great deal of experience in Arctic exploration.

ID
Inside diameter—of a pipe or tube; initials used in specifying pipe sizes, e.g. 3½-inch ID; also OD, outside diameter, e.g. 5-inch OD.

IDIOT STICK
A shovel or other digging tool not requiring a great deal of training to operate.

IDLER GEAR OR WHEEL
A gear so called because it is usually located between a driving gear and a driven gear, transmitting the power from one to the other. It also transmits the direction of rotation of the driving to the driven gear. Without the idler or the intermediate gear, the driving gear by directly meshing with the driven gear reverses the direction of rotation. Idler wheels or pulleys are also used for tightening belts or chains or to maintain a uniform tension on them.

IGNITION MAGNETO
An electric current generator used on stationary engines in the field. A magneto is geared to the engine and, once the engine is started either by hand cranking or by a battery starter, the magneto continues to supply electric current for the ignition system. Current is produced by an armature rotating in a magnetic field created by permanent magnets.

IGNORANT END
The heaviest end of a tool or piece of equipment to be carried or operated.

I.H.P.
Indicated horsepower.

Drilling platform. Note the escape capsule and the commuting helicopter

IMPACT WRENCH

An air-operated wrench for use on nuts and bolts of large engines, valves, and pumps. Impact wrenches have taken the place of heavy end-wrenches and sledgehammers in tightening and loosening large nuts. A small version of the impact wrench is the air-operated automobile lug-wrench used at modern service stations and garages.

INCENTIVE PRICING

Pricing above the going market price for a product that may be more costly to produce. For example, gas found at great depths, geopressurized gas, coal-seam gas (q.v.) may receive incentive pricing if it qualifies under the Natural Gas Policy Act of 1978 and the Federal Energy Regulatory Commission regulations. Incentive pricing is often the difference between producing a natural resource and not producing because of the high cost of production.

INCORPOREAL RIGHTS

Having no material body or form. Said of easements, bonds, or patents. Rights that have no physical existence but that authorize certain activities or interests.

INCLINOMETER

An instrument used downhole to determine the degree of deviation from the vertical of a well bore at different depths; a drift indicator. There are several types of drift indicators; one is the acid bottle inclinometer (q.v.), another is a plumb bob encased in a small steel tube which on signal punches a hole in a paper disk. When the instrument is retrieved, the distance from the center of the disk the hole was punched by the free-swinging plumb bob indicates the degree of drift or deviation from the vertical hole is being drilled.

INDEPENDENT PRODUCER

(1) A person or corporation that produces oil for the market, having no pipeline system or refinery. (2) An oil-country entrepreneur who secures financial backing and drills his own well; an independent operator. Independent operators and small producing companies are credited with finding most of the new oil fields. Once discoveries are made it is the large oil companies that do most of the development work. Independents often lease and drill on small parcels of land, land either overlooked by the majors or thought not worth fooling with until a discovery is made.

INDICATED HORSEPOWER

Calculated horsepower; the power developed within the cylinder of an engine which is greater than the power delivered at the drive shaft by the amount of mechanical friction which must be overcome. *See* Brake Horsepower. An engine's horsepower is calculated by using the bore, stroke, revolutions per minute, and the number of cylinders.

INDICATOR PASTE, GASOLINE

A viscous material applied to a steel gauge line or gauge pole that changes color when it comes in contact with gasoline, making it easy for the gauger to read the height of gasoline in the tank.

INDICATOR PASTE, WATER
A paste material applied to a steel gauge line or wooden gauge pole that changes color when immersed in water. It is used to detect the presence of water in a tank of oil.

INDUSTRIAL GAS
Gas purchased for resale to industrial users; interruptible gas (q.v.).

INFILL DRILLING
Wells drilled to fill in between established producing wells on a lease; a drilling program to reduce the spacing between wells in order to increase production from the lease.

INFLUENT
The flow of liquids or gas into a vessel or equipment. *See* Effluent.

INFORMATION CONSOLE, DRILLER
A bank of indicators, counters, and display dials showing weight of the drillstring, weight on the drill bit, mud pump speed, mud pressure, engine speed, etc., to keep the driller informed of all aspects of the drillng operation.

Information console

INHIBITORS
A substance that slows down a chemical reaction. An inhibitor's role is the reverse of a catalyst's. Inhibitors are sometimes used to interfere with a chemical reaction somewhere along the process train.

INLAND BARGE RIG
A drilling rig mounted on a barge-like vessel for use in shallow water or swampy locations. Barge rigs are not self-propelled and must be towed or pushed by a towboat. In addition to all necessary drilling equipment, such barges have crew quarters.

Submersible drilling barge

INGAA
Interstate Natural Gas Association of America.

IN-LINE EQUIPMENT
Pumps, separators, heat exchanges integral to a process or processing chain; in the line, not auxiliary or only supporting.

INNAGE GAUGE
A measure of the quantity of oil in a tank calculated on the basis of the depth of oil in the tank; the most common method of gauging a tank. *See* Outage Gauge.

INNOVATOR'S ROYALTY
A type of overriding royalty paid to the person instrumental in bringing a company to a concession from a foreign government; British: a fixer's royalty. *See* Overriding Royalty.

INPUT WELL
A well used for injecting water or gas into a formation in a secondary recovery or pressure maintenance operation.

IN SITU COMBUSTION
A technique used in some locations for recovering oil of low gravity and

Input well

high viscosity from a field when other primary methods have failed. Essentially, the method involves the heating of the oil in the formation by igniting the oil (burning it in place) and keeping the combustion alive by pumping air downhole. As the front of burning oil advances, the heat breaks down the oil into coke and light oil. And as the coke burns, the lighter, less viscous oil is pushed ahead to the well bores of the producing wells.

INSPECTION PLATE
A flat metal plate fitted with a gasket and bolted over an opening in the gearbox of a pump or the crankcase of an engine. By removing the plate an inspection of the gears or crank and connecting-rod bearings can be made. On large, multicylinder engines, inspection windows are large enough to permit a mechanic to enter the crankcase to inspect or "change out" a bearing.

INTANGIBLE DRILLING COSTS
Expenditures made by an operator for labor, fuel, repairs, hauling, and supplies used in drilling and completing a well for production. Intangible costs include also the construction of derricks, tanks, pipelines on the lease, buildings, and preparation of the drillsite, but do not include the material or equipment. The rule of thumb is, do the items for which expenditures were made have salvage value? If not they qualify under the tax laws as intangible drilling costs.

INTANGIBLES
Short for intangible drilling costs (q.v.).

INTEGRATED OIL COMPANY
A company engaged in all phases of the oil business, i.e., production, transportation, refining, and marketing; a company that handles its own oil from wellhead to gasoline pump.

INTEGRATION, HORIZONTAL
See Horizontal Integration; also Vertical Integration.

INTEREST IN AN OIL OR GAS WELL
See Operating Interest; also Working Interest.

INTERFACE
The point or area where two dissimilar products or grades of crude oil meet in a pipeline as they are pumped, one behind the other. In a packed or full-line pumping under pressure the integrity of the interface is well maintained. In simple English, this means that there is surprisingly little mixing of the two batches of product or types of crude oil as it moves through the line. Products are kept "pure" up to specification by drawing off the few barrels of interface-mix into a "slop tank."

INTERMEDIATE STRING
See Casing.

INTERRUPTIBLE GAS
A gas supply, usually to industrial plants and large commercial firms, that can be curtailed or interrupted during emergencies or supply shortages in order to maintain service to domestic customers.

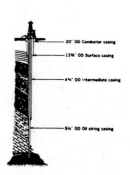

20" OD Conductor casing

13⅜" OD Surface casing

8⅝" OD Intermediate casing

5½" OD Oil string casing

Intermediate string

INTERSTATE OIL COMPACT
A compact between oil-producing states negotiated and approved by Congress in 1935, the purpose of which is the conservation of oil and gas by the prevention of waste. The Compact provides no power to coerce but relies on voluntary agreement to accomplish its objectives. Originally, there were six states as members; today, there are nearly 30.

IOCC
Interstate Oil Compact Commission.

IOSA
International Oil Scouts Association.

IPAA
Independent Petroleum Association of America.

ISO-
A prefix denoting similarity. Many organic substances, although composed of the same number of the same atoms, appear in two, three, or more varieties or isomers which differ widely in physical and chemical properties. In petroleum fractions there are many substances that are similar, differing only in specific gravity, for example, isooctane, isobutane, isopentane, and many other isomers.

ISOMERIZATION
A refinery process for converting chemical compounds into their isomers, i.e., rearranging the structure of the molecules without changing their size or chemical composition.

ISOMERS
Compounds having the same composition and the same molecular weight but differing in properties.

ISOPACHOUS MAP
A geological map; a map that shows the thickness and conformation of underground formations; used in determining underground oil and gas reserves.

ISOPENTANE
A high-octane blending stock for automotive gasoline.

ISOTHERMAL
At constant temperature. When a gas is expanded or compressed at a constant temperature, the expansion or compression is isothermal. Heat must be added to expanding gas and removed from compressing gas to keep it isothermal.

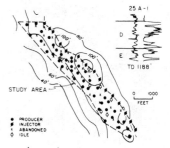

Isopachous map *(after Gates and Brewer)*

J

J-4 FUEL
A designation for highly refined kerosene used as fuel for jet engines.

JACK
An oil well pumping unit powered by a gasoline engine, electric motor, or rod line from a central power. The pumping jack's walking beam provides the up and down motion to the well's pump rods.

1890 pipeline gang using pipe tongs and jack board

Jackknife rig

Jack-up rig

Jeeping

JACK BOARD
A wood or metal prop used to support a joint of line pipe while another joint is being screwed into it. Jack boards have metal spikes inserted at intervals to support the pipe at different levels.

JACKET, OFFSHORE PLATFORM
See Platform Jacket.

JACKKNIFE RIG
A mast-type derrick whose supporting legs are hinged at the base. When the rig is to be moved, it is lowered or laid down intact and transported by truck.

JACK RABBIT
A device that is put through casing or tubing before it is run to make certain it is the proper size inside and outside; a drift mandrel.

JACKSHAFT
An intermediate shaft in the power train. Jack shafts usually are relatively short and often are splined.

JACK-UP RIG
A barge-like, floating platform with legs at each corner that can be lowered to the sea bottom to raise or "jack up" the platform above the water. Towed to location offshore, the legs of the jack-up rig are in a raised position, sticking up high above the platform. When on location, the legs are run down hydraulically or by individual electric motors.

JAM NUT
A nut used to jam and lock another nut securely in place; the second and locking nut on a stud bolt. After the first nut is threaded and tightened on a stud, a second nut is tightened down on the first nut to prevent it from working loose.

JARS
A tool for producing a jarring impact in cable-tool drilling, especially when the bit becomes stuck in the hole. Cable-tool jars (part of the drillstring) are essentially a pair of elongated, interlocking steel links with a couple of feet of "play" between the links. When the drilling line is slacked off, the upper link of the jars moves down into the lower link. When the line is suddenly tightened the upper link moves upward engaging the lower link with great force that usually frees the stuck bit. *See* Bumper Sub, Fishing.

JEEPING
Refers to the operation of inspecting pipe coating with the aid of electronic equipment. An indicator ring is passed over the pipe which carries an electric charge. If there is a break or holiday (q.v.) in the protective coating, a signal is transmitted through the indicator ring to an alarm.

JERKER
A line which connects the bandwheel crank to the drilling cable. As the crank revolves, the drilling line is jerked (pulled up and released suddenly) providing an up and down motion to the spudding tools on a cable tool rig.

JET FUEL
A specially refined grade of kerosene used in jet propulsion engines.

JET MIXER (CEMENT)
A device consisting of a hopper to which a water supply under pressure is connected. Sacks of cement are opened and dumped one at a time into the hopper. The high-pressure water is jetted through the lower part of the hopper, mixing with the dry cement to form a slurry for pumping downhole to cement the casing in a well or for a squeeze job. *See* Squeeze a Well.

Modified jet mixers

JET SLED
An underwater trenching machine for burying a pipeline below the sea floor. The patented jet sled straddles the pipeline and scours out the sea-bed material ahead and beneath the line with a series of high-pressure jets of sea water. The power is supplied by a series of high-pressure pumps aboard an accompanying jet barge. The jetted water, at 1,200 psi, is directed ahead and below the line and literally cuts a ditch in the sea floor into which the line is laid.

JETTED-PARTICLE DRILLING
A method of drilling in hard rock formations using steel pellets forced at high velocity from openings in the bottom of the drill bit. The jetted particles are used with air drilling. The small steel pellets after striking and chipping the rock are returned to the surface by the force of the returning drilling air, along with the rock which has been pulverized. Jetted particle drilling has been used more or less experimentally and has not had wide acceptance.

JETTING
Injecting gas into a subsurface formation for the purpose of maintaining reservoir pressure.

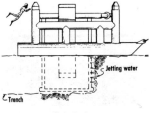

Jet sled

JETTING THE PITS
A method of removing cuttings, drilled rock fragments from the bottom of the working mud pits. This is done with a suction hose that derives its partial vacuum from a jet-nozzle arrangement, a type of venturi (q.v.) through which "clean" drilling mud is pumped at high pressure and velocity. At the waist of the venturi the stream's velocity is increased even more and the pressure at this point is substantially reduced, creating enough vacuum to draw in, through an attached hose, the cuttings from the bottom of the working pit in the manner of a vacuum cleaner. As the chips are drawn into the hose, they are discharged in the reserve pits.

JETTY
A pier (q.v.).

JOCKEY
An experienced and proficient driver of large trucks or earth-moving equipment.

JOINT
A length of pipe, casing, or tubing usually from 20 to 30 feet long. On drilling rigs, drillpipe and tubing are run the first time (lowered into the hole) a joint at a time; when pulled out of the hole and stacked in the rig, they are usually pulled two, three, or four at a time depending upon the height of the derrick. These multiple-joint sections are called stands (q.v.).

JOINT ADVENTURE
See Joint Venture.

Jetty

JOINT VENTURE

A business or enterprise entered into by two or more partners. Joint venture leasing is a common practice. Usually the partner with the largest interest in the venture will be the operator. *See* Consortium.

JOURNAL

That part of a rotating shaft that rests and turns in the bearing; the weight-bearing segment of the shaft.

JOURNAL BOX

A metal housing that supports and protects a journal bearing. *See* Journal.

JUG

(1) Colloquial term for geophones. (2) Colloquial term for the vertical caverns, shaped like a vinegar jug, leached out of subsurface salt formations for the storage of liquefied petroleum gases and other petroleum products. *See* Salt Dome Storage.

JUG HUSTLER

One who carries and places geophones in seismic work. Geophones are strung along the ground over an area where seismic shots are to be made by jug hustlers.

JUMBO BURNER

Jumbo burner *(Courtesy Otis)*

A flare used for burning waste gas produced with oil when there is no ready market or the supply of gas is too small or temporary to warrant a pipeline. A special kind of jumbo burner is used on offshore drilling platforms to burn oil and gas when a well is being tested or in the event of an emergency. Out on the water there is no place to put the oil during a test or when a well is allowed to blow to clean the hole so the oil has to be burned. The big burner, mounted on a boom or an extension of the platform deck, is equipped with air and water jets around the perimeter of the burner nozzle. When oil is burned the air and water jets are turned on which results in the complete combustion of the stream of oil. Without the high-pressure air and water jets to aerate and supply oxygen the oil would not burn completely; some of the oil would fall to the water below and present a real hazard. Jumbo burner or forced-draft burner.

JUMBOIZING

A technique used to enlarge an oil tanker's carrying capacity by cutting the vessel in two amidships and inserting a section between the halves.

JUNK BASKET

A type of fishing tool used to retrieve small objects lost in the borehole or down the casing, such objects as small slips, drilling cones off the bit, tools, etc. The basket is lowered into the hole and by the turbulence set up by pumping of the drilling mud the lost object is washed into the basket.

JUNK MILLS

Drill bits with specially hardened, rough cutting surfaces to grind and pulverize downhole "junk" material or nonretrievable tools or equipment such as millable packers and the like. After the junk has been ground or broken up into small pieces, the pieces can be circulated to the surface by the drilling mud or bypassed by the regular drillstring.

K

K
The abbreviation for kilo, one thousand. In certain employment ads, notably petroleum industry ads, the letter K is used instead of three zeros in giving salary ranges, e.g., 25K to 60K, also $25K - $60K. To the ad writer this is scientific shorthand meant to catch the eye of the no-nonsense engineer or technical person.

KELLY COCK
A blowout preventer built inside a three-foot section of steel tubing inserted in the drillstring above the kelly. A kelly cock is also inserted in the string below the kelly joint in some instances.

KELLY HOSE
See Mud Hose.

KELLY JOINT
The first and the sturdiest joint of the drill column; the thick-walled, hollow steel forging with two flat sides and two rounded sides that fit into a square hole in the rotary table which rotates the kelly joint and the drill column. Attached to the top of the kelly or grief stem (q.v.) are the swivel and mud hose.

KELLY SAFETY VALVE
See Kelly Cock.

KELLY SAVER SUB
See Kelly Valve, Lower.

KELLY SPINNER
A mechanism attached to the swivel that spins the kelly joint in and out of the first joint of drillpipe after the kelly has been broken out, unloosened. The spinner saves time in unscrewing and again in screwing in when a joint of drillpipe must be added to the string,

KELLY VALVE, LOWER
An automatic valve attached to the lower end of the kelly joint that opens and closes by mud pump pressure. The purpose of the valve is to prevent the mud in the kelly joint from pouring out on the derrick floor each time the kelly is disconnected from the drillpipe. When the mud pump is stopped, the kelly valve automatically closes. After a joint of drillpipe is added to the string and the kelly is made up tight, the pumps are started and the mud pressure opens the kelly valve and drilling resumes. The automatic valve saves valuable mud, keeps the rig floor dry, and speeds up the job of making a connection.

KEROGEN
A bituminous material occurring in certain shales which yield a type of oil when heated. *See* Kerogen Shales.

KEROGEN SHALES
Commonly called oil shales, kerogen shales contain material neither petroleum nor coal but an intermediate bitumen material with some of the

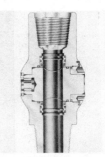

Kelly cock

Kelly hose *(Courtesy B.F. Goodrich)*

Kelly joint and bushing

Kerogen shale
semiworks project

properties of both. Small amounts of petroleum are usually associated with kerogen shales but the bulk of the oil is derived from heating the shale to about 660°F. Kerogen is identified as a pyrobitumen.

KEROSENE, RAW
Kerosene-cut from the distillation of crude oil, not treated or "doctor tested" to improve odor and color.

KEY
(1) A tool used in pulling or running sucker rods of a pumping oil well; a hook-shaped wrench that fits the square shoulder of the rod connection. Rod wrenches are used in pairs; one to hold back-up and the other to break out and unscrew the rod. (2) A slender metal piece used to fasten a pulley wheel or gear onto a shaft. The key fits into slots (keyways) cut in both the hub of the wheel and the shaft.

KEY BED
The stratum chosen for contouring or making an isopachous map. If, as it is hoped, other strata conform to the key bed, then an accurate contouring of the key bed will indicate the subsurface strata.

KEYSEAT
A section of the well bore deviating abruptly from the vertical causing drilling tools to hang up; a shoulder in the borehole.

KEYSEATING
A condition downhole when the drill collar or another part of the drillstring becomes wedged in a section of crooked hole, particularly a dogleg which is an abrupt deviation from the vertical or the general direction of the hole being drilled.

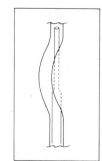

Keyseating drill collars in
crooked hole

KEYWAY
A groove or slot in a shaft or wheel to hold a key (q.v.).

KICK
Pressure from downhole in excess of that exerted by the weight of the drilling mud, causing loss of circulation. If the gas pressure is not controlled by increasing the mud weight, a kick can violently expel the column of drilling mud resulting in a blowout.

KICKING DOWN A WELL
A primitive method of drilling a shallow well using manpower (leg power). In oil's very early days, a pole made from a small tree was used to support the drilling line and bit in the hole. The driller with his foot in a stirrup attached to the line would kick downward causing the pole to bend and the bit to hit the bottom of the hole. The green sapling would spring back, lifting the bit ready for another "kick" by the driller.

KICKOUT CLAUSE
In some purchase contracts for oil and gas a clause that permits the purchaser, under certain conditions, to renegotiate the contract. Usually the conditions concern pricing or market availability.

KICK, WATERFLOOD
See Waterflood Kick.

Kicking down a well with
spring poles (Courtesy
API)

KIER, SAMUAL M.

In the early 1850s, Kier was skimming crude oil from the water of his salt wells in Pittsburgh, Pa., and selling it as Kier Rock Oil, a medicinal cure-all. Soon he had more oil than he could peddle in bottles so he became interested in refining. With the assistance of J.C. Booth, a Philadelphia chemist who designed a crude, coal-fired still, Kier began refining kerosene. By 1859 and the advent of Drake's well, there were nearly a hundred small, one-vessel refineries around the country making kerosene for use in a new lamp that had been invented.

KILL AND CHOKE LINES

Lines connected to the blowout preventer stack through which drilling mud is circulated when the well has been shut in because excessive pressure downhole has threatened a blowout. Mud is pumped through the kill line and is returned through the choke line, bypassing the closed valves on the BOP. When the mud has been heavied up to overcome downhole pressure, drilling can proceed.

KILL A WELL

To overcome downhole pressure in a drilling well by the use of drilling mud or water. One important function of drilling mud is to maintain control over any downhole gas pressures that may be encountered. If gas pressure threatens to cause loss of circulation or a blowout, drilling mud is made heavier (heavied-up) by the addition of special clays or other material. *See* Kick.

KILLER WELL

A directional well drilled near an out-of-control well to "kill" it by flooding the formation with water or mud. Wells that have blown out and caught fire are often brought under control in this manner if other means fail.

KNOCK-OFF POST

A post through which a rod line moves as it operates a pumping jack. When the well is to be hung off (shut down), a block is inserted between the rod-line hook and the knock-off post which interrupts the line's forward movement putting slack in the line so that the hook may be disengaged.

KNOCKOUT

A tank or separator vessel used to separate or "knock out" water from a stream of oil.

KNOWLEDGE BOX

The drilling crew's name for the place the driller keeps his orders and reports; smart box.

KNUCKLE BUSTER

A wrench so worn or of such poor quality that it will not hold when under the strain of heavy work.

KNUCKLE JOINT

A universal joint (q.v.); a type of early drilling tool hinged on a movable joint so that the drill could be deflected at an angle from the vertical.

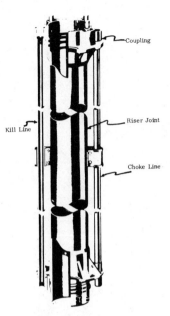

Kill and choke lines attached to marine riser

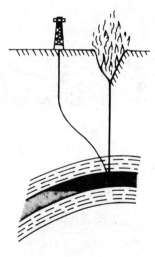

Killer well

KORT NOZZLE
A type of ship's propeller that rotates within a cylindrical cowling which concentrates the thrust of the propeller. This produces a nozzle effect as the water is jetted from the cowling. Kort nozzles are installed on some tugboats and drilling-tender vessels to increase their maneuverability and response.

L

LACT
Lease Automatic Custody Transfer. *See* Automatic Custody Transfer.

LANDED COST (OF OIL)
The cost of a barrel of imported oil offloaded at a U.S. port. Landed cost includes all foreign taxes and royalties plus cost of transportation.

LANDING CASING
Lowering a string of casing into the hole and setting it on a shoulder of rock at a point where the diameter of the borehole has been reduced. The beginning of the smaller diameter hole forms the shoulder on which the casing is landed.

LANDMAN
A person whose primary duties are managing an oil company's relations with its landowners. Such duties include securing of oil and gas leases, lease amendments, and other agreements.

LANDOWNER ROYALTY
A share of the gross production of the oil and gas on a property by the landowner without bearing any of the cost of producing the oil or gas. The usual landowner's royalty is one-eighth of gross production.

LAP-WELDED PIPE
Line pipe or casing made from a sheet of steel which is formed on a mandrel. The two edges, tapered to half normal thickness, are lapped over and welded. *See* Seamless Pipe.

LATERAL LINES
Pipelines that tie into a trunk line; laterals are of smaller diameter and are laid as part of a gathering system or a distribution system. In an oil field, laterals bring oil or gas from individual leases or tank batteries to the booster station and the trunk line.

LAY BARGE
A shallow-draft, barge-like vessel used in the construction and laying of underwater pipelines. Joints of line pipe are welded together and played out over the stern of the barge as it is moved ahead. Lay barges are used in swampy areas, in making river crossings, and in laying lines to offshore installations.

LAY-DOWN RACK
A storage area for tubing and drillpipe that are removed from a well and laid down rather than set back and racked vertically in the derrick.

Pipe lay barge

Lay-down rack

LAY DOWN THE TUBING
To pull the tubing from the well, a joint at a time, and remove it from the derrick floor to a nearby horizontal pipe rack. As each joint is unscrewed from the string, the lower end of the joint is placed on a low cart and pulled out to the rack as the driller lowers the pipe which is held up by the elevators.

LAY TONGS
See Pipe Tongs.

LAZY BENCH
A bench on which workers, when not working, may rest. A perch from which a work operation may be observed by workers or loafers.

LAZY BOARD
A stout board with a handle used to support the end of a pipeline while another length of pipe is screwed into it. On small lines, the man operating the lazy board or "granny" board usually handles the back-up wrench which holds one joint of pipe firm while another joint is being screwed in.

1890 pipeline gang using lazy board and pipe tongs

LB/LB
Pound per pound. In a refining process, the ratio of ingredients to be mixed or introduced to the process.

LBS-H₂O/MMSCF
Pounds of water per million standard cubic feet (MMSCF) of natural gas. The designation of water content for large volumes of gas. *See* PPM/WT.

LCCV
Large crude-carrying vessels; tankers from 100,000 to 500,000 deadweight tons capable of transporting 2.5 to 3.5 million barrels of oil in one trip. Cruising speed of LCCVs is 12 to 18 knots; overall length, about 1,200 feet; draft when fully loaded, more than 80 feet.

LCCV

LEAD LINES
Lines through which production from individual wells is run to a lease tank battery.

LEAN GAS
Natural gas containing little or no liquefiable hydrocarbons. *See* Wet Gas.

LEAN OIL
The absorbent oil in a gasoline absorption plant from which the absorbed gasoline fractions have been removed by distillation. Before distillation to remove gasoline fractions, the oil is referred to as "fat oil" (q.v.).

LEASE
(1) The legal instrument by which a leasehold is created in minerals. A contract that, for a stipulated sum, conveys to an operator the right to drill for oil or gas. The oil lease is not to be confused with the usual lease of land or a building. The interests created by an oil-country lease are quite different from a realty lease. (2) The location of production activity; oil installations and facilities; location of oil field office, tool house, garages.

LEASE BROKER

A person whose business is securing leases for later sale in the hope of profit. Lease brokers operate in areas where survey or exploration work is being done.

LEASE CONDENSATE

Liquid hydrocarbons produced with natural gas and separated from the gas at the well or on the lease. *See* Condensate.

LEASE HOUND

Colloquial term for a person whose job is securing oil and gas leases from landowners for himself or a company for which he works. *See* Landman.

LEASE LINES

Gathering lines on a lease; usually small-diameter (2 to 4-inch) pipelines that carry production from the lease wells to a central tank battery; lead lines.

LEASE TANKS

A battery of two or more 100 to 500-barrel tanks on a lease that receive the production from the wells on the lease. Pipeline connections are made to the lease tanks for transporting the oil to the trunk line and thence to the refinery.

Oil derrick legs *(Courtesy Parker Drilling)*

LEFT-HAND THREAD

A pipe or bolt thread cut to be turned counterclockwise in tightening. Most threads are right-hand, cut to be tightened by turning clockwise. Nipples with one kind of thread on one end, another on the other end, are referred to as "bastard (q.v.) nipples."

LEGAL SUBDIVISION

Forty acres; one-sixteenth of a section (square mile).

LEGS, OIL DERRICK

The four corner-members of the rig, held together by sway braces and girts.

LENS

A sedimentary deposit of irregular shape surrounded by impervious rock. A lens of porous and permeable sedimentary rock may be an oil-producing area.

LESSEE

The person or company entitled, under a lease, to drill and operate an oil or gas well.

Lift pump *(Courtesy KOBE)*

LIFO − FIFO − FILO

Last in first out; first in first out; first in last out. Acronyms that designate the sequence of movement in and out or the handling of crude oil and products in inventory or held in storage.

LIFTING

(1) Refers to tankers and barges taking on cargoes of oil or refined product at a terminal or transshipment point. (2) Producing an oil well by mechanical means; pump, compressed air, or gas.

LIFTING COSTS
The costs of producing oil from a well or a lease.

LIGHT CRUDE
Crude oil that flows freely at atmospheric temperatures and has an API gravity in the high 30s and 40s; a light-colored crude oil. *See* Heavy Crude Oil.

LIGHT ENDS
The more volatile products of petroleum refining; e.g. butane, propane, gasoline.

LIGHT PLANT
An early-day term for an installation on a lease or at a company camp (q.v.) that provided electricity for lighting and small appliances. The light plant often was simply a belt-driven D.C. generator run off one of the engines at a pipeline pumping station or a pumping well's engine. The lights "surged" with the power strokes of the engines and went out when an engine "went down," but the lights were far better than gas lights—or none at all.

LIME
Colloquial for limestone.

LINE FILL AND LINE PACK
Line fill: The amount of gas or oil or product required to fill a new line before deliveries can be made at take-off points or the end of the line. Line pack: The volume of gas or barrels or oil maintained in a trunk pipeline at all times in order to maintain pressure and provide uninterrupted flow of gas or oil. There are millions of barrels of oil and billions of cubic feet of gas in the country's pipelines at all times.

LINE, GAS SALES
Merchantable natural-gas line from a lease or offshore production-processing platform carrying gas that has had water and other impurities removed; a line carrying pipeline gas (q.v.).

LINE LIST
Instructions to the pipeline construction crews building a line across the land of many property owners. The instructions list all owners, the length of line across each property, and include any special restrictions such as "keep all gates closed and in good repair" and "avoid at all costs damaging large trees." The right-of-way man helps make up the line list.

LINE, OIL SALES
Merchantable crude-oil line from a lease or offshore production-processing platform carrying oil that has had water and other impurities removed; a line transporting pipeline oil, (q.v.).

LINE-PACK GAS
Gas maintained in a gas transmission line at all times to maintain pressure and effect uninterrupted flow of gas to customers at take-off points.

LINER
In drilling, a length of casing used downhole to shut off a water or gas formation so drilling can proceed. Liners are also used to case a "thief

zone" (q.v.) where drilling fluid is being lost in a porous formation. A liner is also a removable cylinder used in reciprocating pumps and certain types of internal-combustion engines; a sleeve.

Internal lineup clamps

LINEUP CLAMPS

A device that holds the ends of two joints of pipe together and in perfect alignment for welding. Lineup clamps operate on the outside of the pipe and are used on smaller diameter line pipe. Large-diameter pipe—20 to 36-inch and over—are aligned by internal, hydraulically operated mandrel-like devices.

LINKAGE

A term used to describe an arrangement of interconnecting parts—rods, levers, springs, joints, couplings, pins—that transmit motion, power, or exert control.

LIQUEFIED NATURAL GAS (LNG)

Natural gas that has been liquefied by severe cooling (−160°C.) for the purpose of shipment and storage in high-pressure cryogenic tanks. To transform the liquid to a useable gas, the pressure is reduced and the liquid is warmed.

LIQUEFIED PETROLEUM GAS

Butane, propane, and other light ends (q.v.) separated from natural gasoline or crude oil by fractionation or other processes. At atmospheric pressure, liquefied petroleum gases revert to the gaseous state.

LPG storage tanks

LIQUID HYDROCARBONS

Petroleum components that are liquid at normal temperatures and atmospheric pressure.

LITER

A metric unit of volume: 1.057 U.S. quarts; 61.02 cubic inches.

LITHOLOGY

(1) The study and identification of rocks. (2) The character of a rock formation.

LITTLE BIG INCH PIPELINE

A 20-inch products pipeline built from East Texas to the East Coast during World War II to solve the problem caused by tanker losses as the result of submarine warfare. After the war, the line was sold to a private gas transmission company.

LIVE OIL

Crude oil which contains dissolved natural gas when produced. A flowing well always has dissolved gas, as it is the gas pressure which pushes the oil out of the porous formation to the well bore and up to the surface. Pumping wells may produce oil with a small amount of gas entrained in the production.

LNG carrier

LNG

See Liquefied Natural Gas.

LNGC
Liquefied natural gas carrier; a specially designed oceangoing vessel for transporting liquefied natural gas. (See illustration, p. 124)

LOADING ARMS
Vertical standpipes with swivel-jointed arms that extend to a tanker or barge's deck connections for loading crude oil or products.

Loading arms

LOADING RACK
An elevated walkway that supports vertical filling lines and valves for filling tank cars from the top. Tank cars used to be filled individually with a type of articulated loading arm attached to the filling line riser. The operator walked from car to car on the walkway checking his flow lines. When a tank car was topped out, the hatch was closed and sealed. For a more efficient way, see Tank Train.

LOAD OIL
Oil of any kind put back into a well for any purpose; e.g. hydraulic fracturing, shooting, or swabbing.

LOCAL DRAINAGE
The movement of oil or gas toward the well bore of a producing well. See Drainage.

Loading rack (Courtesy Diamond Alkali)

LOCATION
The well site; the place where a well is to be drilled or has been drilled; a well spacing unit, e.g. "Two locations south of the discovery well . . ."

LOCATION DAMAGES
Compensation paid by an operator to the owner of the land for damages to the surface or to crops during the drilling of a well. Mud pits must be dug, a surface leveled for tanks and rig, and access roads built, so there are always some location damages to be paid.

LOGGING UNIT
Well service wireline equipment for downhole well surveys. The spool on which the wire is wound is powered by a small engine to reel in the thousands of feet of wire lowered into the hole with the logging tool.

Logging unit minus wireline spool

LOG ROAD
See Cord Road.

LOG, SAMPLE
See Sample Log.

LONG STRING
See Production String.

LOOPING A LINE
The construction of a pipeline parallel to an existing line, usually in the same right-of-way, to increase the throughput capacity of the system; doubling a pipeline over part of its length, with the new section tied into the original line.

LOOSE-VALVE TREE
The designation for a Christmas tree or production tree nippled up or made

Log road

Loose-valve tree

Lowboy *(Courtesy Elder-Oilfield)*

Lowering in

up with individual valves as contrasted to solid-block tree valves, i.e., two or more valves made in one compact steel block. A stacked, loose-valve tree.

LOSE RETURNS
Refers to a condition in which less drilling mud is being returned from downhole than is being pumped in at the top. This indicates that mud is being lost in porous formations, crevices, or a cavern.

LOSS OF CIRCULATION
A condition that exists when drilling mud pumped into the well through the drillpipe does not return to the surface. This serious condition results from the mud being lost in porous formations, a crevice or a cavern penetrated by the drill.

LOWBOY
A low-profile, flat-bed trailer with multiple axles (6 to 10) for transporting extra-heavy loads over relatively short distances. The many wheels and axles serve to spread the weight of the trailer and its load over a large area to avoid damaging streets and highways. The low bed makes it easier to load and unload the heavy equipment it was designed to move.

LOWER IN
To put a completed pipeline in the ditch. This is done with side-boom tractors that lift the pipe in slings and carefully lower it into the ditch. The slings are made of layers of heavy canvas or rubber-impregnated fabric so as not to scuff or break the anticorrosion pipe coating which is applied to all buried pipelines.

LPG
Liquefied petroleum gas; LP-gas; "bottled gas"; butane, propane and other light ends (q.v.) separated from natural gasoline or crude oil by fractionation or other refining processes. At atmospheric pressure, liquefied petroleum gases revert to the gaseous state. Liquid butane and propane or a mixture of the two are used extensively in areas where there is no natural gas service. When the valve on the "bottle" or small tank of butane, for example, is opened, releasing the pressure on the liquid, a small quantity of liquid "boils" or turns to a gas and can then be used just as natural gas for cooking or heating.

LP-GAS DRIVE
The injection of high-pressure enriched gas or an LPG slug to effect the miscible displacement of oil. *See* Tertiary Recovery.

LUBE OIL
Short for lubricating oil or lubricant. Also lube and lubes.

LUBRICATING OIL, MULTIGRADE
Specially formulated lubricating oil that flows freely in cold weather, and in the heat of engine operation, maintains sufficient viscosity or "body" to properly lubricate the engine; e.g. 10-30 SAE.

LUBRICATION SYSTEM, GRAVITY SPLASH
A type of lubrication system for relatively slow-moving machinery. The crankcase of a pump, for example, contains the lube oil. As the crankshaft

turns, the crank throws and connecting rods splash through the reservoir of oil creating a "storm" of lubrication for all bearings inside the crankcase.

LUBRICATOR, OIL
A small, box-like reservoir containing a number of gear-operated pumps. The individual pumps, working in oil, measure out a few drops at a time into small, copper lines that distribute the lubricant to the bearings.

LUBRICATOR, MUD
A temporary hookup of pipes and valves for introducing additional, heavy drilling mud into the well bore to control gas pressure. Through one or two joints of large-diameter casing attached atop the wellhead, the heavy mud is fed into the well bore, against pressure, as through a lubricator.

LUCAS, CAPT. ANTHONY F.
It was Capt. Lucas' Spindletop gusher in 1901 (75,000 bbls/day) that ushered in the modern oil age of large oil companies. John H. Galey and James M. Guffey owned the Spindletop gusher located near Beaumont, Texas.

LURGI PROCESS
A process for the commercial gasification of coal which originated in Germany.

LXT UNIT
A low-temperature separator; a mechanical separator which uses refrigeration obtained by expansion of gas from high pressure to low pressure to increase recovery of gas-entrained liquids.

Oil lubricator *(Courtesy Lincoln)*

M

MAGNETO
See Ignition Magneto.

MAGNETOMETER
An instrument for measuring the relative intensity of the earth's magnetic effect. Used to detect rock formations below the surface; an instrument used by geophysicists in oil exploration work.

MAIN LINE
Trunk line; a large-diameter pipeline into which smaller lines connect; a line that runs from an oil-producing area to a refinery.

MAKE A HAND
To be a good worker.

MAKE IT UP
To screw a pipe or threaded connection tight by the use of a wrench.

Making up a connection

MAKE-UP TORQUE
The power necessary to screw a joint of pipe into another sufficiently tight to hold and not loosen under working conditions.

MAKING HOLE
Progress in drilling a well, literally.

Male connection

Automated production
manifold

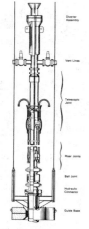

Marine riser system

MALE CONNECTION
A pipe, rod, or coupling with threads on the outside circumference.

MANHOLE
A hole in the side of a tank or other vessel through which a man can enter. Manholes have fitted covers with gaskets that are kept bolted in place when the tank is in use.

MANIFEST
A document issued by a shipper invoicing oil or products transported by a ship; a bill of lading.

MANIFOLD
An area where pipelines entering and leaving a pumping station or tank farm converge and that contains all valves for controlling the incoming and outgoing streams.

MAN RACK
A portable "doghouse" or cab mounted on a flat-bed truck for transporting pipeline workers to and from the job.

MAP, BASE
A map that contains latitude and longitude lines, land and political boundaries, rivers, lakes, and major cities.

MAP, RELIEF
A model of an area in which variation in the surface is shown in relief; a three-dimensional model of a surface area.

MAP, SURVEY
A map containing geologic information of the surface and/or the subsurface.

MAP, TOPOGRAPHIC
A map which shows in detail the physical features of an area of land, including rivers, lakes, streams, roads.

MARGINAL STRIKE
A discovery well on the border line between what is considered a commercial and a noncommercial well; a step-out well that may have over-reached the pool boundary.

MARGINAL WELL
A low-producing well, usually not subject to allowable regulations.

MARINE OIL
Petroleum found by wells offshore or on the continental shelf.

MARINE RISER SYSTEM
A string of specially designed steel pipes that extends down from a drillship or floating platform to the subsea wellhead. Marine risers are used to provide a return fluid-flow conductor between the well bore and the drill vessel and to guide the drillstring to the wellhead on the ocean floor. The riser is made up of several sections including flexible joints and a telescoping joint to absorb the vertical motion of the ship caused by wave action.

MARINE WHITE GASOLINE
Gasoline made for camp stoves, lanterns, blow torches, boat motors. Marine white contains no tetraethyl lead or other additives that could clog the needle valves of gasoline appliances.

MARSH BUGGY
A tractor-like vehicle whose wheels are fitted with extra-large rubber tires inflated with air for use in swamps. The great, balloon-like tires are 10 or 12 feet high and two or three feet wide providing buoyancy as well as traction in marshland. The marsh buggy is indispensable in exploration work in swampy terrain.

Marsh buggy

MASS-FLOW GAS METER
A gas meter that registers the quantity of gas in pounds which is then converted to cubic feet. Mass-flow meters, which are somewhat more accurate than orifice meters, are used in many refineries where large volumes of gas are consumed.

MAST
A simple derrick made of timbers or pipe held upright by guywires; a sturdy A-frame used for drilling shallow wells or for workover; a gin pole.

MASTER GATE
A large valve on the wellhead used to shut in a well if it should become necessary.

Drilling mast

MASTER BUSHING
The large bushing that fits into the rotary table of a drilling rig into which the kelly bushing fits. When the kelly bushing is lifted out of the master bushing, tapered slips are then inserted around the drillpipe to hold it securely while another joint is added to the drillstring.

MATING PARTS
Two or more machine or equipment parts made to fit and/or work together, e.g. piston and cylinder, pump plungers and liners or sucker rod box and pin.

MAT STRUCTURE
The steel platform placed on the sea floor as a rigid foundation to support the legs of a jack-up drilling platform.

Master bushing with kelly bushing above

MAT-SUPPORTED DRILLING PLATFORM
A self-elevating (jack-up) offshore drilling platform whose legs are attached to a metal mat or substructure that rests on the sea floor when the legs are extended.

MATTOCK
A tool for digging in hard earth or rock. The head has two sharpened steel blades; one is in the shape of a pick, the other the shape of a heavy adz.

MAXIMUM EFFICIENCY RATE (MER)
Taking crude oil and natural gas from a field at a rate consistent with "good production practice," i.e. maintaining reservoir pressure, controlling water, etc.; also the rate of production from a field established by a state regulatory agency.

MCF AND MMCF

Thousand cubic feet; the standard unit for measuring volumes of natural gas. MMCF is one million cubic feet.

MEA

Short for monoethanolamine, an organic base used in refining operations to absorb acidic gases in process streams. Also DEA, diethanolamine, another common organic base.

MEASURE, UNITS OF

LENGTH

1 Centimeter	= 0.3937 inches	= 0.0328 feet
1 Meter	= 39.37 inches	= 1.0936 yards
1 Kilometer	= 0.6213 miles	= 3,280 feet
1 Foot	= 0.3048 meters	
1 Inch	= 2.54 centimeters	
1 Mil	= 0.001 inch	

SQUARE MEASURE

1 Sq. Centimeter =	0.1550 sq. inches	
1 Sq. Meter =	1.196 sq. yards	=10.784 sq. feet
1 Sq. Kilometer =	0.386 sq. miles	
1 Sq. Foot =	929.03 sq. centimeters	
1 Sq. Mile =	2.59 sq. kilometers	
1 Sq. Inch =	1 million sq. mils	

MER

Maximum efficiency rate (of production) (q.v.).

MERCAPTANS

Chemical compounds, containing sulfur, present in certain refined products that impart objectionable odor to the product.

MERCHANTABLE OIL

Oil (crude) of a quality as to be acceptable by a pipeline system or other purchaser; crude oil containing no more than one percent BS&W (q.v.).

MERCURY NUMBER

A measure of the free sulfur in a sample of naphtha. Mercury is mixed with a sample and shaken, and the degree of discoloration in the sample is compared with a standard to determine the mercury number.

METAMORPHISM

Changes in rock induced by pressure, heat, and the action of water that results in a more compact and highly crystalline condition.

METER CHART

A replaceable paper chart for recording pressure or flow for a 24-hour period. As the chart revolves on its spindle, an inked pen traces the variations in pressure or volume.

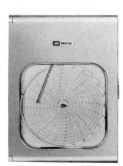

Meter chart *(Courtesy Acco)*

METHANE
The simplest saturated hydrocarbon; a colorless flammable gas; one of the main constituents of illuminating gas.

METHANE-RICH GAS PROCESS
See MRG Process.

METHANOL
Methyl alcohol; a colorless, flammable liquid derived from methane (natural gas).

METRIC SYSTEM CONVERSION

Inches	×	0.0254	=	meters
Feet	×	0.305	=	meters
Miles	×	1609.00	=	meters
Miles	×	1.609	=	kilometers
Millimeters	×	0.03937	=	inches
Centimeters	×	0.3937	=	inches
Meters	×	39.37	=	inches
Meters	×	3.281	=	feet
Kilometers	×	0.621	=	miles
Sq. Centimeters	×	0.155	=	sq. inches
Sq. Meters	×	10.764	=	sq. feet
Cu. Centimeters	×	0.061	=	cu. inches
Liters	×	0.2642	=	gallons
Gallons	×	3.78	=	liters

METRIC SYSTEM PREFIXES

Micro	= one millionth		Hecto	= one hundred
Milli	= one one-thousandth		Kilo	= one thousand
Centi	= one hundredth		Myria	= ten thousand
Deci	= one tenth		Mega	= one million
Deca	= ten			

METRIC TON
A unit of weight equal to 1,000 kilograms or 2,264.6 pounds. A metric ton of oil is 6.5 to 8.5 barrels depending upon the oil's gravity. A good approximation is 7.5 barrels of oil is one metric ton. In Europe and the Middle East, production and refining throughput figures are expressed in tons of crude or products instead of barrels as in the U. S.

MICELLAR-POLYMER FLOODING
See Micellar-Surfactant Flooding.

MICELLAR-SURFACTANT FLOODING
A tertiary recovery technique; a method of recovering additional crude oil from a field depleted by conventional means including repressuring and waterflooding. Micellar-surfactant drive or flooding involves injecting water mixed with certain chemicals into the producing formation. The chemical

solution reduces the surface tension of the oil clinging or adhering to the porous rock thus "setting the oil free" to be pumped out with the flooding solution. Such a project may have various names, e.g. micellar; micellar-polymer; soluble-oil; petroleum sulfonate.

MICROBALLOONS
A foam blanket that floats on the liquid in storage tanks to reduce losses from evaporation. The blanket is composed of billions of hollow, balloon-like plastic spheres containing a sealed-in gas—usually nitrogen. The spheres are almost microscopic in size. When poured in sufficient quantity on top of crude oil or refined products in a tank, they spread across the surface forming a dense layer that is effective in reducing evaporation.

MICRON
A unit of measure equal to one-thousandth of a millimeter. Fines (q.v.) and other low-gravity solids in drilling mud are described as being so many microns in size (10 microns, for example) and must be removed from the circulating mud by the use of a desilting device.

MID-CONTINENT CRUDE
Oil produced principally in Kansas, Oklahoma, and North Texas.

MIDDLE DISTILLATES
The term applied to hydrocarbons in the so-called middle range of refinery distillation; e.g. kerosene, light diesel oil, heating oil, and heavy diesel oil.

MIDNIGHT REQUISITION
Obtaining material without proper authority; borrowing unbeknown to the "lender"; swiping for a "good" cause.

MILL
To grind up; to pulverize with a milling tool (q.v.).

MILLABLE
Said of material used downhole, i.e., packers, bridges, and plugs, "soft" enough to be bored out or pulverized with milling tools.

MILLIDARCY
A unit of permeability of a rock formation; one-thousandth of a darcy. *See* Darcy.

MILLING
Cutting a "window" in a well's casing with a tool lowered into the hole on the drillstring.

MILLING TOOL
A grinding or cutting tool used on the end of the drill column to pulverize a piece of downhole equipment or to cut the casing.

MILL SCALE
A thin layer or incrustation of oxide which forms on the surface of iron and steel when it is heated during processing. Pipelines must be cleaned of mill scale before being put in service carrying crude oil, gas or products. This is done by running steel-bristle pigs and scrapers.

Milling tools

MINERAL SPIRITS
Common term for naphthas (solvents), those used for dry cleaning and paint thinners.

MINIMUM TENDER
The smallest amount of oil or products a pipeline will accept for shipment. Regulations set minimum tender amounts a common carrier pipeline is required to take into its system and pump to destination.

MINISEMI
A scaled-down semisubmersible drilling platform built for service in relatively shallow water.

MISCIBLE FLOOD
A secondary or tertiary oil recover method in which two or more formation-flooding fluids are used, one behind the other. For example, CO_2 may be injected into the formation followed by waterflooding. *See* Tertiary Recovery.

MIST
Small, almost microscopic droplets of water entrained in natural gas. Such gas must be treated to remove the water before it will be accepted by a gas transmission pipeline.

MMBTU, $
So many dollars per million BTU; a pricing formula in some gas purchase contracts which is tied directly to formulas involving prices paid for No. 2 fuel oil at specific locations in the U.S.

MMBTU/HR
Million BTU (British thermal units) per hour; rating used for large industrial heaters and other large thermal installations such as furnaces and boilers.

MOBILE PLATFORM
A self-contained, offshore drilling platform with the means for self-propulsion. Some of the larger semisubmersible drilling platforms are capable of moving in the open sea at five to seven knots.

MOCK-UP
A full-sized structural model built accurately to scale for study and testing of an installation to be used or operated commercially. For deep-water, offshore work mock-ups are made to simulate conditions in subsea well-head chambers and sea-floor work areas.

MODULE
An assembly (q.v.) that is functional as a unit, and can be joined with other units for increasing or enlarging the function; for example, a gas-compressor module; an electronic or hydraulic module.

MON
Motor Octane Number; the measure of a gasoline's antiknock qualities, whether or not it will knock or ping in an engine with a given compression ratio. Motor octane number of a gasoline is determined by test engines run under simulated conditions of load and speed. *See* Octane Rating.

Monkey board

Multibuoy mooring
system

MONEY LEFT ON THE TABLE
A phrase referring to the difference between the high and the second highest bid made by operators or companies when bidding on Federal or state oil leases. For example: high bid, $1,000,000; second-highest bid, $750,000, money left on the table, $250,000.

MONKEY BOARD
A colloquial and humorous reference to the tubing board (q.v.) high in the derrick.

MONKEY WRENCH
An adjustable, square-jawed wrench whose adjusting screw collar is located on the handle, and whose head can be used as a hammer; a crude wrench suitable for mechanical work of the roughest kind.

MONOCLINE
A geological term for rock strata that dip in one direction. When the crest of an anticline (q.v.) is eroded away, a partial cross section of the strata making up the fold is exposed at the earth's surface and the undisturbed lower flanks form what are called monoclines.

MONOPOD DRILLING PLATFORM
A type of offshore drilling platform with a single supporting leg. The design of the monopod makes it effective in Arctic regions where thick, moving bodies of ice present serious problems for more conventional platforms.

MOONPOOL
The opening in a drillship through which drilling operations are carried on; the moonpool or drillwell is usually located amidship, with the derrick rising above.

MOORING SYSTEM, MULTIBUOY
See Multibuoy Mooring System; also SBM; Single-buoy Mooring.

MOOSE AND GOOSE MEN
A humorous and somewhat sarcastic term for conservation (Environmental Protection Agency) people who, by law, can shut down a drilling well or a construction project to allow a rare or endangered species of bird to incubate her eggs unmolested or migrating or mating moose to go about their important business without being disturbed.

MOPE POLE
A lever; a pry pole usually made by cutting a small tree; used on pipeline construction as an adjunct to the jack board and in lowering the pipeline into the ditch.

MORMON BOARD
A broad, reinforced sled-like board with eye bolts on each end and a handle in the center. Used to backfill a pipeline ditch using a team of horses or a tractor pulling the board forward and a workman pulling it back into position for another bite.

MORNING REPORT
The report the tool pusher or drilling supervisor makes each morning after assembling the drilling reports of the drillers under his supervision. The

report includes depths reached at the end of each tour, footage drilled, mud records, formations penetrated, bit weights, rotary speeds, cores taken, pump speeds and pressures, and other pertinent information of the past 24 hours of operation.

MOTOR SPIRIT
A highly volatile fraction in petroleum refining; an ingredient of motor gasoline.

MOUSEHOLE
A hole drilled to the side of the well bore to hold the next joint of drillpipe to be used. When this joint is pulled out and screwed onto the drillstring, another joint of drillpipe is made ready and slipped into the mouse hole to await its turn. *See* Rat Hole.

Mousehole

MRG PROCESS
Methane Rich Gas Process. MRG is a patented process (Japan Gasoline Co.) to make synthetic natural gas from propane. Liquid propane is hydro-desulfurized and gasified with steam at temperatures between 900° and 1000°F. The resulting gas mixture is methanated, scrubbed to remove CO_2, dried, cooled, and fed to distribution lines.

MUD
See Drilling Mud.

MUD BARREL
A small bailer used to retrieve cuttings from the bottom of a cable tool drilling well.

MUD CAKE
See Filter Cake.

MUD COOLING TOWER
In drilling in or near a geothermal reservoir, the drilling mud becomes superheated and must be cooled to avoid flashing or vaporizing of the liquid (water or oil) in the mud stream at the surface. Cooling also reduces the thermal stress on the drill string.

Mud hog

MUD CUP
A device for measuring drilling mud density or weight; a funnel-shaped cup into which a measured quantity of mud is poured and allowed to run through, against time.

MUD ENGINEER
One who supervises the preparation of the drilling mud, tests the physical and chemical properties of the slurry, and prepares reports detailing the mud weight and additives used. A drilling fluid specialist.

MUD HOG
A mud pump; a pump to circulate drilling mud in rotary drilling; slush pump.

MUD HOSE
The flexible, steel-reinforced, rubber hose connecting the mud pump with the swivel and kelly joint on a drilling rig. Mud is pumped through the mud hose to the swivel and down through the kelly joint and drillpipe to the bottom of the well.

Mud hose *(Courtesy B.J. Goodrich)*

Mud line

Mud pit

Mud pump *(Courtesy Gardner s Denver)*

Mud tanks

MUD LINE
The sea or lake bottom; the interface between a body of water and the earth.

MUD LOG
A progressive analysis of the well-bore cuttings washed up from the bore hole by the drilling mud. Rock chips are retrieved with the aid of the shale shaker (q.v.) and examined by the geologist.

MUD-MOTOR DRILLING
See Turbodrilling.

MUD PITS
Excavations near the rig into which drilling mud is circulated. Mud pumps withdraw the mud from one end of a pit as the circulated mud, bearing rock chips from the borehole, flows in at the other end. As the mud moves to the suction line, the cuttings drop out leaving the mud "clean" and ready for another trip to the bottom of the borehole. *See* Reserve Pit.

MUD PUMP
A large, reciprocating pump that circulates drilling mud in rotary drilling. The duplex (two-cylinder) or triplex (three-cylinder) pump draws mud from the suction mud pit and pumps the slurry downhole through the drillpipe and bit and back up the borehole to the mud settling pits. After the rock cuttings drop out in the settling pit, the clean mud gravitates into the suction pit where it is picked up by the pump's suction line. In rotary drilling there are at least two mud pumps, sometimes more. In case of a breakdown or other necessary stoppages, another pump can be immediately put on line.

MUDSCOW
A portable drilling-mud tank in the shape of a small barge or scow used in cable-tool drilling when relatively small amounts of mud were needed or in a location when a mud pit was not practical. Also, a conveyance, a kind of large sled for transporting pipe and equipment into a marshy location. The mudscow is pulled by a crawler-type tractor which would not bog down as would a wheeled vehicle.

MUD TANKS
Portable metal tanks to hold drilling mud. Mud tanks are used where it is impractical to dig mud pits (q.v.) at the well site.

MUD UP
In the early days of rotary drilling and before the advent of accurate well logging, producible formations could be mudded up (plastered over) by the sheer weight of the column of drilling mud, so said the cable-tool men who were skeptical of the newfangled drilling method. Mudding up occurs also in pumping wells. The mud may be from shaley portions of the producing formation, from sections of uncased hole, or the residue of drilling mud.

MUD VALVE, AUTOMATIC
See Kelly Valve, Lower.

MULE SKINNER
Forerunner to the truck driver; a driver of a team or span of horses or mules hitched to an oil field wagon. Unhitched from the wagon, the team was

used to pull, hoist, and do earthwork with a slip or Fresno (q.v.). The "skinner" got his name from the ability to skin the hair off a mule's rump with a crack of the long reins he used, appropriately called butt lines.

MULLET
Humorous and patronizing reference to an investor with money to put into the drilling of an oil well with the expectation of getting rich; a sucker; a person who knows nothing about the oil business or the operator with whom he proposes to deal.

Mule skinner and team

MULTIBUOY MOORING SYSTEM
A tanker loading facility with five or seven mooring buoys to which the vessel is moored as it takes on cargo or bunkers (q.v.) from submerged hoses that are lifted from the sea bottom. Submarine pipelines connect the pipeline-ended manifold to the shore.

MULTIPAY WELL
See Multiple Completion.

MULTIPLE COMPLETION
The completion of a well in more than one producing formation. Each production zone will have its own tubing installed, extending up to the Christmas tree. From there the oil may be piped to separate tankage. *See* Dual Completion.

MULTIPLIER
A device or linkage for increasing (or decreasing) the length of the stroke or travel of a rod line furnishing power for pumping wells on a lease. A beam which oscillates on a fulcrum and bearing to which is attached the rod line from the power source (central power, q.v.) and a rod line to the pumping well. By varying the distance from the fulcrum of the two rod-line connections, the travel of the well's rod line can be lengthened or shortened to match the stroke of the well's pump.

Multibuoy mooring system

N

NACE
National Association of Corrosion Engineers.

NAMEPLATE RATING
The manufacturers' ratings as to speed (rpm), working pressure, horsepower, type of fuel, voltage requirement, etc., printed or stamped on the makers' nameplates attached to pumps, engines, compressors, or electric motors. To ensure proper and lasting performance of machines and equipment, nameplate ratings are always heeded.

NAMING A WELL
See Well Naming.

NAPHTHA
A volatile, colorless liquid obtained from petroleum distillation; used as a solvent in the manufacture of paint, as dry-cleaning fluid, and for blending with casinghead gasoline in producing motor gasoline.

NAPHTHENE-BASE CRUDE OIL
Asphalt-base crude (q.v.).

NATIONAL PETROLEUM RESERVE—ALASKA
An area west of Prudhoe Bay field and south of Point Barrow containing millions of acres set aside and held in reserve for national security purposes. *See* Naval Petroleum Reserves.

NATIVE GAS
Gas originally in place in an underground formation as opposed to gas injected into the structure.

NATURAL GAMMA RAY LOGGING
A procedure in which gamma rays naturally given off or emitted by rock formations, cut through by the well's borehole, are measured. A radiation detector is lowered into the hole and picks up gamma rays emitted by the rock. The signals are transmitted to a recording device at the surface. *See* Gamma Ray Logging.

Natural gas well

NATURAL GAS
Gaseous forms of petroleum consisting of mixtures of hydrocarbon gases and vapors, the more important of which are methane, ethane, propane, butane, pentane, and hexane; gas produced from a gas well.

NATURAL GASOLINE
Drip gasoline (q.v.); a light, volatile liquid hydrocarbon mixture recovered from natural gas. A water-white liquid similar to motor gasoline but with a lower octane number. Natural gasoline, the product of a compressor plant or gasoline plant, is much more volatile and unstable than commercial gasoline because it still contains many lighter fractions that have not been removed.

NATURAL GAS, UNCONVENTIONAL
See Unconventional Natural Gas.

NAVAL PETROLEUM RESERVES
Areas containing proven oil reserves which were set aside for national defense purposes by Congress in 1923. The Reserves, estimated to contain billions of barrels of crude oil, are located in Elk Hills and Buena Vista, California; Teapot Dome, Wyoming; and on the North Slope in Alaska.

NEEDLE VALVE
A valve used on small, high-pressure piping where accurate control of small amounts of liquid or gas is desired. The "tongue" of the valve is a rod that tapers to a point and fits into a seat which permits fine adjustments as when used with pressure gauges.

NEOPRENE
A rubber-like product derived from petroleum and compounded with natural rubber to produce a substance highly resistant to chemicals and oils. Neoprene, first called polychloroprene, was discovered by W. Carothers, Ira Williams, A. Collins, and J. Kirby of the DuPont research laboratory.

NET PROFITS INTEREST
A share of gross production from a property, measured by the net profits

from the operation of the property. Such an interest is carved out of the working interest and represents an economic interest in the oil and gas produced from the property. Sometimes referred to as net royalty.

NET REVENUE INTEREST
A fractional share of the working interest not required to contribute to, nor liable for, any part of the expense of drilling and completing the first well on the property or lease. Net revenue is income from a property after all costs, including taxes, royalties, and other assessments, have been paid.

NEUTRAL STOCK
Lubricating oil stock that has been dewaxed and impurities removed and can be blended with bright stock (q.v.) to make good lube oil; one of the many fractions of crude oil that, owing to special properties, is ideal as a blending stock for making high-quality lube oil.

NEW OIL
For the purposes of price regulation under the Emergency Petroleum Allocation Act of 1973, new oil is the production from a property in excess of production in 1972; all subsequent production from a property producing in 1972. *See* Old Oil.

N.G.A.
Natural Gas Act. An Act of Congress which empowers the Federal Power Commission to set prices and regulate the transportation of natural gas.

NGPA
Natural Gas Policy Act of 1978.

N.G.P.A.
Natural Gas Processors Association, successor to the Natural Gasoline Association of America.

NIPPLE
A short length of pipe with threads on both ends or with weld ends.

NIPPLE CHASER
The material man who serves the drilling rig; the person who makes certain all supplies needed are on hand.

NIPPLE UP
To put together fittings in making a hook up; to assemble a system of pipe, valves, and nipples as in a Christmas tree.

NOBLE METAL (CATALYST)
A metal used in petroleum refining processes that is chemically inactive with respect to oxygen.

NOISE LOG
A sound detection system inside a logging tool designed to pick up vibrations caused by flowing liquid or gas downhole. The device is used to check the effectiveness of a squeeze job (q.v.), to estimate the gas flow from perforated formations, etc.

NOMINATIONS
(1) The amount of oil a purchaser expects to take from a field as reported to

a regulatory agency that has to do with state proration. (2) Information given to the proper agency of the Federal government or a state relative to tracts of offshore acreage a person or company would like to see put up for bid at a lease sale.

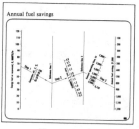

Nomograph

NOMOGRAPH
A device used by engineers and scientists for making rapid calculations; a graph that enables one, with the aid of a straightedge, to find the value of a dependent variable when the values of two or more independent variables are given.

NONDRILLING LEASE
A lease which grants the lessee the customary rights relative to oil and gas under the acreage but provides that a well shall not be drilled on the property. Under such circumstances, production from beneath the property requires that any drilling be done on other land. Nondrilling leases are usually granted where surface installations or activities make the drilling of a well impractical or incompatible.

NONFERROUS
Containing no iron; nonferrous tools, valves, or rods are made of other metal or combination of metals, e.g., brass, copper, bronze, spent uranium, or tungsten. Nonsparking tools are made of nonferrous metals, usually brass or bronze, because they are softer and will not give off sparks when struck against another piece of metal.

NONOPERATING INTEREST (IN A WELL)
An interest in an oil or gas well bearing no cost of development or operation; the landowner's interest; landowner's royalty (q.v.).

NONOPERATOR
The working-interest owner or owners other than the one designated as operator of the property; a "silent" working-interest owner.

NONSPARKING TOOLS
Hand tools made of bronze or other nonferrous alloys for use in areas where flammable oil or gas vapors may be present.

NPR
Naval Petroleum Reserves (q.v.).

NPR–A
National Petroleum Reserve–Alaska

N.P.R.A.
National Petroleum Refiners Association.

NPSH
Net positive suction head.

NPT
National pipe thread; denotes standard pipe thread.

"N" STAMP
Designates equipment qualified for use in nuclear installations: pipe, fittings, pumps, valves, etc.

NUMBER 2 FUEL
Furnace oil; also Two oil, distillate fuel.

NUT CUTTING, DOWN TO THE
The crucial point; the vital move or decision; a "this is it" situation.

O

OAPEC
Organization of Arab Petroleum Exporting Countries.

OBJECTIVE DEPTH (OF A WELL)
The depth to which a well is to be drilled. Drilling contracts often state that the hole shall be drilled to a specified depth or to a certain identifiable formation, whichever comes first, e.g., "to 5,500 feet or the Skinner sand, the objective depth."

OBO VESSEL
A specially designed vessel for carrying ore and crude, both in bulk form. The first oil and bulk ore tanker/carrier was launched in 1966 and used in handling relatively small cargoes of oil and ore.

OCAW
Oil, Chemical and Atomic Workers Union, a labor organization representing a large number of the industry's refinery and other hourly workers.

OCS
Outer Continental Shelf (q.v.).

OCTANE RATING
A performance rating of gasoline in terms of antiknock qualities. The higher the octane number the greater the antiknock quality; e.g., 94 octane gasoline is superior in antiknock qualities to a gasoline of 84 octane.

OD
Outside diameter of pipe; OD and ID (inside diameter) are initials used in specifying pipe sizes, e.g., 4½-inch OD, 8⅝-inch ID.

ODORANT
A chemical compound added to natural gas to produce a detectable, unpleasant odor to alert householders should they have even a small leak in the house piping. Odorants are used also in liquids or gases being stored or transported to detect leaks.

OFFLOADING
Another name for unloading; offloading refers more specifically to liquid cargo—crude oil and refined products.

OFFSET WELL
(1) A well drilled on the next location to the original well. The distance from the first well to the offset well depends upon spacing regulations and whether the original well produces oil or gas. (2) A well drilled on one tract of land to prevent the drainage of oil or gas to an adjoining tract where a well is being drilled or is already producing.

A replica of Drake's derrick and engine house *(Courtesy Pennsylvania Historical and Museum Comm.)*

OFFSHORE "WELL NO. 1"
The first offshore well (out of sight of land) was drilled on November 14, 1947, in the Gulf of Mexico, 43 miles south of Morgan City, Louisiana. By 1979, more than 20,000 wells had been drilled offshore.

OFF THE SHELF
Said of products or equipment that are ready and waiting at a supplier's warehouse and can be taken "off the shelf" and shipped immediately. Refers also to techniques and procedures that have been perfected and are ready to be employed on some job.

OIC
Oil Information Committee of the American Petroleum Institute (API).

OIL
Crude petroleum (oil) and other hydrocarbons produced at the wellhead in liquid form; includes distillates or condensate recovered or extracted from natural gas.

OIL ANALYZER, NET
A well testing installation that separates the oil flow and water content of individual wells on a lease. The analyzer automatically determines net oil and net water in a liquid stream. This information is important on leases where the production of individual wells (perhaps with different royalty owners) is to be commingled in the lease tanks or the pipeline gathering system.

OIL BONUS
An oil payment (q.v.) reserved by the lessor (usually the landowner) in addition to the cash bonus and royalty payment he is entitled to receive. The cash bonus is the money paid by the lessee to the landowner (the lessor) for the granting of an oil and gas lease. The landowner's royalty traditionally is one-eighth of the gross production from the well or the lease, if there is more than one well.

OIL BROKER
One who acts as a go-between in the domestic or international crude oil market. A broker will find a market for a quantity of crude or product not committed by long-term contract. Just as readily, he will come up with oil for someone who wishes to buy. Brokers perform a useful function in the oil business by being knowledgeable about the industry's supply and demand situation. He is the unobtrusive link between buyer and seller, independent producer and small refiner. For his services, the broker receives either a flat fee or a percentage of the deal he helps consummate.

OIL COUNTRY TUBULAR GOODS
Well casing, tubing, drillpipe, drill collars, and line pipe.

OILER
The third man at a pumping station in the old days. The normal shift-crew on a large gathering or mainline station was the station engineer, the telegraph operator-assistant engineer, and the oiler whose job included feeling the engine and pump bearings, keeping the wick oilers full and dripping properly, and cleaning and mopping the station floors.

OIL FINDER

A wry reference to a petroleum geologist. However, geologists maintain that they do not find oil but instead locate or identify formations that in their opinion are favorable to the accumulation of oil. It takes the drill to find oil.

OIL IMPORT TICKET

A license issued by an agency of the Federal government to refiners to buy certain amounts of crude oil shipped in from abroad.

OIL IN PLACE

Crude oil estimated to exist in a field or a reservoir; oil in the formation not yet produced. Oil in place is the province of the reservoir engineer who works with many factors — kind of formation, how thick, its porosity and permeability, wells drilled, well spacing, reservoir pressure, and other information — to arrive at the best scientific estimate of how much crude oil or gas remains in a given field or reservoir. It may take years to prove his estimate was correct — or that it missed the mark.

OIL-MIST SYSTEM

A lubricating system which pneumatically conveys droplets of a special oil from a central source to the points of application. An oil-mist system is economical in its use of lubricant and efficient on many types of antifriction applications.

CIL PATCH

A term referring broadly to the oil field, to areas of exploration, production, and pipelining.

OIL PAYMENT

A share of the oil produced from a well or a lease, free of the costs of production. An oil payment or overriding royalty (q.v.) may be conveyed to another party by the owner of a larger interest, for example the owner of the seven-eighths working interest. It may be granted to a bank to pay off a loan, or to someone else for other considerations. An oil payment is in fact a slice of the royalty from the well and comes to the owner of the royalty at specified times, monthly, quarterly, or annually, and is free of any costs or assessments for operating the well or lease.

OIL POOL

An underground reservoir or trap containing oil. A pool is a single, separate reservoir with its own pressure system so that wells drilled in any part of the pool affect the reservoir pressure throughout the pool. An oil field may contain one or more pools.

OIL RING

A metal ring that runs on a horizontal line shaft in the bearing well which has a supply of lube oil. As the ring slowly rotates through the well of oil, it deposits oil on the shaft. Oil rings are generally made of brass and are used on relatively slow-moving shafts.

OIL ROYALTY

The lessor's or landowner's share of oil produced on his land. The customary ⅛ royalty can be paid in money or in oil. In some instances, another fraction of production is specified as royalty.

OIL RUN
(1) The production of oil during a specified period of time. (2) In pipeline parlance, a tank of oil gauged, tested, and put on the line; a pipeline run. *See* Run Ticket.

OIL SANDS BITUMEN
A heavy, petroleum-like substance extracted from oil sands. Bitumen is defined as "any of various mixtures of hydrocarbons together with their non-metallic derivitives"; asphalts and tars.

OIL SHALE
Kerogen shale (q.v.).

Oil shale semiworks project

OIL SPILL
A mishap that permits oil to escape from a tank, an offshore well, an oil tanker, or a pipeline. Oil spill has come to mean oil on a body of water where even small amounts of oil spread and become highly visible.

OIL-SPILL BOOM
Any of various devices or contraptions to contain and prevent the further spread of oil spilled on water until it can be picked up. A curtain-like device deployed around or across the path of a drifting oil spill. The curtain is weighted on the bottom edge to hold it a foot or two below the surface and has floats on the upper edge to hold the curtain a foot or more above the surface. Once surrounded, the oil is sucked up by a vacuum cleaner-like suction pump.

Oil spill boom

OIL STRING
See Production String.

OIL WELL PUMP
See Pumping Unit, et seq.

OIL WELL PUMP, GRABLE
See Grable Oil Well Pump.

OLD OIL
For the purposes of price regulation under the Emergency Petroleum Allocation Act of 1973, old oil is production from a property up to the 1972 level of production. Any production in excess of this amount from a property is new oil (q.v.).

ONE-THIRD FOR ONE-QUARTER
A term used by independent oil operators who are selling interests in a well they propose to drill. An investor who agrees to a one-third for one-fourth deal will pay one-third of the cost of the well to casing point and receive one-fourth of the well's net production.

ON-LINE PLANT (GAS)
Gas processing plant located on or near a gas transmission line which takes gas from the trunk line for processing—stripping, scrubbing, drying—and returns the clean, dry gas to the line.

ON STREAM
Term used for a processing plant, a refinery or a pumping station that is operating. To bring on stream is to start up a plant or an operation.

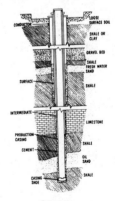

Oil string

ON THE BEAM
Refers to a well on the pump, operated by a walking beam instead of a pumping jack.

ON THE LINE
(1) Said of a tank of oil whose pipeline valve has been opened and the oil is running into the line. (2) A pumping unit that has been started and is pumping on the pipeline.

ON THE PUMP
A well that is not capable of flowing and is produced by means of a pump.

OOG
Office of Oil and Gas.

OOIP
Original oil in place.

OP-DRILLING SERVICE
Optimization drilling, a consulting service first developed by American Oil Company that makes available to operators of drilling rigs technical, geological, and engineering information gathered from wells drilled in the same area. Included is advice on mud programs, bits, drill speed, pressures, as well as consultation with drilling experts.

OPEC
Organization of Petroleum Exporting Countries (q.v.).

OPEN FLOW
The production of oil or gas under wide-open conditions; the flow of production from a well without any restrictions (valves or chokes) on the rate of flow. Open flow is permitted only for testing or clean-out. Good production practice nowadays is to produce a well under maximum efficiency rate conditions (q.v.).

OPEN HOLE
An uncased well bore; the section of the well bore below the casing; a well in which there is no protective string of pipe.

OPERATING INTEREST
An interest in oil and gas that bears the costs of development and operation of the property; the mineral interest less the royalty. See Working Interest.

OPERATOR
An actuating device; a mechanism for the remote operation and/or control of units of a processing plant. Operators usually are air or hydraulically actuated. Their main use is for opening and closing stops and valves.

OPERATOR, PLANT OR STATION
A worker who is responsible for the operation of a small plant or a unit of a larger plant during his working shift. In the old days, an operator was a telegrapher at a pumping station who sent reports on pumping rates, tank gauges, and line pressures to the dispatcher (q.v.).

ORGANIC SUBSTANCE
A material that is or has been part of a living organism. Oil, although

Operator or actuator
(Courtesy Fisher)

classified as a mineral, is an organic substance derived from living organisms.

ORGANIZATION OF PETROLEUM EXPORTING COUNTRIES
Oil producing and exporting countries in the Middle East, Africa, and South America that have organized for the purpose of negotiating with oil companies on matter of oil production, prices, future concession rights.

ORIFICE METER
A measuring instrument that records the flow rate of gas, enabling the volume of gas delivered or produced to be computed.

O-RING
A circular rubber gasket used in flanges, valves, and other equipment for making a joint pressure tight. O-rings in cross section are circular and solid.

OR LEASE
One of the two most common forms of oil and gas leases; the other is the unless lease (q.v.). Both types of leases are granted for a primary term, five years for example, and "so long thereafter as oil and gas are produced." In an or lease, the lessee promises to drill on or before the first anniversary date or do something else: pay rental, forfeit the lease, etc. The delay rental clause (q.v.) of an or lease is often written as follows: "Lessee agrees to begin a well on said premises within one year of date hereof or thereafter pay lessor as rental $_____ each year in advance to the end of this term or until said well is commenced, or this grant is surrendered as stipulated herein."

Orifice meter

O & S
Over and short (q.v.).

OSHA
Occupational Safety and Health Administration. A government agency that sets and enforces working condition standards relative to on-the-job safety and health. A result of the Occupational Safety and Health Act of 1971.

OTTO-CYCLE ENGINE
A four-stroke cycle gas engine; the conventional automobile engine is an Otto-cycle engine, invented in 1862 by Beau de Rochas and applied by Dr. Otto in 1877 as the first commercially successful internal combustion engine. The four strokes of the Otto cycle are intake, compression, power, and exhaust.

OUTAGE GAUGE
A measure of the oil in a tank by finding the distance between the top of the oil and the top of the tank and subtracting this measurement from the tank height. Outage gauging is used on large storage tanks in which there may be several feet of heavy sediment in the bottom preventing the plumb bob on the gauge line from touching bottom. By measuring the distance from the top of the tank whose height is known, the height of the oil from the bottom can be easily arrived at. *See* Innage Gauge.

OUTCROP
A subsurface rock layer or formation that, owing to geological conditions,

appears on the surface in certain locations. That part of a strata of rock that comes to the surface.

OUTER CONTINENTAL SHELF (OCS)
"All submerged land (1) which lies seaward and outside the area of lands beneath the navigable waters as defined in the Submerged Lands Act (67 Stat. 29) and (2) of which the subsoil and seabed appertain to the U.S. and are subject to its jurisdiction and control."

OUTPOST WELL
A well drilled in the hope of making a long extension to a partly developed field; an ambitious, giant-stride, step-out well that hopes to pick up the same pay zone as that encountered in the field.

OVER AND SHORT (O&S)
In a pipeline gathering system O&S refers to the perennial imbalance between calculated oil on hand and the actual oil on hand. This is owing to contraction, evaporation, improper measuring of lease tanks, and losses through undetected leaks. Oil is paid for on the basis of the amount shown in the lease tanks. By the time this oil is received at the central gathering station, the amounts invariably are short which represent a loss to the pipeline system.

OVERHEAD
A product or products taken from a processing unit in the form of a vapor or a gas; a product of a distillation column.

OVERRIDE
See Overriding Royalty.

OVERRIDE SYSTEM
A backup system; controls that take over should the primary system of controls fail or be taken out for adjustment or repair; a redundancy built in for safety and operational efficiency.

OVERRIDING ROYALTY
An interest in oil and gas produced at the surface free of any cost of production; royalty in addition to the usual landowner's royalty reserved to the lessor. A 1/16 override is not unusual.

OVERSHOT
A fishing tool; a specially designed barrel with gripping lugs on the inside that can be slipped over the end of tubing or drillpipe lost in the hole. An overshot tool is screwed to a string of drillpipe and lowered into the hole, and over the upper end of the lost pipe. The lugs take a friction grip on the pipe which can then be retrieved.

OVER-THE-DITCH COATING
Doping and wrapping line pipe above the ditch just before it is lowered in. Most line pipe is coated and wrapped in the pipe yard and then transported to the right-of-way and strung. Over-the-ditch coating has the advantage of minimizing scuffing or other damage to the coating suffered through moving and handling.

Coating and wrapping over the ditch

OXYACETYLENE WELDING

The use of a mixture of oxygen and acetylene in heating and joining two pieces of metal. When the weld edges of the two pieces are molten, metal from a welding rod is melted onto the molten puddle as the welder holds the tip of the rod in the flame of the torch. Oxygen and acetylene are used also in cutting through metal. The intense heat generated at the tip of the cutting torch (about 3,500°F) literally melts away the metal in the area touched by the flame. *See* Welding Torch.

P

PACKAGE PLANT

A facility at a refinery where various refined products are put in cartons and boxes ready for shipment. Waxes, greases, and small-volume specialty oils are boxed in a package plant.

PACKED-HOLE ASSEMBLY

A drill column containing special tools to stabilize the bit and keep it on a vertical course as it drills. Included among the tools are stabilizer sleeves (q.v.), square drill collars (q.v.), and reamers. Packed-hole assemblies are often used in "crooked-hole country."

PACKER

An expanding plug used in a well to seal off certain sections of the tubing or casing when cementing, acidizing, or when a production formation is to be isolated. Packers are run on the tubing or the casing, and when in position can be expanded mechanically or hydraulically against the pipe wall or the wall of the well bore.

PACKING

Any tough, pliable material—rubber or fiber—used to fill a chamber or "gland" around a moving rod or valve stem to prevent the escape of gas or liquid; any yielding material used to effect a pressuretight joint. Packing is held in place and compressed against a moving part by a "follower," an adjustable element of the packing gland.

PACKING GLAND

A stuffing box; a chamber that holds packing material firmly around or against a moving rod or valve stem to prevent the escape of gas or liquid. An adjustable piece that fits into the gland to compress the packing against the moving part is called a "follower" and can be screwed into the gland or forced into the gland by nuts on stud bolts.

PALYNOLOGY

The science that deals with the study of live and fossil spores and with pollen grains and other microscopic plant structures. As palynology concerns oil prospecting, particularly stratigraphic problems, the science involves age dating rocks and determining the environment when sedimentary formations were laid down. This can be observed from well borehole cuttings, cores, and surface outcrop samples; also microscopic analysis of source rock samples and other basic geochemical studies.

Packer *(Courtesy Otis)*

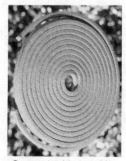

Square-braided packing

A complete offshore operation with a reel barge, drilling platform, producing platform, and tender.

PARAFFIN
A white, odorless, tasteless, and chemically inert waxy substance derived from distilling petroleum; a crystalline, flammable substance composed of saturated hydrocarbons.

PARAFFIN-BASE CRUDE
Crude oil containing little or no asphalt materials; a good source of paraffin, quality motor lubricating oil, and high-grade kerosene; usually has lower nonhydrocarbon content than an asphalt-base crude.

PARTICIPATION
A type of joint venture between a host country and an international oil company holding concession rights in that country. Participation may be voluntary on the part of the oil company or as the result of coercion by the host country.

PARTICIPATION CRUDE OIL
See Buy-back Crude Oil.

PARTICULATE MATTER
Minute particles of solid matter—cinders and fly ash—contained in stack gases.

PAY HORIZON
The subsurface, geological formation where a deposit of oil or gas is found in commercial quantities.

PAY OUT
The recovery from production of the costs of drilling, completion, and equipping a well. Sometimes included in the costs is a prorata share of lease costs.

PAY STRING
The pipe through which a well is produced from the pay zone. Also called the "long string" because only the pay string of pipe reaches from the wellhead to the producing zone.

PAY ZONE
See Pay Horizon.

PCV
Positive Crankcase Ventilation (q.v.).

PEAK-SHAVING LNG PLANT
A liquefied natural gas plant that supplies gas to a gas pipeline system during peak-use periods. During slack periods the liquefied gas is stored. With the need for additional gas, the liquid product is gasified and fed into the gas pipeline.

PEAPICKER
An inexperienced worker; a green hand; boll weevil.

PEMEX
Petroleos Mexicanos, the state-owned Mexican oil company.

PENDULUM DRILL ASSEMBLY
A heavily weighted drill assembly using long drill collars and stabilizers to help control the drift from the vertical of the drill bit. The rationale for the weighted drill assembly is that, like a pendulum at rest, it will resist being moved from the vertical and will tend to drill a straighter hole.

PENNSYLVANIA-GRADE CRUDE OIL
Oil with characteristics similar to the crude oil produced in Pennsylvania from which superior quality lubricating oils are made. Similar-grade crude oil is also found in West Virginia, eastern Ohio, and southern New York state.

PERFORATING
To make holes through the casing opposite the producing formation to allow the oil or gas to flow into the well. Shooting steel bullets through the casing walls with a special downhole "gun" is a common method of perforating.

PERFORATING GUN
A special tool used downhole for shooting holes in the well's casing opposite the producting formation. The gun, a steel tube of various lengths, has steel projectiles placed at intervals over its outer circumference, perpendicular to the gun's long axis. When lowered into the well's casing on a wire line opposite the formation to be produced, the gun is fired electrically, shooting numerous holes in the casing which permit the oil or gas to flow into the casing.

PERFORATIONS, SHAPED-CHARGE
See Shaped-charge Perforation.

PERMAFROST
The permanently frozen layer of earth occurring at variable depths in the Arctic and other frigid regions.

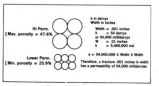

Permeability

PERMEABILITY
A measure of the resistance offered by rock to the movement of fluids through it. Permeability is one of the important properties of sedimentary rock containing petroleum deposits. The oil contained in the pores of the rock cannot flow into the well bore if the rock in the formation lacks sufficient permeability. Such a formation is referred to as "tight." *See* Porosity.

PERSUADER
An oversized tool for a small job; an extension added to the handle of a wrench to increase the leverage.

PESA
Petroleum Equipment Suppliers' Association.

PETROCHEMICALS
Chemicals derived from petroleum; feedstocks for the manufacture of a variety of plastics and synthetic rubber.

PETROCHEMISTRY
A word derived from petroleum and chemistry; the science of synthesizing substances derived from crude oil, natural gas, and natural gas liquids.

Petrochemical plant

PETROFRACTURING
A process in which a mixture of oil, chemicals, and sand is pumped under high pressure into an oil-bearing formation penetrated by the well bore. This produces cracks and fissures in the formation to improve the flow of oil. *See* Hydraulic Fracturing.

PETROLEUM
In its broadest sense, the term embraces the whole spectrum of hydrocarbons—gaseous, liquid, and solid. In the popular sense, petroleum means crude oil.

PETROLEUM RESERVES, NAVAL
See Naval Petroleum Reserves.

PETROLEUM RESERVES, STRATEGIC
See Strategic Petroleum Reserves.

PETROLEUM ROCK
Sandstone, limestone, dolomite, shale, and other porous rock formations where accumulations of oil and gas may be found.

PETROLEUM SULFONATE FLOODING
See Micellar-surfactant Flooding.

PETROLEUM TAR SANDS
Native asphalt, solid and semisolid bitumen, including oil-impregnated rock or sand from which oil is recoverable by special treatment. Processes have been developed for extracting the oil, referred to as synthetic crude.

A tar sand plant

PETROLOGIST, SEDIMENTARY
A specialist in petrology; a geologist who studies the origin, history, occurrence, structure, and chemical composition of sedimentary rocks; also a specialist in the acoustical properties of rocks who often works with geophysicists in determining the presence of oil and gas in sedimentary formations.

PETROLOGY
The science that deals with the origin, history, occurrence, structure, chemical composition, and classification of rocks.

PH (pH)
A symbol used in expressing both acidity and alkalinity on a scale whose values run from 0 to 14, with 7 representing neutrality; numbers less than 7, increasing acidity; greater than 7, increasing alkalinity.

PHOTOMETRIC ANALYZER
A device for detecting and analyzing the changes in properties and quantities of a plant's stack gases. The analyzer, through the use of electronic linkage, automatically sounds a warning or effects changes in the stack emissions when they exceed predetermined levels.

PHYSICAL DEPLETION
The exhausting of a mineral deposit or a petroleum reservoir by extraction or production.

A pier

Pipeline pig

Pig trap

Pileless platform or
gravity structure

PICKLE

A cylindrical weight (two to four feet in length) attached to the end of a hoisting cable, just above the hook, for the purpose of keeping the cable hanging straight and thus more manageable for the person using the wire line.

PIER

A walkway-like structure built on piling out from shore, a distance over the water for use as a landing place or to tie up boats; a jetty.

PIG

A cylindrical device (three to seven feet long) inserted in a pipeline for the purpose of sweeping the line clean of water, rust, or other foreign matter. When inserted in the line at a "trap," the pressure of the oil stream behind it pushes the pig along the line. Pigs or scrapers (q.v.) are made with tough, pliable discs that fit the internal diameter of the pipe, thus forming a tight seal as they move along cleaning the pipe walls.

PIG

To run or put a pig or scraper through a pipeline; to clean the line of rust, mill scale, corrosion, paraffin and water.

PIG TRAP

A scraper trap (q.v.).

PILELESS PLATFORM

A concrete offshore drilling platform of sufficient weight to hold the structure firmly in position on the sea bottom. Referred to as a "gravity structure," the platform is constructed on shore and then floated and towed to location where it is "sunk" by flooding its compartments. Some platforms of this type have oil storage facilities within the base of the structure. *See* Gravity Structure; *also* Tension-leg Platform.

PILING, DRILLED-IN

Piling that is inserted into holes drilled by special large-diameter bits. In this operation the piles are cemented in to achieve more stability. Drilled-in piling is often used to secure platform jackets to the ocean floor. *See* Drilling and Belling Tool.

PILLOW TANKS

Pliable, synthetic rubber and fabric fuel "tanks" that look like giant pillows. Pillow tanks, first used by the military to store fuel, are now in service at remote locations to store diesel fuel, gasoline, and lube oil until steel tankage can be erected. Easily deployable, the rubber pillows can be filled by tank truck or air shuttle, and when no longer needed they may be emptied, folded up and taken to another location. (See illustration, p. 155)

PILOT MILL

A type of junk mill (q.v.) with a tapered center projection below the cutting surface of the bit to guide or pilot the bit into the open end of a piece of junk or a tool to be milled out downhole.

PILOT PLANT

A small model of a future processing plant, used to develop and test processes and operating techniques before investing in a full-scale plant.

PINCHING A VALVE
Closing a valve part way to reduce the flow of liquid or gas through a pipeline. *See* Cracking a Valve.

PINCH-OUT
The disappearance or "wedging out" of a porous, permeable formation between two layers of impervious rock. The gradual, vertical "thinning" of a formation, over a horizontal or near-horizontal distance, until it disappears.

PIPE FITTER
One who installs and repairs piping, usually of small diameter. An "oil patch plumber" according to pipeliners who traditionally work with large-diameter pipe.

PIPE FITTINGS
See Fittings.

PIPE LAY BARGE
See Lay Barge.

PIPELINE CAT
A tough, experienced pipeline construction worker who stays on the job until is is flanged up and then disappears—until the next pipeline job. A hard-working, impermanent construction hand; a boomer.

PIPELINE DELUMPER
A motor-driven chopping machine that is flanged into a pipeline to break up any solid material that may have found its way into the fluid stream. The electric motor furnishes power for the chopper blades. Delumpers are used for the most part on coal slurry pipelines.

PIPELINE GAS
Gas under sufficient pressure to enter the high-pressure gas lines of a purchaser; gas sufficiently dry so that liquid hydrocarbons—natural gasoline, butane, and other gas liquids—usually present in natural gas will not condense or drop out in the transmission lines.

PIPELINE GAUGER
See Gauger.

PIPELINE INSPECTION SPHERE
A manned bathysphere for inspection of offshore pipelines or to investigate the underwater terrain, the sea floor, for a proposed route for laying a pipeline. The diving sphere is lowered to the sea floor by a boom and tackle extending from the deck of a work boat or diving tender equipped with support systems.

PIPELINE OIL
Clean oil; oil free of water and other impurities so as to be acceptable by a pipeline system.

PIPELINE PATROL
The inspection of a pipeline for leaks, washouts, and other unusual conditions by the use of light, low-flying aircraft. The pilot reports by radio to ground stations of any unusual condition on the line.

Pillow tanks being filled

Pipe lay barge

Pipeline delumper

A manned bathysphere

PIPELINE PRORATIONING
The refusal by a purchasing company or a pipeline to take more oil than it needs from the producer by limiting pipeline runs from the producer's lease; an informal practice in the days of overproduction when market conditions were unsatisfactory or when the pipeline system lacked storage space. Also referred to as purchaser prorationing.

PIPELINER
One who does pipeline construction or repair work: welders, ditching machine operators, cat drivers, coating and wrapping machine operators, connection men; broadly, anyone who is involved in the building, maintenance, and operation of a pipeline system.

Pipeline

PIPELINE RIDER
One who covers a pipeline by horseback looking for leaks in the line or washed-out sections of the right of way. The line rider has been replaced by the pipeline patrol using light planes or, for short local lines, by the pickup truck and the man on foot.

PIPELINE SPREAD
See Spread.

PIPELINE WELDING
In pipeline welding, the bevelled ends of two joints are brought together and aligned by clamps. Welders then lay on courses of weld metal called passes or beads designated as : (1) stringer bead, (2) hot pass, (3) third pass or hot fill (for heavy-wall pipe), (4) filler pass, and (5) final or capping pass.

PIPE MILL, PORTABLE
See Portable Pipe Mill.

Pipeline spread

PIPE SLING
A stirrup-like sling made of heavy belting material used on the winch line of boom cats for lifting, handling, and lowering in of pipe. Fabric slings are used to prevent scarring or damaging the pipeline's protective coating.

PIPE STRAIGHTENER
A heavy, pipeyard press equipped with hydraulically powered mandrels for taking the kinks and bends out of pipe. The replaceable mandrels come in all sizes, 2″ to 12″.

PIPE TONGS
Long-handled wrenches that grip the pipe with a scissors-like action; used in laying a screw pipeline. The head (called the butt) is shaped like a parrot's beak and uses one corner of a square "tong key," held in a slot in the head, to bite into the surface of the pipe in turning it.

PITCH
Asphalt; a dark brown to black bituminous material found in natural beds, also produced as a black, heavy residue in oil refining. *See* Brea.

Pipeline sling

PITCHER PUMP
A small hand pump for very shallow water wells. Looking much like a large, cast-iron cream pitcher, the pitcher pump is built on the order of the "old

town pump" with one exception. The pitcher pump's handle, working on a fulcrum, does not have a string of pump rods attached. Water is pumped by the suction created by a leather cup and valve arrangement in the throat or lower body of the pump together with a foot valve 20 feet or so down in the tubing. A simple and elegantly fundamental pumping machine.

PIT LINERS
Specially formulated plastic sheeting for lining earthen or leaking concrete pits to prevent seepage of oil or water into the ground.

PITMAN
The connecting piece between the crank on a shaft and another working part. On cable tool rigs, the pitman transmits the power from the bandwheel crank to the walking beam (q.v.).

Pitman

PITOT TUBE
A measuring device for determining the gas-flow rates during tests. The device consists of a tube with a ⅛-inch inside diameter inserted in a gas line horizontal to the line's long axis. The impact pressure of the gas flow at the end of the tube compared to the static pressure in the stream are used in determining the flow rate.

PITTING (OF PIPE)
Line pipe corroded in such a manner as to cause the surface to be covered with minute, crater-like holes or pits. *See* Anode; *also* Rectifier Bed.

PLANT OPERATOR
An employee who runs plant equipment, makes minor adjustments and repairs, and keeps the necessary operating records.

PLASTIC FLOW
The flow of liquid (through a pipeline) in which the liquid moves as a column; flowing as a river with the center of the stream moving at a greater rate than the edges which are retarded by the friction of the banks (or pipe wall). *See* Turbulent Flow.

PLAT
A map of land plots laid out according to surveys made by a Government Land Office showing section, township, and range; a grid-like representation of land areas showing their relationship to other areas in a state or county.

PLAT BOOK
A book containing maps of land plots arranged according to Township and Range for counties within a state. *See* Plat.

PLATFORMATE
High-octane gasoline blending stock produced in a catalytic reforming unit, commonly known as a platformer (q.v.).

PLATFORM BURNER
See Forced-draft Burner.

Platform burner

PLATFORMER
A catalytic reforming unit which converts low-quality straight-chain paraffins or naphthenes to low-boiling, branched-chain paraffins or aromatics of

Jacket piling

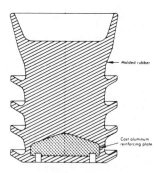

Top plug

Molded rubber

Cast aluminum reinforcing plate

Bottom plug

higher octane; a refinery unit that produces high-octane blending stock for manufacture of gasoline.

PLATFORM JACKET
A supporting structure for an offshore platform consisting of large-diameter pipe welded together with pipe braces to form a four-legged stool-like structure (stool without a seat). The jacket is secured to the ocean floor with piling driven through the legs. The four-legged offshore platform is then slipped into legs of the jacket and secured with pins and by the weight of platform and equipment.

PLATFORM TREE
A production Christmas tree (q.v.) on an offshore platform; an assembly of control and production valves, used on offshore platforms, through which wells are produced.

PLEM
Pipeline-end manifold; an offshore, submerged manifold connected to the shore by pipelines that serve a tanker loading station of the multibuoy mooring type (q.v.).

PLENUM
A room or enclosed area where the atmosphere is maintained at a pressure greater than the outside air. Central control rooms at refineries are usually kept at pressures of a few ounces above the surrounding atmosphere to prevent potentially explosive gases from seeping into the building and being ignited by electrical equipment. Some offshore drilling and production platforms are provided with plenums as a safety measure. *See* Acoustic Plenum.

PLUG
To fill a well's borehole with cement or other impervious material to prevent the flow of water, gas, or oil from one strata to another when a well is abandoned; to screw a metal plug into a pipeline to shut off drainage or to divert the stream of oil to a connecting line; to stop the flow of oil or gas.

PLUG BACK
To fill up the lower section of the well bore to produce from a formation higher up. If the well has been cased, the casing is plugged back with cement to a likely formation and then perforated (q.v.). *See* Bridge plug.

PLUGGING A WELL
To fill up the borehole of an abandoned well with mud and cement to prevent the flow of water or oil from one strata to another or to the surface. In the industry's early years, wells were often improperly plugged or left open. Modern practice requires that an abandoned well be properly and securely plugged.

PLUG VALVE
A type of quick-opening pipeline valve constructed with a central core or "plug." The valve can be opened or closed with one-quarter turn of the plug; a stop.

PLUNGER
The piston in the fluid end of a reciprocating pump. *See* Plunger pump.

PLUNGER PUMP
A reciprocating pump in which plungers or pistons, moving forward and backward or up and down in cylinders, draw in a volume of liquid and, as a valve closes, push the fluid out into a discharge line.

POGO PLAN
A plan for financing oil and gas exploration developed for offshore exploration. The form of the plan is usually corporate, the investors receiving shares of stock in the corporation and other securities.

POINT MAN
The member of a pipeline tong crew who handles the tips (the points) of heavy pipe-laying tongs. He is the "brains" of the crew as he keeps his men pulling and "hitting" in unison and in time with the other tong crews working on the same joint of screw pipe.

POLISHED ROD
A smooth brass or steel rod that works through the stuffing box or packing gland of a pumping well; the uppermost section of the string of sucker rods, attached to the walking beam of the pumping jacket.

POLYETHYLENE
A petroleum-derived plastic material used for packaging, plastic housewares, and toys. The main ingredient of polyethylene is the petrochemical gas ethylene.

POLYMERIZATION
A refining process of combining two or more molecules to form a single heavier molecule; the union of light olefins to form hydrocarbons of higher molecular weight; polymerization is used to produce high-octane gasoline blending stock from cracked gases.

PONTOONS
The elements of a floating roof tank that provide buoyance; airtight, metal tanks that float on the fluid and support the moveable deck structure of the roof.

PONY RODS
Sucker rod made in short lengths of 2′ to 8′ for use in making up a string of pumping rods of the correct length to connect the polished rod of the pumping jack. Pony rods are screwed into the top of the string just below the polished rod (q.v.).

POOL
See Oil Pool.

POOLING
The bringing together of small, contiguous tracts, resulting in a parcel of land large enough for the granting of a well permit under applicable spacing regulations. Pooling is often erroneously used for unitization (q.v.). Unitization describes a joint operation of all or some significant portion of a producing reservoir.

POOLS, SALT-DOME/SALT-PLUG
See Salt-dome Pools.

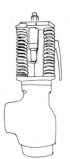

Pop-off valve

POP-OFF VALVE
See Relief Valve.

POPPET VALVE
A type of check valve installed in a riser or a downhole packer to prevent fluid from rising vertically in the pipe or the well bore. A spring-loaded vertical valve that permits downward flow as fluid pressure opens the valve. Pressure from below moving upward is blocked by the valve's clapper held shut by spring tension.

POPPING
The discharge of natural gas into the atmosphere; a common practice in the 1920s and 1930s, especially with respect to sour gas and casinghead gas. After the liquid hydrocarbons were extracted, the gas was "wasted" as there was no ready market for it.

PORCUPINE
A cylindrical steel drum with steel bristles protruding from the surface; a super, pipe-cleaning pig for swabbing a sediment-laden pipeline.

POROSITY
The state or quality of being porous; the volume of the pore space expressed as a percent of the total volume of the rock mass; an important property of oil-bearing formations. Good porosity indicates an ability to hold large amounts of oil in the rock. And with good permeability (q.v.), the quality of a rock that allows liquids to flow through it readily, a well penetrating the formation should be a producer.

PORTABLE PIPE MILL
A very large, self-propelled "factory on wheels" that forms, welds, and lays line pipe in one continuous operation. The pipe is made from rolls of sheet steel (skelp) shaped into a cylindrical form, electric welded, tested, and strung out behind the machine as it moves forward.

POSITIVE CRANKCASE VENTILATION SYSTEM
A system installed on automobiles manufactured after 1968 to reduce emissions from the engine's crankcase. The emissions—oil and unburned gasoline vapors— are directed into the intake manifold and from there they mix with the gasoline to be burned.

POSITIVE-DISPLACEMENT PUMP
A pump that displaces or moves a measured volume of liquid on each stroke or revolution; a pump with no significant slippage; a plunger or rotary pump.

POSSUM BELLY
A metal box built underneath a truck bed to hold pipeline repair tools— shovels, bars, tongs, chains, and wrenches.

POSTED PRICE
The price an oil purchaser would pay for crude oil of a certain API gravity and from a particular field or area. Once literally posted in the field.

POTS (PUMP VALVE)
See Valve Pots.

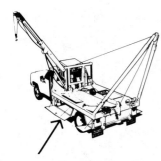

Possum belly

POUR POINT
The temperature at which a liquid ceases to flow; or at which it congeals.

POUR-POINT DEPRESSANT
A chemical agent added to oil to keep it flowing at low temperatures.

POWER
See Central Power.

POWER SYSTEMS, DRILLING RIG
See Drilling Rig, Electric; *also* Drilling Rig, Mechanical.

POWER TAKEOFF
A wheel, hub, or sheave that derives its power from a shaft or other driving mechanism connected to an engine or electric motor; the end of a power shaft designed to take a pulley.

POWER TONGS
An air or hydraulically powered mechanism for making up and breaking out joints of drillpipe, casing or tubing. After a joint is stabbed, the power tongs are latched onto the pipe which is screwed in and tightened to a predetermined torque.

PPM
Parts per million; a measure of the concentration of foreign matter in air or a liquid.

PPM/VOL
Parts per million (of water) in a given volume of natural gas. *See also* LBS-H_2O/MMSCF.

PPM/WT
Parts per million (of water) in a given weight of gas; used to express water content in a small amount of gas. *See also* LBS-H_2O/MMSCF.

PRAIRIE-DOG PLANT
A small, basic refinery located in a remote area; a "distillation system" (q.v.) which is very small, temporary refinery (200 to 1,000 barrels a day) set up at a remote drilling site to make diesel fuel and low-grade gasoline from available crude oil for the drilling engines and auxiliary equipment such as a light-plant engine, welding unit, etc.

PRESSURE GAUGE
Used on gas or liquid lines to make instantly visible the pressure in the lines. Some gauges have damping devices to protect the delicate mechanisms from the transient pulsations of pressure.

PRESSURE MAINTENANCE
See Repressuring Operation.

PRESSURE SNUBBER
A pulsation dampener (q.v.).

PRESSURE VESSEL
A cylindrical or spherical tank so constructed as to hold a gas or a liquid under pressure. Pressure vessels are used to hold air for air-actuated

Old central power plant

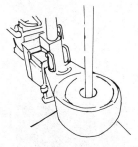
Power tongs

Pressure gauge
(Courtesy Wika)

Pressure snubber

valves, air-starting of engines, and other pneumatic applications. In a refinery or chemical plant, pressure vessels are integral parts of the processing chain where feedstock is subjected to both heat and pressure as part of the refining process.

PRIMARY PRODUCTION

Primary recovery; production from a reservoir by natural energy in the reservoir resulting in flowing wells, or wells on the pump with the oil flowing freely by gravity into the well bore.

PRIMARY TERM

The period of time an oil and gas lease is to run or be valid. When a lease's primary term expires, the lease must be renewed, if possible, or the interest in the property reverts automatically to the lessor or landowner. *See* Or Lease; *also* Delay Rental.

PRIME MOVER

The term describes any source of motion; in the oil field it refers to engines and electric motors; the power source. Prime mover is also applied to large four-wheel-drive trucks or tractors.

PRIVATE BRAND DEALER

A gasoline dealer who does not buy gasoline from a "major" supplier, but retails under the brand name of an independent supplier or his own brand name.

PROCESSING PLATFORM

A production platform (q.v.).

Processing platform

PROCESS STEAM

Steam produced in a refinery's or chemical plant's boilers to heat a process stream or for use in a refining process itself. Of the energy used in the U.S., a large percentage is consumed in the production of process steam. Petrochemical plants are important users of superheated, high-pressure steam.

PRODUCED WATER

Salt water produced from the oil from a well. When water and oil are mixed in the production stream they go into a gun barrel or other type of oil/water separator. The oil goes to the lease tanks, the water to an evaporation pit or, where there are large volumes of water, it is pumped into a salt water disposal well.

PRODUCING PLATFORM

An offshore structure with a platform raised above the water to support a number of producing wells. In offshore operations, as many as 60 wells are drilled from a single large platform by slanting the hole at an angle from the vertical away from the platform. When the wells are completed, the drilling equipment is removed and the platform is given over to operation of the producing wells.

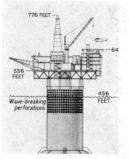

Producing platform

PRODUCT GAS

End product gas; gas resulting from a special manufacturing process; synthetic natural gas.

PRODUCT IMPORT TICKET
A license issued by an agency of the Federal government to a refiner or marketer to import products from abroad.

PRODUCTION ISLAND
An island made by dredging up material from the bottom of a lake or the ocean bottom to support one or more producing wells. Production islands are constructed in shallow water, close to shore, and are usually cheaper to build than steel production platforms. And with a lower profile, the islands are less offensive to the esthetic eye. Also, an island can be landscaped to hide the pumping wells and other equipment from view.

Production island
(Courtesy Conoco)

PRODUCTION PACKER
An expandable plug-like device for sealing off the annular space between the well's tubing and the casing. The production packer is run as part of the tubing string, inside the casing; when lowered to the proper depth, the packer is mechanically or hydraulically expanded and "set" firmly against the casing wall isolating the production formation from the upper casing while permitting the oil or gas to flow up the tubing.

PRODUCTION PAYMENT LOAN
A loan that is to be repaid out of the production of a well or a lease. It is a common practice in the oil country to borrow money on a producing well to finance further development of a lease. To pay off the loan the operator "carves out" a royalty payment to the lending institution from his seven-eighths working interest. This overriding royalty conveyed to the bank or other lender is free and clear of any costs of production or maintenance. *See* Carved-Out Interest, *also* Overriding Royalty.

Production platform

PRODUCTION PLATFORM
An offshore structure built for the purpose of providing a central receiving point for oil produced in an area of the offshore. The production platform supports receiving tanks, treaters, separators, and pumping units for moving the oil to shore through a submarine pipeline.

PRODUCTION SKID
A prefabricated oil and gas production unit assembled on a base or skid on the shore and transported to an offshore platform by one or more derrick barges. After the skid has been lifted into position and secured to the platform it is connected to the flow lines of the offshore wells it is to serve, and begins its function of receiving, separating, treating, storing, and pumping the oil and gas to shore stations. *See* Production Platform.

PRODUCTION STRING
The casing set just above or through the producing zone of a well. The production string is the longest and smallest diameter casing run in a well. It reaches from the pay zone to the surface.

PRODUCTION TAX
See Gross Production Tax.

PRODUCTION TREE
See Christmas Tree.

Production tree

PRODUCT LUBRICATED

Describes a pump whose bearings are lubricated by the liquid it is pumping. The pump is constructed with channels and wells that fill with product and in which the bearings or other moving parts run. Product lubricated equipment, needless to say, handles only clean liquids, i.e., various kinds of oils with lubricating qualities.

PRODUCT YIELD (AVERAGE)

From a 42-gallon barrel of crude oil the average yield is as follows: gasoline, 49.6%; jet fuel and kerosene, 6.6%; gas oil and distillates, 21.2%; residual fuel oil, 9.3%; lubricating oils, 7.0%; other products, 6.3%. With modern-day refining methods, these product percentages can be changed depending upon market demand.

PROFIT-SHARING BIDS

A type of bidding for federal and sometimes state oil leases in which a relatively small cash bonus is paid for the lease acreage plus a share in the net profits should the lease prove to be commercially productive. In some instances bidders have offered a 75 to 90 percent share in net profits for an especially promising parcel. This type of bidding substantially reduces the front-end cost for an operator but extends the payout time for his wells.

PROPANE

A petroleum fraction; a hydrocarbon, gaseous at ordinary atmospheric conditions but readily converted to a liquid. When in a liquid state propane must be stored in a high-pressure metal container. Propane is odorless, colorless, and highly volatile. It is used as a household fuel "beyond the gas mains."

Variety of propping agents *(Courtesy Pan American Petroleum)*

PROPPANTS

Material used in hydraulic fracturing (q.v.) for holding open the cracks made in the formation by the extremely high pressure applied in the treatment; the sand grains, beads, or other miniature pellets suspended in the fracturing fluid that are forced into the formation and remain to prop open the cracks and crevices permitting the oil to flow more freely.

PROPRIETARY DATA

Information on subsurface, geological formations gathered or purchased from a supplier of such data by an operator and kept secret; land and offshore reconnaissance surveys from seismic, magnetic, and gravity studies that are privately owned.

PRORATIONING

Restriction of oil and gas production by a state regulatory commission, usually on the basis of market demand. Prorationing involves allowables which are assigned to fields, and from fields to leases, and then allocated to individual wells.

PROTECTIVE STRING

A string of casing used in very deep wells and run on the inside of the outermost casing to protect against the collapsing of the outer string from high gas pressure encountered.

Proppants, close-up view *(Courtesy Amoco)*

PROVEN RESERVES
Oil which has been discovered and determined to be recoverable but is still in the ground.

P.S.I.A.
Pounds per square inch absolute; pressure measurement which includes atmospheric pressure.

P.S.I.G.
Pounds per square inch gauge (as observed on a gauge).

PULLED THREADS
Stripped threads; threads on pipe or tubing damaged beyond use by too much torque or force used in making up the joint.

PULLING MACHINE
A pulling unit (q.v.).

Pulling rods

PULLING RODS
The operation of removing the pumping or sucker rods from a well in the course of bringing up the bottom-hole pump for repairs or replacement. Rods must also be pulled when they have parted downhole. The rods above the break are pulled in a normal manner; the lower section must first be retrieved with a "fishing tool" (q.v.).

PULLING THE CASING
Removing the casing from the hole after abandoning the well. Prior to plugging the well with mud and cement, as much of the casing as can be pulled is retrieved. It is rare that all the casing can be removed from the hole. Often part of the string must be cut off and left in the hole.

PULLING TOOLS
Taking the drillpipe and bit out of the hole. If the tools are to be run again (put back in the hole), the drillpipe is unscrewed in two or three-joint sections (stands) and stacked in the derrick. *See* Doubles.

Pulling unit

PULLING UNIT
A portable, truck-mounted mast equipped with winch, wire lines, and sheaves, used for pulling rods or well workover.

PULL ONE GREEN
To pull a drill bit from the hole before it is worn out; to pull a bit before it is necessary.

PULL ROD
Shackle rod (q.v.).

PULL/ROD LINE
See Shackle Rod.

PULSATION DAMPENER
Various devices for absorbing the transient, rhythmic surges in pressure that occur when fluid is pumped by reciprocating pumps. On such pumps air chambers (q.v.) are installed on discharge lines, which act as air cushions. To protect pressure gauges and other instruments from the incessant pounding, fine-mesh, sieve-like disks are placed in the small tubing or

Pulsation dampener

Duplex pump *(Courtesy Gaso)*

Beam-balanced pumping unit

Crank-balanced unit *(Courtesy Lufkin Industries, Inc.)*

piping to which the gauge is attached; this arrangement "filters out" much of the surging which can damage delicate gauges.

PUMP, CASING

A sucker-rod pump designed to pump oil up through the casing instead of the more common method of pumping through tubing. A casing pump is run into the well on the sucker rods; a packer (q.v.) on top or bottom of the pump barrel provides packoff or seal between the pump and the wall of the casing at any desired depth. Oil is discharged from the pump into the casing and out the wellhead.

PUMP, DOUBLE-ACTING

See Double-Acting Pump.

PUMP, DOUBLE-DISPLACEMENT

A type of downhole rod pump which has plungers placed in tandem and operated simultaneously by the pump rods.

PUMP, DUPLEX

A two-cylinder reciprocating plunger pump.

PUMPER

A person who operates a well's pumping unit, gauges the lease tanks, and keeps records of production; a lease pumper.

PUMP, GEAR

A type of rotary pump made with two sets of meshing gears. When rotated on their shafts in the pump housing, fluid is taken in the suction port and forced out the discharge port. As the gears rotate, they mesh in a rolling action like an old-fashioned clothes wringer. Gear pumps, like other rotary pumps, efficiently handle small volumes of fluid, often of high viscosity, at high pressures.

PUMPING, BACKSIDE

An arrangement that permits one prime mover (electric motor or engine) to operate two pumping wells. The hook up is such that the downstroke load on one well counterbalances the upstroke load of the other well. *See also* Central Power.

PUMPING UNIT

A pump connected to a source of power; an oil well pumping jack; a pipeline pump and engine.

PUMPING UNIT, BEAM-BALANCED

An oil well pumping unit that carries its well-balancing weights on the walking beam on the end opposite the pump rods. The weights are usually in the form of heavy iron plates added to the walking beam until they balance the pull or weight of the string of pumping rods.

PUMPING UNIT, CRANK-BALANCED

An oil well pumping unit that carries its counterweights on the two cranks that flank the unit's gear box. The string of pump rods is balanced by adding sufficient extra iron weights to the heavy cranks. The walking beam on this type unit is short and is not used as a balancing member.

PUMP, JERKER

A single-barrel, small-volume plunger pump actuated by the to-and-fro motion of a shackle-rod line and an attached counterweight. The jerker pumps on the pull stroke of the rod line; it takes in fluid (the suction stroke) as the counterweight pulls the plunger back from the pumping stroke. Jerkers pump small volumes but can buck high pressure.

PUMP OFF

To pump a well so rapidly that the oil level falls below the pump's standing valve; to pump a well dry, temporarily.

PUMP, POSITIVE-DISPLACEMENT

See Positive-displacement Pump.

PUMP, ROD

A class of downhole pumps in which the barrel, plunger, and standing valve are assembled and lowered into the well through the tubing. When lowered to its pumping position, the pump is locked to the tubing to permit relative motion between plunger and barrel. The locking device is a hold-down, and consists either of cups or a mechanical, metal-to-metal seal.

PUMP, ROD-LINE

An oil well pump operated by a shackle-rod line; a pumping jack. *See* Rocker.

PUMP, SCREW

A small-volume, rotary pump for handling viscous or abrasive liquids. The pumping element is an Archimedes screw housed in a sturdy, cylindrical body, and powered by an electric motor.

PUMP, SIMPLEX

A one-cylinder steam pump used in refineries and processing plants where extra or excess steam is available. Simplex pumps are simple, direct-acting pumps with the steam piston connected directly to the pump's fluid plunger.

PUMP SPECIFICATIONS

A plunger pump designated as 6 x 12 duplex is a two-cylinder pump whose cylinders are 6 inches in diameter with a stroke of 12 inches. A pump with replaceable liners (cylinders) may carry a specifications plate that reads: 4-6 x 10. This pump can be fitted with liners and pistons from 4 inches to 6 inches in diameter; its stroke is 10 inches.

PUMP, SUBMERSIBLE

A bottom-hole pump for use in an oil well when a large volume of fluid is to be lifted. Submersible pumps are run by electricity and, as the name implies, operate below the fluid level in the well.

PUMP, TRAVELING-BARREL

A downhole pump, operated by rods, in which the barrel moves up and down over the plunger, instead of the plunger reciprocating in the barrel as in more conventional pumping devices.

PUMP, TRIPLEX

See Triplex Pump.

Jerker pump *(Courtesy Gaso)*

Screw pump

Triplex pump *(Courtesy Gaso)*

Turbine pump

PUMP, TUBING

A class of downhole pumps in which the barrel of the pump is an integral part of the tubing string. The barrel is installed on the bottom of the string of tubing and is run into the well on the tubing string. The plunger assembly is lowered into the pump barrel on the string of pump rods.

PUMP, TURBINE

A type of centrifugal pump driven by a direct-connected electric motor; commonly used to aerate large settling ponds.

PUP JOINT

A joint of pipe shorter than standard length; any short piece of usable line pipe.

PUP JOINTS, API

Short sections of well tubing made to American Petroleum Institute standards. Pup joints come in different lengths to make up a string of tubing of the proper length, from the bottom of the well to the tubing hanger in the wellhead. Made at the pipe mill under controlled conditions, the short joints are of the same quality as the rest of the tubing.

PURCHASER PRORATIONING

See Pipeline Prorationing.

PUSHER, TOOL

See Tool Pusher; *also* Gang Pusher.

PUT ON THE PUMP

To install a pumping unit on a well. Some wells are pumped from the time they are brought in or completed; others flow for a time (sometimes for many years) and then must be put on the pump.

PVC

Polyvinylchloride; a tough, durable, petroleum-derived plastic that can be extruded or molded and is used for pipe, fittings, light structural members. PVC is highly resistant to salt water and chemicals.

PYROBITUMEN

See Kerogen and Kerogen Shales.

PYROMETER

An instrument for measuring very high temperatures, beyond the range of mercury thermometers. Pyrometers use the generation of electric current in a thermocouple (q.v.) or the intensity of light radiated from an incandescent body to measure temperatures.

Q

Quarter-turn ball valve
(Courtesy Marpac)

QUAD

Short for quadrillion; 1,000 trillion.

QUARTER-TURN VALVE

A plug valve, ball valve, or butterfly valve. A valve made with a plug or sphere with a full-bore opening on the horizontal axis that can be opened or

closed with a quarter or 90°C. turn of the handle. A butterfly valve with its disk that rotates on a shaft or trunion in the valve body also is opened and closed with a quarter turn of the handle.

QUENCH OIL
A specially refined oil with a high flash point (q.v.) used in steel mills to cool hot metal.

QUITCLAIM
An instrument or document releasing a certain interest in land owned by a grantor at the time the agreement takes effect. The key phrase of a quit-claim is: " . . . to release, remise, and forever quitclaim all right, title, and interest in the following described land."

R

RABBIT
A plug put through lease flow lines for the purpose of clearing the lines of foreign matter, water, and to test for obstructions. *See* Pig.

RACKING BOARD
A platform high in the derrick, on well-service rigs, where the derrick man stands when racking tubing being pulled from the well. *See also* Tubing Board.

Racking board

RACK PRICING
Selling to petroleum jobbers or other resellers at f.o.b. at the refinery, with the customer picking up pipeline or other transportation charges. The price of petroleum products at the refinery loading rack; cash and carry at the refinery's loading dock.

RAFFINATE
In solvent-refining practice, raffinate is that portion of the oil being treated that remains undissolved and is not removed by the selective solvent.

RAINBOW
(1) The irridescence (blues, greens, and reds) imparted to the surface of water by a thin film of crude oil. (2) The only evidence of oil from an unsuccessful well— "Just a rainbow on a bucket of water."

RAM
A closure mechanism on a blowout preventer stack; a hydraulically operated type of valve designed to close in a well as with a conventional valve or to close on tubing or drillpipe and maintain high-pressure contact. Hydraulic pistons push the valve's rams against the pipe, one ram on each side, making a pressuretight contact.

RAM, SHEAR
A closure mechanism on a well's blowout preventer stack fitted with chisel-like jaws that are hydraulically operated. When the ram is closed on the pipe the jaws or blades cut the pipe, permitting the upper section to be removed from the BOP stack.

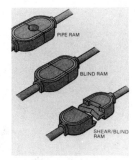

Ram elements

R&D
Research and Development; often used to denote a function up to the stage where the commercial potential of a process or technology can be evaluated. *See* Pilot Plant.

RANGE OIL
Kerosene-type product used in old kerosene stoves or cooking ranges.

RATABLE TAKE
(1) Production of oil and/or gas in such quantities that each landowner whose property overlies a producing formation will be able to recover an equitable share of the oil and/or gas originally in place beneath his land. (2) Production in accordance with allowables set by a state regulatory commission. (3) In some states, common carriers (q.v.) and common purchasers of gas and oil are prohibited from discriminating in favor of one supplier over another.

RAT HOLE
(1) A slanted hole drilled near the well's borehole to hold the kelly joint when not in use. The kelly is unscrewed from the drillstring and lowered into the rat hole as a pistol into a scabbard. (2) The section of the borehole that is purposely deviated from the vertical by the use of a whipstock (q.v.).

RAW GAS
Gas straight from the well before the extraction of the liquefied hydrocarbons (gasoline, butane); wet gas.

RAW MIX
A stream of mixed components: butane, propane, hexane, and others; the product of gas processing plants that is sent on to fractionating plants for the separation of the various components. *See* Field Butanes.

REAMER
A tool used to enlarge or straighten a borehole; a milling tool used to cut the casing downhole. Reamers are run on the drillstring and are built with cutting blades or wheels that can be expanded against the walls of the hole.

REBOILER
A refinery heater that reheats or reboils a part of a process stream drawn off a distilling column and then is reintroduced to the column as a vapor. Reboiling is a process of reworking a part of the charge in a distilling column to ensure more complete fractionating.

RECIPROCATING PUMP
A pump with cylinders and pistons or plungers for moving liquids through a pipeline; a plunger pump (q.v.). The pistons or plungers move forward and backward alternately drawing in fluid into the cylinders through the suction valves and discharging the liquid through discharge valves into a pipeline. Reciprocating pumps are used extensively in the field and at refineries for moving crude oil and products. They handle relatively small volumes but do so at high pressures. Large volumes of oil as are moved in trunk or main lines are pumped with large high-speed centrifugal pumps (q.v.).

RECLAIMED OIL
Lubricating oil which, after a period of service, is collected, re-refined, and sold for reuse.

RECTIFIER BED
A source of electric current for protection against corrosion of pipelines, tanks, and other metal installations buried or in contract with the earth. Using a source of AC electric current, the rectifier installation converts the AC to DC (direct current) and allows the DC to flow into the metal to be protected. By reversing the flow of electric current, the corrosion is inhibited. Metal corrosion is a chemical action which produces minute quantities of current that normally flows away from the metal into the ground.

RECYCLING (GAS)
Injecting gas back into a formation to maintain reservoir pressure so as to produce a larger percentage of oil from the formation.

REDUCED CRUDE OIL
Crude oil that has undergone at least one distillation process to separate some of the lighter hydrocarbons. Reducing crude lowers its API gravity.

REEF
A type of reservoir trap composed of rocks, usually limestone, made up of the skeletal remains of marine animals. Reef reservoirs are often characterized by high initial production which falls off rapidly, requiring pressure maintenance techniques to sustain production.

REEL BARGE
A pipe-laying barge equipped with a gigantic reel on which line pipe up to 12 inches in diameter is spooled at a shore station. To lay the pipe, it is unspooled, run through straightening mandrels, inspected, and paid out over the stern of the barge in the manner of a hawser.

REELED TUBING
A well-service tool used in well workovers. The one-inch or so flexible tubing is carried on a larger spool mounted on a specially equipped truck. The tubing is inserted in the well through the wellhead valves and is used basically for flushing out the well and reestablishing a circulating path.

REENTRY
To reestablish contact with the well's borehole in offshore waters, after having moved off location because of weather or other reasons halting drilling operations. A notable example of reentering was that of the Deep Sea Drilling Program by the Scripps Institution of Oceanography when the crew of the drillship *Glomar Challenger* reentered the hole nine times while drilling in 14,000 feet of water in the Atlantic. *See* Acoustic Reentry.

REEVING A LINE
To string up a tubing or other line in preparation for hoisting; to run a line from the winch up and over a sheave in the crown block and down to the derrick floor.

REFINER-MARKETER
A marketer of gasoline and/or heating oils who operates his own refinery.

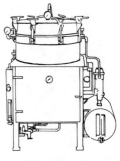

Oil reclaimer

Reel barge

Reid vapor test gauge
(Courtesy Weksler)

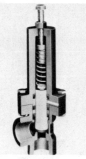

Relief valve

Relief well drilling

REFINERY, SKID-MOUNTED
A small, basic refining unit that is transportable by lowboy trailer to locations where low-grade or straight-run gasoline and diesel fuel are needed and a source of crude oil is available. For example, such a midget "distillation system" can be trucked to a remote drilling site and can supply fuel for diesel drilling engines and gasoline for auxiliary equipment.

REFORMING PROCESSES
The use of heat and catalysts to effect the rearrangement of certain of the hydrocarbon molecules without altering their composition appreciably; the conversion of low-octane gasoline fractions into higher octane stocks suitable for blending into finished gasoline; also the conversion of naphthas to obtain more volatile product or higher octane number.

REGENERATOR
A refinery vessel into which inactive or spent catalyst is pumped to regenerate it, to burn off the coating of carbon or coke. Air at a temperature of 1,100°F. is mixed with the spent catalyst, causing the oxidation of the carbon leaving the catalyst clean and regenerated.

REID VAPOR PRESSURE
A measure of volatility of a fuel, its ability to vaporize. Reid vapor pressure, the specific designation, is named after the man who designed the test apparatus for measuring vapor pressure (q.v.).

RELEASED OIL
Under the Emergency Petroleum Allocation Act of 1973, released oil is old oil production equal to any volume of new oil produced. Unlike old oil, released oil could be sold at free market prices.

RELIEF VALVE
A valve that is set to open when pressure on a liquid or gas line reaches a predetermined level; a pop-off valve.

RELIEF WELL
A directional well drilled near an out-of-control or burning well to kill the well by flooding the formation with water or drilling mud; a well drilled as close as possible or prudent to an out-of-control well and into the same formation in order to vent off or relieve the flowing pressure of the blowout so that the wild well may be brought under control. In some instances more than one relief well is drilled to reduce the flow of the blowing well. A killer well (q.v.).

RENTAL, DELAY
Payment of a sum of money by lessee to the lessor to delay the drilling of a well.

REPEATER STATION
An electronic installation, part of a surveillance and control system for offshore or other remote production operations.

REPRESSURE GAS
Gas purchased for injection into an underground formation, a reservoir, for maintaining reservoir pressure. *See* Recycling (Gas).

REPRESSURING OPERATION
The injection of fluid into a reservoir whose pressure has been largely depleted by producing wells in the field. This secondary recovery technique used to increase the reservoir pressure in order to recover additional quantities of oil. *See* Service Well.

RE-REFINED OIL
Reclaimed oil (q.v.).

RESERVE PIT
An excavation connected to the working mud pits of a drilling well to hold excess or reserve drilling mud; a standby pit containing already-mixed drilling mud for use in an emergency when extra mud is needed.

RESERVOIR
A porous, permeable sedimentary rock formation containing quantities of oil and/or gas enclosed or surrounded by layers of less permeable or impervious rock; a structural trap; a stratigraphic trap (q.v.).

RESERVOIR ENGINEER
A petroleum engineer; one who advises production people on matters relating to petroleum reservoirs: estimating and determining effects of reservoir pressure drops, gas and water encroachment, changes in gas-oil ratios, rates of production, and feasibility of secondary and tertiary recovery programs.

RESERVOIR MODELER
A reservoir engineer or geologist who, by various means, simulates petroleum reservoirs. Using data from wells in the area, seismic information, test-hole findings, cores, and rock samples, the modeler projects and expands his information beyond what is known and provable into the realm of the conjectural. This is accomplished with inferences based on an assumed continuity of the data in hand. The work of the reservoir modeler is important in producing a field at the maximum efficient rate (MER) (q.v.). It is necessary also in projects such as waterflooding, thermal recovery of oil, and hydraulic fracturing.

RESERVOIR PRESSURE
The pressure at the face of the producing formation when the well is shut in. It is equal to the shutin pressure (at the wellhead) plus the weight in pounds of the column of oil in the hole. The hydrostatic pressure exerted by a column of oil 5,000 feet high, for example, would be several thousand pounds. In a flowing well, the reservoir pressure would be sufficient to overcome the pressure of the hydrostatic head.

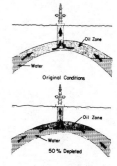

RESERVOIR ROCKS
Sandstone, limestone, and other permeable formations into which petroleum, formed eons ago, migrated and accumulated in reservoirs.

RESERVOIR, WATER-DRIVE
See Water-drive Reservoir.

Water-drive reservoir

RESID MARKET
The market for residual oils; black oils market.

173

RESIDUALS
A term used to describe oils that are "leftovers" in various refining processes; heavy black oils used in ships' boilers and in heating plants.

RESIDUE GAS
Gas that remains after processing in a separator or a plant to remove liquids contained in the gas when produced. *See* Tail Gas.

RESIDUUM
What is left after crude oil has been refined to extinction; a heavy, black, tar-like substance remaining after all useable fractions have been distilled off. The "bottom of the barrel," literally.

RETROFITTING
To modify or add to an engine, item of equipment, or operating plant something new for the sake of efficiency, better performance, or increased safety. To *retro* (go back) and fit or make a change or refinement in the original item of equipment or plant, e.g., "The Ft. Lewis gas plant was retrofitted with automation . . . ." After years of hand operation the plant was modernized and made more efficient.

RETROGRADE GAS CONDENSATE
A liquid hydrocarbon (condensate) formed in deep formations as the reservoir pressure is reduced through production of natural gas. As the pressure is reduced, the gas condenses to form a liquid instead of the usual pattern of liquid changing to gas. Hence the term "retrograde gas condensate." As liquefaction occurs, the formation rock is "wet" by the condensate which is then not as recoverable as when it was in a gaseous state.

REVERSE CIRCULATION
A technique used in fishing for "junk" in the bottom of the well's borehole. A junk basket (q.v.) is lowered into the hole, just above the junk to be retrieved, and through ports in the sides of the basket the drilling mud is jetted to the bottom of the hole and back into the open end of the tool, washing the junk back up into the junk basket.

REVERSE OSMOSIS
A process used in the industry for removing salt and other contaminants from water. The process uses the phenomenon of osmosis, the diffusion through a semipermeable membrane of a solvent leaving behind the solute or dissolved substance. In reverse osmosis, the solvent (water) diffuses through the man-made membrane leaving the salt and other contaminants behind.

REWORKING A WELL
To restore production where it has fallen off substantially or ceased altogether; cleaning out an accumulation of sand and silt from the bottom of the well. In addition to removing or washing out sand and silt accumulations, the well may be hydraulically fractured to open new cracks and fissures in the formation. Or, if conditions warrant, a squib shot (q.v.), a small charge of nitroglycerine, is detonated in the bottom of the hole.

RHABDOMACY
The "science" of divination by rods, wands, and switches. *See* Doodlebug.

RHEOLOGY
The science that treats of the flow of matter. Rheology in drilling refers to the makeup and handling of a drilling mud circulation system; drilling mud control and characteristics.

RICH GAS
Natural gas containing significant amounts of liquefiable hydrocarbons, i.e., casinghead gasoline, butane, propane, etc.; wet gas.

RIG
(1) A drilling rig (q.v.). (2) A large tractor-trailer.

RIG BUILDER
(1) A person whose job is to build or (in a modern context) to assemble a derrick. Steel derricks are erected by bolting parts together. (2) Originally, a person who built derricks on the spot out of rig timbers and lumber on which he used crosscut saws, augers, axes, hammers, and the adz to fit the wood to his pattern.

Rig

RIG-DOWN
To prepare to move the drilling rig and associated tools and equipment to another location or to storage, to stack the tools; to disassemble the mud system, disconnect the engines, lay the derrick down (a jackknife or other portable rig), fill the mud pits, and load up the pipe and fittings and other equipment ready for transport to another well site.

RIGHT OF WAY
(1) A legal right of passage over another person's land. (2) The strip of land for which permission has been granted to build a pipeline and for normal maintenance thereafter. The usual width of right of way for common carrier pipelines (q.v.) is 50 feet.

RIGHT-OF-WAY GANG
A work crew that clears brush, timber, and other obstructions from the right of way. The crew also installs access gates in fenced property. *See* Dress-up Crew.

RIGHT-OF-WAY MAN
A person who contacts landowners, municipal authorities, government agency representatives for permission to lay a pipeline through their property or through the political subdivision. He also arranges for permits to cross navigable waterways, railroads, and highways from the proper authorities.

RIG MANAGER
One who supervises all aspects of offshore rig operation. Large semi-submersibles, anchored miles at sea with hundreds of workers, are much like a small town engaged in drilling a well in hundreds of feet of water. The rig manager is the resident boss of this floating microcosm.

RIG, PUSH-DOWN
A drilling rig which is a modification of rigs used by the mining industry and for drilling water wells. The drillpipe is supported within an A-frame, with the rotary and its pipe-turning mechanism on top of the first joint of drillpipe 30 feet or so up in the A-frame. As the drillpipe and bit are rotated, the pipe

is pushed downward hydraulically until the first joint is in the hole and the rotary is at floor level. A second joint is then added and the rotary is raised to the top to turn and push down on the second joint. A push-down rig has a conventional mud system, but the rig is practical only for drilling holes to about 3,500 feet in relatively soft formations.

RIG REGISTER
A roster of offshore drilling equipment—jackups, semisubmersibles, drill-ships, platforms, tenders and drilling barges—deployed around the world. The register, a modern Jane's *Fighting Ships* as it were, was introduced by the *Petroleum Engineer* magazine. It is kept current and lists the vessel's or platform's depth capability, equipment, whether self-propelled or towed, and other pertinent information.

RIG, SS CLASS 2000
See SS Class 2000 Rig.

RIG TIMBERS
Large-dimension wooden beams used to support the derrick, drilling engines, or other heavy equipment; heavy, roughcut timbers used in the trade by rig builders when derricks were built rather than assembled.

RIG-UP
To make preparations to drill; to get all equipment in place ready to make hole: dig the cellar and mud pits; set up the derrick, reeve the lines; set engines and pumps; connect the lines of the mud system and set auxiliary equipment. Also have necessary bits, tubular goods, valves, rams, and fittings on hand.

RISER
(1) A pipe through which liquid or gas flows upward. (2) In offshore drilling by semisubmersible, jackup, fixed platform, or drillship, a riser is the casing extending from the drilling platform through the water to the sea bed through which drilling is done. *See* Marine Riser System.

RIVER CLAMPS
Heavy steel weights made in two halves bolted on screw pipe at each collar to strengthen the joints and keep the line lying securely on the river bottom or in a dredged trench.

RIVET BUSTER
An air-operated (pneumatic) chisel-like tool for cutting off rivet heads. Used by tankies when tearing down an old tank or other vessel put together with rivets.

ROCK A WELL
To agitate a "dead" well by alternately bleeding and shutting in the pressure on the casing or tubing so that the well will start to flow.

ROCK, CLASTIC
One of the categories of sedimentary rock; a so-called secondary rock which consists of particles that are fragments of preexisting rocks. They may range in size from "blocks the size of boxcars down to colloids so fine as to remain in suspension almost indefinitely." The three classes of clastic

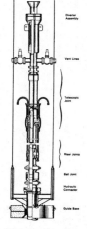

Riser system for floating rig

sedimentary rocks are sandstone, conglomerate, and shale and mud-stone.

ROCKERS

A counterbalance installed on a shackle-rod line, operating a pumping jack, to pull the rod line back after its power stroke. Rod lines can only pull, so on the return stroke the line is kept taut by a counterbalance. Rockers often are in the shape of a box or crate filled with rocks. One edge of the box is attached to a fulcrum bearing on which it moves back and forth like a rocking chair.

ROCK HOUND

A geologist; a humorous but affectionate colloquialism for a person who assiduously pursues rock specimens in a search for evidence of oil and gas deposits.

ROCK, IGNEOUS

Rocks that have solidified from a molten state deep in the earth. Those rocks that have reached the surface while still molten are called lavas; they can form volcanic cones or spread out in flows or sheets, they can be forcibly thrust up between beds of other kinds of rocks in what are called sills, or they can fill crevices and then solidify as "dikes." Rocks that have solidified deep beneath the earth's crust are referred to as plutonic, from the Greek god of the lower regions, Pluto. Granite is an example of plutonic rock.

ROCK, METAMORPHIC

Rocks formed by the metamorphosis of other rocks. When either igneous or sedimentary rocks are subjected to enough heat, pressure, and chemical action, their character and appearance are changed. These factors act to cause recrystallization of the minerals of the rock. Granite may become gneisses or schists; sandstones become quartzites; shales become slates; limestone becomes marble.

ROCK PRESSURE

An early-day term for a well's shutin or wellhead pressure when all valves are closed and the pressure is observed at the surface.

ROD

(1) Sixteen and one-half feet; the unit of measure used in buying certain types of pipeline right of way. (2) A sucker rod; an engine's connecting rod; a piston rod.

RODDAGE FEE

The fee paid to a landowner for the easement of a pipeline right of way across his property. Right of way is measured in rods (16½ feet) hence the term roddage fee. *See* Right of Way.

ROD HANGER

A rack with finger-like projections on which rods are hung when pulled from the well; a vertical rack for hanging lengths of pumping rods.

ROD JOB

See Pulling Rods.

Rod job

Rod-line pump

ROD LINE
See Shackle-rod Line.

ROD-LINE PUMP
See Pump, Jerker.

ROD PUMPS
See Pump, Rod.

ROLL A TANK
To agitate a trank of crude oil with air or gas for the purpose of mixing small quantities of chemical with the oil to break up emulsions or to settle out impurities.

ROLLER BIT
The rock-cutting tool on the bottom of the drillstring made with three or four shanks welded together to form a tapered body. Each shank supports a cone-like wheel with case-hardened teeth that rotate on steel bearings.

ROLLERS, CASING AND TUBING
A steel tubular device for opening up and reconditioning buckled, dented, or collapsed casing and tubing in the hole. The long, steel tool with a tapered end has a series of rollers. The tool is forced into the damaged pipe and, as it is pushed down and rotated by the drillstring, the series of rollers forces the damaged pipe open and restores it to its original diameter and roundness.

ROLL IN
To include the cost of new facilities, service, and supply as part of the overall cost of operating a company for the benefit of all customers served by a pipeline or other common carrier; to roll in the cost of new supplies and facilities for the purpose of arriving at a new rate structure.

ROLLING PIPE
Turning a joint of screw pipe into the coupling of the preceding joint by the use of a rope looped once around the pipe and pulled by a rope crew. This procedure was used on larger diameter line pipe—10 and 12-inch—to make up the connection rapidly before the tongs were put on the pipe for the final tightening.

RON
Research octane number; a measure of a gasoline's antiknock quality determined by tests made on engines running under moderate conditions of speed and load. MON, motor octane number, is a measure of gasoline's antiknock characteristics determined by tests under more severe conditions of load and speed. *See* Octane Rating.

ROOF ROCK
A layer of impervious rock above a porous and permeable rock formation that contains oil and gas.

ROOT RUN
The first course of metal laid on by a welder in joining two lengths of pipe or other elements of construction; the stringer bead (q.v.).

ROPE SOCKET
A device for securing the end of a steel cable into a connecting piece—a clevis, hook or chain. A metal cup or socket (like a whip socket) into which the cable end is inserted and which then is filled with molten lead or babbitt (q.v.).

ROTARY BUSHING
The metal casting that fits into the master bushing of the rotary table on a drilling well, and through which the kelly joint moves downward as drilling procedes. The kelly bushing is turned by the rotary table and the bushing rotates the kelly and drillstring.

ROTARY PUMP
A positive displacement pump consisting of rotary elements—cams, screws, gears, or vanes—enclosed in a case; employed, usually, in handling small volumes of liquid at either high or low pressures. Because of the close tolerances in the meshing of the gears or cams, rotary pumps cannot handle liquids contaminated with grit or abrasive material without suffering excessive wear or outright damage.

Rotary pump *(Courtesy Roper Pump Co.)*

ROTARY REAMER
A rock-cutting tool inserted in the drill column just above the drill bit for the purpose of keeping the hole cut to full diameter. Often in drilling deep, hard-rock formations the bit will become worn or distorted, thus cutting less than a full hole. The following reamer trims the hole wall, maintaining full diameter.

ROTARY RIG
A derrick equipped with rotary drilling equipment, i.e., drilling engines, draw works, rotary table, mud pumps, and auxiliary equipment; a modern drilling unit capable of drilling a borehole with a bit attached to a rotating column of steel pipe.

ROTARY SLIPS
See Slips.

Rotary rig

ROTARY TABLE
A heavy, circular casting mounted on a steel platform just above the derrick floor with an opening in the center through which the drillpipe and casing must pass. The table is rotated by power transmitted from the draw works and the drilling engines. In drilling, the kelly joint fits into the square opening of the table. As the table rotates, the kelly is turned, rotating the drill column and the drill bit.

ROUGHNECKS
Members of the drilling crew; the driller's assistants who work on the derrick floor, up in the derrick racking pipe, tend the drilling engines and mud pumps, and on "trips" operate the pipe tongs breaking out or unscrewing the stands of drillpipe.

ROUND-POINT SHOVEL
A digging tool whose blade is rounded and tapers to a point in the center of the cutting edge. A long-handled shovel, standard equipment for digging ditches by hand.

Rotary table *(Courtesy U.S. Steel)*

Making a trip

ROUND TRIP
Pulling the drillpipe from the hole to change the bit and running the drillpipe and new bit back in the hole. On deep wells, round trips or "a trip," as it is more commonly called, may take 24 hours, three 8-hour shifts.

ROUSTABOUT
A production employee who works on a lease or around a drilling rig doing manual labor.

ROYALTY
A share of the minerals (oil and gas) produced from a property by the owner of the property. Originally, the right or prerequisite of the king to receive a percentage of the gold or silver taken from the mines of his realm. Today, the sovereign is the landowner who traditionally receives 12½ percent or one-eighth of the oil and gas produced from his land. This is the basic form of royalty. But as you will note there are variations and "refinements" to the concept of what belongs to the king. *See* Landowner's Royalty.

ROYALTY BIDDING
An uncommon practice of bidding on Federal leases by offering a high royalty interest to the government on any production discovered on the tract in lieu of the traditional cash bonus. Royalty interests as high as 70 and 80 percent of gross production have been offered. The advantages to a company bidding royalty interests instead of cash could be a savings in millions of dollars of front money. In case the lease is unproductive, the company is out only the cost of the well and any seismic or other exploratory expenses.

ROYALTY BONUS
Describes an overriding royalty or oil payment reserved by the lessor, the landowner. Usually any consideration received or promised to a lessor on the execution of a lease in excess of the customary one-eighth royalty is called a bonus or royalty bonus.

ROYALTY, COMPENSATORY
Payments to royalty owners as compensation for loss of income which they may suffer due to the failure of the operator to develop a lease properly.

ROYALTY, FEE
The lessor's share of oil and gas production; landowner's royalty traditionally one-eighth of gross production free of any cost.

ROYALTY, FIXED-RATE
Royalty calculated on the basis of a fixed rate per unit of production, without regard for the actual proceeds from the sale of the production.

ROYALTY, GUARANTEED
The minimum amount of royalty income a royalty owner is to receive under the lease agreement, regardless of his share of actual proceeds from the sale of the lease's production.

ROYALTY, INNOVATOR'S
See Innovator's Royalty.

A rig floor. Note the draw works, the kelly joint, the bushing, and the slips

ROYALTY INTEREST, TERM

A royalty interest not in perpetuity but for a definite period of time. Most royalty interests are created for a fixed period and so long thereafter as oil and gas are produced. But there are such interests that run only for a specified, fixed length of time with no qualifying "thereafter" clause.

ROYALTY, LANDOWNER'S

A share of the gross production of the oil and gas on a property by the landowner without bearing any of the cost of producing the oil or gas. The usual landowner's royalty is one-eighth of gross production.

ROYALTY OIL

Oil owned by the government, Federal, state or local. Oil on the Outer Continental Shelf and on Federal land is royalty oil; also oil on land owned by a state or a municipality is royalty oil.

ROYALTY, SHUT-IN

Payment to royalty owners under the terms of a mineral lease which allows the operator or lessee to defer production from a well which is shut in for lack of a market or pipeline connection.

ROYALTY, SLIDING-SCALE

Royalty paid to the Federal government on oil and gas production from a government lease, usually offshore, which varies from the normal $16\frac{2}{3}$ percent up to 50 percent of the value of the production. As the value of production increases the percentage of royalty also increases to a maximum of 50 percent.

RUN

A transfer of crude oil from a stock tank on a production lease to a pipeline gathering system for transportation to the buyer's facilities; running oil from a tank into a pipeline for delivery to a purchaser.

RUNNING THE TOOLS

Putting the drillpipe, with the bit attached, into the hole in preparation for drilling.

RUN TICKET

A record of the oil run from a lease tank into a connecting pipeline. The ticket is made out in triplicate by the gauger and witnessed by the lease owner's representative, usually the pumper. The run ticket, an invoice for oil delivered, shows opening and closing gauge, API gravity, and temperature, tank temperature, and BS&W. The original of the ticket goes to the purchaser, a copy goes to the pumper and to the gauger.

RUPTURE DISK

A thin, metal plug or membrane in a fitting on a pressure line made so as to blow out or rupture when the pressure exceeds a predetermined level; a safety plug. See Soft Plug.

S

SADDLE
A clamp, fitted with a gasket, for stopping the flow of oil or gas from holes or splits in a pipeline; a device for making temporary repairs to a line. The clamp conforms to the curve of the pipe and is held in place by U-bolts that fit around the pipe and extend through the clamp.

SADDLE BEARING
A broad, heavy bearing located on top of the Samson post to support the walking beam on a cable tool drilling rig or an oil well pumping jack.

SAE
Society of Automotive Engineers.

SAE NUMBER
A classification of lubricating oils in terms of viscosity only. A standard established by the Society of Automotive Engineers. SAE 20; SAE 10W-30, multiviscosity lubricating oil (q.v.).

SAFETY VALVE
See Relief Valve.

SALT-BED STORAGE
Thick formations or underground layers of salt in which cavities are mined or leached out with super-heated water for the storage of pertroleum products, e.g., heating oils, butane, propane, and other LP-gases.

SALT DOME
A subsurface mound or dome of salt. Two types of salt domes are recognized: the piercement and nonpiercement. Piercement domes thrust upward into the formations above them causing faulting; nonpiercement domes are produced by local thickening of the salt beds and merely lift the overlying formations to form an anticline (q.v.).

SALT-DOME/SALT-PLUG POOL
Structural or stratigraphic traps (q.v.) associated with rock-salt intrusions; pools formed by the intrusion of underlying salt formations into overlying porous and permeable sedimentary layers creating traps favorable to the presence of oil and gas.

SALT-DOME STORAGE
Cavities leached out of underground salt formations by the use of superheated water for the storage of petroleum products, especially LP-gases.

SALT PLUG
See Salt Dome.

SAMPLE
Cuttings of a rock formation broken up by the drill bit and brought to the surface by the drilling mud. Rock samples are collected from the shale shaker (q.v.) and examined by the geologist to identify the formation, the type of rock being drilled.

Vertical safety valves on high-pressure gas line

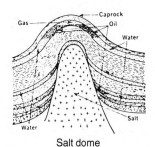

Salt dome

SAMPLE BAG
A small cotton bag with a drawstring to hold rock cutting samples. Each bag with its sample is tagged with identifying information, well name, lease, location, depth at which cuttings were taken, etc.

SAMPLE LOG
A record of rock cuttings as a well is being drilled, especially in cable-tool drilling. The cuttings, brought to the surface by the bailer, are saved and the depth where obtained is recorded. This record shows the characterisitics of various strata drilled through.

SAMSON POST
A heavy, vertical timber that supports the well's walking beam (q.v.). On a cable-tool rig, the samson post is located just ahead of the band wheel. The walking beam rests on the samson post on a broad saddle bearing and moves up and down like a child's teeter-totter.

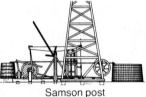
Samson post

SAND
Short for sandstone; one of the more prolific sedimentary rock formations. In informal usage, other sedimentary rocks are referred to as "sands."

SAND BODY
A sand or sandstone formation defined by upper and lower layers of impervious rock. A sandstone formation, sometimes in the shape of a lens, sandwiched between two impervious layers of rock; a geologic trap favorable to the accumulation of oil and gas.

SAND CONTROL
A technique for coping with sand from unconsolidated (loose, unpacked) formations that migrate (drift or wash) into downhole pumping equipment or into the borehole. *See* Gravel Packing.

SANDED UP
A well clogged by sand that has drifted and washed into the well bore from the producing formation by the action of the oil.

SAND LINE
A wire line (cable) used on a drilling rig to raise and lower the bailer or sand pump in the well bore. Logging devices and other lightweight equipment are also lowered in the hole on the sand line.

SAND PUMP
A cylinder with a plunger and valve arrangement used for sucking up the pulverized rock, sand, and water from the bottom of the well bore. More effective than a simple bailer. Shell pump; sludger.

SAND REEL
A small hoisting drum on which the sand line is spooled and used to run the bailer or sand pump on a cable-tool rig. The sand reel is powered by contact with the band wheel (q.v.).

SANDS
Common terminology for oil-bearing sandstone formations. (Oil is also found in limestone, shales, dolomite, and other porous rock.) In informal or loose usage, other sedimentary rocks are referred to as "sands."

SAND SEPARATOR
A device for removing "drilled solids," pulverized rock and sand, from drilling mud. The sand separator is used in addition to the shale shaker (q.v.) and by removing most of the abrasive material reduces wear on mud pumps and bits.

SATELLITE PLANT
A facility that supports the main processing plant; a plant that derives its feedstock or raw material from the main processing unit.

SATELLITE PLATFORM
Production platform (q.v.).

SATS GAS PLANT
A refiner's term for the part of the refinery that processes gas streams carrying saturates (q.v.) to be stripped out of the gases.

SATURATES
Components of refinery-process gas streams: methane, ethane, propane, butanes, and others. Saturates is a synonym for hydrocarbons whose carbon atoms are "saturated with hydrogen atoms." These gas streams are further refined in a facility called by refinery engineers the sats gas plant.

SATURATION
(1) The extent to which the pore space in a formation contains hydrocarbons or connate water (q.v.). (2) The extent to which gas is dissolved in the liquid hydrocarbons in a formation.

SATURATION PRESSURE
The pressure at which gas begins to be released from solution in oil. *See* Bubble Point.

SAYBOLT SECONDS
See Seconds Saybolt Furol; *also* Seconds Saybolt Universal.

SBM SYSTEM
Single-buoy mooring system (q.v.).

SBR
Initials for synthetic butadiene rubber, the main ingredients of which are derived from petroleum. SBR is used in the manufacture of tires, hose, shoes, other heavy-duty products.

SCAT (WELDING) RIG
A rack that carries welding generators, gas bottles (CO_2), and spools of welding wire along the pipe being welded. The rig is powered by a small diesel engine. Automatic welding heads ("bugs") are moved ahead on the pipe as the joints are welded.

SCHEDULER
A person in an oil-dispatching office who plans the future movement of batches of crude oil or product in a pipeline system, keeping batches separated and making arrangements for product input and downstream deliveries. *See* Dispatcher.

Sand separator

Satellite platform

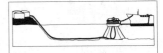
Single-buoy mooring system

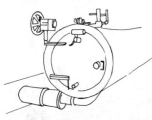

Pipeline scraper
*(Courtesy Oil States
Rubber Co.)*

Scraper trap

Screw conveyor

Screw pump

SCHLUMBERGER (Slum-ber-jay)

Trade name of a pioneer electrical well-surveying company. In many areas it is common practice to speak of an electric well log as a "slumberjay" even though the log was made by another company.

SCOURING

The erosion or washing away of the sand/clay covering of a buried subsea pipeline. Scouring caused by sea currents is a serious problem for undersea lines. Excessive scouring causes spanning, the hanging of a section of the line one to several feet off bottom. If allowed to go uncorrected the pipeline welds crack or the pipe ruptures from its unsupported weight. Subsea lines are inspected for scouring and spanning by side-scan sonar devices or by diver inspection.

SCOUT

A person hired by an operator or a company to seek out information about activities of drilling wells in an area, survey data, drilling rates and depths, and well potentials.

SCOUTING A WELL

Gathering information, by all available means, about a competitor's well—the depth, formations encountered, well logs, drilling rates, leasing, and geophysical reports.

SCRAPER, PIPELINE

A pig; a cylindrical, plug-like device equipped with scraper blades, wire brushes, and toothed rollers used to clean accumulations of wax, rust, and other foreign matter from pipelines. The scraper is inserted in the line at a "trap" (q.v.) and is pushed along by the pressure of the moving column of oil.

SCRAPER TRAP

A facility on a pipeline for inserting and retrieving a scraper or "pig." The trap is essentially a "breech-loading" tube isolated from the pipeline by valves. The scraper is loaded into the tube like a shell into a shotgun; a hinged plug is closed behind it, and line pressure is then admitted to the tube *behind* the scraper. A valve is opened ahead of the scraper and it is literally pushed into the line and moved along by the oil pressure.

SCREW CONVEYOR

A mechanism for moving dry, solid material—pelletized plastics, sulfur, cement, etc.—from one location to another by means of a helix or screw rotating in a cylindrical conduit. Archimedes thought of it first.

SCREW PUMP

A rotary pump made with one, two, or three screws or spiral members. When rotated on their shafts, the screws closely mesh and take in fluid at the suction end of the pump and force it out the discharge port in a continuous stream. Screw pumps, which are small, usually are driven by electric motors but can be hooked up to gas engines. Screw pumps and other types of rotary pumps are used in refineries and chemical plants to handle highly viscous fluids and as transfer pumps for small volumes of liquid at high pressures.

SCRUBBING
Purifying a gas by putting it through a water or chemical wash; also the removal of entrained water. Natural gas when it is produced or when it flows from the well under pressure usually contains impurities, traces of other gases and microscopic droplets of water, as well as liquid hydrocarbons. Before it can be accepted by a gas transmission line, a trunk line, the gas must be stripped of the liquid hydrocarbons, scrubbed and dried out.

SCRUBBING PLANT
A facility for purifying or treating natural gas for the removal of hydrogen sulfide or other impurities.

SEALINES
Submarine pipelines; lines laid on the ocean floor from offshore wells to a production platform and to receiving stations onshore.

SEALS
Thin strips of metal, imprinted with serial numbers, used to "seal" a valve in an open or closed position. The metal strip has a locking snap on one end into which the free end is inserted, locking it securely. Seals are used on tanks in a battery to prevent the undetected opening or closing of a valve.

SEAMLESS PIPE
Pipe made without an axial seam; pipe made from a billet or solid cylinder of hot steel and "hot-worked" on a mandrel into a tubular piece without a seam. *See* Lap-Welded Pipe.

SEA TERMINAL
An offshore loading or unloading facility for large, deep-draft tankers. The terminal is served by filling lines from shore or by smaller, shallow-draft vessels.

Sea terminal

SECONDARY RECOVERY
The extraction of oil from a field beyond what can be recovered by normal methods of flowing or pumping; the use of waterflooding, gas injection, and other methods to recover additional amounts of oil.

SECONDS SAYBOLT FUROL (SSF)
A measurement of the viscosity of a heavy oil. Sixty cubic centimeters of an oil are put in an instrument known as a "Saybolt viscosimeter" and permitted to flow through a standardized orifice in the bottom at a specified temperature. The seconds required to flow through is the oil's viscosity, its SSF number. *See also* Seconds Saybolt Universal.

SECONDS SAYBOLT UNIVERSAL (SSU)
A measurement of the viscosity of a light oil. A measured quantity of oil—usually 60 cubic centimeters—is put in an instrument known as a "Saybolt viscosimeter" and permitted to flow through an orifice in the bottom at a specified temperature. The number of seconds required for the flow-through is the oil's SSU number, its viscosity.

SECTION MILL
A downhole cutting tool made with expandable arms used to cut sections out of the casing in the hole. The mill is attached to the end of the drillstring and lowered into the hole to the point where the casing is to be cut. The

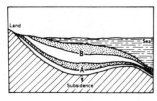

Sedimentation process

Seismic thumper

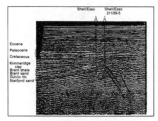

The record made by
a seismometer

cutter arms are then expanded, either hydraulically or mechanically, against the casing wall. As the drillpipe is rotated, the cutters do their work.

SECTION OF LAND
One square mile; 640 acres; sixteen 40-acre plots.

SEDIMENTARY BASIN
An extensive area (often covering thousands of square miles) where substantial amounts of unmetamorphized sediments occur. Most sedimentary basins are geologically depressed areas (shaped like a basin). The sediment is thickest in the interior and tends to thin out at the edges. There are many kinds of such basins, but it is in these formations that all the oil produced throughout the world has been found.

SEDIMENTARY ROCK
Rock formed by the laying down of matter by seas, streams, or lakes; sediment (mineral fragments, animal matter) deposited in bodies of water through geologic ages. Limestone, sandstone, shale are sedimentary rocks.

S.E.G.
The Society of Exploration Geophysicists, a professional organization of geophysicists engaged in exploration for oil and gas.

SEISMIC SEA STREAMER
A cable, trailed from a geophysical vessel, towing a series of hydrophones along the sea floor recording seismic "signals" from underwater detonations. As the vessel moves slowly ahead, harmless electronic or air detonations are set off which are reflected from rock formations beneath the sea floor and picked up by the sensitive, sound-detecting hydrophones. *See* Geophone.

SEISMIC SHOT HOLE
See Shot Hole.

SEISMIC THUMPER
See Vibrator Vehicle.

SEISMOGRAM
The record produced by a seismographic survey.

SEISMOGRAPH
A device that records vibrations from the earth. As used in the exploration for oil and gas, a seismograph records shock waves set off by explosions detonated in shot holes (q.v.) and picked up by geophones (q.v.).

SEISMOGRAPHIC SURVEY
Geophysical information on subsurface rock formations gathered by means of a seismograph (q.v.); the investigation of underground strata by recording and analyzing shock waves artificially produced and reflected from subsurface bodies of rock.

SEISMOMETER
A device for receiving and recording shock waves set off by an explosion or other seismic sources and reflected by underground rock formations; a seismograph which measures the movements of the ground.

SEIZE
To stick together, as two pieces of metal that have become hot from excessive friction as one piece moves relative to the other; to bond or adhere, as a piston to a cylinder from heat and pressure.

SEMIDIESEL
A misnomer for a diesel-cycle engine whose compression is not high enough to create sufficient heat to ignite the injected fuel when starting cold. Semidiesels or, more correctly, hot-head or hot-plug diesels, are equipped with a plug that extends into the firing chamber heated by a torch or by electricity to assist in the ignition of the diesel fuel until the engine is running and up to operating temperature; a small, low-compression diesel engine. *See* Hot-Plug Engine.

SEMISUBMERSIBLE
A large, floating drilling platform with a buoyant substructure, part of which is beneath the surface of the water. Semisubmersibles are virtually self-contained, carrying on their main and lower decks all supplies and personnel for drilling and completing wells in hundreds of feet of water and miles from shore. Some of the huge platforms are self-propelled and are capable of moving at 6 to 8 knots. As they often drill in waters too deep for conventional chain and cable anchors, they maintain their position over the borehole by the use of thrusters, jets, or Kort nozzles controlled by onboard computers. Some of the largest floaters are designated SS-2000 with 18,000-ton displacement, a 2,000-ton deck-load capacity, and are capable of drilling in 2,000 feet of water in severe weather and sea conditions.

Semisubmersible drilling platform

SEPARATOR
A pressure vessel (either horizontal or vertical) used for the purpose of separating well fluids into gaseous and liquid components. Separators segregate oil, gas, and water with the aid, at times, of chemical treatment and the application of heat.

SEPARATOR GAS
Natural gas separated out of the oil by a separator at the well.

SEPARATOR, LOW-TEMPERATURE (GAS)
See LXT Unit.

SEPARATOR, SAND
See Sand Separator; *also* Decanting Centrifuge.

SERVICE TOOLS
A variety of downhole equipment used in drilling, completion, and workover of oil and gas wells; so-called wireline tools such as logging, sampling, temperature and pressure gauging; fishing, fracturing, acidizing and shooting are some of the service tools and procedures provided by the numerous service companies that perform specialized work in the oil field.

Separator

SERVICE WELL
A nonproducing well used for injecting water or gas into the reservoir or producing formation in pressure maintenance or secondary recovery programs; also a salt-water disposal well.

SERVO
Short for servomechanism (q.v.).

SERVOMECHANISM
An automatic device for controlling large amounts of power with a small amount of force. An example of a servomechanism is the power-steering on an automobile. Any small force on the steering wheel activates a hydraulically powered mechanism that does the real work of turning the wheels.

SERVOMOTOR
A power-driven mechanism that supplements a primary control operated by a comparatively small force. *See* Servomechanism.

SETBACK
The space on the derrick floor where stands of drillpipe or tubing are "set back" and racked in the derrick. Offshore drilling platforms often list the stand capacity of their setbacks as an indication of their pipe-handling capability and capacity. On transportable, mast-type derricks used on land, setbacks are outside the derrick proper.

SET CASING
To cement casing in the hole. The cement is pumped downhole to the bottom of the well and forced up a certain distance into the annular space between casing and the rock wall of the drillhole. It is then allowed to harden, thus sealing off upper formations that may contain water. The small amount of cement in the casing is drilled out in preparation for perforating (q.v.) to permit the oil to enter the casing. The decision to set casing (or pipe) is an indication that the operator believes he has a commercial well.

SETTLED PRODUCTION
The lower average production rate of a well after the initial flush production (q.v.) tapers off; the production of a well that has ceased flowing and has been put on the pump.

SEVEN SISTERS
A term applied to the seven large international oil companies: Exxon, Texaco, Gulf, Standard of California, and Mobile of the U.S.; and British Petroleum and Royal Dutch Shell, the two overseas sisters. It is said that these seven companies control a major portion of production and refinery runs in the Free World. The term was first used by Enrico Mattei, then head of the Italian government oil company Ente Nazionale Idrocarburi.

SEVERANCE TAX
A tax levied by some states on each barrel of oil or each thousand cubic feet of gas produced. Production tax.

SHACKLE ROD
Jointed steel rods, approximately 25 feet long and ¾ to 1 inch in diameter, used to connect a central power (q.v.) with a well's pumping unit or pumping jack. Shackle-rod lines are supported on metal posts (usually made of 2-inch line pipe) topped with wooden guide blocks which are lubricated with a heavy grease.

SHAKE OUT
To force the sediment in a sample of oil to the bottom of a test tube by whirling the sample at high speed in a centrifuge machine. After the sample has been whirled for three to five minutes, the percent of BS&W (sediment and water) is read on the graduated test tube.

SHALE
A type of sedimentary rock composed of fine particles of older rock laid down as deposits in the water of lakes and seas. Most shales are compacted mud and consequently do not contain oil or gas in commercial quantities.

SHALE OIL
Oil obtained by treating the hydrocarbon kerogen found in certain kinds of shale deposits. When the shale is heated the resulting vapors are condensed and then treated in an involved process to form what is called shale oil or synthetic oil.

SHALE SHAKER
A vibrating screen for sifting out rock cuttings from drilling mud. Drilling mud returning from downhole carrying rock chips in suspension flows over and through the mesh of the shale shaker leaving the small fragments of rocks which are collected and examined by the geologist for information on the formation being drilled.

Shale shaker *(Courtesy Sweco)*

SHALES, KEROGEN
See Kerogen Shales.

SHAPED-CHARGE PERFORATION
A perforation technique using shaped explosive charges instead of steel projectiles to make holes in casing. Quantities of explosives are made in special configurations and detonated at the bottom of the hole against the casing wall to make the perforations.

SHARPSHOOTER
A spade; a narrow, square-ended shovel used in digging. Sharpshooters are one of the pipeliner's digging tools used for squaring up a ditch or the sides of a bell hole (q.v.).

SHAVE-TAILS
A skinner's (q.v.) term for his mules.

SHEAR PIN
A retaining pin or bolt or screw designed to shear or give way before damage can be done to the item of equipment it is holding in place. A common use for a shear pin is to secure a propeller to a shaft. Should the propeller strike an obstruction, the pin will shear, preventing damage to the shaft or other parts of the power train. In other applications, shear pins or screws are used in downhole tools or equipment to hold a part in position until the tool is landed or in place. Then when thrust or torque is applied, the pin or screw shears, permitting an element of the tool to assume a predetermined attitude.

SHEAVE
A grooved pulley or wheel; part of a pulley block; a sheave can be on a

Sheave

fixed shaft or axle (as in a well's crown block) (q.v.) or in a free block (as in block and tackle).

SHEET IRON
Galvanized, corrugated sheet metal used for roofing, garages, and other more or less temporary buildings. Because a sheet iron or corrugated iron building is relatively inexpensive and easy to assemble, this kind of construction is common on oil leases.

SHELL PUMP
See Sand Pump.

SHIMS
Thin sheets of metal used to adjust the fit of a bearing or to level a unit of equipment on its foundation. For fitting a bearing, a number of very thin (.001 to .030-inch) shims are put between the two halves of the bearing (between the box and cap). Shims are added or removed until the bearing fits properly on the journal.

SHIRTTAILS
Colloquial term for the structural members or shanks of a drill bit that anchor the cutting wheels; the frame of the bit below the threaded pin.

SHOCK ABSORBER
A spring-loaded slip joint run in the drillstring just above the bit to absorb vibrations and dynamic bit forces while drilling.

Shock absorber

SHOESTRING SAND
Narrow strands of saturated formation that pinch out (q.v.) or are bounded by less permeable strata that contain no oil or gas. Shoestring sands sometimes occur or have the shape of a lense several miles long and are completely saturated with oil or gas or both. The more common type of shoestring sand follows the bed of an ancient creek or river and is narrow and sinuous, completely hemmed in or trapped by impervious layers of rock or shale.

SHOOT A WELL, TO
To detonate an explosive charge in the bottom of a well to fracture a "tight" formation in an effort to increase the flow of oil. *See* Well Shooter.

SHOOTER
See Well Shooter.

SHOOTING LEASE
An agreement granting permission to conduct a seismic or geophysical suvey. The lease may or may not give the right to lease the land for oil or gas exploration.

SHORT TANK
A colloquial term for a lease tank not full to the top which the lease owner wants run. Some small independent operators want a gauger to run a short tank at the end of the month to get some revenue from their stripper lease to pay the expenses of the lease. It requires as much time and work to run a short tank as a full one, so gaugers are sometimes reluctant to work a short, half-, or third-full stock tank.

SHORT TRIP

Pulling the drillstring partway out of the hole. Short trips may be necessary to raise the drill up into the protective string of casing to avoid having the drillstring stuck in the hole by a cave-in or sloughing of the wall of the borehole below the protective casing.

SHOTGUN TANK

A tall, slender tank for separating water and sediment from crude oil. On a small-production or stripper lease, the shotgun serves as a water and sediment knockout vessel. As the oil and water are pumped into the tall column, the water and sediment settles to the bottom; the oil frees itself from most of the water and floats on top of the water where it gravitates through a take-off line into the stock tanks. The water is drawn off through a "syphon" which may be adjusted to take off only the water.

SHOT HOLE

A small-diameter hole, usually drilled with a portable, truck-mounted drill, for "planting" explosive charges in seismic operations.

SHOT POINT

The shot hole (q.v.), the point at which a detonation is to be made in a geophysical survey.

Shot hole

SHOW OF OIL

A small amount of oil in a well or a rock sample; a show of oil usually signifies the well will not be a commercial producer.

SHRINK FIT

An extremely tight fit as the result of "shrinking" one metal part around another. A heated part is placed around a companion piece, and as the heated part cools, a shrink fit results. Conversely, an expansion fit may be made by cooling a part (a valve-seat insert, for example) to extremely low temperature with "dry ice" and placing the part in position. As it returns to normal temperature, a tight "expansion fit" will result.

SHUT DOWN—SHUT IN WELL

There is a great difference between a shut down and a shut in. A well is shut down when drilling ceases which can happen for many reasons: failure of equipment; waiting on pipe; waiting on cement; waiting on orders from the operator, etc. A well is shut in when its wellhead valves are closed, shutting off production. A shut in well often will be waiting on tankage or a pipeline connection.

SHUT IN

To close the valves at the wellhead so that the well stops flowing or producing; also describes a well on which the valves have been closed.

SHUTIN PRESSURE

Pressure as recorded at the wellhead when the valves are closed and the well is shut in. To allow the pressure to build up to its peak, the well is shut in for 24 hours or even longer. This permits the gas to move out of the formation toward the borehole and build its head of pressure. *See* Reservoir Pressure.

Shutin gas well
*(Courtesy Pan American
Petroleum)*

Side boomcats

Sidetrack drilling from a
workover rig

Silencers on stationary
engines

SHUTIN ROYALTY

Payments made when a gas well, capable of producing in paying quantities, is shut in for lack of market for the gas. This type of royalty or some form of rental is usually required to prevent termination of the lease.

SIDE BOOMCATS

See Boomcats.

SIDE-DOOR ELEVATORS

Casing or tubing elevators (q.v.) with a hinged latch that opens on one side to permit it to be fastened around the the pipe and secured for hoisting.

SIDETRACKING

Drilling of another well beside a nonproducing well and using the upper part of the nonproducer. A method of drilling past obstructions in a well, i.e., lost tools, pipe or other material blocking the hole. This can be done with the use of a whipstock (q.v.), a downhole tool which forces the drill bit to drill at a slight angle from the vertical. By beginning the deviation of the hole several hundred feet above the junk in the original hole, the new hole will bypass the obstruction and may be taken on down to the pay zone.

SIDEWALL CORE/SAMPLE

A sample of rock taken from the wall of the well's borehole.

SIDEWALL TAP OR COCK

A small-diameter valve inserted in the wall of a tank or other vessel for drawing samples or bleeding off pressure.

SIGHT GLASS

A glass tube in which the height of a liquid in a tank or pressure vessel may be observed. The glass tube is supported by fittings that extend through the vessel wall thus allowing the fluid in the tank to assume a corresponding level in the glass.

SIGHT PUMP

An "antique" gasoline dispensing system in which the gasoline was pumped by hand into a ten-gallon glass tank atop the pump in plain sight of the customer. When the glass cylinder had been pumped full, the attendant opened the valve on the filling hose which permitted the gasoline to gravitate into the vehicle's tank. Gravity pump.

SIGMA

Society of Independent Gasoline Marketers of America.

SILENCER

A large, cylindrical vessel constructed with an arrangement of baffles, ports, and acoustical grids to muffle the exhaust noises of stationary engines. The exhaust is piped into the silencer or muffler by piping the same size or larger than the exhaust port on the engine. The rushing, "exploding" exhaust gases enter the silencer by being ducted through the labyrinth which dissipates much of the noise. There are silencers also for the intake and exhaust ports of large blowers.

SINGLE-BUOY MOORING SYSTEM

An offshore floating platform (20 to 35 feet in diameter) connected to

pipelines from the shore for loading or unloading tankers. The SBM system is anchored in deep water thus permitting large tankers to offload or "lift" cargo in areas where it is impractical to build a loading jetty or the close-in water is too shallow for deep-draft vessels.

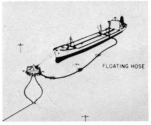

Single-buoy mooring

SINGLE-POINT MOORING
Single-buoy mooring (q.v.).

SITTING ON A WELL
The vigil of the geologist, the operator, and other interested parties who literally sit waiting for the well's drill to bore into what is expected to be the producing formation. The geologist examines the cuttings brought up by the drilling mud to ascertain just when the pay zone is penetrated. On a "big well," a very good well, everyone knows when the pay is reached; on small or marginal wells, the geologist may be the only one who recognizes it.

SIZING SCRAPER
A cylindrical, plug-like tool that is pushed or pulled through a length of pipe to test for roundness. In casing and tubing which is to be run in a well, roundness is very important. There are a number of downhole tools which must be run, e.g. packers, swabs, pumps, etc., and if the pipe were out of round the tools and equipment would not go down. Also, out-of-round pipe is more vulnerable to collapsing pressures encountered in the hole.

Skidding the rig

SKID
Squared, wooden timbers used to support line pipe while it is being welded; any rough-cut lumber used to move or support a heavy object.

SKIDDING THE RIG
Moving the derrick from one location to another without dismantling the structure; transporting the rig from a completed well to another location nearby. The use of skids (heavy timbers), rollers, and a truck or tractor. Transportable folding or jackknife rigs are seldom skidded; they are folded down to a horizontal position and moved on a large, flatbed truck.

Skid-mounted unit

SKID-MOUNTED
Refers to a pumping unit or other oil field equipment that has no permanent or fixed foundation but is welded or bolted to metal runners or timber skids. Skid-mounted units are usually readily movable by pulling as a sled or by hoisting onto a truck.

SKID TANK
A product-dispensing tank mounted on skids or runners. Can be pulled or carried on a truck.

SKIMMER
A type of oil-spill clean-up device propelled over the water that sucks or paddles oil into a tank. The skimmer is mounted on a skiff or small boat. As the boat is maneuvered to where the oil has been corralled by an oil-spill boom (q.v.) an engine and suction pump on board takes the oil off the top of the water much like a vacuum cleaner, and pumps it into a tank on board.

Skimmer

Mule skinner and his
team

Slant-hole drilling

Slant rig

Pipeline sling

SKIMMING PLANT
(1) A topping plant (q.v.); (2) A facility built alongside a creek or small stream to catch and skim off oil that, in the early days in some fields, was turned into creeks or accidentally discharged from lease tanks or from broken pipelines.

SKINNER
See Mule Skinner.

SLAB PATCH
A metal patch made out of a section of pipe welded over a pitted or corroded section of pipeline. *See* Half Sole.

SLANT-HOLE TECHNIQUE
A procedure for drilling at an angle from the vertical by means of special downhole drilling tools to guide the drill assembly in the desired direction. Slant holes are drilled to reach a formation or reservoir under land that can not be drilled on, such as a town site, beneath a water-supply lake, a cemetery or industrial property where direct, on-site drilling would be impractical or unsafe. Slant holes also are drilled to flood a formation with water or mud to kill a wild or burning well. *See* Killer Well.

SLANT RIG
A drilling derrick designed to drill from offshore platforms at angles of 20 to 35 degrees from the vertical. The slant rig, canted from the vertical, has a companion structure for racking the drillpipe vertically when coming out of the hole on a trip. The rig's traveling equipment—block, hook, swivel, and kelly joint—moves up and down on rails which are an integral part of the derrick. With a slant rig it is possible to reach farther out from a drill platform, particularly in relatively shallow water, than with a conventional rig using directional drililng techniques.

SLANT WELL
Directional well.

SLEEVE FITTING
A collar or nipple that is slipped over a length of pipe to repair a leak caused by a split or corrosion. When the sleeve is in place, the ends are welded to the pipe beyond the damaged section.

SLIDE VALVES
Very large, box-like valves for flues and stacks. Made from sheet steel, the valves are mechanically or hydraulically operated.

SLING, PIPELINE
A wide, rubber and fabric sling for lowering-in or handling coated and wrapped pipe. The slings, at the end of the boom cat's hoisting lines, are used to minimize scuffing or damaging the pipelines anticorrosion coating.

SLIM-HOLE DRILLING
A means of reducing the cost of a well by drilling a smaller diameter hole than is customary for the depth and the types of formations to be drilled through. A slim hole permits the scaling down of all phases of the drilling and completion operations, i.e., smaller bits, less powerful and smaller rigs (engines, pumps, draw works), smaller pipe and less drilling mud.

SLIP
A horse-drawn, earthmoving scoop. The slip has two handles by which the teamster guides the metal scoop into the ground at a slight angle to skim off a load of earth. Teams and slips were used to dig slush pits and build tank dikes before the days of the bulldozer. A full slip would hold about one-half cubic yard.

SLIP JOINT
A special sleeve-like section of pipe run in the drillstring to absorb the vertical motion of a floating drilling platform caused by wave action; a heave compensator (q.v.).

SLIP LOAD
The weight of the string of drillpipe, tubing or casing suspended in the drillhole by the slips (q.v.). When making a trip, coming out of the hole with a drillstring 10,000 feet long, for example, the traveling block and the hook will be lifting hundreds of thousands of pounds. But when one stand is above the rotary table and is being unscrewed to be set back into the derrick's pipe rack, the full weight of the remaining string of pipe is held by the slips. This is a slip load, and quite a load it is.

SLIPS
Wedge-shaped pieces of metal that fit into a bushing in the rotary table to support the string of tubing, drillpipe or casing suspended in the hole. Slips have a bail or handle on them so they may be lifted and set into the bushing, around the pipe, by the floormen to hold the pipe from slipping back into the hole when a stand is being unscrewed or screwed onto the string. Whenever the elevators (and the hook and traveling block) do not have a hold on the string of pipe, the slips are supporting it.

Slips inserted into rotary table bushing

SLIP STICK
An engineer's slide rule; a log-log rule; an instrument consisting of a ruler and a medial slide graduated with logarithmic scales used for rapid calculations.

SLOP TANK
(1) On a products pipeline, a tank where off-specification products or interface mix is stored. (2) At a marine terminal, a tank for holding the oil/water mix from a vessel that has washed down its compartments. (3) Any vessel used for retaining contaminated oil or water until it can be properly disposed of.

Casing slips

SLUDGE
An oleo-like substance caused by the oxidation of oil or by contamination with other material; a thick, heavy emulsion containing water, carbon, grit, and oxidized oil.

SLUDGER
See Sand Pump.

SLUG
A measured amount of liquid injected into a pipeline; a batch; a pipeline scraper or pig.

Slug

Slush pump

A flare

Snatch block with hook

SLUGGING
(1) Intermittent flow in a pipeline. When gas and oil are pumped in the same line, the oil will accumulate in low places until sufficient gas pressure builds behind it to push it out forcibly as a slug. (2) A small slug of acid pumped into a pumping well to open up the formation as part of a well workover operation.

SLURRY PIPELINE
A pipeline whose primary service is carrying a mixture of crushed solids in a water or oil medium. The common use of the term refers to a pipeline carrying pulverized coal in water. A pipeline is the cheapest and most efficient form of transportation for liquids. In recent years the pumping of small-particle solids, notably coal in water, has gained favor with shippers who are attracted by the pipeline's economics and safety, as well as environmental acceptance.

SLUSH-PIT LAUNDER
A wooden or metal square-sided conduit or sluice box where the bailer is dumped, the water, mud, and rock chips flushing down the launder into the slush pit. This device, a cousin to the launder used in washing ore from a mine, is part of a cable-tool drilling scene.

SLUSH PUMP
Mud pump (q.v.).

SMOKELESS FLARE
A specially constructed vertical pipe or stack for the safe disposal of hydrocarbon vapors or, in an emergency, process feed that must be disposed of. Smokeless flares are equipped with steam jets at the mouth of the stack to promote the complete combustion of the vented gases. The jets of steam induce greater air flow and cool the flame resulting in complete combustion without smoke or ash.

SMOKE POINT
One of the specifications on jet-engine fuel. Kerosene or jet fuel with a low smoke point is not as desirable as fuel with a high smoke point. Hydrotreating the fuel reduces the smoke or gives it a higher smoke point. This is not a contradiction, as it appears. The high and low smoke points indicate the high and low points on the wick of a testing device made like an old-fashioned kerosene lamp. The higher the wick can be turned up while burning the sample of jet fuel without producing smoke, the cleaner burning it is; thus the high smoke point.

SNAP GRABBER
A member of a work gang who manages to find easy jobs to keep himself busy while the heavy work is being done by his companions. A fully occupied loafer.

SNATCH BLOCK
A block whose frame can be unlatched to insert a rope or wire line; a single-sheave block used more often for horizontal pulling than for hoisting with A-frame or mast. Snatch block is so named because it opens to grasp or snatch the bight of a rope or cable. For those who may have forgotten their rope lore, a bight is the middle of a slack rope or a loop in a rope being held with its two ends together.

SNG
Synthetic natural gas; gas manufactured by various processes from coal, tar sands, or kerogen shales (q.v.). Substitute natural gas.

SNOW-BANK DIGGING
Colloquial expression for the relatively soft, easy drilling in sand, shales, or gumbo.

SNUB
To check a running line by taking a turn around a post or fixed object; to take up and hold fast the slack in a line; to secure or hold on object from moving with an attached rope turned around an anchoring piece.

SNUBBERS
An ingenious rig-up of lines and blocks to push down on joints of pipe that must be put into the well through the blowout preventer stack against very high well pressure. With a special hookup, the upward pull of the rig's traveling block and hook is transmitted to lines and a yoke that push down on a joint of drillpipe, forcing it by the packing of the rams in the BOP stack while the rams are holding the well pressure leaktight. After a number of joints of pipe are forced in (the joints are screwed together), their weight equals the upthrust of the well's pressure so the snubbers may be removed and the remainder of the pipe put in through the BOP without being pushed.

SNUBBING
A procedure for servicing wells that are under pressure. Tubing, packers, and other downhole tools are withdrawn from the well through a stack of rams (valve-like devices that close around pipe or tubing being withdrawn and seal off the well pressure). As each joint of tubing is withdrawn, it is unscrewed.

SOFT PLUG
A safety plug in a steam boiler, soft enough to give way or blow before the boiler does from excessive high pressure; the plug in an engine block that will be pushed out in case the cooling water in the block should freeze, thus preventing the ice from cracking the block.

SOFT ROPE
Rope made of hemp, sisal, jute, or nylon, as distinguished from wire rope which is a steel cable.

SOFTWARE
The collection of programs used in a particular application for use in a computer. Tapes, cards, disk packs containing programs designed for a process or series of processes.

SOLENOID
An electrical unit consisting of a coil of wire in the shape of a hollow cylinder and a moveable core. When energized by an electric current, the coil acts as a bar magnet, instantly drawing in the moveable core. A solenoid on an automobile's starting mechanism causes the starter-motor gear to engage the toothed ring on the vehicle's flywheel, turning the engine. Solenoids are used for opening and closing quick-acting, plunger-type valves, as those on washing machines and automatic dishwashers.

An LNG carrier, the *Lachmer Louisiana*

SOLUBLE OIL FLOODING
See Micellar-surfactant Flooding.

SOLUTION GAS
Natural gas dissolved and held under pressure in crude oil in a reservoir.
See Solution-gas Field.

SOLUTION-GAS FIELD
An oil reservoir deriving its energy for production from the expansion of the
natural gas in solution in the oil. As wells are drilled into the reservoir, the
gas in solution drives the oil into the well bore and up to the surface.

SOLVENT
A liquid capable of absorbing another liquid, gas, or solid to form a homo-
geneous mixture; a liquid used to dilute or thin a solution.

SONIC INTERFACE-DETECTOR
A pipeline sensing "probe" for detecting the approach of a product interface by identifying the change in sound velocities between the two products being pumped. The electronic device has a probe inserted through the
wall of the pipeline, protruding into the fluid stream. The probe picks up the
variations in sound velocities, and through the proper linkage can give an
audible alarm or actuate valves when the interface arrives.

SOUP
Nitroglycerine used in "shooting" a well. Nitro in its pure form is a heavy,
colorless, oily liquid made by treating glycerin with a mixture of nitric and
sulfuric acids. It is usually mixed with absorbents for easier handling. Nitro,
when used in well shooting, is put in tin "torpedos," 4 to 6 inches in diameter, and lowered into the well on a line. The bottom of each torpedo can is
made to nest in the top of the preceding one, so as many cans as necessary for the shot can be lowered in and stacked up. Nitro is measured in
quarts; the size of the shot depends upon the thickness and hardness of
the formation to be fractured.

SOURCE ROCKS
Sedimentary formations where nearly all the world's petroleum has been
found. Nearly 60 percent of the world's petroleum reserves are in sandstone; the other 40 percent are in limestone, dolomite, et al.

SOUR GAS
Natural gas containing chemical impurities, notably hydrogen sulfide (H_2S)
or other sulfur compounds that make it extremely harmful to breathe even
small amounts; a gas with a disagreeable odor resembling that of rotten
eggs.

SOUR PRODUCTS
Gasolines, naphthas, and refined oils which contain hydrogen sulfide
(H_2S) or other sulfur compounds. Sourness is directly connected with odor.

SOUR-SERVICE TRIM
A designation by manufacturers of oil field fittings and equipment that their
products have finishes resistant to corrosion by hydrogen sulfide (H_2S)
and other corrosive agents in "sour" oil and gas. *See* Sour Gas.

Before well spacing
was controlled

SPACERS AND WASHES

Specially formulated fluids for removing drilling mud from a well's borehole just ahead of the cement in a downhole cementing job. It is essential to a good cement job that the mud be removed and the wall of the hole be clean to ensure a good bond between cement and the wall. Spacers are thick fluids which displace the drilling mud ahead of the cement in a slug or piston-like manner, owing to the fluid's high viscosity and weight differential. Washes are much thinner fluids which separate the drilling mud from the cement being pumped downhole and simultaneously remove the coating of mud left on the formations. This is accomplished through a combination of turbulent and surfactant action.

SPACING PATTERN

Geographic subdivision established by government authority, usually state, defining the number of acres to be allotted to each well drilled in a reservoir. This is a conservation measure for it is generally agreed that increased recovery from a reservoir is not a function of the number of wells drilled. One oil well on 40 acres is a general rule, in some states. But there are many exceptions. Gas wells, one or two to a section (640 acres), depending upon well depth, producing formation, and other factors.

SPEARS

Fishing tools for retrieving pipe or cable lost in the borehole. Some spears resemble harpoons with fixed spurs, others have retractable or releasing-type spurs.

SPHERICAL BLOWOUT PREVENTER

A large, barrel-shaped well closure mechanism attached to the top of the well's casing. Its purpose is to close around the drillpipe in the event of a severe gas kick or threatened blowout. When the preventer's closing mechanism is hydraulically actuated, pressure is applied to a piston which moves upward, forcing the packing element to extend into the well bore and around the drillpipe in a pressure-tight seal. Should the spherical preventer be damaged or for some reason not hold pressure, rams in the BOP stack below can also be closed on the pipe to hold the pressure until the well is killed (the pressure is equalized) by the injection of heavier drilling mud.

SPHEROID

As it applies to the industry, a spheroid is a steel storage tank in the shape of a sphere flattened at both "poles," designed to store petroleum products, mainly LP-gases, under pressure. *See* Hortonsphere.

SPIDERS

The hinged, latching device attached to the elevators (the hoisting arms that lift pipe and casing in the derrick). An elevator spider is a unit attached to the travelling-block hook for hoisting pipe, casing, and tubing out of the hole and lowered in. The spider is manually locked around a length of tubing just below the tool joint. Some advanced types of elevator spiders are air operated.

SPINDLETOP

The name of the gusher brought in by Capt. Anthony Lucas, near Beaumont, Texas, in 1902. The well, the first important producer ever drilled with

Spherical blowout
preventer

Spindletop well

rotary tools, blew in (literally) and produced at the rate of 75,000 to 100,000 barrels a day.

SPINNING CHAIN
A light chain used by the drilling crew on the derrick floor when running and pulling tubing and drillpipe. After a joint has been "broken" or loosened by the pipe tongs, the spinning chain is given several turns around the pipe and, when the chain is pulled, the pipe is rotated counterclockwise and quickly unscrewed. In adding a joint or a stand of pipe, the pipe is stabbed into the tool joint or the collar and then spun up by the spinning chain. To make the joint tight the tongs are latched on and torque is applied to the satisfaction of the driller.

SPINNING TONGS
See Spinning Wrench.

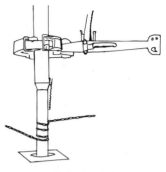

Spinning chain

SPINNING WRENCH
An air-operated drillpipe or tubing wrench used in place of the spinning chain (q.v.) and the winch-powered tongs. After the joint of pipe is stabbed into the tool joint or collar, the air wrench is latched on and spins the pipe in and applies a measured or fixed amount of torque, thus ensuring the equal tightening of all joints.

SPIN-UP
To screw one stand of drillpipe or tubing rapidly into another with a spinning chain (q.v.). After making up the joint in this manner, the heavy pipe tongs are applied to make the joint tight.

Spinning wrench
(Courtesy Eckel)

SPLASH ZONE
The area where waves of ocean or lake strike the support members of offshore platforms and production installations; the water line. The splash zone is particularly subject to corrosion because of the action of both salt and water and air.

SPLIT-LEVEL DRILLING RIG
See Drilling Rig, Split-Level.

SPLIT SLEEVE
A type of pipeline repair clamp made in two halves that bolt together to form a pressuretight seal over a hole or split in the pipe. Split sleeves also are made to enclose leaking valves and flanges until they can be permanently repaired.

Spin-up

SPM
Strokes per minute; indicates the speed or pumping rate of reciprocating pumps.

SPONGE OIL
A type of lean oil (q.v.) used in refinery absorber columns to absorb light petroleum fractions or a lighter lean oil that has vaporized in an upstream process.

SPONSON
An air chamber along the sides of a barge or small ship to increase buoyancy and stability. Sponsons are used on crane barges for additional

Split-level drilling rig

buoyancy and to minimize listing when heavy, off-side lifts are being made with the crane.

SPOOL PIECE
A short section of piping specially cut to join the ends of two pipelines lying at unusual attitudes to each other in tight, difficult-to-reach places. In undersea work, spool pieces are used to connect a seabed flow line to a platform riser, or two undersea lines. Spool pieces are difficult to measure and cut because of the pitch and yaw angle of the pipes to be joined. Spool pieces may either be a simple nipple with the ends cut at the proper angles or they may include a valve or other fittings.

SPOOLS, CASING AND TUBING
Short-length castings, flanged on both ends, used in Christmas tree assemblies to separate and support the various valves in the stack. Spools act as spacers for the valves in the blowout preventer.

SPOT CHARTER TANKER RATES
The cost per ton to move crude oil product by tanker from one port to another on a one-time basis, as compared to long-term charter rates. Spot charter rates fluctuate widely with demand and availability of tonnage.

SPOT MARKET SALES
(1) The term applied to sales of crude oil or products on a one-time basis and usually at prices above the going rate or world prices. Often these sales are arranged by an oil broker (q.v.) who can obtain certain quantities of oil for a price and for a one-time sale. (2) Sales of domestic crude oil by major producers to independent refiners from the majors' temporary overproduction or surplus. These spot sales usually are intermittent and often at prices somewhat below the posted prices.

SPREAD
A contractor's men and equipment assembled to do a major construction job, a "spread" may be literal, as the men and equipment are strung out along the right of way for several miles. On well workover, or other jobs, the spread is a concentration of the equipment for the work.

SPREAD BOSS
The person in charge of men and equipment on a large pipeline or other construction project; the stud duck.

SPRING LOADED
Refers to an item of equipment, machinery, or valve incorporating one or more springs to effect an action or motion. A spring (spiral, coil, or leaf) which, when compressed, exerts a pressure or force against whatever is compressing it equal to the compressive force. This stored-up energy of the compressed spring is used to close a valve after being opened by a momentary greater force (a pop-off or relief valve); a machine's working part to assume its original position after being acted upon for a split instant by a larger force, e.g., the instantaneous closing of an automobile's exhaust and intake valves after being opened by the engine's push rods and rocker arms.

Pipeline spread

SPUDDER
The name for a small, transportable cable-tool drilling rig. Spudders are used in shallow-well workovers for spudding in or bringing in a rotary-drilled well.

SPUD
To start the actual drilling of a well. The first section of the hole is drilled with a large-diameter spudding bit down several hundred feet to accommodate the surface pipe which may be 8 to 20 inches in diameter, depending upon the depth the well is ultimately to be drilled, its TD. After the surface casing is lowered in to the hole, it is firmly cemented in. The hole below the surface pipe is of smaller diameter and is drilled with a conventional bit which can penetrate harder formations, even rock. The surface pipe serves three important functions: it shuts out shallow water formations, prevents caving or sloughing of unconsolidated (loosely packed) upper layers of soil, sand, and shales, and acts as a foundation or anchor for all subsequent drilling and completion work.

SQUEEZE A WELL
A technique to seal off with cement a section of the well bore where a leak or incursion of water or gas occurs; forcing cement to the bottom of the casing and up the annular space between the casing and the wall of the borehole to seal off a formation or plug a leak in the casing; a squeeze job.

SQUEEZE JOB
See Squeeze a Well.

SQUIB SHOT
A small charge of nitroglycerin set off in the bottom of a well as part of a workover operation (q.v.). After cleaning out a well, freeing the producing interval of sand and silt, a small explosive charge may be set off to "wake up the well."

SS-2000 RIG CLASS
The designation for the class of semisubmersible drilling platforms (the largest built to date: 1979) which are of 18,000-ton displacement; 2,000-ton deck-load capacity; and capable of drilling in 2,000 feet of water.

SSU & SSF
Seconds Saybolt Universal and Seconds Saybolt Furol (q.v.).

STABBER
(1) A pipeline worker who holds one end of a joint of pipe and aligns it so that it may be screwed into the collar of the preceding joint. Before the days of the welded line, the pipeline stabber worked only half a day because of the exhausting nature of his work. (2) On a pipe-welding crew, the stabber works the line-up clamps or line-up mandrel. (3) On a drilling rig, the floorman (roughneck) centers the joint of pipe being lowered into the tool joint (q.v.).

STABBING BOARD
A platform 20 to 40 feet up in the derrick used in running casing. The derrick man stands on the stabbing board and assists in guiding the threaded end of the casing into the collar of the preceding joint that is hanging in the slips in the rotary table.

Stabbing a joint of
drillpipe

STABILIZER SLEEVE

A bushing the size of the borehole inserted in the drill column to help maintain a vertical hole, to hold the bit on course. The bushing or sleeve can be the fixed or rotating type with permanent or replaceable wings or lugs. (The lugs protrude from the body of the sleeve, making contact with the wall of the hole.)

STAB-IN CEMENTING (OF A WELL)

A method of cementing large-diameter casing in the borehole in which cement is pumped down through the drillpipe. The drillpipe is landed in a special casing shoe at the bottom of the casing. When the drillpipe is locked into the casing shoe, pumping of the cement downhole begins. When the cement works its way up the outside of the casing, filling the annular space, and reaches the surface, cement pumping is stopped and water and drilling mud are started down the pipe behind the cement. This displaces the cement to the bottom of the tubing. Stab-in cementing uses less cement than pumping down the casing and minimizes contamination at the cement/drilling mud interface.

Stands of pipe in the derrick

STACK THE TOOLS

Pulling the drillpipe and laying it down (stacking outside the derrick) in preparation for skidding or dismantling the derrick. If the rig is transportable, it is folded down and made ready to move.

STALKS

Colloquialism for joints of line pipe, tubing, or drillpipe.

STANDARD CUBIC FOOT OF GAS

The volume of gas contained in one cubic foot of space at a pressure of 14.65 pounds per square inch absolute and a temperature of 60°F. Volumes of gas are bought and sold corrected to the standard pressure and temperature.

STANDARD PUMPING RIG

A conventional pumping unit consisting of an engine or electric motor operating a walking beam which raises and lowers the sucker rods in an up and down pumping action.

STANDARD TOOLS

See Cable Tools.

STANDBY RIG TIME

Payment made during the period of time when the drilling rig is shut down awaiting a decision from the lease owner and other interested parties whether or not drilling is to continue.

STAND OF PIPE

A section of drillpipe or tubing (one, two, or three—sometimes four joints) unscrewed from the string as a unit and racked in the derrick. The height of the derrick determines the number of joints that can be unscrewed in one "stand of pipe." *See* Doubles.

Standpipe

STANDPIPE

The pipe that conveys the drilling mud from the mud pump to the swivel (q.v.). The standpipe extends part way up the derrick and connects to the

mud hose which is connected to the gooseneck (a curved pipe) of the swivel.

STARVE A PUMP
To have insufficient suction head at the pump's intake connection. A pump whose capacity or pumping rate is greater in volume than the volume of fluid being fed into it is being "starved," which can cause cavitation (q.v.) particularly in rotary and centrifugal pumps. *See* Suction Head, Net Positive.

STATIC LINE
(1) A wire or line to drain off or ground static electricity that may have built up from friction in a vehicle or its contents; a grounding line for gasoline transports to prevent arcing or static charges when unloading. (2) A line used to actuate part of a device or mechanism when the line is pulled, as the static line which is attached to a jumper's parachute and to the plane to open the chute. (3) A guy wire; an anchored line for stabilizing a pole, rig, or A-frame. Any wire or line; one end anchored to a deadman (q.v.), the other end attached to an upright construction for support.

STATION KEEPING
See Dynamic Stationing.

STEAM FLOODING
A secondary or tertiary oil recovery method in which superheated, high-pressure steam is injected into an oil formation to heat the oil, to reduce its viscosity so it will separate from the oil sand and drain into the well bore. The water from the cooled and condensed steam is pumped out of the well with the oil and separated at the surface. *See* Heavy Oil Process (HOP).

STEAMING PLANT
See Treating Plant.

STEAM PUMP
A reciprocating pump that receives its power from high-pressure steam. Steam is piped into the pump's steam chest and from there it is admitted to the power cylinder where it acts upon the pump's power pistons, driving them to and fro as the steam valves open and close. The fluid end of the pump is driven by the steam pistons. *See* Pump, Simplex.

STEAM SOAK
See Steam Flooding.

STEAM TRAP
A device on a steam line designed to trap air and water condensate and automatically bleed the air and drain the water from the system with a minimum loss of steam pressure.

STEEL REEF
Refers to the artificial "reefs" fromed by the substructures of offshore drilling and production platforms that attract a variety of marine life from barnacles and algae to many kinds of fish.

STEEL STORAGE
Refers to the storage of crude oil and products in above-ground steel

Mobile steam generators
for steam flooding

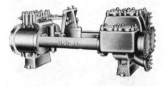

Steam pump *(Courtesy
Union Steam Pump)*

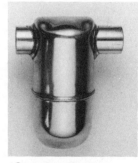

Steam trap *(Courtesy
Armstrong)*

Storage tank under
construction

tanks. In the 1920s and 30s vast tank farms of steel tankage were filled with crude oil. The market could not absorb all of the country's production but oil still flowed from the wells as each operator or company tried to produce as much oil as possible—and store it in 55,000-barrel tanks, which was the going size in those days. In a few years, however, prorationing, the restriction of production by state regulatory bodies, came into effect sharply reducing the amount of oil stored on tank farms, where it was set afire by lightning and suffered great losses through evaporation. Under prorationing, the oil stayed in the ground until the market could handle it.

STEP-OUT WELL
A well drilled adjacent to a proven well but located in an unproven area; a well located a "step out" from proven territory in an effort to determine the boundaries of a producing formation.

STICK-ELECTRODE WELDING
Electric-arc welding in which the welding rod or electrode is hand-held as compared to automatic welding. *See* Gas Welding.

Stick-electrode welding

STILE
Steps made for walking up and over a fence or other obstruction. Made in the shape of the letter A, stiles are used on farms and fenced leases to get to the other side without going through a gate. *See* Cattle Guard.

STILL, PIPE
A type of distillation unit in which oil to be heated passes through pipes or tubes in the form of a flat coil, similar to certain kinds of heat exchangers. There are two main chambers in a pipe still: one where the oil is preheated by flue gases (the convection chamber), the other, the radiant-heat chamber where oil is raised to the required temperature. No distillation or fractionation takes place in the still proper. The hot oil is piped to a bubble tower or fractionation tower where the oil flashes or vaporizes. The vapors are then condensed into a liquid product.

STILL, SHELL
The oldest and simplest form of a distillation still; a closed vessel in which crude oil is heated and the resulting vapors conducted away to be condensed into a liquid product.

STIMULATION
The technique of getting more production from a downhole formation. Stimulation may involve acidizing, hydraulic fracturing, shooting or simply cleaning out to get rid of and control sand.

STINGER
The pipe guide at the laying-end of a lay barge (q.v.). On a reel-type lay barge where the coiled pipe is straightened before being laid over the end of the barge, the stinger controls the conformation of the pipe as it leaves the barge.

Stinger

STOCK AND DIES
A device for making threads on the end of a joint of pipe or length of rod; an adjustable frame holding a set of steel dies or cutting teeth that is clamped over the end of the pipe to be threaded. When properly aligned the dies are

rotated clockwise in the frame, cutting away excess metal, leaving a course of threads.

STOCK TANK
(1) A lease tank into which a well's production is run. (2) Colloquial term for a cattle pond, particularly in the Southwest.

STOP
A common term for a type of plug valve used on lease tanks and low-pressure gravity systems.

Stock tanks

STOP-AND-WASTE VALVE
A type of plug valve that when in a closed position drains the piping above or beyond it. When the valve is turned a quarter turn to shut it off, a small port or hole in the valve body is uncovered, permitting water above the valve to drain out, preventing a freeze up in cold weather. Stop-and-waste valves are used mainly on small-diameter water piping.

STOPCOCK
A type of plug valve usually installed on a small-diameter piping; pet cock.

STOPCOCKING
Shutting in wells periodically to permit a buildup of gas pressure in the formations and then opening the wells for production at intervals.

STOPPEL
A plug inserted in a pipeline to stop the flow of oil while repairs are being made; a specially designed plug inserted in a pipeline through the use of a tapping machine (q.v.). In making a pipeline repair, cutting out a short section of the pipe, for example, the pump is shut down and the line is drained as completely as possible. But when the pipe is cut some oil may still be draining from a higher level of the line. To block this drainage a stoppel or plug is put in. On small-diameter lines, 12-inch or smaller, mud is packed in the end of the line to dam up the drainage. When the line is repaired, the pump is started and the mud plug is pushed along inside the line. It soon disintegrates and will end up either in a storage tank or at a scraper trap. On large-diameter lines, an inflatable rubber sphere is inserted in the cut line and inflated with compressed air. The sphere effectively fills the line and makes a good dam. When the repair is made, pumping is resumed and the inflated sphere is pushed ahead to take-off point just as a batching sphere (q.v.).

STORAGE JUG
The name applied to underground salt cavities for the storage of LP-gases and other petroleum products. Jug-shaped cavities are leached or washed out of salt beds using super-heated water under pressure. The resulting underground caverns, some are 100 feet in diameter and 900 feet deep, are ideal storage wells for petroleum products. See Salt-bed Storage.

STORAGE, SALT-BED
See Salt-bed Storage.

STORE LEASE
A preprinted lease form (bought at the store) with blanks to be filled in by the parties to the lease.

STORM CHOKE

A safety valve installed in the well's tubing below the surface to shut the well in when the flow of oil reaches a predetermined rate. Primarily used on offshore, bay, or townsite locations, the tubing valve acts as an automatic shut-off in the event there is damage to the control valve or the Christmas tree. Should the flow valves be damaged or be torn away by a storm or other cause, the well is wide open, flowing full stream. It is at this moment that the high-pressure stream activates the storm choke, shutting the well in completely.

STORM CONDITIONS, HUNDRED-YEAR

See Hundred-year Storm Conditions.

STOVE OIL

A light fuel oil or kerosene used in certain kinds of wickless-burner stoves.

STOVEPIPE METHOD (LAYING PIPE)

Adding one joint at a time (as in building a stovepipe) in laying an offshore pipeline from a weld-and-lay barge. In contrast, reel-barge pipe laying is done by unreeling a spool of pipe over the stern of the reel barge, over the stinger (q.v.) and onto the sea floor. On some of the largest reel barges 12,000 feet of 10-inch pipe can be carried on the massive reel and payed out like a giant hawser as the barge moves through the water at about a mile an hour. A reel barge can lay as much pipe in an hour or so as can be welded and laid from a conventional lay barge in a day. This capability is very important where "weather windows" (q.v.) may be of short duration as in the North Sea or in the extremely hostile environment of arctic waters.

STRADDLE PLANT

See On-line Plant.

STRAIGHT RUN

Refers to a petroleum product produced by the primary distillation of crude oil; the simple vaporization and condensation of a petroleum fraction, without the use of pressure or catalysts.

STRAINER, POT

An inline strainer used to catch and hold debris being pumped through a pipeline in a products line, a refinery, or processing plants. The strainer is flanged and is bolted into a pipeline.

STRAIN GAUGE

Any of various devices that measure the deformation of a structural element, pipe, or cable subject to loads. *See* Transducer.

STRAKE, HELICAL

The helical band of metal or durable plastic attached to tall metal smokestacks or vent stacks. The helical strakes, like giant grapevines entwining the stack, reduce the oscillation of the structure caused by the wind. Strakes are designed to protrude from the circumference of the stack a distance equal to about 10 percent of the stack's diameter.

STRAPPING

Measuring a tank with the use of a steel tape to arrive at its volume;

Strainer *(Courtesy Andale)*

strapping involves measuring the circumference at intervals, top to bottom; height, steel thickness, and computing deadwood (q.v.). Tank tables (q.v.) are made from these measurements.

STRATEGIC PETROLEUM RESERVES
Crude oil stored in underground formations and sealed caverns as a fuel reserve in the event of a national emergency or a prolonged oil embargo by foreign suppliers. The caches of crude are located in various areas across the country.

STRATIGRAPHIC TEST HOLE
A hole drilled to gather information about a stratigraphic formation, the general character of the rocks, their porosity, and permeability.

STRATIGRAPHIC TRAP
A type of reservoir (q.v.) capable of holding oil or gas, formed by a change in the characteristics of the formation—loss of porosity and permeability, or a break in its continuity—which forms the trap or reservoir.

STRATIGRAPHY
Geology that deals with the origin, composition, distribution, and succession of rock strata.

STRAW IN THE CIDER BARREL
To have a well in a producing reservoir; or to have an interest in a well in a producing field.

STREAM DAY
An operating day on a process unit as opposed to a calendar day. Stream day includes an allowance for regular downtime. In computing a plant's throughput on a daily basis, calendar days would include time not on the line and therefore give a distorted result. Stream day computations include only days the plant was on stream, ignoring regular downtime for turnarounds or for other reasons.

STRIKE
(1) The angle of inclination from the horizontal of an exposed strata of rock. (2) A good well; to make a strike is to find oil in commercial quantities; a hit.

STRINGER BEAD
A welding term that refers to the first bead or course of molten metal put on by the welder as two joints of line pipe are welded together. *See* Pipeline Welding.

STRINGING PIPE
Placing joints of pipe end to end along a pipeline right of way in preparation for laying, i.e., screwing or welding the joints together to form a pipeline. On large-diameter pipelines two joints are welded together in a "doubling yard" (q.v.), an area convenient to a large pipeline construction project where the pipe is unloaded from railway flatcars, coated and wrapped and two joints welded together. After "doubling" the sections of pipe are hauled to the job and strung along the right of way.

STRIP CHART
In lieu of the circular chart for recording gas flow through an orifice meter,

Stringing pipe

strip charts are sometimes used. Strip charts, as long as 35 to 40 feet, need not be changed more than once a month if the operator desires. Also, the speed at which the long chart moves through the meter is adjustable so the recording of fluctuations in gas flow may be spread out, permitting more accurate readings.

STRIPPER
An oil well in the final stages of production; a well producing less than 10 barrels a day. Most stripper wells are pumped only a few hours a day. In 1978 there were nearly 400,000 stripper wells in the U.S. producing 20 percent of the country's oil.

STRIPPER TOWER, SOUR-WATER
A refinery vessel, a tower for the physical removal of contaminants from "sour water," water from knockout drums, condensates from accumulators, and other processing units, before it undergoes biological treatment or is discharged in the plant's waste-water system.

STRIPPER WHEEL
A hand wheel which is attached to the upper rod in a string of sucker rods in the well to unscrew them.

STRIPPING JOB
See Pulling Rods.

STRIPPING PLANT
See Gasoline Plant.

STRIPPING THE PIPE
The job of removing drillpipe or tubing from a well under pressure, while maintaining control of the well. The pipe is "stripped" by withdrawing it, a "stand" at a time, through a wellhead plug equipped with a hydraulic closure mechanism (ram) that maintains pressure contact with the pipe being withdrawn.

STRIPPING THE WELL
To pull the rods and tubing from the well at the same time. The tubing must be "stripped" over the rods, a joint at a time.

STRIP
To disassemble; to dismantle for the purpose of inspection and repair; to remove liquid components from a gas stream. *See also* Stripping the Pipe.

STUCK PIPE
Refers to drillpipe stuck in the well's borehole from one or more of several possible causes: the formation above the drill bit caves or sloughs off, filling an intervals of the hole; keyseating (q.v.); junk or a foreign object wedged against the pipe; sand or shale packed around the tool joints; or a section of shale, absorbing water from the drilling mud, swelling sufficiently to reduce the size of the hole.

STRUCTURAL TRAP
A type of reservoir containing oil and/or gas formed by movements of the earth's crust which seal off the oil and gas accumulation in the reservoir forming a trap. Anticlines, salt domes, and faulting of different kinds form structural traps. *See* Stratigraphic Trap.

Stripping job

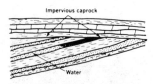

Oil trapped under a caprock

STRUCTURE
Subsurface folds or fractures of rock layers which may form a reservoir capable of holding oil or gas.

STRUCTURE CONTOUR MAP
See Contour Map.

STUB IN
To attach a line (usually of smaller diameter) to an existing line, manifold or vessel and make the connection by cutting a hole in the existing installation and welding on a nipple or other fitting.

STUB LINE
An auxiliary line attached to an existing line by use of a tap saddle (q.v.) or by welding on a nipple or other fitting.

STUD DRIVER
A mechanical device for driving or screwing stud bolts into a bored and threaded hole; a wrench-like device attached to one threaded end of a stud bolt without damaging the threads. When torque is applied, the other end of the bolt screws into the hole. Simple stud drivers are hand held. But for large jobs they can be adapted for impact wrenches (q.v.), drill-press or air motors.

STUD DUCK
Top man; the big boss.

STUFFING BOX
A packing gland; a chamber or "box" to hold packing material compressed around a moving pump rod or valve stem by a "follower" to prevent the escape of gas or liquid.

SUB
A short length of tubing containing a special tool to be used down hole; a section of steel pipe used to connect parts of the drill column which, because of difference in thread design, size or other reason, cannot be screwed together; an adapter.

SUBMERSIBLE BARGE PLATFORM
A type of drilling rig mounted on a barge-like vessel used in shallow coastal waters. When on location, the vessel's hull is submerged by flooding its compartments leaving the derrick and its equipment well above the water line.

SUBMERSIBLE PUMP
See Pump, Submersible.

SUBSEA COMPLETION SYSTEM
A self-contained unit resembling a bathysphere to carry men to the ocean bottom to install, repair, or adjust wellhead connections. One type of modular unit is lowered from a tender and fastened to a special steel, wellhead cellar. Men work in a dry, normal atmosphere. The underwater wellhead system was developed by Lockheed Petroleum Services Ltd. in cooperation with Shell Oil Company.

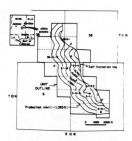

Structure contour map
(Courtesy Mace)

Stuffing box on pumping well

Submersible barge platform

Subsea inspection vessel
(Courtesy Comex)

Substructure

Sucker rod, pin end

SUBSTRUCTURE

The sturdy platform upon which the derrick is erected. Substructures are from 10 to 30 feet high and provide space beneath the derrick floor for the blowout preventer valves.

SUCKER ROD

Steel rods that are screwed together to form a "string" that connects the pump inside a well's tubing downhole to the pumping jack on the surface; pumping rods.

SUCKER-ROD GUIDES

Small washer-like devices attached to a pumping well's sucker rods to center the rods in the tubing as the rods move up and down. The guides prevent excessive wear of the tubing and the rods as well.

SUCKER RODS, HOLLOW

In certain applications, slim-hole pumping, hollow sucker rods are used, serving the dual purpose of rod and production tubing in the same string. Traveling-barrel pumps are most often used with hollow-rod pumping. The rods are attached to the cage or pull tube (traveling barrel); the pump is installed in the seating nipple or a packer-type pump anchor is used.

SUCKER-ROD SCRAPERS

Perforated disks attached to the string of sucker rods of a pumping well to prevent the buildup of paraffin on the inside of the tubing. As the rods move up and down, the perforated disks (several to each rod) scrape off the paraffin attempting to coat and then build up on the tubing, reducing the amount of oil that can be pumped from the well.

SUCTION HEAD, NET POSITIVE

The hydrostatic head; the height of the column of liquid required to ensure that the liquid is above its bubble point (q.v.) pressure at the impeller eye of a centrifugal pump. If a pump requires 10 feet of net positive suction head to fill properly and prevent cavitation, then the minimum liquid level above the pump's immediate intake connection should be 12 feet. The additional two feet of liquid level are needed to overcome the friction of connecting piping.

SUITCASE ROCK

Any formation that indicates further drilling is impractical. Upon hitting such a formation, drilling crews traditionally pack their suitcases and move on to another site.

SUPERCHARGE

To supply air to an engine's intake or suction valves at a pressure higher than the surrounding atmosphere. *See* Supercharger.

SUPERCHARGER

A mechanism such as a blower or compressor for increasing the volume of air charge to an engine over that which can normally be drawn into the cylinders through the action of the pistons on the suction strokes. Superchargers are operated or powered by an exhaust-gas turbine in the engine's exhaust stream.

SUPERPORT
A terminal or oil-handling facility located offshore in water deep enough to accommodate the largest, deep-draft oil tankers.

SUPERTANKER
The largest crude oil carrier yet designed.

SUPPLY-BOAT MOORING SYSTEM (SBMS)
A type of sea terminal for tankers and supply boats featuring a single leg securely fixed to the ocean floor with a truss-like yoke which attaches to the bow of the vessel being loaded. Loading lines are supported by the yoke which is hinged to the boat allowing free articulation to accommodate any kind of sea condition during loading. The leg of the mooring system is equipped with a universal joint and is able to rotate as the ship weather-vanes.

312,000-dwt
supertanker, the
Universe Ireland

SURFACE PIPE
See Casing.

SURFACTANT FLOODING
See Micellar-surfactant Flooding.

SURGE TANK
A vessel on a flow line whose function is to receive and neutralize sudden, transient rises or surges in the stream of liquid. Surge tanks often are used on systems where fluids flow by heads (q.v.) owing to entrained gas.

SURVEYOR'S CHAIN
A measuring instrument; a chain of 100 links, each link equaling 7.92 inches.

SURVEYOR'S TRANSIT
A telescope mounted on a calibrated base, on a tripod, for measuring horizontal as well as vertical angles; a theodolite. A transit is commonly used by surveyors for running levels.

SURVEY STAKES
Wooden markers driven into the earth by a survey crew identifying the boundaries of a right of way, the route of a pipeline, or a well location. Survey stakes may bear notations indicating elevation or location.

SUSPENDED DISCOVERY
An oil or gas field that has been identified by a discovery well but is yet to be developed.

SUSPENSE MONEY
The term applied to revenue or money collected by a regulated gas pipeline company after filing a rate increase, which is subject to an obligation to refund the money to purchasers if the regulatory agency, controlling such increases, fails to approve the increase. Escrow money.

SWAB
(1) To clean out the borehole of a well with a special tool attached to a wire line. Swabbing a well is often done to start it flowing. By evacuating the fluid contents of the hole the hydrostatic head is reduced sufficiently to permit the oil in the formation to flow into the borehole, and if there is

enough gas in solution the well may flow for a time. (2) To glean as much information as possible from a person; to pump someone for facts he would not normally volunteer.

SWAG
A downward bend put in a pipeline to conform to a dip in the surface of the right of way, or to the contours of a ravine or creek; a sag.

SWAGE
A heavy, steel tool, tapered at one end, used to force open casing that has collapsed downhole in a well.

SWAGE NIPPLE
An adapter; a short pipe fitting, a nipple, that is a different size on each end, e.g. 2-inch to 3-inch; 2-inch to 4-inch.

SWAMPER
A helper; the person who assists a truck driver load and unload and helps take care of the vehicle.

SWAY BRACES
The diagonal support braces on a rig structure. Along with the horizontal girts, sway braces hold the legs (the corner members) of the rig in place.

SWEET
Having a good odor; a product testing negative to the "doctor test"—free of sulfur compounds.

SWEET CRUDE
Crude oil containing very little sulfur and having a good odor.

SWEET GAS
Natural gas free of significant amounts of hydrogen sulfide (H_2S) when produced.

SWEPT-FREQUENCY EXPLOSION
A type of controlled explosion used in seismic work in which a string of small detonations are set off in sequence instead of the more conventional single, large explosion. In oil and gas exploration, swept-frequency explosions are a vibration or shock source in conducting seismographic surveys.

SWING CHECK
A check valve (q.v.).

SWING JOINT
A combination of pipe fittings that permits a limited amount of movement in the connection without straining the lines, flanges, and valves.

SWING LINE
A suction line inside a storage or a working tank that can be raised or lowered by a wire line attached to a hand winch mounted on the outside of the tank. By raising the swing line above the level of water and sediment in the tank, only the clean oil is pumped out.

Sway braces *(Courtesy Parker Drilling)*

SWING MAN
One whose job is working in place of other employees on their days off. In a refinery or pump station operating 24 hours a day, there are three shifts of workers, and a swing shift. The swing shift covers the days off of the other workers, so the swing man works two day shifts, two evening shifts, and one graveyard or hoot-owl shift. The other graveyard shift is worked either by another swing man or another plant worker who is not a regular shift worker.

SWING SHIFT
See Swing Man.

SWITCHER
A person who works on an oil lease overseeing the filling of lease stock tanks. When a tank is full he switches valves, turning the production into other tanks. A switcher works on a lease with flowing production; if the lease had only pumping wells, he would be called a pumper.

SWIVEL
A part of the well-drilling system; a heavy, steel casting equipped with a bail—held by the hook of the traveling block—containing the wash pipe, gooseneck, and bearings on which the kelly joint hangs and rotates; the heavy link between the hook and the drillstring onto which the mud hose is attached; an item of "traveling" equipment.

Swivel *(Courtesy B.F. Goodrich)*

SYNCLINE
A bowl-shaped geological structure usually not favorable to the accumulation of oil and gas. Stratigraphic traps (q.v.) are sometimes encountered in synclines. *See* Anticline.

SYNFUEL
Short for synthetic gas or oil (q.v.).

SYNTHANE PLANT
A coal-to-gas pilot plant operated by the Energy Research and Development Administration in Pennsylvania. Designed to produce 1.2 MMCFD of pipeline gas, designated as synthane, synthetic methane.

SYNTHETIC GAS
Commercial gas made by the reduction or gasification of solid hydrocarbons: coal, oil shale and tar sand. *See* Gasification.

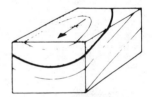

Plunging syncline *(Courtesy Petex)*

SYNTHETIC OIL
A term applied to oil recovered from coal, oil shales, and tar sands (q.v.).

T

TACK WELD
Spot weld temporarily joining two joints of pipe to hold them in position for complete welding.

TAIL
To carry the light end of a load; to extricate a vehicle from a ditch or mud.

TAIL CHAIN
The short length of chain, with a hook attached, on the end of a winch line.

TAILS ENDS
In a distillation column at a refinery, tail ends are the overlapping ends of the distillation curves of two products. For example, when naphtha and kerosene are being distilled, the end point (q.v.) of naphtha is about 325°F. but the initial boiling point of kerosene is about 305°F. So before naphtha reaches its end point, kerosene has begun to boil or vaporize. This unavoidable overlap results in tail ends; the high end of one product and low end of a closely related product.

TAIL GAS
Residue gas from a sulfur recovery unit; any gas from a processing unit treated as residue.

TAILING-OUT RODS
Unscrewing and stacking rods horizontally outside the derrick. As a rod is unscrewed, a worker takes the free end and, as the elevator holding the other end is slacked off, he "walks" the rod to a rack where it is laid down.

TAILINGS
Leftovers from a refining process; refuse material separated as residue.

Tallying the pipe

TALLYING THE PIPE
In setting pipe, casing a well, it is important to keep tab on the footage of pipe run in the hole. So before lowering a joint it is carefully tallied (measured) so the operator, by counting the number of joints run, knows to the foot where the bottom of the casing is downhole.

TALUS
Rock fragments at the base of a cliff, sometimes forming a slope of chips and larger fragments one-fourth to one-third the way up the face of the disintegrating rock cliff.

TANDEM
A heavy-duty, flat-bed truck with two closely coupled pairs of axles in the rear; a ten-wheeler.

TANK
(1) Cylindrical vessel for holding, measuring, or transporting liquids. (2) Colloquial for small pond; stock tank.

TANK BATTERY
See Battery.

TANK, BULLET
See Bullet Tanks.

TANK BOTTOMS
Oil-water emulsion mixed with free water and other foreign matter that collect in the bottoms of stock tanks and large crude storage tanks. Periodically, tank bottoms are cleaned out by physically removing the material or by the use of chemicals which separate oil from water permitting both to be pumped out.

TANK DIKE
A mound of earth surrounding an oil tank to contain the oil in the event of a rupture in the tank, a fire, or the tank running over.

TANKER RATES, SPOT CHARTER
See Spot Charter Tanker Rates.

TANKER TERMINAL
A jetty or pier equipped to load and unload oil tankers. *See* Sea Terminal.

TANK FARM
A group of large riveted or welded tanks for storage of crude oil or product. Large tank farms cover an area of a square mile or more.

TANK MIXER
Motor-driven propeller installed on the shell of a storage tank to stir up and mix tank sediments with the crude. The propeller shaft protrudes through the shell, with the motor mounted on the outside. Turbulence created by the prop thrust causes the BS&W to remain suspended in the oil as it is pumped out.

TANK TABLES
A printed table showing the capacity in barrels for each one-eighth inch or one-quarter inch of tank height, from bottom to the top gauge point of the tank. Tank tables are made from dimensions furnished by tank strappings (q.v.). *See* Strapping.

TANK TRAIN
A new concept in the rail shipment of crude oil, products, and other liquids developed by General American Transportation (GATX). "Tank Train" tank cars are interconnected which permits loading and unloading of the entire train of cars from one trackside connection. This arrangement does away with the need for the conventional loading rack (q.v.), and vapors from the filing operation can be more easily contained. *See* Densmore, Amos.

TAP
A cutting tool for making threads in a drilled hole in metal, wood, or other hard material. A slightly tapered, bolt-like threaded tool that is forcibly screwed into a bored hole, cutting away some of the metal, forming internal threads. For cutting external threads on a bolt or pipe, dies are used (q.v.).

TAPER MILL
A type of junk mill (q.v.); an elongated, tapered grinding and pulverizing bit (tapered from several inches to one or two inches in diameter) whose surface had been hardfaced with superhard, durable cutting material.

TAPPED OR FLANGED CONNECTIONS
Indicates the two types of pump or process unit connections available from suppliers. Tapped is an internally threaded (female) connection into which an externally threaded piece may be screwed; a flanged connection is one furnished with a screw or weld flange.

TAPPING AND PLUGGING MACHINE
A device used for cutting a hole in a pipeline under pressure. A nipple, with a full-opening valve attached, is welded to the line. The tapping machine is

Tanker terminal

Tank farm

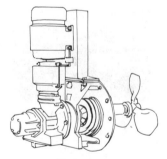
Tank mixer

Tank train

Tapping and plugging
machine

screwed onto the valve and, working through the open valve, bores a hole in the line. The tapping drill is withdrawn, the valve is closed, and the tapping machine is unscrewed from the valve. A connection can then be made to the pipeline at the valve.

TAPS
Trans-Alaska Pipeline System; a large-diameter pipeline built from the oil-rich North Slope of Alaska to the warm-water port of Valdez on the state's south shore. The 48-inch, 800-mile pipeline was completed in the summer of 1977. Its designed throughput is 1.2 million barrels a day. The Valdez crude oil terminal covers 1,000 acres and is one of the world's largest. The terminal has four loading berths for tankers up to 200,000 deadweight tons displacement.

TAP SADDLE
A type of pipeline clamp with a threaded hole in one of the two halves of the bolt-on clamp for use when a pipeline is to be tapped; to have a hole made in it for drawing off gas or liquid. Tap saddles are used on field lines, 2″ to 10″; for tapping larger lines, nipples are welded to the pipe and a tapping machine (q.v.) is used.

TARIFF
A schedule of rates or charges permitted a common carrier or utility; pipeline tariffs are the charges made by common carrier pipelines for moving crude oil or products.

TAR SANDS
See Petroleum Tar Sands.

TATTLETALE
A geolograph; a device to record the drilling rate or rate of penetration during an eight-hour tour (q.v.).

TBA
Among marketing department people, TBA stands for tires, batteries, and accessories.

Tar sand plant

T.D.
Total depth. Said of a well drilled to the depth intended.

TEAMING CONTRACTOR
A person who furnished teams of horses and mules and oil field wagons for construction and earth work in the oil fields. Some large teaming contractors in the early days kept stables with 600 teams (1,200 horse and mules). In the days of dirt roads in the booming oil fields, the horse and wagon was the most dependable mode of transportation.

TEAMSTER
See Mule Skinner.

TEAPOT DOME
Part of the Naval Petroleum Reserves set aside by Congress in 1923. Teapot Dome in Wyoming was the center of controversy and scandal in the 1920s during the presidency of Warren G. Harding.

Teamster and team

TECTONIC MAP
A geological map; structural map showing the folding and faulting of sub-surface formations.

TELEGRAPH
A device for the remote control of a steam drilling engine on a cable-tool rig. The "telegraph" consisted of a wire or a small cable running between the pulleys, one at the driller's stand, the other mounted on the steam valve of the engine. By turning his wheel, the driller regulated the speed of the engine by opening or closing the steam valve.

TELEGRAPHER'S BUG
An automatic Morse code sending machine operated by pressure from the telegrapher's thumb and forefinger. The advantage of the bug is that it makes dots in rapid succession by a slight pressure on the thumb lever; dashes are made one at a time with the forefinger. A popular, patented bug is the Vibroplex which has a beetle on the nameplate, hence the name.

TELEGRAPH KEY
A Morse code sending instrument made with a spring loaded lever on a fulcrum. When the lever is depressed, the brass lever or key makes contact with a fixed terminal, closing the electric circuit which energizes two small coils into magnets. The magnets draw down a small bar on the telegraph sounder, making a dot or a dash sound depending upon the length of time (split seconds) contact is made by the telegrapher. Dots are short, dashes are slightly longer.

TEMPERATURE BOMB
A device used downhole to measure bottom-hole and circulating temperatures on a drilling well. One technique involves attaining a temperature-sensitive probe in a protective sleeve attached to a carrier mounted on the drillpipe.

TEMPERATURE CONVERSION
(°F. to °C.) °C. = 5/9 (°F. − 32°); (°C. to °F.) °F. = 9/5 (°C.)+ 32°.

TEMPERATURE LOG
Recording temperature variations downhole by the use of an electrode containing a length of platinum wire that readily assumes the temperature of drilling mud, gas, or water leaking into the hole. One important use of the logging device is to determine the location of cement in the annular space between casing and well bore after a cement job. The curing or hardening cement gives off heat which alters the flow of electric current observable at the surface.

TEMPER SCREW
A device on the cable of a string of cable tools that permits the driller to adjust tension on the drilling line. A temper screw is made in the general form of a turnbuckle (q.v.).

TEMPLATE PLATFORM
An offshore platform whose supporting legs fit into a frame previously constructed and anchored to the sea floor. The platform, constructed onshore, is taken out to location by a crane barge where it is set into the frame.

Floater-type subsea template

Jack-up rig (left) and
drilling tender (right)

Tensioner system

TENDER
(1) A permit issued by a regulatory body or agency for the transportation of oil or gas. (2) A barge or small ship serving as a supply and storage facility for an offshore drilling rig; a supply ship.

TENDERS
A quantity of crude oil or refined product delivered to a pipeline for transportation. Regulations set the minimum amount of oil that will be accepted for transportation.

TENSIOMETER
A gauge attached to a cable or wire rope to detect the tension being applied. From two positions on a section of the cable a sensitive gauge measures the stretch and twist of the cable under load, indicating the tension on a scale; a strain gauge.

TENSIONER SYSTEMS
Tensioner systems are installed on deepwater floating drilling platforms to maintain a constant tension on the marine riser (q.v.). Two types of systems are used: the deadweight system and the pneumatic system. Tensioning systems serve the dual purpose of compensating for the vertical motion of the drilling vessel or platform and maintaining a constant tension or lifting force on the riser.

TENSION-LEG PLATFORM
A semisubmersible drilling platform held in position by multiple cables anchored to the ocean floor. The constant tension of the cables makes the platform immune to heave, pitch, and roll caused by wave action, conditions that affect conventional semisubmersibles.

TERTIARY RECOVERY
The third major phase of crude oil recovery. The primary phase is flowing and finally pumping down the reservoir until it is "depleted" or no longer economical to operate. Secondary recovery usually involves repressuring or simple waterflooding. The third or tertiary phase employs more sophisticated techniques of altering one or more of the properties of crude oil, e.g., reducing surface tension. This is accomplished by flooding the formation with water mixed with certain chemicals that "free" the oil adhering to the porous rock so it may be taken into solution and pumped to the surface. See Micellar-surfactant Flooding.

TEST COUPONS
Small samples of materials—metals, alloys, coatings, plastics and ceramics—which are subjected to heat, cold, pressure, humidity and other conditions of stress to test durability and performance under simulated operating conditions.

TESTING, HYDROSTATIC
See Hydrostatic Testing.

TEST SET (TELEPHONE)
A lineman's portable equipment for testing the circuit on a telephone line. The test set includes a hand-cranked telephone instrument whose lead wires are clipped to the phone lines when the lineman wants to call in to the switchboard.

TETRAETHYL LEAD
A lead compound added, in small amounts, to gasoline to improve its antiknock quality. Tetraethyl lead (TEL) is manufactured from ethyl chloride which is derived from ethylene, a petrochemical gas.

TEXAS DECK
The top deck of a large semisubmersible drilling platform. The upper deck of any offshore drilling rig that has two or more platform levels.

TEXAS TOWER
A radar or microwave platform supported on caissons anchored to the ocean floor. The tower resembles an offshore drilling platform in the Texas Gulf, hence the name.

THEODOLITE
A surveyor's transit (q.v.).

THERMAL CRACKING
A refining process in which heat and pressure are used to break down, rearrange, or combine hydrocarbon molecules. Thermal cracking is used to increase the yield of gasoline obtainable from crude oil.

THERMAL OXIDIZERS
A large, cylindrical furnace, with refractory lining and banks of burners at various levels, for burning refinery gases before they are vented to the flare tower (q.v.).

THERMOCOUPLE
A pyrometer; a temperature-measuring device used extensively in refining. The thermocouple is based upon the principle that a small electric current will flow through two dissimilar wires properly welded together at the ends, when one junction is at a higher temperature than the other. The welded ends are known as the "hot junction" which is placed where the temperature is to be measured. The two free ends are carried through leads to the electromotive force detector, known as the "cold junction." When the hot junction is heated, the millivolts can be measured on a temperature scale.

Thermowell thermocouple *(Courtesy Honeywell)*

THERMOMETRIC HYDROMETER
A hydrometer (q.v.) which has a thermometer as an integral part of the instrument to show the temperature of the liquid. This is of first importance as the density or API gravity varies with the temperature. Hydrometers used by pipeline gaugers are thermometric hydrometers.

THIEF
A metal or glass cylinder with a spring-actuated closing device that is lowered into a tank to obtain a sample of oil, or to the bottom of the tank to take a column of heavy sediment. The thief is lowered into the tank on a line that when jerked will trip the spring valve enabling the operator to obtain a sample at any desired level.

Thief

THIEF HATCH
An opening in the top of a tank large enough to admit a thief and other oil-sampling equipment.

Thribbles racked outside
a deep rig

Throwing the chain

800-hp tunnel thruster

Thumper

THIEFING A TANK

Taking samples of oil from different levels in a tank of crude oil and from the bottom to determine the presence of sediment and water with the use of a thief (q.v.).

THIEF ZONE

A very porous formation downhole into which drilling mud is lost. Thief zones, which also include crevices and caverns, must be sealed off with a liner or plugged with special cements or fibrous clogging agents before drilling can resume.

THIRD-GENERATION HARDWARE

Equipment developed from earlier, less sophisticated models or prototypes; the latest in the evolution of specialized equipment.

THIXOTROPIC

The property of certain specially formulated cement slurries—used in cementing jobs downhole—that causes them to "set," become rigid when pumping ceases. But when force is again applied (pumping is resumed) the cement again becomes a pumpable slurry. This procedure may be repeated until the predetermined setting time of the cement is reached.

THREAD PROTECTOR

A threaded cap or lightweight collar screwed onto the ends of tubular goods (pipe, casing, and tubing) to protect the threads from damage as the pipe is being handled.

THRIBBLES

Drillpipe and tubing pulled from the well three joints at a time. The three joints make a stand (q.v.) of pipe that is racked in the derrick. Two-joint stands are doubles; four-joint stands are fourbles.

THROWING THE CHAIN

Wrapping the spinning chain (q.v.) around the drillpipe in preparation for running the pipe up or backing it out. Crew members become proficient at throwing the chain in such a way as to put several wraps on the pipe with one deft motion.

THRUSTERS

Jets or propellers on large tankers, drillships, and deepwater drilling platforms that provide a means to move the vessel sideways—at right angles to the ship's normal line of travel—when docking or in maintaining position in water too deep for conventional anchors. See Dynamic Stationing.

THUMPER

See Vibrator Vehicle.

TIDELANDS

Land submerged during high tide. The term also refers to that portion of the continental shelf between the shore and the boundaries claimed by states. The Federal government now has the right to produce oil and gas from this area of the continental shelf.

TIE-IN

An operation in pipeline construction in which two sections of line are connected: a loop tied into the main line; a lateral line to a trunk line.

TIGHT GAS
Natural gas produced from a tight formation, one that will not give up its gas readily or in large volumes. The production of tight gas is more costly and therefore less attractive to producers owing to the need for fracturing, acidizing, and other expensive treatments to free the gas from the relatively impermeable formations. In view of these constraints, such gas has been given an incentive price of 150 percent of the price of gas from new, conventional onshore gas wells by the Natural Gas Policy Act of 1978.

TIGHT HOLE
A drilling well about which all information—depth, formations encountered, drilling rate, logs—is kept secret by the operator.

TIN HAT
The metal, derby-like safety hat worn by all workers in the oil fields, refineries, and plants to protect their heads.

TLP
Term-limit pricing; an agreement on price between a supplier and a wholesaler or jobber that runs for a specified length of time.

TOE BOARD
The enclosure at toe height around a platform or on a catwalk to prevent tools or other objects on the platforms from being kicked off accidentally.

TO LAY OFF AN INTEREST
To sell off a portion of one's interest in a well to another person to reduce the financial loss should the well be noncommercial or dry. For example, an investor who has a 30 percent interest in a well to be drilled may lay off five or 10 percent of his interest for cash he needs to minimize his risk or to reduce his "exposed position."

TOLUENE
An aromatic hydrocarbon resembling benzene but less volatile and flammable. It is used as a solvent and as an antiknock agent in gasoline.

TONGS, DRILLPIPE AND CASING
Large wrenches for making up or breaking out (tightening or loosening) joints of pipe or casing. The tongs are counterbalanced because of their weight and size. When a joint of pipe is to be tightened, the two tongs (one for applying torque, the other for backup) are swung toward the pipe by the floormen, and upon making contact with the pipe a latching device clamps the tongs onto the pipe. A chain and rope lanyard attached to the handle of the torque tongs is pulled by a friction turn or two around the cat head (operated by the driller). The backup tongs are anchored by a rope or chain secured to a substantial rig member.

TON OF CRUDE OIL
A ton of crude oil is 6.5 to 8.5 barrels, depending on the oil's specific gravity. For rough approximation, 7.5 barrels equals a metric ton or long ton; 1,000 kilograms or 2,204.6 pounds.

TOOL DRESSER
In cable-tool drilling, a worker who puts a new cutting edge on a drill bit that is worn or blunted. Like a blacksmith, the tool dresser heats the bit in a

Tin hat

Drillpipe tongs
*(Courtesy Peter Bawden
Drilling Ltd.)*

225

charcoal fire and using a hammer draws out the metal into a sharp, chisel-like cutting edge.

TOOLIE
A tool dresser (q.v.) on a cable-tool rig.

TOOL JOINT
Heavy-duty, threaded joints specially designed to couple and uncouple drillpipe into "stands" (q.v.) of such length that they can be racked in the derrick. Intermediate couplings between the tool joints are made with regular pipe collars.

Tool joints

TOOL JOINT LEAK DETECTOR
A hydraulic testing device which is clamped around a tool joint after it is made up tight in the drillstring and before it is lowered in the hole. The leak detector puts a 1,000-psi pressure or more on the outside circumference of the joint and holds the pressure for a few seconds. The smallest leak in the connection is indicated on a gauge by a drop in pressure.

TOOL PUSHER
A supervisor of drilling operations in the field. A tool pusher may have one drilling well or several under his direct supervision. Drillers are directed in their work by the tool pusher.

TOP OUT
To finish filling a tank; to put in an additional amount that will fill the tank to the top.

TOPPED CRUDE OIL
Oil from which the light ends (q.v.) have been removed by a simple refining process.

TOPPING PLANT
An oil refinery designed to remove and finish only the lighter constituents of crude oil, such as gasoline and kerosene. In such a plant the oil remaining after these products are taken off is usually sold as fuel oil.

TOPS
The "tops" in a refinery operation are the fractions or products distilled or flashed off at the top of a tower or distillation unit.

TORPEDO
An explosive device used in shooting (q.v.) a well. The well-shooting torpedo was invented and used by Col. E. A. L. Roberts, a Civil War veteran, in 1865. The first torpedoes used black powder as an explosive; later, nitroglycerin was substituted for the powder.

TORQUE
A turning or twisting force; a force that produces a rotation or torsion, or tends to.

TORQUE CONVERTER
An item of hydraulic equipment which is installed between the prime movers (drilling engines, for example) and the driven components (mud pumps and rotary) to transmit a smooth, continuous flow of power. The torque

converter absorbs or cushions the pulsations, the transient, uneven surges of torque in the power train.

TORQUE WRENCH
A tool for applying a turning or twisting motion to nuts, bolts, pipe or anything to be turned and which is equipped with a gauge to indicate the force or torque being applied. Torque wrenches are useful in tightening a series of bolts or nuts with equal tension, as on a flange or engine head.

Torque wrench

TORREY CANYON
An oil tanker that ran aground off the coast of England causing the largest, most costly, and most publicized oil spill up to that time. The mishap touched off reactions that put oil-spill pollution in the international spotlight. The *Torrey Canyon* ran aground March 18, 1967.

TORSION BALANCE
A delicate instrument used by early-day geophysical crews to measure the minute variations in magnetic attraction of subsurface rock formations. As the differences in attraction of the subsurface features were plotted over a wide area, the geophysicist had some idea as to where sedimentary formations that might contain oil were located in relation to non-sedimentary rocks. The torsion balance has been superceded by the less complicated (to use) gravity meter or gravimeter. *See* Gravimeter and Magnetometer.

TOUR
A work period; a shift of work, usually eight hours, performed by drilling crews, pump station operators, and other oil field personnel. In the field, "tour" is pronounced tower, to rhyme with sour; a trick.

TOWER HAND
A member of the drilling crew who works up in the derrick; derrick man.

TRACER LINES
Small-diameter tubing paralleling and in contact with process or instrumentation piping in a refinery or other plant to provide heat or cooling for the fluid or gases in transit. More often tracer lines carry steam. In the field, larger diameter tracer lines are used to heat low-gravity, viscous crude oils so they may be pumped. *See also* Heat Tape.

TRACT BOOK
A record book maintained by the district land offices of the Bureau of Land Management (BLM), listing all entries affecting described land.

TRACTOR FUEL
A low-octane fuel, less volatile than motor gasoline, used in low-compression farm tractors.

TRADER
One who deals in bulk petroleum or products both domestic and foreign; one who operates in the international oil market, arranging for supplies and trading surpluses of one product for others; an oil broker.

TRANSDUCER
A device or instrument actuated by power from one kind of system and in turn supplies power to another system. A classic example of a transducer

Pressure transmitter

is the telephone receiver which is actuated by electric power and supplies acoustic power to the atmosphere. A form of transducer is an air or hydraulic system that will actuate an electric system by pressure on a contact switch. Another and true form of transducer is the thermocouple (q.v.) wherein heat on two dissimilar pieces of metal will create a small, measureable electric current.

TRANSITE PIPE
A patented, composition pipe for handling corrosive liquids and salt water.

TRANSITION FITTINGS
When using plastic pipe in the field or at a plant, it is usually necessary to make connection with steel tank fittings or a pipeline. If so, special transition fittings, made with one end acceptable to the plastic pipe and the other end a standard thread end or weld end, are installed.

TRANSPONDER
A radio or other electronic device that, upon receiving a designated signal, emits a signal of its own.

TRANSSHIPMENT TERMINAL
A large, deepwater terminal where crude oil and products are delivered by "supertanker" (LCCV) (q.v.) and transshipment of product is by smaller tankers. Such terminals have large storage capacities and high-volume unloading facilities to accommodate the mammoth vessels that carry more than two million barrels of oil each trip.

TRAP
A type of geological structure that retards the free migration of oil and concentrates the oil in a limited space. A mass of porous, permeable rock which is sealed on top and down both flanks by nonporous, impermeable rock thus forming a trap. *See* Anticline.

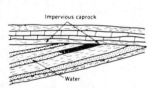

Fault trap

TRAVELING BLOCK
The large, heavy-duty block hanging in the derrick and to which the hook is attached. The traveling block supports the drill column and "travels" up and down as it hoists the pipe out of the hole and lowers it in. The traveling block may contain from three to six sheaves depending upon the loads to be handled and the mechanical advantage necessary. The wire line from the hoisting drum on the draw works runs to the derrick's crown block and down to the traveling block's sheaves.

TRAYED COLUMNS
Any of several kinds of vertical, cylindrical refining or processing columns fitted with internal, horizontal trays or baffles over which charge stock flows from top to bottom in a vaporization or absorption process. *See* Bubble Cap Trays.

Traveling block *(Courtesy B.F. Goodrich)*

TREATER, ELECTROSTATIC
An oil treater that uses AC and DC electrical force fields to cause the water droplets in the oil-water emulsion to come together, coalesce, and then drop out by gravity. The patented dehydrator uses some heat in its process, particularly on low-gravity crude oils.

TREATING PLANT
A facility for heating oil containing water, emulsions, and other impurities and with the addition of chemicals causing the water and oil to separate. The water and other foreign matter settle to the bottom of the tank and are then drawn off.

TREE SAVER
A patented, mandrel-like piping made to slip into and through the valves and connecting spools of a Christmas tree when a well is to be stimulated, acidized, or hydrofracked under high pressure. The mandrel or inner sleeve takes the pressure, protecting the tree both from the high pressure and any corrosive or abrasive fluids during the stimulation operation.

Trencher

TRENCHER
A ditching machine; a large, self-propelled machine with digging buckets fixed to an "endless" chain belt or circular frame that, when rotated, scoops out a ditch to predetermined width and depth.

TRICK
See Tour.

TRICKLE-CHARGED BATTERY
A storage battery, usually for standby, emergency service, kept charged by a small amount of current from a primary electrical source. Should the main source of power fail, the battery, fully charged, is ready for use.

TRIM, SOUR SERVICE
See Sour Service Trim.

Ferris-wheel-type trencher

TRIP
See Round Trip.

TRIP GAS
High-pressure gas encountered in drilling deep wells that can cause serious problems of control when the tools are pulled out of the hole in making a trip. The driller must exercise extreme care to prevent loss of control or a blowout. Sufficient mud must be in the hole to provide the hydrostatic head necessary to contain the downhole gas pressure. Sometimes, in order to come out of the hole under high-pressure conditions, the crew must resort to stripping the pipe (q.v.), i.e., removing the drillstring through the well's stack of control valves, the blowout preventer, on the wellhead.

TRIPLEX PUMP
A reciprocating pump with three plungers or pistons working in three cylinders. The triplex pump discharges fluid more evenly than a duplex or two-plunger pump, as it has a power stroke every one-third or a revolution of the crankshaft compared to every half revolution for the duplex pump.

Triplex pump *(Courtesy Gaso Pump)*

TRIPPING DOUBLES
An expression meaning pulling the drillpipe out of the hole (or going in) in two-joint stands (q.v.). Tripping doubles requires one-third more pipe connections to make up and break out by the floor men than if they were tripping "thribbles," three-joint stands. Handling thribbles calls for a large, tall derrick, as three-joint stands are 90 feet high and can present problems in windy areas.

Making a trip

Tube bundle

Tubing board

Tubing spool *(Courtesy Cameron Iron Works)*

TRIPPING THE BIT
Removing the bit from the hole and running it in again. (In removing the bit, the drillpipe must be pulled a stand at a time in order to reach the bit.) *See* Round Trip.

TRUNK LINE
Main line (q.v.).

TRUNNION VALVE
A type of butterfly valve whose orifice is opened and closed by a disk rotating on trunnions or pins seated in the valve body. Trunnion valves are opened and closed by a quarter turn of the handle.

TUBE BUNDLE
The name given to the tubes in the core of a heat exchanger (q.v.). The tubes or pipes, all the same length, are spaced equidistance apart in parallel rows and are supported by perforated endplates thus forming a "bundle."

TUBE STILL
A pipe still (q.v.).

TUBE TURN
A weld or flanged fitting in the shape of a U used in construction of manifolds, exchanger bundles, and other close pipe work.

TUBING ANCHOR
A downhole, packer-like device run in a string of tubing that clamps against the wall of the casing. The tubing anchor prevents the "breathing" of the tubing, the cyclic up and down movement of the lower section of tubing as the well is pumped by a rod pump.

TUBING AND CASING ROLLERS
A downhole tool for reconditioning buckled, dented or collapsed well tubing or casing. The tool is lowered into the hole, entering the small, deformed diameter of the damaged pipe. As the cylindrical tool is forced lower and rotated it pushes out dents and restores the pipe to its original diameter.

TUBING BOARD
A small platform high in the derrick where a "derrick man" (a member of the drilling crews who is not affected with acrophobia) stands to rack drillpipe or tubing as it is being pulled and set back (q.v.).

TUBING HEAD
The top of the string of tubing with control and flow valves attached. Similar in design and function to the casinghead, the tubing head supports the string of tubing in the well, seals off pressure between casing and the inside of the tubing, and provides connections at the surface to control the production of gas or oil.

TUBING PUMPS
See Pump, Tubing.

TUBING SPOOL
A heavy, forged-steel fitting that is flanged to the casinghead and into

which the tubing hangers fit; an element of the above-ground well completion hookup.

TUBULAR GOODS
Refers to drillpipe, casing, well tubing, and line pipe; a generic term for any steel pipe used in the oil fields.

TUNDRA
A vast area in the Arctic lying between the permanent ice cap and the more southerly forested region. Even in the warmest months of summer, the subsoil remains frozen, the top few inches supporting only limited vegetation.

TURBINE METER, LIQUID
A mechanism inserted into a liquid flow line that measures volumetric flow rate and total flow. The meter is constructed with vanes on a spindle inside a housing that can be flanged into a flow line. The movement of liquid through the meter exerts a force on the curved vanes, causing the spindle to turn, as on a water wheel. The spindle is connected to a counter and readout mechanism, showing rate of flow and total daily or monthly throughput.

TURBINE PUMP
See Pump, Turbine.

TURBODRILLING
A type of rotary drilling in which a fluid-drive turbine (motor) is placed in the drillstring just above the drill bit. The mud pressure from the pumps at the surface pumping mud down through the drillpipe turns the turbine which rotates the drill bit. The drillpipe does not rotate as in conventional drilling, hence there is no kelly joint being turned by the rotary table.

TURBOCHARGER
A centrifugal blower driven by an engine's exhaust-gas turbine to supercharge the engine. To supercharge (q.v.) is to supply air to the intake of an engine at a pressure higher than the surrounding atmosphere.

TURBULENT FLOW
The movement of liquid through a pipeline in eddies and swirls which tends to keep the column of liquid "together" rather than running like a river with the center of the stream moving faster than the edges. *See* Plastic Flow.

TURNAROUND
The planned, periodic inspection and overhaul of the units of a refinery or processing plant; the preventive maintenance and safety check requiring the shutting down of a refinery and the cleaning, inspection and repair of piping and process vessels.

TURNBUCKLE
A link with a screw thread at one end and a swivel at the other; a right-and-left screw link used for tightening a rod, a guy wire, or stay.

TURNKEY CONTRACT
A contract in which a drilling contractor agrees to furnish all materials and labor and do all that is required to drill and complete a well in a workman-

Turbine meter (*Courtesy Rockwell*)

Motor-driven turbine pump

Turbodrill

Hood-and-eye turnbuckle (*Courtesy Crosby*)

like manner. When on production, he "delivers" it to the owner ready to "turn the key" and start the oil running into the lease tank, all for an amount specified in the contract.

TURNKEY
A verb made from the adjective "turnkey," to perform a complete job as under a turnkey contract (q.v.); to take over and perform all necessary work of planning, procurement, construction, completion, and testing of a project before turning it over to the owner for operation.

TURNKEY WELL
A well drilled under a turnkey contract (q.v.).

TURNTABLE, ROTARY DRILLING
See Rotary Table.

TURRET-MOORED ICE-DRILLING BARGE
A drilling barge of new concept developed by Dome Petroleum Ltd. for use in Arctic waters where floating or moving ice is a danger to conventional drillships or barges. The new barge has a 16-anchor mooring system attached to a swivel directly beneath the drilling derrick. At the approach of advancing ice on the barge's beam, the vessel weathervanes until its bow is headed into the ice flow. This maneuver reduces the tension on the mooring lines to a small fraction of that on a vessel moored in a fixed position.

TURTLEBACK
A two-part clamp for joining lengths of shackle rod (q.v.). The connector is in the general configuration of an English walnut; the two halves are held together by a bolt and nut.

TVD
Total vertical depth. TVD is always less than a well's total depth (TD) because of the inevitable deviation from the vertical of the well bore.

TWIST A TAIL
To bring pressure to bear in order to speed up a job or to get action from someone who is suspected of dragging his feet.

TWO-CYCLE ENGINE
An internal combustion engine that produces one power stroke for each revolution of the crankshaft; intake, compression, power and exhaust stroke are accomplished in one revolution.

TWO-STAGE COMPRESSOR
Two-stage identifies a type of compressor that intakes gas and compresses or raises the pressure in the first chamber of the compressor and passes the gas into the second-stage chamber where it is further compressed, raising the pressure to the required level.

Turntable (Courtesy U.S. Steel)

U

U-BOLT
A bolt in the shape of a U, both ends of which are threaded. A follower or saddle piece fits over the threaded ends and is held in place by nuts. U-bolts or U-clamps are used to hold two ends of wire lines together or to make a loop in a length of wire cable by turning back the running part (the loose end) on the standing part of the cable and clamping them together.

ULTRASONIC ATOMIZER
A development in burners for heating oils in which high-frequency sound waves are focused on the stream of fuel, forming a spray of microscopic fuel droplets. The resulting intimate mixture of fuel and air makes for greater combustion efficiency.

UNASSOCIATED GAS
Natural gas occurring alone, not in solution or as free gas with oil or condensate. *See* Associated Gas.

UNBRANDED GASOLINE
Gasoline sold by a major refiner to jobbers and other large distributors without bearing the name of the refiner.

UNCOMFORMITY
The surface that separates two rock units. If the rock layers on either side (top and bottom) are parallel, it is a parallel unconformity; if they lie at an angle to each other, it is an angular unconformity. For example, a layer of sandstone is lying on a layer of limestone. Where the two dissimilar formations touch or meet, this surface is an unconformity; the upper layer does not conform to the lower layer or vice versa.

UNCONVENTIONAL NATURAL GAS
The term applied to natural gas so difficult and expensive to produce that the sources have been bypassed in favor of more easily obtainable supplies. Such sources are to be found in tight sandstone reservoirs in the western and southwestern states, in certain shales in the Appalachian Basin, and in geopressurized reservoirs along the Gulf coast. Geologists have known of these sources for many years but, because of the low prices for conventional, more cheaply producible conventional gas, the unconventional gas supplies have remained untouched. Also, to get at these marginal sources advances in technology have to be developed, and at great cost.

UNDERGROUND STORAGE
In certain areas of the country where there are underground caverns petroleum and products are stored for future use. All caverns are not suitable; some are not naturally sealed and would permit the stored oil to leak into subsurface water sources. *See* Salt-dome Storage.

UNDERREAM
To enlarge the size of the borehole of the well by the use of an underreamer (q.v.), a tool with expanding arms or lugs that, when lowered into the hole, can be released at any depth to ream the hole with steel or insert cutters.

Cavern hewn for oil storage *(Courtesy Shell)*

Often the borehole, which has penetrated the producing formation, is underreamed to enlarge the exposed area of the hole and increase the flow of oil into the hole. A belling tool, a type of underreamer, is sometimes used to excavate a bell-shaped hole just below the production string of tubing or casing for the same purpose, to enlarge or increase the face area of the hole.

UNDERREAMER

A type of drilling tool used to enlarge the diameter of the borehole in certain downhole intervals. The underreamer is made with expandable arms fitted with cutters. When in position the expandable arms are released and the cutters chew away the rock to enlarge the hole. When the reamer is pulled from the hole the arms fold in toward the body of the tool.

UNIBOLT COUPLING

A patented coupling or flange for joining two lengths of pipe. The two mating halves of the coupling have tapered shoulders. When torque is applied to the two halves by a single bolt, drawing the bolt lugs together, the coupling is tightened. Unibolt couplings are for medium-diameter piping and take up less space than conventional multibolt flanges.

UNITIZATION

A term denoting the joint operation of separately owned producing leases in a pool or reservoir. Unitization makes it economically feasible to undertake cycling, pressure maintenance, or secondary recovery programs. With the knowledge that a pool or a reservoir is a unit or an entity, with its own pressure system and a continuous oil-bearing strata, unitization was the logical arrangement to maintain as long as possible the productive life of the pool. In such an arrangement, each lease bears its prorata share of the expense of any project undertaken, as well as a share of the production. For another type of joint venture see Pooling, which is not the same as unitization.

UNIT OPERATOR

Head well puller; the man in charge of the pulling unit crew that does routine subsurface work on producing wells, e.g., cleaning out, changing pumps, pulling rods and tubing.

UNIVERSAL

A shaft coupling able to transmit rotation to another shaft not directly in line with the first shaft; a moveable coupling for transmitting power from one shaft to another when one shaft is at an angle to the other's long axis.

UNLESS CLAUSE

The clause in an unless lease that provides for the termination of the lease interest unless the lessee commences drilling or pays rental during the primary term of the lease. See Delay Rental.

UNLESS LEASE

A type of lease in general use; the other common type is the or lease (q.v.). There is no single form of the unless lease, but it is known as the unless lease because of the wording of the delay rental clause (q.v.) which usually takes the following form: "If no well is commenced on said land on or before the date hereof, this lease shall terminate as to both parties unless the

Aerial view of the Sweeney refinery

lessee on or before that date shall pay or tender to lessor the sum of _____ dollars ($ _____) which shall operate as rental and cover the privilege of deferring the commencement of a well for 12 months from said date."

UNMANNED STATION
A pipeline pumping station that is started, stopped, and monitored by remote control. Through telecommunication systems, most intermediate booster stations on large trunk lines are unmanned and remotely controlled from the dispatcher's office.

UPDIP WELL
A well located high on a structure where the oil-bearing formation is found at a shallower depth.

UPSET TUBING
Tubular goods that are "upset" are made thicker in the area of the threads in order to compensate for the metal cut away in making threads. In the manufacture of casing and drillpipe, the additional metal is usually put on the inside, but in well tubing, especially the smaller sizes, the thickening is on the outside. This is known as exterior-upset tubing.

U.S.G.S.
U.S. Geological Survey; an agency of the Federal government that, among its many services and duties, regulates the placement of wells in Federal offshore lands.

V

VACUUM DISTILLATION
Distillation under reduced pressure (less than atmospheric) which lowers the boiling temperature of the liquid being distilled. This technique with its relatively low temperatures prevents cracking or decomposition of the charge stock. For example water which boils at 212 degrees F. under the atmospheric pressure of 14.7 lbs./sq. in. boils at 102 degrees at a pressure of one pound. So vacuum distillation saves in refinery fuel costs and may prevent the breaking down or changing of molecules which might occur at higher distillation temperatures.

VACUUM FLASHER
A refinery vessel; a large-diameter column where charge stock is distilled at less than atmospheric pressure. The pressure in some flasher vessels is less than one-third atmospheric—4 or 5 pounds per square inch. At this reduced pressure, lighter fractions of the heavy charge stock will flash off or vaporize. The lower the pressure, the lower the boiling point for all liquids.

VACUUM STILL
A refining vessel in which crude oil or other feedstock is distilled at less than atmospheric pressure.

VACUUM TAR
See Asphalt.

Vacuum still *(Courtesy ARCO)*

VALVE AND SEAT
On a reciprocating pump, the seat is firmly held in the body of the pump, the suction and discharge cavity; the valve, held in place and guided by its stem, moves up on the suction stroke of the pump. On the discharge stroke it closes or seats itself.

VALVE, MULTIPLE-ORIFICE
A patented orifice valve with two orifice plates or disks in pressuretight contact. One disk can be rotated through 90°. For full flow through the valve, the orifices in the two disks are in perfect alignment. To reduce the flow, the moveable disk is rotated a certain number of degrees which partially covers the orifice in the fixed disk, thus restricting the flow through the valve.

VALVE, NEEDLE
See Needle Valve.

VALVE, PACKLESS
A special kind of valve that uses a welded bellows rather than soft packing around the valve stem. The stem of the packless valve does not rotate; it is raised and lowered into the valve body by a connecting stem outside the fluid cavity. Packless or packingless valves usually are for small-diameter piping (one-quarter to 2-inch) and are used on piping carrying hazardous or toxic fluids or gases and for high-pressure steam.

VALVE, PILOT
A small relief valve that, through a linkage of pressure piping, controls the opening of a larger relief or safety valve. A pilot valve is usually employed to modulate or dampen the action of a larger valve as it opens to relieve the system pressure.

VALVE POTS
The wells in the body of a reciprocating (plunger) pump where the suction and discharge valves are located. Valve pots are on the fluid end of the pump, and are covered and sealed by heavy, threaded plugs or metal caps bolted over the top of the pots.

VANE PUMP
A type of rotary pump designed to handle relatively small volumes of liquid products: gasoline and light oils as well as highly viscous fluids.

VAN SYCKEL, SAMUEL
The man who invented and successfully operated the first crude oil pipeline. The line was two-inch and ran from Pithole City, Pa., to a railroad five miles away. It pumped 81 barrels the first day, thus sounding the knell for the teamster and his wagonload of oil barrels.

VAPOR LOCK
A condition that exists when a volatile fuel vaporizes in an engine's fuel line or carburetor preventing the normal flow of liquid fuel to the engine. To handle gas lock or vapor lock the gas must be bled off the system by removing a line or loosening a connection, or the lines and carburetor cooled sufficiently to condense the gas back to a liquid.

Valve and seats
(Courtesy Woolley)

VAPOR PRESSURE

The pressure exerted by a vapor held in equilibrium with its liquid state. Stated inversely, it is the pressure required to prevent a liquid from changing to a vapor. The vapor pressure of volatile liquids is commonly expressed in pounds per square inch absolute (q.v.) at a temperature of 100°F. For example, butane has a vapor pressure of 52 psia at 100°F.

VAPOR RECOVERY UNIT

A facility for collecting and condensing vapors of volatile products being loaded into open tanks at refineries, terminals, and service stations. The vapors are drawn into a collecting tank and by pressure and cooling are condensed to a liquid. VR units significantly reduce air pollution by petroleum vapors.

VAPOR TENSION

See Vapor Pressure.

V-BELT

A type of "endless" V-shaped belt used in transmitting power from an engine's grooved drive pulley to the grooved sheave of a pump, compressor, or other equipment. The V-belts, bigger and tougher versions of the automobile fan belt, are used in sets of from two to twenty belts depending upon the size of the drive pulley.

V-DOOR

The opening in the derrick opposite the draw works used for bringing in drillpipe and casing from the nearby pipe racks.

VENTURI METER

An instrument for measuring the volume of flowing gases and liquids. It consists of two parts—the tube through which the fluid flows and a set of indicators which show the pressures, rate of flow, or quantity discharged. The tube, in the shape of an elongated hourglass, is flanged into a pipeline carrying the fluid. The effect of the tube is to increase the velocity and decrease the pressure at the point where the tube's diameter is reduced. The relationship between the line pressure and the pressure at the narrow "waist" of the tube is used in computing the rate of flow.

VERTICAL INTEGRATION

Refers to the condition in which a company produces raw material, transports it, refines or processes it, and markets the product, all as one integrated operation. Specifically, an oil company is said to be vertically integrated when it finds and produces oil and gas; transports it in its own pipelines; refines it; and markets its products under its brand name. According to the critics of the industry, this is not in the country's best interest. *See* Horizontal Integration.

VERTICAL-MOORED PLATFORM

A buoyant drilling-producing platform moored to the sea floor by flexible risers cemented into the seabed. Wells are drilled through the risers by conventional methods and completed at the platform deck. When all wells are drilled and completed the VMP becomes a producing platform. The buoyancy of the platform exerts sufficient tension on its mooring systems to stabilize it in all kinds of weather.

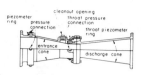

Venturi tube *(Courtesy Honeywell)*

VESSEL, EXPLORATION
See Exploration Vessel.

VIBRATOR VEHICLE
A specially designed tractor-like vehicle used to produce shock waves for geophysical and seismic surveys. The vehicle incorporates a hydraulically operated hammer or "thumper" that strikes the ground setting off shock waves which are reflected from subsurface rock formations and recorded by seismic instruments at the surface.

Seismic vibrator vehicle

VIBROPLEX
A patented automatic telegraph sending machine; a telegrapher's bug (q.v.).

VIBROSEIS
Producing seismic shockwaves by the use of "thumpers" or vibrator vehicles.

VICTAULIC COUPLING
A patented pipe coupling made in two halves that wrap around the grooved ends of two lengths of pipe and are forced together by bolts. Before the halves of the coupling are put in position, a rubber ring is placed over the junction of the two lengths of pipe. When the coupling is tightened with the two bolts, the rubber is compressed, making a pressuretight connection.

Vibroseis

VISCOSITY
One of the physical properties of a liquid, i.e., its ability to flow. It happens that the more viscous an oil, for example, the less readily it will flow, so the term has an inverse meaning—the lower the viscosity, the faster the oil will flow. Motor oil with a viscosity of SAE 10 flows more readily than oil with a viscosity of SAE 20. *See* Seconds Saybolt.

VISCOSITY INDEX
An arbitrary scale used to show the changes in the viscosity of lubricating oils with changes in temperatures.

VLCC
Very large crude carrier; a crude oil tanker of 160,000 deadweight tons or larger, capable of transporting one million barrels or more.

Victaulic coupling

VLPC
Very large product carriers (oceangoing tankers).

V.O.I.C.E.
Voluntary Oil Industry Communications Effort; part of a full-scale advertising/information program conducted by the American Petroleum Institute to tell the oil industry's story to the public. V.O.I.C.E. or VOICE is part of the program in which speakers from the industry appear before interested groups to tell oil's story.

VOLATILITY
The extent to which gasoline or oil vaporizes; the ease with which a liquid is converted into a gaseous state.

VOLUME TANK

A small cylindrical vessel connected to a gas line in the oil field to provide an even flow of gas to an engine and to trap liquids that may have condensed in the gas line. The volume tank is usually located a few feet away from the gas engine's fuel line takeoff. One reason for the volume tank, besides its being a small catch vessel for water and drip gasoline (q.v.), is that gas from the lease is piped to each gas engine through a small pipe, often one inch, and as the engine intakes gas the small pipe would not provide a steady and continuous enough supply.

VUG

Large pits or cavities found in certain types of sedimentary rocks.

VUGULAR-TYPE ROCK

Rock with large pits or cavities in its structure. Limestone, which often contains pits and cavities, is an example of a vugular-type sedimentary rock.

W

WAGON DRILLS

A battery of pneumatic rock drills mounted on a wagon. A type of trailer pulled along a pipeline right of way to drill holes in rock for the placement of dynamite charges to break up the rock so it can be removed by trenching equipment or bulldozers.

WALKING BEAM

A heavy timber or steel beam supported by the samson post that transmits power from the drilling engine via the band wheel, crank, and pitman (q.v.) to the drilling tools. The walking beam rocks or oscillates on the samson post, imparting an up-and-down motion to the drilling line or to the pump rods of a well.

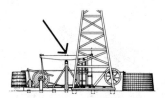

Walking beam

WALL CAKE

See Filter Cake.

WALL CLEANER

A scraping or cutting device attached to the lower joints of casing in the string for the purpose of cleaning the wall of the borehole in preparation for cementing. There are numerous types of scratching, raking, and cutting devices designed to remove the clay sheet or "filter cake" deposited by the circulating drilling mud. Mechanically cleaning the walls frees the production formation from the caked mud and also enlarges the hole diameter through the production zone making for more efficient oil flow into the well bore.

WALL SCRAPER

See Wall Cleaner.

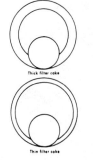

Thick filter cake

Thin filter cake

Pipe stuck in wall cake

WALL STICKING (OF DRILLPIPE)

A condition downhole when a section of the drillstring becomes stuck or hungup in the deposit of filter cake on the wall of the borehole. Also referred to as "differential sticking," an engineer's term.

WARM UP
To hammer a pipe coupling so as to loosen the threaded connection. Repeated pounding with a hammer literally warms the connection as well as "shocking" corroded threads so they can be unscrewed.

WASHERS AND SPACERS
See Spacers and Washers.

WASTE
Wiping material; cotton waste; tangled skein of cotton thread from a textile mill used in engine rooms or on "dirty" jobs to wipe up oil and grease.

WASTING ASSET
A material (usually mineral property) whose use results in depletion; a non-replaceable mineral asset; oil, gas, coal, uranium, and sand.

WATER-CONING
The encroachment of water in a well bore in a water-drive reservoir owing to an excessive rate of production. The water below the oil moves upward to the well bore through channels, fissures, and permeable streaks leaving the oil sidetracked and bypassed.

WATER DRIVE
The force of water under immense pressure below the oil formation that, when the pressure is released by drilling, drives the oil to the surface through the well bore. The drilling of a well in water-drive reservoir (q.v.) is comparable to opening a valve on a closed vessel whose contents are under pressure. As soon as the valve is opened (the well is drilled into the reservoir) the pressure forces some of the contents of the vessel into and through the valve opening. Water drives are effective means of moving oil out of the formation to the well bore. Recoveries from water-drive fields are high when care is taken to maintain reservoir pressures or energy as long as possible by good production practices.

WATER-DRIVE RESERVOIR
An oil reservoir or field in which the primary natural energy for the production of oil is the from edge, or bottom water in the reservoir. Although water is only slightly compressible, the expansion of vast volumes of it beneath the oil in the reservoir will force the oil to the well bore.

WATERFLOODING
One method of secondary recovery (q.v.) in which water is injected into an oil reservoir to force additional oil out of the reservoir rock and into the well bores of producing wells.

WATERFLOOD KICK
The first indication of increased crude oil production as the result of a waterflood project. In such an operation, the massive and forcible injection of water into a reservoir, it may be a year or longer before there is a kick, a measurable increase in the field's production rate.

WATERFLOOD PROGRAM, FIVE-SPOT
See Five-spot Waterflood Program.

WATER LOSS IN DRILLING MUD
See Filtration-loss Quality of Mud.

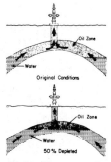

Water-drive reservoir

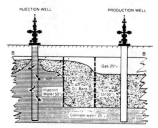

Waterflooding process
(after Clark, Courtesy
SPE)

Waterflood facilities

WATER PRESSURE
Pounds per square inch = height of the water column in feet $\times$ 0.434; e.g. 10-foot column of water $\times$ 0.434 = 4.34 pounds per square inch of pressure.

WATER TABLE
The level of ground water in the earth; the surface below which all pores of the ground are filled with water.

WEATHERED CRUDE
Crude oil which has lost an appreciable quantity of its volatile components owing to natural causes (evaporation) during storage and handling. In the early days when great quantities of crude were held above ground in steel storage tanks, on large tank farms, there were significant losses of the "light ends" which lowered the crude oil's gravity.

WEATHERING
The old practice of allowing highly volatile products such as natural gasoline (q.v.) to stand in tanks vented to the atmosphere to "weather off," to lose some of the more volatile fractions before being pumped into a pipeline. This wasteful procedure is no longer permitted.

WEATHER WINDOW
The period of time between storms when relatively calm weather prevails. In offshore work, particularly in the stormy North Sea, weather windows are often of short duration so work must be planned to take advantage of the brief intervals of good weather. This is true also for exploration work and supply operations in the Arctic.

Weight indicator and driller's control console

WEEVIL-PROOF
Refers to tools and equipment simple to operate or to assemble; fittings and equipment parts impossible for a boll weevil, a green hand, to put together improperly.

WEIGHBRIDGE
A facility to measure the contents of rail tank cars loading LP-gas at a refinery or terminal. The rail cars are moved onto the scales and loading is done while the cars are on the weighbridge (scales). When the tank car is filled, the flow is automatically shut off and a ticket for the net weight is printed simultaneously.

WEIGHT INDICATOR
A large scale-like instrument suspended a few feet above the derrick floor that constantly displays the total weight of the drillstring in the hole as the well is being drilled. By observing the indicator, the driller can tell at a glance the weight of the string and, just as importantly, the weight or downward pressure on the drill bit.

WEIGHTING MATERIAL (DRILLING MUD)
Small pellets or particles of inert, nonabrasive material such a barite (barium sulfate) added to drilling mud to increase its unit weight per gallon.

Weld-end fittings
(swage, ell, flange, tee)

WELD-END FITTINGS
Nipples, flanges, valves, and plugs without threads, but with plain, beveled ends that can be welded to nonthreaded, plain-end pipe. For proper weld-

ing, the ends of both fittings and pipe are beveled to provide a V-shaped groove for the courses of welding metal.

WELDER'S HELPER
A person who assists the welder. His most important job is to brush the weld with a wire brush to dislodge rust and scale. He also keeps the welder supplied with rods and holds or turns the piece being welded.

WELD, FILLET
An electric or oxyacetylene weld joining two pieces of metal whose ends overlap; a weld which fills in the angular or concave junction of two overlapping pieces; a strip weld.

Welding-bottle gauges

WELDING BOTTLES
Steel cylinders of oxygen and acetylene gas used in oxyacetylene or gas welding. The oxygen bottle is green and taller and smaller in diameter than the black acetylene bottles.

WELDING-BOTTLE GAUGES
A type of small, adjustable-flow regulator screwed onto oxygen and acetylene gas bottles to regulate the flow of gases to the welding torch (q.v.).

WELDING "BUG"
An automatic electric welding unit; specifically, the welding head that contains the welding-wire electrode and moves on the pipe's circumference on an aligned track-like guide. Used in welding large-diameter line pipe.

WELDING, CO$_2$-SHIELDED
A semiautomatic technique of field welding that has the advantage of making welding a hydrogen-free operation, thus eliminating hydrogen cracking of the weld metal; an inert gas-shielded welding process; electric welding in which the molten metal being laid down is blanketed by CO_2 to protect it from active gases making contact with the molten surface.

Welding "bug"

WELDING GOGGLES
Dark, safety glasses used by oxyacetylene welders and welder's helpers to protect their eyes from the intense light of the welding process and from the flying sparks. *See* Welding Hood.

WELDING HOOD
A wraparound face and head shield used by electric welders. The hood has a dark glass window in the face shield; the hood tilts up, pivoting on the headband.

WELDING MACHINE
A self-contained electric generating unit composed of a gasoline engine direct-connected to a DC generator which develops current for electric welding. Welding machines or units are skid-mounted and transportable by dragging or by truck.

WELDING TORCH
An instrument used to produce a hot flame for welding; a hand-held, tubular device connected by hoses to a supply of oxygen and acetylene and equipped with valves for regulating the flow of gases to the "tip" or welding nozzle. By opening the valves to permit a flow of the two gases from the tip,

Pipeline welding. Note goggles and hood.

the torch is ignited and then, with adjustments of the valves, a hot (3,500°F.) flame results.

WELD, WET

A weld made underwater "in the wet" without the use of a "dry-box" as in hyperbaric (extreme pressure) welding (q.v.).

WELL

A hole drilled or bored into the earth, usually cased with metal pipe, for the production of gas or oil. Also, a hole for the injection under pressure of water or gas into a subsurface rock formation. *See* Service Well.

WELL COMPLETION

The work of preparing a newly drilled well for production. This is a costly procedure and includes setting and cementing the casing, perforating the casing, running production tubing, hanging the control valves (nippling up the production tree, i.e., Christmas tree), connecting the flow lines, and erecting the flow tanks or lease tanks.

WELLHEAD CELLAR

Access capsule (top) and wellhead cellar (bottom)

An underwater "cellar" directly below the floating platform that contains the wellhead connectors, the "wet tree," the blowout preventer valves, and the flow lines and control valves. In a platform completion, the wellhead cellar may contain the blowout preventer stack and the production riser connector. In dry, one-atmosphere completions on the sea floor, the wellhead cellar is a chamber attached to the wellhead foundation and contains the marine wellhead, Christmas tree, flow lines and valves. This type of completion chamber permits workers to make adjustments or repairs in the dry at a pressure of one atmosphere, 14.7 psi.

WELLHEAD PRESSURE

The pressure exerted by a well's oil or gas at the casinghead or the wellhead when all valves have been closed for a period of time, usually 24 hours. The pressure is shown on a gauge on the wellhead. *See* Reservoir Pressure.

WELL JACKET

A structure built around a completed offshore well to protect its production tree (the valves and piping protruding above the surface of the water) from damage by boats or floating debris. The structure is equipped with navigational warning lights and other devices to signal its position at night or in a fog.

WELL NAMING

4,000 psi wellhead pressure

The naming of a well follows a long-standing, logical practice. First is the name of the operator or operators drilling the well; then the land owner from whom the lease was obtained; and last the number of the well on the lease or the block. For example Gulf Oil drills on a lease acquired from Dorothy Doe, and it is the first well on the lease. The name will appear in the trade journals as Gulf Doe No. 1, sometimes Gulf 1 Doe. A lease from the state of Oklahoma by Gulf and the name would appear on a sign at the well as: Gulf State No. 1 and perhaps followed by its location—SW NW 2–29s–13w (SW quarter of NW quarter (40 acres) of Section 2, Township 29 south, Range 13 west).

WELL PERMIT
The authorization to drill a well issued by a state regulatory agency. In some states, the law requires an operator to obtain permission before drilling so that well spacing may be enforced. Also, a permit will not be issued unless the well site is the proper size. If the well site is determined to be too small it may have to be united with other contiguous tracts in a pooling arrangement, so as to end up with a well-size plot of land.

WELL PLATFORM
An offshore structure with a platform above the surface of the water that supports the producing well's surface controls and flow piping. Well platforms often have primary separators and well-testing equipment. See Producing Platform.

WELL PROGRAM
The step by step procedure for drilling, casing, and cementing a well. A well program includes all data necessary for the tool pusher (q.v.) and drilling crews to know: formations to be encountered, approximate depth to be drilled, hole sizes, bit types, sampling and coring instructions, casing sizes, and methods of completion—or abandonment if the well is dry.

WELL SHOOTER
A person who uses nitroglycerin and other explosives to shoot a well, to fracture a subsurface rock formation into which a well has been drilled. The shooter lowers the explosive into the well bore on a wire line. When the explosive charge has been landed at the proper depth, it is detonated electrically.

WELL STIMULATION
See Stimulation.

WELL SYMBOLS (OIL MAPPING SYMBOLS)
Symbols used on a map to indicate the kind of well and its condition—successful, dry, abandoned, etc.

WET GAS
Natural gas containing significant amounts of liquefiable hydrocarbons. Gas from a well producing wet gas or rich gas is stripped of its liquid or liquefiable components in a stripping or gasoline plant to remove natural gasoline, butane, pentane, and other light hydrocarbons. After the gas stream has been "dried out" it is acceptable to a gas transmission line.

WET JOB
Pulling tubing full of oil or water. As each joint or stand (q.v.) is unscrewed, the contents of the pipe empties onto the derrick floor, drenching the roughnecks (q.v.). The tubing is standing full of fluid because the pump valve on the bottom of the tubing is holding and will not permit the fluid to drain out as it is being hoisted out of the hole.

WET STRING
See Wet Job.

WHIPSTOCK
A tool used at the bottom of the borehole to change the direction of the

Well platform

API STANDARD OIL-MAPPING SYMBOLS

Location	O
Abandoned location	erase symbol
Dry hole	-O-
Oil well	●
Abandoned oil well	-●-
Gas well	O
Abandoned gas well	-O-
Distillate well	◑
Abandoned distillate well	-◑-
Dual completion—oil	◉
Dual completion—gas	◎
Drilled water-input well	⊘w
Converted water-input well	◗w
Drilled gas-input well	⊘g
Converted gas-input well	◗g
Bottom-hole location (× indicates bottom of hole. Changes in well status should be indicated as in symbols above.)	O····×
Salt-water disposal well	⊖swd

Standard well symbols
(Courtesy API)

Wet job

Whirley crane barge
(Courtesy R. Reece
Barret Assoc.)

drilling bit. The whipstock is, essentially, a wedge that crowds the bit to the side of the hole causing it to drill at an angle from the vertical.

WHIPSTOCK ANCHOR
A downhole tool used to prevent the downward movement of a whipstock (q.v.) during milling or sidetracking. Essentially, the tool is a plug run and set permanently in the casing just below the projected window to be cut in the casing wall. The anchor isolates the casing and the formations below the window and keeps the whipstock from moving, forcing the milling tool to cut the casing at the intended point. In sidetracking, the anchor directs the drill bit at an angle from the vertical to sidetrack or drill a new hole around an obstruction, a fish (q.v.), or to correct the direction of the bit in a severely deviated hole.

WHIRLEY
The name applied to a full-revolving crane for offshore duty. Other barge-mounted cranes revolve 180°, over the stern and over both sides of the vessel.

WHITAKER SYSTEM
A patented system for protection of crews working on offshore drilling platforms, semisubmersibles, and other structures. The heart of the system is a survival capsule into which offshore crew members can retreat in the event of fire or explosion or other disaster and lower themselves to the water. The capsule is self-propelled and provides food, water, First Aid supplies for 28 persons. Large offshore structures have several survival capsules that hang from davits at various locations on the platform. *See* Brucker Survival Capsule.

Whitaker system

WHITE CARGO
Clean cargo; a term to describe distillate—gasoline, kerosene, heating oils—carried by tankers.

WHITE, DR. ISRAEL CHARLES
The "father of petroleum geology." Dr. White, in the 1880s, brought about the transition from superstition and "creekology" to scientific geological methods. He was a poor West Virginia boy who grew up to become world famous as the discoverer of the anticlinal or structural theory of oil accumulation.

WICK OILER
A lubricator for large, slow-moving crank bearings. Oil is fed from a small canister, a drop at a time, onto a felt pad or "wick." As the crank turns beneath the wick, a scraper on the crank makes contact with the wick taking a small amount of oil.

WILDCATTER
A person or company that drills a wildcat well; a person held in high esteem by the industry, if he is otherwise worthy; an entrepreneur to whom taking financial risks to find oil is the name of the game.

WILDCAT WELL
A well drilled in an unproved area, far from a producing well; an exploratory well in the truest sense of the word; a well drilled out where the wildcats prowl and "the hoot owls mate with the chickens."

Wildcat well

WINCH
A device used for pulling or hoisting by winding rope or cable around a power-driven drum or spool.

WINDLASS
A winch; a steam or electric-driven drum, with a vertical or horizontal shaft, for raising a ship's anchor.

WIND-LOAD RATING
Drilling derricks not only are rated as to lifting or load capacity (up to two million pounds) but for wind load, resistance to the wind trying to blow them over when their racks are full of drillpipe or tubing. The biggest, sturdiest rigs are designed to withstand 100 to 130 mph wind gusts.

WIRE LINE
A cable made of strands of steel wire; the "lines" used on a drilling rig. Also the small-diameter steel cable used in well logging.

WIRE-LINE TOOLS
Special tools or equipment made to be lowered into the well's borehole on a wire line (small-diameter steel cable), e.g. logging tools, packers, swabs, measuring devices, etc.

WIRE-LINE TRUCK
A service vehicle on which the spool of wire line is mounted for use in downhole wire-line work. To run a wire-line tool, the service truck is backed up to a position near the rig floor and by an arrangement of small guide pulleys, the wire line with a tool attached is lowered into the hole. When the downhole work is finished, the line is reeled in by the truck-powered reel.

WIRE ROPE
A rope made of braided or twisted strands of steel wire; a cable. *See* Soft Rope.

WIRE WELDING
Electric welding with a continuous wire electrode instead of the more common hand-held electrode or welding rod. Wire welding is used in automatic electric welding on pipelines to lay down filler beads or hot passes after the joints of pipe have been joined by tack welds and stringer beads with the use of hand-held welding rods.

WITCHING
See Doodle Bug.

WOC TIME
Waiting-on-cement time; the period between the completion of the actual cementing operations and the drilling out of the hardened cement plug in the casing at the bottom of the well.

WOODCASE THERMOMETER
A thermometer used by gaugers in taking the "tank temperature," the temperature of the oil in the tank as contrasted to the temperature of the sample oil to be tested. The thermometer is encased in a wood frame to which a line may be attached for lowering the thermometer into the oil.

Rope mooring winch

Wire-line trailer
(Courtesy Mathey)

Skid-mounted wireline
unit

Work boat

WORK BOAT
A boat or self-propelled barge used to carry supplies, tools, and equipment to job sites offshore. Work boats have large areas of clear deck space which enable them to carry a variety of loads. *See* Drilling Tender.

WORKING GAS
Gas stripped of all liquid hydrocarbons and used in gas-lift production of crude oil; lift gas.

WORKING INTEREST
The operating interest under an oil and gas lease. The usual working interest consists of seven-eighths of the production subject to all the costs of drilling, completion, and operation of the lease. The other one-eighth of production is reserved for the lessor or landowner. *See* Landowner's Royalty.

WORKING INTEREST, FULL-TERM
A working interest which lasts as long as oil and gas are produced, as distinguished from a limited-term working interest which may terminate before the depletion of the well. Or the interest may terminate at a specified time or after the production of a certain amount of oil and gas.

WORKING PRESSURE
The pressure at which a system or item of equipment is designed to operate. Normal pressure for a particular operation. Pumps, valves, pipe are, however, designed and manufactured with a wide margin of safety, extra strength for protection of personnel and equipment in the event that, owing to some emergency, the working pressure is exceeded for a short period.

WORKING TANK
A terminal or main-line tank pumped into and out of regularly; a tank that is "worked" as contrasted to a storage tank not regularly filled and emptied.

WORKOVER
Operations on a producing well to restore or increase production. Tubing is pulled and the casing at the bottom of the well is pumped or washed free of sand that may have accumulated. In addition to washing out sand and silt that has clogged the face of the formation, a workover may also include an acid treatment, hydrofracking, or a "squib shot" (q.v.), a small shot of nitroglycerin to "wake up the well."

WORM GEAR
A type of pinion gear mounted on a shaft, the worm gear meshing with a ring gear; a gear in the shape of a continuous spiral or with the appearance of a pipe thread, often used to transmit power at right angles to the power shaft.

WRINKLE PIPE
To cut threads on a piece of pipe in order to make a connection.

WRIST PIN
The steel cylinder or pin connecting the rod to the engine's or pump's piston. The wrist pin is held in the apron or lower part of the piston by a friction fit and a circular, spring clip. The upper end of the connecting rod is fitted with a lubricated bushing which permits the rod to move on the pin. A piston pin.

Workover crew pulling tubing

X-Y-Z

XYLENE
An aromatic hydrocarbon; one of a group of organic hydrocarbons (benzene, toluene, and xylene) which form the basis for many synthetic organic chemicals. *See* BTX.

ZEOLITIC CATALYST
Catalyst formulations that contain zeolite (any of various hydrous silicates, a mineral) for use in catalytic cracking units.

ZONE
An interval of a subsurface formation containing one or more reservoirs; that portion of a formation of sufficient porosity and permeability to form an oil or gas reservoir.

ZONE ISOLATION
A method of sealing off, temporarily, a producing formation while the hole is being deepened. A special substance is forced into the formation where it hardens allowing time for the well bore to be taken on down. After a certain length of time, the substance again turns to a liquid unblocking the producing formation.

ZONE OF LOST CIRCULATION
An interval in a subsurface formation so porous or cut with crevices and fissures that the drilling mud is lost in the pores, cracks, or even a cavern, leaving none to circulate back to the surface.

Zone of lost circulation
(Courtesy Dresser Magcobar)

D&D

Standard Oil Abbreviator

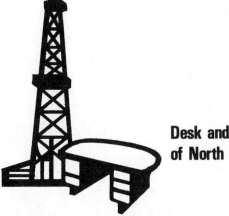

Desk and Derrick Clubs
of North America

A

A/	Acidized with
AA	After acidizing, as above
A&A	Adjustments and Allowances
ab	Above
ABC	Audit Bureau of Circulation
abd	Abandoned
abd loc	Abandoned location
abd-gw	Abandoned gas well
abd-ow	Abandoned oil well
abdogw	Abandoned oil & gas well
abrsi jet	Abrasive jet
ABS	Acrylonitrile butadiene styrene rubber
absrn	Absorption
abst	Abstract
abt	About
abun	Abundant
abv	Above
ac	Acid, acidizing
ac	Acres, acre, acreage
AC	Alternating current
AC	Austin Chalk
acct	Account, account of, accounting
accum	Accumulative
acd	Acidize, acidizing, acidized
A-Cem	Acoustic cement
acfr	Acid fracture treatment
ac-ft	Acre feet
ACM	Acid-cut mud
acrg	Acreage
ACSR	Aluminum conductor steel reinforced
ACT	Automatic custody transfer
ACW	Acid-cut water
AD	Authorized depth
add	Additive
addl	Additional
adj	Adjustable
adm	Administrative, administration
ADOM	Adomite
ADP	Automatic data processing
adpt	Adapter
adspn	Adsorption
advan	Advanced
AF	Acid frac, after fracture
AFE	Authorization for expenditure
affd	Affirmed

afft	Affidavit
AFP	Average flowing pressure
aggr	Aggregate
aglm	Agglomerate
AIR	Average injection rate
Alb	Albany
alg	Algae
alk	Alkalinity
akly	Alkylate, alkylation
allow	Allowable
alm	Alarm
alt	Alternate
amb	Ambient
amor	Amorphous
amort	Amortization
amp	Ampere
amph	Amphipore
Amph	Amphistegina
amp hr	Ampere hour
amt	Amount
anal	Analysis, analytical
ang	Angle, angular
Angul	Angulogerina
anhy	Anhydrite, anhydritic
anhyd	Anhydrous
ANYA	Allowable not yet available
AOF	Absolute open flow potential (gas well)
app	Appears, appearance
appd	Approved
appl	Appliance, applied
applic	Application
approx	Approximate(ly)
apr	Apparent (ly)
apt	Apartment
aq	Aqueous
AR	Acid residue
A/R	Accounts receivable
Ara	Arapahoe
arag	Aragonite
Arb	Arbuckle
arch	Architectural
Archeo	Archeozoic
aren	Arenaceous
afg	Argillaceous, argillite
ark	Arkose(ic)
Arka	Arkadelphia
arm	Armature
arnd	Around
ARO	At rate of
arom	Aromatics

AS	Anhydrite stringer, after shot
ASAP	As soon as possible
asb	Asbestos
asbr	Absorber
asgmt	Assignment
Ash	Ashern
asph	Asphalt, asphaltic
assgd	Assigned
assn	Association
assoc	Associate (d) (s)
asst	Assistant
assy	Assembly
astn	Asphaltic stain
AS&W ga	American Steel & Wire gauge
at	Atomic
AT	All thread, acid treat(ment), after treatment
At	Atoka
ATC	After top center
ATF	Automatic transmission fluid
atm	Atmosphere, atmospheric
att	Attempt(ed)
atty	Attorney
at wt	Atomic weight
aud	Auditorium
Aus	Austin
auth	Authorized
auto	Automatic, automotive
autogas	Automotive gasoline
aux	Auxiliary
AV	Annular velocity, Aux Vases sand
av	Aviation
avail	Available
AVC	Automatic volume control
avg	Average
avgas	Aviation gasoline
AW	Acid water
AWG	American Wire Gauge
awtg	Awaiting
az	Azimuth
aztrop	Azeotropic

B

B/	Base, bottom of given formation (i.e., B/Frio)
BA	Barrels of Acid
Ball	Balltown sand
bar	Barite(ic)
bar	Barometer or barometric
BAR	Barrels acid residue
Bar	Barlow Lime
Bark Crk	Barker Creek
Bart	Bartlesville
base	Basement (Granite)
bat	Battery
BAT	Before acid treatment
Bate	Bateman
BAW	Barrels acid water
BAWPD	Barrels Acid Water Per Day
BAWPH	Barrels Acid Water Per Hour
BAWUL	Barrels acid water under load
B&B	Bell and bell
BB fraction	Butane-butane fraction
B/B	Back to Back, Barrels per barrel
BB	Bridged back
BBE	Bevel both ends
B.Bl	Base Blane
BBL	Barrel, barrels
B & CB	Beaded and centre beaded
BC	Barrels of condensate, bottom choke
BCF	Billion cubic feet
BCFD	Billion cubic feet per day
BCPD	Barrels condensate per day
BCPH	Barrels condensate per hour
BCPMM	Barrels condensate per million
BD	Barrels of distillate, budgeted depth
B/D	Barrels per day
B/dry	Bailed dry
bd	Board
BDA	Breakdown acid
Bd'A	Bois d'Arc
BDF	Broke (break) down formation
bd ft	Board foot; board feet
BD-MLW	Barge deck to mean low water
BDO	Barrels diesel oil
BDP	Breakdown pressure
BDPD	Barrels distillate per day
BDPH	Barrels distillate per hour
BDT	Blow-down test
B.E.	Bevelled end
Be	Berea
Be	Baume
Bear R	Bear River
bec	Becoming

Beck	Beckwith	bldg	Building, bleeding, bleeding gas
Bel	Beldon		
Bel C	Belle City	bldg drk	Building derrick
Bel F	Belle Fourche	bldg rds	Building roads
Belm	Belemnites	bldo	Bleeding oil
Ben	Benoist (Bethel) sand	bldrs	Boulders
Ben	Benton	blg	Bailing
Bent	Bentonite, bononitic	Blin	Blinebry
bev	Bevel, beveled, as for welding	BL/JT	Blast joint
		blk	Black, block
b & f	Ball and flange	Blk Lf	Black Leaf
bf	Buff	Blk Li	Black Lime
BF	Barrels fluid	blk lnr	Blank liner
BFO	Barrels frac oil	blnd	Blend/blended/blending
BFPD	Barrels fluid per day	blndr	Blender
BFPH	Barrels fluid per hour	BLO	Barrels load oil
BFW	Barrels formation water, boiler feed water	blo	Blow
		BLOR	Barrels load oil recovered
B/H	Barrels per hour	Blos	Blossom
BHA	Bottom-hole assembly	BLOYR	Barrels load oil yet to recover
BHC	Bottom-hole choke	blr	Bailer
BHFP	Bottom-hole flowing pressure	B. Ls	Big Lime
BHL	Bottom-hole location	blts	Bullets
BHM	Bottom-hole money	BLW	Barrels load water
BHN	Brinell hardness number	BM	Barrels mud, bench mark, Black Magic (mud)
B Hn	Big Horn		
BHP	Bottom-hole pressure	B/M	Bill of material
bhp	Brake horsepower	BMEP	Braked mean effective pressure
bhp-hr	Brake horsepower hour		
BHPC	Bottom-hole pressure, closed (See also SIBHP and BHSIP)	BMI	Black malleable iron
		bmpr	Bumper
BHPF	Bottom-hole pressure, flowing	bn	Brown
BHPS	Bottom-hole pressure survey	bnd	Band(ed)
B/hr	Barrels per hour	bndry	Boundary
BHSIP	Bottom-hole shut-in pressure	bnish	Brownish
BHT	Bottom-hole temperature	BNO	Barrels new oil
Big.	Bigenerina	bnz	Benzene
Big. f.	Bigenerina floridana	BO	Barrels oil, backed out (off)
Big. h.	Bigenerina humblei	BOCD	Barrels oil per calendar day
Big. nod.	Bigenerina nodosaria	BOCS	Basal Oil Creek sand
B. Inj.	Big Injun	BOD	Barrels oil per day
bio	Biotite	Bod	Bodcaw
bit	Bitumen, bituminous	BOE	Bevel one end
bkdn	Breakdown	BOE	Blow out equipment
bkr	Breaker	Bol.	Bolivarensis
BL	Barrels load	Bol. a.	Bolivina a.
B/L	Bill of lading	Bol. flor.	Bolivina floridana
bl	Blue	Bol. p.	Bolivina perca
BL&AW	Barrels load & acid water	Bonne	Bonneterre
bld	Bailed, blind (flange)	BOP	Blowout preventer
		BOPCD	Barrels oil per calendar day

BOPD	Barrels oil per day	bsmt	Basement
BOPH	Barrels oil per hour	BSPL	Base plate
BOPPD	Barrels oil per producing day	BSUW	Black sulfur water
BOS	Brown oil stain	BSW	Barrels salt water
bot	Bottom	BSWPD	Barrels salt water per day
BP	Back Pressure, boiling point, bridge plug, bull plug, bulk plant	BSWPH	Barrels salt water per hour
		BT	Benoist (Bethel) sand
		BTDC	Before top dead center
BP	Bearpaw, Base Pennsylvanian	btm	Bottom
		btm chk	Bottom choke
BPCD	Barrels per calender day	btmd	Bottomed
BPH	Barrels per hour	btry	Battery
BPLO	Barrels of pipeline oil	BTU	British thermal unit
BPLOPD	Barrels of pipeline oil per day	btw	Between
BPM	Barrels per minute	BTX (unit)	Benzene, toluene, xylene (Unit)
BP Mix	Butane and propane mix		
BPV	Back pressure valve	Buck	Buckner
BPWPD	Barrels per well per day	Buckr	Buckrange
BR	Building rig, building road	Bul. text.	Buliminella textularia
brach	Brachiopod	Bull W.	Bullwaggon
brec	Breccia	bunr	Burner
brg	Bearing	Burg	Burgess
Brid	Bridger	butt	Buttress thread
brit	Brittle	BV/WLD	Beveled for welding
B. Riv.	Black River	BW	Barrels of water, boiled water, butt weld
brk	Break (broke)		
brkn	Broken	BW/D	Barrels of water per day
brkn sd	Broken sand	BW ga	Birmingham (or Stubbs) iron wire gauge
brksh	Brackish (water)		
brkt(s)	Brackett(s)	BWL	Barrels water load
brn or br	Brown	BWOL	Barrels water over load
Brn Li	Brown lime	BWPD	Barrels of water per day
brn sh	Brown shale	BWPH	Barrels of water per hour
Brom	Bromide	bx	Box(es)
brtl	Brittle		
bry	Bryozoa		
B & S	Bell and spigot		
BS	Basic sediment, bottom sediment, bottom settlings, Bone Springs		

C

BS&W	Basic sediment & water	C	Center (land description), Centrigrade temp. scale
B/S	Bill of sale, base salt		
B/SD	Barrels per stream day (refinery)	C/	Contractor (i.e., C/John Doe)
		c	Coarse(ly)
BSE	Bevel small end	C/A	Commission agent
BSFC	Brake specific fuel consumption	C&A	Compression and absorption plant
bsg	Bushing	C to C	Center to center
B&S ga	Brown and Sharpe gauge	C to E	Center to end
bskt	Basket	C to F	Center to face
B slt	Base of the salt	Cadd	Caddell

CAG	Cut across grain	CCU	Catalytic Cracking Unit
cal	Caliper survey, calorie, calcite, calcitic, caliche	ccw	Counterclockwise
		CD	Contract depth, calendar day
Calc	Calcium, calcareous, calcerenite, calculate(d)	CDM	Continuous dipmeter survey
		Cdr Mtn	Cedar Mountain
calc OF, COF	Calculated open flow (potential)	cdsr	Condensor
		Cdy	Cody (Wyoming)
calc gr	Calcium base grease	cell	Cellar, cellular
calc	Calceneous	cem	Cement(ed)
Calv	Calvin	CEMF	Counter electromotive forces
Cam.	Camerina	Ceno	Cenozoic
Camb	Cambrian	cent	Centralizers
Cane R	Cane River	centr	Centrifugal
Cany	Canyon	ceph	Cephalopod
Cany Crk	Canyon Creek	Cert	Ceratobulimina eximia
CAOF	Calculated absolute open flow	CF	Casing flange
		Cf	Cockfield
cap	Capacity, capacitor	CF	Cubic feet, clay filled
Cap	Capitan	CFBO	Companion flanges bolted on
Car	Carlile	CFG	Cubic feet gas
carb	Carbonaceous	CFGPD	Cubic feet gas per day
carb tet	Carbon tetrachloride	CFGH	Cubic feet of gas per hour
Carm	Carmel	CFM	Cubic feet per minute
Casp	Casper	CFOE	Companion flange one end
cat	Catalyst, catalytic, catalog	CFR	Cement friction reducer
CAT	Carburetor air temperature, catalog, catalyst, catalytic	CFS	Cubic feet per second
		c-gr	Coarse-grained
Cat	Catahoula	CG	Corrected gravity, center of gravity
Cat ckr	Catalytic cracker		
Cat Crk	Cat Creek	cg	Coring
cath	Cathodic	cglt	Conglomerate, conglomeritic
caus	Caustic	cgs	Centimeter-gram-second system
cav	Cavity		
CB	Counterbalance (pumping equip.), core barrel, changed(ing) bits	CH	Casinghead (gas)
		C/H	Cased hole
		ch	Chert, choke
cc	Cubic centimeter	chal	Chalcedony
CC	Carbon copy, casing cemented (depth), closed cup	Chapp	Chappel
		Char	Charles
C & C	Circulating and conditioning	Chatt	Chattanooga shale
C-Cal	Contact caliper	chem	Chemical, chemist, chemistry
CCHF	Center of casinghead flange	chem prod	Chemical products
Cck	Casing choke	Cher	Cherokee
CCL	Casing collar locator	Ches	Chester
CCLGO	Cat cracked light gas oil	CHF	Casinghead flange
CCM	Condensate-cut mud	CHG	Casinghead gas
CCP	Critical compression pressure	chg	Charge, charged, charging
CCPR	Casing collar perforating record	chng	Change, changed, changing
		Chim H	Chimney Hill
CCR	Conradson carbon residue	Chim R	Chimney Rock
CCR	Critical compression ratio	Chin	Chinle

chit	Chitin(ous)	clyst	Claystone
chk	Choke, chalk	cm	Centimeter
Chkbd	Checkerboard	CMC	Sodium carboxymethylcellulose
chkd	Checked		
chky	Chalky	Cmchn	Comanchean
chl	Chloride(s), chloritic	Cmpt	Compact
chl log	Chlorine log	cm/sec	Centimeters per second
Chou	Chouteau lime	cmt (d)(g)	Cement (ed)(ing)
CHP	Casinghead pressure	cmtr	Cementer
chrm	Chairman	CN	Cetane number
chrome	Chromium	cncn	Concentric
chromat	Chromatograph	cntf	Centrifuge
cht	Chart, chert	cntl	Control(s)
chty	Cherty	cntr	Center(ed), controller, container
Chug	Chugwater		
CI	Cast-iron, contour interval (map)	Cnty	County
		cnvr	Conveyor
Cib	Cibicides	CO	Clean out, cleaning out, cleaned out, crude oil, circulated out
Cib h	Cibicides hazzardi		
CIBP	Cast-iron bridge plug		
CI engine	Compression-ignition engine	c/o	Care of
C.I.F.	Cost, insurance and freight	COBOL	Common Business Oriented Language
Cima	Cimarron		
CIP	Cement in place, closed-in pressure	COC	Cleveland open cup
		Coco	Coconino
cir	Circle, circular, circuit	Cod	Codell
circ	Circulate, circulating, circulation	coef	Coefficient
		COF	Calculated open flow
cir mils	Circular mils	COG	Coke oven gas
Cis	Cisco	COH	Coming out of hole
ck	Check, cake	COL	Colored, column
cksn	Chicksan	Col ASTM	Color, American Standard Test Method
Ck Mt	Cook Mountain		
CL	Carload	Cole J	Cole Junction
C/L	Center line	Col Jct	Coleman Junction
Clag	Clagget	coll	Collect, collected, collection, collecting
Claib	Claiborne		
Clarks	Clarksville	colr	Collar
clas	Clastic	Com	Comanche
Clav	Clavalinoides	Com	Comatula
Clay	Clayton, Claytonville	com	Common
Cleve	Cleveland	comb	Combined, combination
Clfk	Clearfork	coml	Commercial
Cliff H	Cliff House	comm	Community, communitized, communication
CLMP	Canvas-lined metal petal basket		
		comm	Commission, commenced
cin (d) (g)	Clean, cleaned, cleaning	commr	Commissioner
Clov	Cloverly	comp	Complete, completed, completion
clr	Clear, clearance		
clrg	Clearing	Com Pk	Comanche Peak
clsd	Closed	comp nat	Completed natural

compnts	Components	CPC	Casing pressure—closed
compr	Compressor	Cp Colo	Camp Colorado
compr sta	Compressor station	CPF	Casing pressure—flowing
compt	Compartment	CPG	Cost per gallon or cents per gallon
con	Consolidated		
conc	Concentric, concentrate, concrete	cplg	Coupling
		CPM	Cycles per minute
conc	Concretion(ary)	CPS	Cycles per second
conch	Conchoidal	CPSI	Casing pressure shut in
concl	Conclusion	CR	Cold rolled, compression ratio
cond	Condensate, conditioned, conditioning		
		CR	Cow Run
condr	Conductor (pipe)	CR	Cane River
condt	Conductivity	cr (d), (g), (h)	Core, cored, coring, core hole
conf	Confirm, confirmed, confirming		
		CRA	Chemically retarded acid
confl	Conflict	crbd	Crossbedded
cong	Conglomerate(itic)	CRC	Coordinating Research Council, Inc.
conn	Connection		
cono	Conodonts	CR Con	Carbon Residue (Conradson)
consol	Consolidated	cren	Crenulated
const	Constant, construction	Cret	Cretaceous
consv	Conserve, conservation	Crin	Crinoid(al)
cont(d)	Continue, continued	Cris	Cristellaria
contam	Contaminated, contamination	crit	Critical
contr	Contractor	crkg	Cracking
contr resp	Contractor's responsibility	Crkr	Cracker
contrib	Contribution	cr moly	Chrome molybdenum
conv	Converse	crn blk	Crown block
Co. Op.	Company-operated	crnk	Crinkled
co-op	Cooperative	Crom	Cromwell
Co. Op. S.S.	Company-operated service stations	crs	Coarse(ly)
		crypto-xln	Cryptocrystalline
coord	Coordinate	cryst	Crystalline
COP	Crude oil purchasing	CS	Cast steel, carbon steel
coq	Coquina	CS	Casing seat
cor	Corner	cs	Centistokes
Corp	Corporation	CSA	Casing set at
corr	Correct (ed) (ion), corrosion, corrugated	cse gr	Coarse grained
		csg	Casing
correl	Correlation	csg hd	Casing head
corres	Correspondence	csg press	Casing pressure
CO & S	Clean out & shoot	csg pt	Casing point
COTD	Cleaned out to total depth	CSL	County school lands, center section line
Cott G	Cottage Grove		
Counc G	Council Grove	CT	Cable tools
CP	Casing point, casing pressure	CTC	Consumer tank car
CP	Chemically pure	ctd	Coated
C & P	Cellar & pits	CTD	Corrected total depth
cp	Centipoise	ctg(s)	Cuttings
CPA	Certified public accountant	CTHF	Center of tubing flange

Ctlmn	Cattleman
ctn	Carton
Ctnwd	Cottonwood
CTP	Cleaning to pits
ctr	Center
CTT	Consumer transport truck
ctw	Coated and wrapped
CTW	Consumer tank wagon
cu	Cubic
CU	Clean up
cu ft	Cubic foot
cu ft/bbl	Cubic feet per barrel
cu ft/min	Cubic feet per minute
cu ft/sec	Cubic feet per second
cu in	Cubic inch
culv	Culvert
cum	Cumulative
cu m	Cubic meter
Cur	Curtis
cush	Cushion
Cut B	Cut Bank
cutbk	Cutback
Cutl	Cutler
Cut Oil	Cutting oil
Cut Oil Act Sul-Dk	Cutting oil-active sulphurized-dark
Cut Oil Act Sul-Trans	Cutting oil-active-sulphurized transparent
Cut Oil Inact Sul	Cutting oil-inactive-sulphurized
Cut Oil Sol	Cutting oil soluble
Cut Oil St Mrl	Cutting oil-straight mineral
cu yd	Cubic yard
CV	Cotton Valley, control valve
cvg(s)	Caving(s)
CW	Continuous weld
cw	Clockwise
C/W	Complete with
C & W	Coat and wrap (pipe)
CWP	Cold working pressure
cwt	Hundred weight
CX	Crossover
Cyc	Cyclamina
Cycl canc	Cyclamina cancellata
cyl	Cylinder
cyl stk	Cylinder stock
cyp	Cypridopsis
Cy Sd	Cypress Sand
Cz	Carrizo

D

d-1-s	Dressed one side
d-2-s	Dressed two sides
d-4-s	Dressed four sides
D-2	Diesel No. 2
DA	Daily allowable
A & A	Dry and abandoned
DAIB	Daily average injection, barrels
Dak	Dakota
Dan	Dantzler
Dar	Darwin
DAR	Discovery allowable requested
dat	Datum
db	Decibel
DB	Drilling break
D & B	Dun & Bradstreet
d/b/a	Doing business as
DBO	Dark brown oil
DBOS	Dark brown oil stains
DC	Delayed coker, direct current, drill collar, dually complete(d), development well—carbon dioxide, diamond core
D & C	Drill and complete
DCB	Diamond core bit
DCLSP	Digging slush pits, digging cellar, or digging cellar and slush pits
DCM	Distillate-cut mud
DCS	Drill collars
D/D	Day to day
dd	Dead
DD	Degree day, drilling deeper
d-d-l-s-l-e	Dressed dimension one side and one edge
d-d-4-s	Dressed dimension four sides
D & D	Desk and Derrick
DDD	Dry desiccant dehydrator
DDT	Dichloro-diphenyl-trichloroethane
DE	Double end
Deadw	Deadwood
deaer	Deaerator
deasph	Deasphalting
debutzr	Debutanizer
dec	Decimal

decr	Decrease (d) (ing)		dim	Dimension, diminish, diminishing
deethzr	Deethanizer		Din	Dinwoody
defl	Deflection		dir	Direct, direction, director
deg	Degonia		dir sur	Directional survey
deisobut	Deisobutanizer		Disc	Discorbis
Dela	Delaware		disc	Discount, discover (y) (ed) (ing)
Del R	Del Rio			
delv	Delivery, delivered, deliverability		Disc grav	Discorbis gravelli
delv pt	Delivery point		disch	Discharge
demur	Demurrage		Disc norm	Discorbis normada
dend	Dendrite(ic)		Disc y	Discorbis yeguaensis
DENL	Density log		dism	Disseminated
dep	Depreciation		disman	Dismantle
depl	Depletion		displ	Displaced, displacement
deprec	Depreciation		dist	Distance, distillate, distillation, district
deprop	Depropanizer			
dept	Department		distr	Distribute (d) (ing) (ion)
desalt	Desalter		div	Division
desc	Description		dk	Dark
Des Crk	Desert Creek		DK Crk	Duck Creek
Des M	Des Moines		D/L	Density log
desorb	Desorbent		dlr	Dealer
desulf	Desulfurizer		DM	Datum, demand meter, dip meter, drilling mud
det	Detail(s), detector			
deterg	Detergent		dml	Demolition
detr	Detrital		dmpr	Damper
dev	Deviate, deviation		dn	Down
Dev	Devonian		dns	Dense
devel	Develop (ed) (ment)		DO	Drill (ed) (ing) out, development oil well
dewax	Dewaxing			
Dext	Dexter		D.O.	Division order
DF	Derrick floor, diesel fuel		D/O	Division Office
DFE	Derrick floor elevation		do	Ditto
DFO	Datum faulted out		DOC	Diesel oil cement, drilled out cement
DFP	Date of first production			
DG	Development gas well, draft gage, dry gas		Doc	Dockum
			doc	Document
DH	Development well—helium		doc-tr	Doctor-treating
DHC	Dry hole contribution		DOD	Drilled out depth
DHDD	Dry hole drilled deeper		dolo	Dolomite(ic)
DHM	Dry hole money		dolst	Dolstone
DHR	Dry hole reentered		dom	Domestic
dia	Diameter		dom AL	Domestic airline
diag	Diagram, diagonal		DOP	Drilled out plug
diaph	Diaphragm		Dorn H	Dornick Hills
dichlor	Dichloride		Doth	Dothan
diethy	Diethylene		Doug	Douglas
diff	Differential, difference, different		doz	Dozen
			DP	Data processing, dew point, drill pipe
dilut	Diluted			

DP	Double pole (switch)
D/P	Drill (ed) (ing) plug
DPDB	Double pole double base (switch)
DPDT	Double pole double throw (switch)
dpg	Deepening
DPM	Drill pipe measurement
dpn	Deepen
DPSB	Double pole single base (switch)
DPST	Double pole single throw (switch)
DPT	Deep pool test
dpt	Depth
dpt rec	Depth recorder
DPU	Drill pipe unloaded
dr	Drain, drive, drum, druse
DR	Development redrill (sidetrack)
Dr Crk	Dry Creek
drk	Derrick
DRL	Double random lenths
drl	Drill
drld	Drilled
drlg	Drilling
drlr	Driller
drng	Drainage
drpd	Dropped
drsy	Drusy
dry	Drier, drying
DS	Directional survey
DS	Drill stem
ds	Dense
dsgn	Design
DSI	Drilling suspended indefinitely
dsl	Diesel (oil)
dsmtl(g)	Dismantle(ing)
DSO	Dead oil show
DSS	Days since spudded
DST	Drill stem test
dstl	Distillate
dstn	Destination
DSU	Development well—sulphur
DT	Drilling time
D/T	Driller's tops
DTD	Drillers total depth
dtr	Detrital
DTW	Dealer tank wagon
Dup	Duperow

dup	Duplicate
Dutch	Dutcher
DV	Differential valve (cementing)
DWA	Drilling with air
dwg	Drawing
dwks	Drawworks
DWM	Drilling with mud
dwn	Down
DWO	Drilling with oil
DWP	Dual (double) wall packer
DWSW	Drilling with salt water
DWT	Dead weight tester
DX	Development well workover
dx	Duplex
dyn	Dynamic

E

E	East
E/2, E/4	East half, quarter, etc.
ea	Each
EAM	Electric accounting machines
Earls	Earlsboro
Eau Clr	Eau Claire
E/BL	East boundary line
ecc	Eccentric
Ech	Echinoid
ECM	East Cimarron Meridian (Oklahoma)
Econ	Economics, economy, economizer
Ect	Ector (county, Tex.)
Ed lm	Edwards lime
EDP	Electronic data processing
Educ	Education
Edw	Edwards
E/E	End to end
EF	Eagle Ford
eff	Effective, efficiency
effl	Effluent
EFV	Equilibrium flash vaporization
e.g.	For example
Egl	Eagle
Eglwd	Englewood
EHP	Effective horsepower
eject	Ejector
E/L	East Line
Elb	Elbert
elec	Electric(al)
elem	Element(ary)

elev	Elevation, elevator	euhed	Euhedral
Elg	Elgin	ev	Electron volts
el gr	Elevation ground	eval	Evaluate
ell(s)	Elbow(s)	evap	Evaporation, evaporite
Ellen	Ellenburger	ev-sort	Even-sorted
Elm	Elmont	EW	Electric weld, exploratory well
EL/T	Electric log tops		
EM	Eagle Mills	E of W/L	East of west line
Emb	Embar	Ex	Exter
EMS	Ellis-Madison contact	exc	Excavation
emer	Emergency	exch	Exchanger
EMF	Electromotive force	excl	Excellent
empl	Employee	exh	Exhaust, exhibit
emul	Emulsion	exist	Existing
encl	Enclosure	exp	Expansion, expense
End	Endicott	expl	Exploratory, exploration
endo	Endothyra	exp plg	Expendable plug
eng	Engine	explos	Explosive
engr(g)	Engineer(ing)	expir	Expiration, expire, expired, expiring
enl	Enlarged		
enml	Enamel	exr	Executor
Ent	Entrada	Exrx	Executrix
E/O	East offset	exst	Existing
EO	Emergency order	ext(n)	Extended, extension
Eoc	Eocene	ext	External
EOF	End of file	Ext M/H	Extension manhole
EOL	End of line	extr	Exterior
EOM	End of month	extrac	Extraction
EOQ	End of quarter	EYC	Estimated yearly consumption
EOR	East of Rockies		
EOY	End of year		
EP	End point, extreme pressure		
Epon	Eponides		
Ep y	Eponides yeguaensis		
eq	Equal, equalizer, equation		
equip	Equipment	°F.	Degree Fahrenheit
equiv	Equivalent	F/	Flowed, flowing
erect	Erection	fab	Fabricate(d)
Eric	Ericson	FAB	Faint air blow
ERW	Electric resistance weld	fac	Facet(ed)
est	Estate, estimate (d) (ing)	FACO	Field authorized to commence operations
ETA	Estimated time of arrival		
el al.	And others	fail	Failure
et con	And husband	Fall Riv	Fall River
eth	Ethane	Farm	Farmington
ethyle	Ethylene	FAO	Finish all over
et seq	And the following	FARO	Flowed(ing) at rate of
et ux	And wife	fau	Fauna
et vir	And husband	FB	Fresh break
Eu	Eutaw	FBH	Flowing by heads
EUE	External upset end	FBHP	Flowing bottom hole pressure

F

FBHPF	Final bottom hole pressure flowing	FIRC	Flow indicating ratio controller
FBHPSI	Final bottom hole pressure shut-in	fis	Fissure
		fish	Fishing
FBP	Final boiling point	fisl	Fissile
FC	Filter cake, float collar	FIT	Formation interval tester
FC	Fixed carbon	fix	Fixture
FCC	Fluid catalytic cracking	FJ	Flush joint
FCP	Flowing casing pressure	FL	Floor, fluid level, flow line, flush
FCV	Flow control valve		
FD	Feed, floor drain	fl/	Flowed or flowing
F-D	Formation density	fl	Fluid
F & D	Flanged and dished (heads), faced and drilled	F & L	Fuels & Lubricants
		FLA	Ferry Lake anhydrite
fdn	Foundation	flat	Flattened
FDL	Formation density log	Flath	Flathead
fdr	Feeder	Fl-COC	Flash Point, Cleveland Open Cup
fed	Federal		
FE/L	Form east line	fld	Failed, feldspar (thic), field
FELA	Federal Employers Liability Act	flex	Flexible
		flg (d) (s)	Flange (d) (s), flowing
Ferg	Ferguson	Flip	Flippen
ferr	Ferruginous	flk	Flaky
fert	Fertilizer	flo	Flow
Fe-st	Ironstone	Flor Fl	Florence Flint
FF	Flat face	flt	Float, fault
FF	Frac finder (Log), full of fluid, fishing for	fltg	Floating
		flu	Flue, fluid
F & F	Fuels & fractionation	fluor	Fluorescence, fluorescent
F to F	Face to face	flshd	Flushed
FFA	Female to female angle	flw (d) (g)	Flowed, flowing
FFG	Female to female globe (valve)	Flwg Pr.	Flowing pressure
		Flwrpt	Flowerpot
FFO	Furnace Fuel Oil	fm	Formation
FFP	Final flowing pressure	FM	Frequency meter, frequency modulation
f-gr	Fine-grained		
F.G.	Fracture gradient	f'man	Foreman
FGIH	Finish going in hole	Fm W	Formation water
FGIW	Finish going in with —	fn	Fine
F/GOR	Formation gas-oil ratio	FNEL	From northeast line
FH	Full hole	FNL	From north line
FHP	Final hydrostatic pressure	fnly	Finely
FI	Flow indicator	fnt	Faint
fib	Fibrous	FNWL	From northwest line
FIC	Flow indicating controller	FO	Farmout, fuel oil, full opening, faulted out
fig	Figure		
FIH	Fluid in hole	FOB	Free on board
fill	Fillister	FOCL	Focused log
filt	Filtrate	F.O.E.	Fuel oil equivalent
fin	Final, finish, finished	FOE-WOE	Flanged one end, welded one end
fin drig	Finished drilling		

FOH	Full open head (grease drum 120 lb)	FS	Feedstock, forged steel, float shoe
fol	Foliated	F/S	Front & side, flange x screwed
Forak	Foraker	F & S	Flanged and spigot
foram	Foraminifera	FSEL	From southeast line
Fort	Fortura	fsg	Fishing
foss	Fossiliferous	FSIP	Final shut-in pressure
FOT	Flowing on test	FSL	From south line
Fount	Fountain	FSP	Flowing surface pressure
Fox H	Fox Hills	FST	Forged steel
FP	Final pressure, flowing pressure, freezing point	FSWL	From southwest line
		FS&WLs	From south and west lines
FPI	Free point indicator	ft	Foot, feet
fpm	Feet per minute	Ft C	Fort Chadbourne
FPO	Field purchase order	ft-c	Foot-candle
fprf	Fireproof	ftg	Fittings, footing, footage
fps	Feet per second	Ft H	Fort Hayes
fps	Foot-pound-second (system)	ft/hr	Feet per hour
FPT	Female pipe thread	ft lb	Foot-pound
FQG	Frosted quartz grains	ft lbs/hr	Foot-pounds per hour
FR	Flow recorder, flow rate, feed rate	ft/min	Feet per minute
		FTP	Final tubing pressure, flowing tubing pressure
fr	Fair, fractional, frosted, from, front		
		Ft R	Fort Riley
FRA	Friction reducing agent	FTS	Fluid to surface
frac (d) (s)	Fracture, fractured, fractures	ft/sec	Feet per second
fract	Fractionation, fractionator	Ft U	Fort Union
frag	Fragment	Ft W	Fort Worth
fran	Franchise	FU	Fill up
Franc	Franconia	Full	Fullerton
FRC	Flow recorder control	furf	Furfural
Fred	Fredericksburg, Fredonia	furn	Furnace
fr E/L	From east line	Furn & fix	Furniture and fixtures
freg	Frequency	Fus	Fuson
Frgy	Froggy	Fussel	Fusselman
fri	Friable	Fusul	Fusulinid
fr N/L	From north line	fut	Future
F-R Oil	Fire-resistant oil	FV	Funnel viscosity
Fron	Frontier	fvst	Favosites
fros	Frosted	FW	Fresh water
FRP	Fiberglass reinforced plastic	FWC	Field wildcat
FRR	Final report for rig	fwd	Forward
fr S/L	From south line	FWD	Four-wheel drive
frs	Fresh	FWL	From west line
frt	Freight	fxd	Fixed
Fruit	Fruitland	f/xln	Finely-crystalline
FRW	Final report for well	FYI	For your information
frwk	Framework		
fr W/L	From west line		
frzr	Freezer		

G

G	Gas
g	Gram
GA	Gallons acid
ga	Gage (d) (ing)
GAF	Gross acre feet
gal	Gallon, gallons
Gall	Gallatin
gal/Mcf	Gallons per thousand cubic feet
gal/min	Gallons per minute
gal sol	Gallons of solution
galv	Galvanized
gaso	Gasoline
gast	Gastropod
GB	Gun barrel
GBDA	Gallons breakdown acid
GC	Gas-cut
g-cal	Gram-calorie
GCAW	Gas-cut acid water
GCD	Gas-cut distillate
GCLO	Gas-cut load oil
GCLW	Gas-cut load water
GCM	Gas-cut mud
GCO	Gas-cut oil
GCPD	Gallons condensate per day
GCPH	Gallons condensate per hour
GCR	Gas-condensate ratio
GCSW	Gas-cut salt water
GCW	Gas-cut water
gd	Good
GD	Glen Dean line
Gdld	Goodland
gd o&t	Good odor & taste
GDR	Gas-distillate ratio
Gdwn	Goodwin
GE	Grooved ends
G egg	Goose egg
gel	Jelly-like colloidal suspension
gen	Generator
genl	General
Geo	Georgetown
Geol	Geologist, geological, geology
GFLU	Good fluorescence
GGW	Gallons gelled water
GHO	Gallons heavy oil
Geop	Geophysics, geophysical
GH	Greenhorn
GI	Gas injection
Gib	Gibson
GIH	Going in hole
gil	Gilsonite
Gilc	Gilcrease
GIW	Gas-injection well
GJ	Ground joint
GL	Gas lift, ground level
G/L	Gathering line
gl	Glassy
glau	Glauconitic, glauconite
Glen	Glenwood
Glna	Galena
GLO	General Land Office (Texas)
Glob	Globigerina
Glor	Glorieta
GLR	Gas-liquid ratio
gls	Glass
glyc	Glycol
gm	Gram
GM	Ground measurement (elevation)
G.M.	Gravity meter
GMA	Gallons mud acid
gm-cal	Gram calorie
G&MCO	Gas & mud-oil cut
g mole	Gram molecular weight
gmy	Gummy
gnd	Grained (as in fine-grained)
gns	Gneiss
GO	Gas odor, gallons oil
G&O	Gas and oil
GOC	Gas-oil contract
G&OCM	Gas and oil-cut mud
GODT	Gas odor distillate taste
Gol	Golconda lime
Good L	Goodland
GOPH	Gallons of oil per hour
GOPD	Gallons of oil per day
GOR	Gas-oil ratio
Gor	Gorham
Gouldb	Gouldbusk
gov	Governor
govt	Government
GP	Gas pay, gasoline plant
G/P	Gun perforate
GPC	Gas purchase contract
GPD	Gallons per day
GPG	Grains per gallon
GPH	Gallons per hour
GPM	Gallons per minute

gpm	Gallons per thousand cubic feet
GPS	Gallons per second
GR	Gamma ray, Glen Rose
gr	Ground, grade, grain, grease
GRA	Gallons regular acid
grad	Gradual, gradually
gran	Granite, granular
Granos	Graneros
Gran W	Granite Wash
grap	Graptolite
gr API	Gravity, °API
grav	Gravity
Gray	Grayson
Grayb	Grayburg
grd	Ground
grdg	Grading
GRDL	Guard log
grd loc	Grading location
G. Riv	Gull River
G Rk	Gas Rock
Grn	Green
grnlr	Granular
Grn Riv	Green River
Grn sh	Green shale
gr roy	Gross royalty
grs	Gross
Gr Sd	Gray sand
grt	Grant (of land)
grtg	Grating
grty	Gritty
grv	Grooved
grvt	Gravitometer
gr wt	Gross weight
gry	Gray
GS	Gas show
GSC	Gas sales contract
GSG	Good show of gas
GSI	Gas well shut-in
gskt	Gasket
GSO	Good show of oil
GSW	Gallons salt water
gsy	Greasy
GTS	Gas to surface (time)
GTSTM	Gas too small to measure
GTY	Gravity
GU	Gas Unit
Guns	Gunsite
GV	Gas volume
gvl	Gravel
GVLPK	Gravel packed

GVNM	Gas volume not measured
GW	Gallons water, gas well
GWC	Gas-water contact
GWG	Gas-well gas
GWPH	Gallons of water per hour
gyp	Gypsum
Gyp Sprgs	Gypsum Springs
gypy	Gypsiferous
Gyr	Gyroidina
Gyr sc	Gyroidina scal
gywk	Graywacke

H

Hackb	Hackberry
Hara	Haragan
Hask	Haskell
Haynes	Haynesville
haz	Hazardous
HB	Housebrand (regular grade of gasoline)
Hberg	Hardinsburg sand (local)
HBP	Held by production
hbr	Harbor
HC	Hydrocarbon
HCO	Heavy cycle oil
HCV	Hand-control valve
HD	High detergent, heavy duty, Hydril
hd	Hard, head
hd li	Hard lime
hdns	Hardness
hdr	Header
hd sd	Hard sand
hdl	Handle
hdwe	Hardware
Heeb	Heebner
hem	Hematite
Her	Herington
Herm	Hermosa
het	Heterostegina
HEX	Heat exchanger
hex	Hexagon(al), hexane
hfg	Hydrofining
HFO	Heavy fuel oil, hole full of oil
HF Sul W	Hole full of sulphur water
HFSW	Hole full of salt water
HFW	Hole full of water
HGCM	Heavily gas-cut mud
HGCW	Heavily gas-cut water

HGOR	High gas-oil ratio
hgr	Hanger
hgt	Height
HH	Hand hole, hydrostatic head
H H P	Hydraulic horsepower
Hick	Hickory
Hill	Hilliard
hily	Highly
hky	Hackly
HLSD	High-level shut-down
HO	Heavy oil, heating oil
hock	Hockleyensis
HOCM	Heavily oil-cut mud
HOCW	Heavily oil-cut water
Hog	Hogshooter
HO&GCM	Heavily oil-and-gas-cut mud
Holl	Hollandberg
Home Cr	Home Creek
hop	Hopper
horiz	Horizontal
Hosp	Hospah
Hov	Hoover
Hox	Hoxbar
HP	High pressure, horsepower, hydraulic pump, hydrostatic pressure
HPF	Holes per foot
HPG	High-pressure gas
hp hr	Horsepower hour
HQ	Headquarters
HRS	Hot-rolled steel
hr	Hour, hours
HRD	High-resolution dipmeter
hrs	Heirs
HSD	Heavy steel drum
ht	Heater treater, heat treated, high temperature, high tension
htr	Heater
HTSD	High-temperature shut-down
Humb	Humblei
Hump	Humphreys
Hun	Hunton
HV	High viscosity
HVI	High viscosity index
hvly	Heavily
hvy	Heavy
HWCM	Heavily water-cut mud
HWP	Hookwall packer
hwy	Highway
HX	Heat exchanger

HYD	Hydril thread, hydraulic
HYDA	Hydril Type A joint
HYDCA	Hydril Type CA joint
HYDCS	Hydril Type CS joint
hydtr	Hydrotreater
Hyg	Hygiene
Hz	Hertz (new name for electrical cycles per second)

I

IAB	Initial air blow
IB	Impression block
IB	Iron body (valve)
IBBC	Iron body, brass core (valve)
IBBM	Iron body, brass (bronze) mounted (valve)
IBHP	Initial bottom-hole pressure
IBHPF	Initial bottom-hole pressure flowing
IBHPSI	Initial bottom-hole pressure shut-in
IBP	Initial boiling point
IC	Iron case
ID	Inside diameter
I.D. Sign	Identification sign
Idio	Idiomorpha
IF	Internal flush
IFP	Initial flowing pressure
Ign	Igneous
IGOR	Injection gas-oil ratio
IHP	Initial hydrostatic pressure
IHP	Indicated horsepower
IHPHR	Indicated horsepower hour
IJ	Integral joint
imbd	Imbedded
immed	Immediate(ly)
Imp	Imperial
imperv	Impervious
Imp gal	Imperial gallon
IMW	Initial mud weight
in	Inch(es)
inbd	Interbedded
inbdd	Inbedded
Inc	Incorporated
incd	Incandescent
incin	Incinerator
incl	Include, included, including
incls	Inclusions

incolr	Intercooler	IPF	Initial potential flowed, initial production flowed(ing)
incr	Increase (d) (ing)		
ind	Induction	IPG	Initial production gas lift
indic	Indicate, indication, indicates	IPI	Initial production on intermitter
indiv	Individual		
indr	Indurated	IPL	Initial production plunger lift
indst	Indistinct	IPP	Initial production pumping
Inf. L	Inflammable liquid	IPS	Initial production swabbing
info	Information	IPS	Iron pipe size
Inf. S	Inflammable solid	IPT	Iron pipe thread
ingr	Intergranular	IR	Injection rate
in. Hg	Inches mercury	Ire	Ireton
inhib	Inhibitor	irreg	Irregular
init	Initial	irid	Iridescent
inj	Injection, injected	IRS	Internal Revenue Service
Inj Pr	Injection pressure	irst	Ironstone
inl	Inland, inlet	IS	Inside screw (valve)
inlam	Interlaminated	ISIP	Initial shut-in pressure (DST), instantaneous shut-in pressure (frac)
in-lb	Inch-pound		
Inoc	Inoceramus		
INPE	Installing, installed, pumping equipment	isom	Isometric
		isoth	Isothermal
ins	Insurance, insulate, insulation	ITD	Intention to drill
in/sec	Inches per second	IUE	Internal upset ends
insp	Inspect, inspected, inspecting, inspection	Ives	Iverson
		IVP	Initial vapor pressure
inst	Install (ed) (ing), instantaneous, institute	IW	Injection well

J

instl	Installation(s)		
instr	Instrument, instrumentation		
insul	Insulate		
int	Interest, interval, internal, interior	J&A	Junked and abandoned
		jac	Jacket
interbd	Interbedded	Jack	Jackson
inter-gran	Intergranular	Jasp	Jasper(oid)
inter-lam	Interlaminated	Jax	Jackson sand
inter-xln	Inter-crystalline	JB	Junk basket, junction box
intgr	Integrator	jbr	Jobber
intl	Interstitial	JC	Job complete
intr	Instrusion	jct	Junction
ints	Intersect	jdn	Jordan
intv	Interval	Jeff	Jefferson
inv	Invert, inverted, invoice	JINO	Joint interest non-operated (property)
inven	Inventory		
invrtb	Invertebrate	jmd	Jammed
I/O	Input/output	jnk	Junk(ed)
IP	Initial potential, initial production, initial pressure	J/O	Joint operation
		JOA	Joint operating agreement
IPA	Isopropyl alcohol	JOP	Joint operating provisions
IPE	Install(ing) pumping equipment	JP	Jet perforated
		JP/ft	Jet perforations per foot

JP fuel	Jet propulsion fuel
JSPF	Jet shots per foot
jt(s)	Joint(s)
Jud R	Judith River
Jur	Jurassic
juris	Jurisdiction
JV	Joint venture
Jxn	Jackson

K

K	Kelvin (temperature scale)
Kai	Kaibab
kao	Kaolin
Kay	Kayenta
KB	Kelly bushing
KBM	Kelly bushing measurement
KC	Kansas City
kc	Kilocycle
kcal	Kilocalorie
KD	Kiln dried, kincald lime
KDB	Kelly drive bushing
KDBE	Kelly drive bushing elevation
KDB-LDG FLG	Kelly drill bushing to landing flange
KDB-MLW	Kelly drill bushing to mean low water
KDB-Plat	Kelly drill bushing platform
Ke	Keener
Keo-Bur	Keokuk-Burlington
kero	Kerosine
ket	Ketone
Key	Keystone
kg	Kilogram
kg-cal	Kilgram calorie
kg-m	Kilogram-meter
Khk	Kinderhood
KHz	Kilohertz (see Hz—Hertz)
Kia	Kiamichi
Kib	Kibbey
Kin	Kinematic
Kin	Kincaid lime
kip	One thousand pounds
kip-ft	One thousand foot-pounds
Kirt	Kirtland
kld	Killed
km	Kilometer
KMA	KMA sand
KO	Kicked off, knock out
Koot	Kootenai

KOP	Kickoff point
Kri	Krider
KV	Kinematic viscosity
kv	Kilovolt
kva	Kilovolt-ampere
kvah	Kilovolt-ampere-hour
kvar	Kilovar; reactive kilovolt-ampere
kvar hr	Kilovar-hour
kw	Kilowatt
KW	Kill (ed) well
kwh	Kilowatt hour
kwhm	Kilowatt-hourmeter

L

l	Liter
L/	Lower, as L/Gallup
/L	Line, as in E/L (East line)
LA	Level alarm, lightening, avvester, load acid
Lab	Labor, laboratory
LACT	Lease automatic custody transfer
lad	Ladder
Lak	Lakota
L/Alb	Lower Albany
lam	Laminated, lamination(s)
La Mte	La Motte
Land	Landulina
Lans	Lansing
Lar	Laramie
LAS	Lower anhydrite stringer
lat	Latitude
Laud	Lauders
Layt	Layton
lb	Pound
LB	Light barrel
lb/ft	Pound per foot
lb-in	Pound-inch
LBOS	Light brown oil stain
lbr	Lumber
lb/sq ft	Pounds per square foot
LC	Lost circulation, long coupling, level controller, lease crude
LC	Lug cover type (5-gallon can)
lchd	Leached
LCL	Less-than-carload lot
LCM	Lost circulation material

L/Cret	Lower Cretaceous	lns	Lense
LCP	Lug cover with pour spout	LO	Load oil, lube oil
LCV	Level control valve	loc	Located, location
LD	Laid down	loc abnd	Location abandoned
ld(s)	Land(s), load	loc gr	Location graded
LDC	Laid down cost	long	Longitude (inal)
LDDCs	Laid (laying) down drill collars	Lov	Lovington, Lovell
LDDP	Laid (laying) down drill pipe	low	Lower
Leadv	Leadville	lp	Loop
Le C	Le Comptom	LP	Low pressure, Lodge pole
LEL	Lower explosive limit	L.P.	Line pipe
Len	Lennep	LP-Gas	Liquefied petroleum gas
len	Lenticular	LPO	Local purchase order
LFO	Light fuel oil	LP sep	Low-pressure separator
lg	Large, length, long, level glass	LR	Level recorder, long radius
		LRC	Level recorder controller
LGD	Lower Glen Dean	lrg	Large
Lg Disc	Large Discorbis	ls	Limestone
Lge	League	LSD	Legal subdivision (Canada), light steel drum
LH	Left hand		
LH/RP	Long handle/round point	lse	Lease
LI	Level indicator	lstr	Lustre
li	Lime, limestone	lt	Light
LIB	Light iron barrel	LT&C	Long threads & coupling
lic	License	LTD	Log total depth
LIC	Level indicator controller	ltd	Limited
Lieb	Liebuscella	ltg	Lighting
lig	Lignite, lignitic	LTL	Less than truck load
LIGB	Light iron grease barrel	ltl	Little
LIH	Left in hole	ltr	Letter
lim	Limit, limonite	LTS unit	Low-temperature separation unit
lin	Linear, liner		
lin ft	Linear foot	LTSD	Low-temperature shut-down
liq	Liquid	LTX unit	Low-temperature extraction unit
liqftn	Liquefaction		
litho	Lithographic	L/Tus	Lower Tuscaloosa
LJ	Lap joint	L U	Lease use (gas)
lk	Leak, lock	lub	Lubricant, lubricate (d) (ing) (ion)
LKR	Locker		
LLC	Liquid level controller	lued	Lueders
LLG	Liquid level gauge	LV	Liquid volume
lm	Lime	LVI	Low viscosity index
LMn	Lower Menard	lvl	Level
Lmpy	Lumpy	Lvnwth	Leavenworth
LMTD	Log mean temperature difference	lwr	Lower
		LW	Load water, lapweld
lmy	Limy		
Lmy sh	Limy shale		
LNG	Liquified natural gas		# M
lngl	Linguloid	M/	Middle
lnr	Liner	m	Meter

MA	Massive Anhydrite, mud acid	McL	McLish
ma	Milliampere	MC Ls	Moore County Lime
MAC	Medium amber cut	McMill	McMillan
mach	Machine	MCO	Mud-cut oil
Mack	Mackhank	mcr-x	Micro-crystalline
Mad	Madison	MCSW	Mud-cut salt water
Mag	Magnetic, magnetometer	MCW	Mud-cut water
maint	Maintenance	MD	Measured depth
maj	Major, majority	md	Millidarcies
mall	Malleable	MDDO	Maximum daily delivery obligation
man	Manual, manifold		
Manit	Manitoban	MDF	Market demand factor
Mann	Manning	mdl	Middle
man op	Manually operated	mdse	Merchandise
Maq	Maquoketa	md wt	Mud weight
mar	Maroon, marine	Mdy	Muddy
March	Marchand	Meak	Meakin
marg	Marginal	meas	Measure (ed) (ment)
Marg	Marginulina	mech	Mechanic (al), mechanism
margas	Marine gasoline	Mech DT	Mechanical down time
Marg coco	Marginulina coco	Med	Median, medium
Marg fl	Marginulina flat	Med	Medina
Marg rd	Marginulina round	Med B	Medicine Bow
Marg tex	Marginulina texana	med FO	Medium fuel oil
Mark	Markham	med gr	Medium-grained
Marm	Marmaton	Medr	Medrano
mass	Massive	Meet	Meeteetse
Mass pr	Massilina pratti	MEG	Methane-rich gas
mat	Matter	MEK	Methyl ethyl keton
math	Mathematics	memo	Memorandum
matl	Material	Men	Menard lime
MAW	Mud acid wash	Mene	Menefee
max	Maximum	MEP	Mean effective pressure
May	Maywood	MER	Maximum efficient rate
MB	Moody's Branch	Mer	Meramec
MB	Methylene blue	merc	Mercury
Mbl Fls	Marble Falls	mercap	Mercaptan
mbr	Member (geologic)	merid	Meridian
MBTU	Thousand British thermal units	Meso	Mesozoic
		meta	Metamorphic
MC	Mud cut	meth	Methane
mc	Megacycle	meth-bl	Methylene blue
MCA	Mud cleanout agent, mud-cut acid	meth-cl	Methyl chloride
		methol	Methanol
McC	McClosky lime	metr	Metric
McCul	McCullough	mev	Million electron volts
MCF	Thousand cubic feet	mezz	Mezzanine
McEl	McElroy	MF	Manifold, mud filtrate
MCFD	Thousand cubic feet per day	M&F	Male and female (joint)
mchsm	Mechanism	MFA	Male to female angle
McK	McKee	mfd	Microfarad, manufactured

mfg	Manufacturing
MFP	Maximum flowing pressure
M&FP	Maximum & final pressure
MG	Multi-grade, motor generator
mg	Milligram
m-gr	Medium-grained
m'gmt	Management
mgr	Manager
MH	Manhole
mh	Millihenry
MHz	Megahertz (megacycles per second)
MI	Malleable iron, mile(s), mineral interest, moving in (equipment)
mic	Mica, micaceous
micfos	Microfossil(iferous)
micro-xin	Microcrystalline
MICT	Moving in cable tools
MICU	Moving in completion unit
Mid	Midway
mid	Middle
MIDDU	Moving (moved) in double drum unit
MIK	Methyl isobutyl ketone
mil	Military, million
mill	Milliolitic
millg	Milling
MIM	Moving in materials
min	Minimum, minute(s), minerals
Minl	Minnelusa
min P	Minimum pressure
Mio	Miocene
MIPU	Moving in pulling unit
MIR	Moving in rig
MIRT	Moving in rotary tools
MIRU	Moving in and rigging up
misc	Miscellaneous
Mise	Misener
MISR	Moving in service rig
Miss	Mississippian
Miss Cany	Mission Canyon
MIST	Moving in standard tools
MIT	Moving in tools
MIU	Moisture, impurities and unsaponifiables (grease testing)
mix	Mixer
mkt	Market(ing)
Mkta	Minnekahta
mky	Milky

ML	Mud logger
ml	Milliliter
m/l	More or less
mld	Milled
mlg	Milling
ml TEL	Milliliters tetraethyl lead per gallon
MLU	Mud logging unit
MLW-PLAT	Mean low water to platform
mly	Marly
MM	Motor medium
mm	Millimeter
MMBTU	Million British thermal units
MMCF	Million cubic feet
MMCFD	Million cubic feet/day
mm Hg	Millimeters of mercury
MMSCFD	Million standard cubic feet per day
mnrl	Mineral
MO	Moving out, motor oil
mob	Mobile
MOCT	Moving out cable tools
MOCU	Moving out completion unit
mod	Moderate(ly), model, modification
modu	Modular
MOE	Milled other end
Moen	Moenkopi
mol	Molas, mollusca, mole
mol wt	Molecular weight
mon	Monitor
MON	Motor octane number
Mont	Montoya
Moor	Mooringsport
MOP	Maximum operating pressure
MOR	Moving out rig
Mor	Morrow
Morr	Morrison
MORT	Moving out rotary tools
Mos	Mosby
mot	Motor
mott	Mottled
MOU	Motor oil units
mov	Moving
Mow	Mowry
MP	Maximum pressure, melting point, multipurpose
MPB	Metal petal basket
MPGR-Lith	Multipurpose grease, lithium base

MPGR-soap	Multipurpose grease, soap base
MPH	Miles per hour
MPT	Male pipe thread
MR	Marine rig, meter run
mrlst	Marlstone
M & R Sta.	Measuring and regulating station
MS	Motor severe
MSA	Multiple service acid
MSP	Maximum surface pressure
mstr	Master
MT	Empty container, macaroni tubing
M/T	Marine terminal
MTD	Measured total depth, mean temperature difference
mtd	Mounted
mtg	Mounting
mtge	Mortgage
mtl	Material
MTP	Maximum top pressure, maximum tubing pressure
mtr	Meter
MTS	Mud to Surface
Mt. Selm	Mount Selman
M. Tus	Marine Tuscaloosa
mtx	Matrix
mudst	Mudstone
mud wt	Mud Weight
musc	Muscovite
mv	Millivolt
M/V	Motor vehicle, motor vessel
Mvde	Mesaverde
MVFT	Motor vehicle fuel tax
MW	Muddy water, microwave
MWD	Marine wholesale distributors
MWP	Maximum working pressure
MWPE	Mill wrapped plain end
Mwy	Midway
mxd	Mixed

N

N/2	North half
N/4	North quarter
NA	Not applicable, not available
Nac	Nacatoch
nac	Nacreous

NAG	No appreciable gas
NALRD	Northern Alberta land registration district
nap	Naphtha
nat	Natural
Nat'l	National
Nav	Navajo
Navr	Navarro
NB	Nitrogen blanket, new bit
Nbg	Newburg
NC	No core, National coarse thread, no change, normally closed
N. Cock	Nonionella Cockfieldensis
NCT	Non-contiguous tract
ND	Non-detergent, not drilling
NDBOPs	Nipple (d) (ing) down blowout preventers
Ndl Cr	Noodle Creek
NDT	Non-destructive testing
NE	Northeast, nonemulsifying agent
NE/4	Northeast quarter
NEA	Non-emulsion acid
NEC	Northeast corner
NEC	National Electric Code
neg	Negative, negligible
NEL	Northeast line
NEP	Net effective pay
neut	Neutral, neutralization
Neut. No.	Neutralization Number
New Alb	New Albany shale
Newc	Newcastle
NF	National Fine (thread), natural flow, no fluorescence, no fluid, no fuel
NFD	New field discovery
NFW	New field wildcat
NG	Natural gas, no gauge
NGL	Ntural gas liquids
NGTS	No gas to surface
NIC	Not in contract
Nig	Niagara
Nine	Ninnescah
Niob	Niobrara
nip	Nipple
nitro	Nitroglycerine
NL	North line
NL Gas	Non-leaded gas
N'ly	Northerly
NMI	Nautical mile

NO	New oil, normally open, number
NO	Noble-Olson
N/O	North offset
nod	Nodule, nodular
Nod Blan	Nodosaria blanpiedi
Nod mex	Nodosaria Mexicana
No Inc	No increase
NOJV	Non-operated joint ventures
nom	Nominal
Non	Nonionella
nonf G	Nonflammable compressed gas
NOP	Non-operating property
nor	Normal
NOR	No order required
no rec	No recovery
noz	Nozzle
NP	Nameplate, notary public, nickle plated, not prorated, no production, not pumping, non-porous
NPD	New pool discovery
npne	Neoprene
NPOS	No paint on seams
NPS	Nominal pipe size
NPT	National pipe thread
NPTF	National pipe thread, female
NPTM	National pipe thread, male
NPW	New pool wildcat
NPX	New pool exempt (nonprorated)
NR	No report, not reported, no recovery, non-returnable, no returns
NRS	Non-rising stem (valve)
NRSB	Non-returnable steel barrel
NRSD	Non-returnable steel drum
NS	No show
NSG	No show gas
NSO	No show oil
NSO&G	No show oil & gas
nstd	Non-standard
N/S S/S	Non-standard service station
NT	Net tons, no time
NTD	New total depth
NTS	Not to Scale
N/tst	No test
NU	Non-upset, nippling up
NUBOPs	Nipple (d) (ing) up blowout preventers

NUE	Non-upset ends
Nug	Nugget
num	Numerous
NVP	No visible porosity
NW	Northwest, no water
NW/C	Northwest corner
NW/4	Northwest quarter
NWL	Northwest line
NWT	Northwest Territories
NYA	Not yet available

O

O	Oil
OA	Overall
OAH	Overall height
Oakv	Oakville
OAL	Overall length
OAW	Old abandoned well
OB	Off bottom
obj	Object
OBM	Oil base mud
OBMO	Outboard motor oil
OBS	Ocean bottom suspension
obsol	Obsolete
OBW & RS	Optimum bit weight and rotary speed
OC	Oil cut, on center, open cup, operations commenced
O/C	Oil change
OCB	Oil circuit breaker
occ	Occasional(ly)
OCM	Oil-cut mud
OCS	Outer continental shelf
OCSW	Oil-cut salt water
oct	Octagon, octagonal, octane
OCW	Oil-cut water
OD	Outside diameter
od	Odor
Odel	O'Dell
OE	Oil emulsion, open end
OEB	Other end beveled
OEM	Oil emulsion mud
OF	Open flow
off	Office, official
off-sh	Off-shore
OFL	Overflush(ed)
OFLU	Oil fluorescence
OFOE	Orifice flange one end
OFP	Open flow potential

O&G	Oil and gas
O&GCM	Oil and gas-cut mud
O&GC SULW	Oil & gas-cut sulphur water
O&GCSW	Oil and gas-cut salt water
O&GCW	Oil and gas-cut water
OGJ	Oil and Gas Journal
O&GL	Oil and gas lease
OH	Open hearth, open hole, over head
O'H	O'Hara
ohm cm	Ohm-centimeter
ohm-m	Ohm-meter
OIH	Oil in hole
Oil Cr	Oil Creek
OIP	Oil in place
ole	Olefin
Olig	Oligocene
ONR	Octane number requirement
ONRI	Octane number requirement increase
OO	Oil odor
ooc	Oolicastic
ool	Oolitic
oom	Oolimoldic
OP	Oil pay, over produced, out post
OPBD	Old plug-back depth
oper	Operate, operations, operator
Operc	Operculinoides
opn	Open (ing) (ed)
opp	Opposite
OPI	Oil payment interest
OPT	Official potential test
optn to F/O	Option to farmout
Or	Oread
Ord	Ordovician
orf	Orifice
org	Organic, organization
orig	Original, originally
Orisk	Oriskany
ORR	Overriding royalty
ORRI	Overriding royalty interest
orth	Orthoclase
Os	Osage
OS	Oil show, overshot
O/S	Out of service, over and short (report), out of stock
OSA	Oil soluble acid
Osb	Osborne
O sd	Oil sand
OSF	Oil string flange

O S & F	Odor, stain & fluorescence
OSI	Oil well shut in
Ost	Ostracod
OSTN	Oil stain
OSTOIP	Original stock tank oil in place
Osw	Oswego
O&SW	Oil and salt water
O&SWCM	Oil & sulphur water-cut mud
OS&Y	Outside screw and yoke (valve)
OT	Open tubing, overtime
OTD	Old total depth
OTE	Oil-powered total energy
otl	Outlet
OTS	Oil to surface
O T & S	Odor, taste & stain
O, T, S & F	Odor, taste, stain & fluorescence
OU	Oil unit
Our	Ouray
ovhd	Overhead
O & W	Oil and water
OWC	Oil-water contact
OWDD	Old well drilled deeper
OWF	Oil well flowing
OWG	Oil-well gas
OWPB	Old well plugged back
OWWO	Old well worked over
ox	Oxidized, oxidation
oxy	Oxygen
oz	Ounce

P

P & A	Plugged & abandoned
PA	Pooling agreement, pressure alarm
PAB	Per acre bonus
Padd	Paddock
Paha	Pahasapa
Pal	Paluxy
Paleo	Paleozoic, paleontology
Palo P	Palo Pinto
Pan L	Panhandle Lime
PAR	Per acre rental
Para	Paradox
Park C	Park City
pat	Patent(ed)
patn	Pattern

pav	Paving	pers	Personnel
Paw	Pawhuska	pet	Petroleum
payt	Payment	Pet	Pettet
PB	Plugged Back	petrf	Petroliferous
PBD	Plugged back depth	petrochem	Petrochemical
PBHL	Proposed bottom hole location	Pet sd	Pettus sd
		Pett	Pettit
pbl	Pebble	PEW	Pipe, electric weld
pbly	Pebbly	PF	Power factor
PBP	Pulled big pipe	P&F	Pump and flow
PBTD	Plugged back total depth	pfd	Preferred
PBW	Pipe, buttweld	PFM	Power factor meter
PBX	Private branch exchange	PFT	Pumping for test
PC	Paint Creek, poker chipped, Porter Creek	PG	Pecan Gap
		PGC	Pecan Gap Chalk
P&C	Personal and confidential	PGW	Producing gas well
pc	Piece	ph	Phase
pcs	Pieces	Ph	Parish
pct	Percent	Phos	Phosphoria
PCV	Pressure control valve, positive crankcase ventilation	PI	Penetration index, Pine Island, pressure indicator, productivity index
PD	Per day, proposed depth, pressed distillate, paid		
		PIC	Pressure indicator controller
PDC	Pressure differential controller	Pic Cl	Pictured Cliff
		pinpt	Pinpoint
PDET	Production department exploratory test	PIP	Pump-in pressure
		piso	Pisolites, pisolitic
PDI	Pressure differential indicator	pit	Pitted
PDIC	Pressure differential indicator controller	PJ	Pump jack, pump job
		pk	Pink
PDR	Pressure differential recorder	pkd	Packed
PDRC	Pressure differential recorder controller	pkg	Packing, package
		pkgd	Packaged
pdso	Pseudo	pkr	Packer
PE	Plain end, pumping equipment	PL	Pipeline, property line
		P & L	Profit & loss
PEB	Plain end beveled	plag	Plagioclase
pebs	Pebbles	Plan. harangensis	Planulina harangensis
pell	Pelletal, pelletoidal		
pen	Penetration, penetration test	Plan. palm.	Planulina palmarie
Pen A.C.	Penetration asphalt cement	P Lar	Post Laramie
penal	Penalize, penalized, penalizing, penalty	plas	Plastic
		platf	Platform
Penn	Pennsylvanian	platfr	Platformer
perco	Percolation	plcy	Pelecypod
perf	Perforate (d) (ing) (or)	pld	Pulled
perf csg	Perforated casing	PLE	Plain large end
perm	Permeable, permeability, permanent	Pleist	Pleistocene
		pl fos	Plant fossils
Perm	Permian	plg	Pulling
perp	Perpendicular	plgd	Plugged

Plio	Pliocene	P&P	Porosity and permeability, porous and permeable
pln	Plan		
plngr	Plunger	ppd	Prepaid
PLO	Pipeline, oil, pumping load oil	ppg	Pounds per gallon
plt	Plant	PPI	Production payment interest
PLT	Pipeline terminal	ppm	Parts per million
plt	Pilot	PPP	Pinpoint porosity
plty	Platy	ppt	Precipitate
PLW	Pipe, lapweld	pptn No	Precipitation Number
PM	Pensky Martins	PR	Polished rod, public relations, pressure recorder
pmp (d)(g)	Pump, pumped, pumping		
PN	Performance Number (Av gas)	pr	Pair
		PRC	Pressure recorder control
pneu	Pneumatic	prcst	Precast
P & NG	Petroleum & natural gas	prd	Period
pnl	Panel	Pre Camb	Pre-Cambrian
PNR	Please note and return	predom	Predominant
P. O.	Post Oak, Pin Oak	prefab	Prefabricated
PO	Pumps off, purchase order, pulled out	prehtr	Preheater
		prelim	Preliminary
po	Phrrhotite	prem	Premium
POB	Plug on bottom, pump on beam	Prep	Prepare, preparing, preparation
Pod.	Podbielniak	press	Pressure
POE	Plain one end	prest	Prestressed
POGW	Producing oil & gas well	prev	Prevent, preventive
POH	Pulled out of hole	PRF	Primary Reference Fuel
pois	Poison	pri	Primary
pol	Polish(ed)	prin	Principal
poly	Polymerization, polymerized	pris	Prism(atic)
poly cl	Polyvinyl chloride	priv	Privilege
polyel	Polyethylene	prly	Pearly
polygas	Polymerized gasoline	prmt	Permit
polypl	Polypropylene	prncpl lss	Principal lessee(s)
PONA	Paraffins, olefins, naphthenes, aromatics	pro	Prorated
		prob	Probable(ly)
Pont	Pontotoc	proc	Process
POP	Putting on pump	prod	Produce (d) (ing) (tion), product (s)
por	Porosity, porous		
porc	Porcelaneous, Porcion	prog	Progress
port	Portable	proj	Project (ed) (ion)
pos	Position, positive	prop	Proportional, propose(d)
poss	Possible(ly)	prot	Protection
pot	Potential	Protero	Proterozoic
pot dif	Potential difference	Prov	Provincial
pour ASTM	Pour point (ASTM Method)	PRPT	Preparing to take potential test
POW	Producing oil well		
POWF	Producing oil well flowing	PR&T	Pull(ed) rods & tubing
POWP	Producing oil well pumping	prtgs	Partings
PP	Pinpoint, production payment, pulled pipe	partly	Partly
		PS	Pressure switch

ps	Pseudo
PSA	Packer set at
PSD	Permanently shut down
PSE	Plain small end
psf	Pounds per square foot
psi	Pounds per square inch
PSI	Profit sharing interest
psia	Pounds per square inch absolute
psig	Pounds per square inch gauge
PSL	Public School Land
PSM	Pipe, seamless
PSW	Pipe, spiral weld
PT	Potential test
pt	Part, partly, point, pint
PTG	Pulling tubing
Pk Lkt	Point Lookout
PTR	Pulling tubing and rods
PTS pot.	Pipe to soil potential
PTTF	Potential test to follow
PU	Picked up, pulled up, pumping unit
purp	Purple
PV	Plastic viscosity, pore volume
PVC	Polyvinyl chloride
pvmnt	Pavement
PVR	Plant volume reduction
PVT	Pressure-volume-temperature
PWR	Power
Pxy	Paluxy
pyls	Pyrolysis
pyr	Pyrite, pyritic
pyrbit	Pyrobitumen
pyrclas	Pyroclastic

Q

Q. City	Queen City
QDA	Quantity discount allowance
qnch	Quench
QRC	Quick ram change
qry	Quarry
Q. Sd	Queen Sand
qt	Quart(s)
qtr	Quarter
qtz	Quartz, quartzite, quartzitic
qtzose	Quartzose
qty	Quantity
quad	Quadrant, quadrangle, quadruple

qual	Quality
quan	Quantity
quest	Questionable
quint	Quintuplicate

R

R	Range, Rankine (temp. scale)
RA	Right angel, radioactive
R/A	Regular acid
rad	Radial, radian, radiological, radius
radtn	Radiation
RALOG	Running radioactive log
Rang	Ranger
RB	Rotary bushing, rock bit
Rbls	Rubber balls
RBM	Rotary bushing measurement
RBP	Retrievable bridge plug
rbr	Rubber
RBSO	Rainbow show of oil
RBSOF	Rubber ball sand oil frac
RBSWF	Rubber ball sand water frac
RC	Rapid curing, remote control, reverse circulation, running casing, Red Cave
RCO	Returning circulation oil
RCR	Ramsbottom Carbon Residue, reverse circulation rig
RD	Rigged down, rigging down
rd	Round, road
RDB	Rotary drive bushing
Rd Bds	Red Beds
RDB-GD	Rotary drive bushing to ground
rdd	Rounded
Rd Fk	Red Fork
Rd Pk	Red Peak
rds	Roads
RDSU	Rigged down swabbing unit
rd thd	Round thread
rdtp	Round trip
R & D	Research and development
reac	Reactor
reacd	Reacidize, reacidizing, reacidized
Reag	Reagan
rebar	Reinforcing bar

reblr	Reboiler	res	Research, reservation, reserve, reservoir, resistance, resistivity, resistor
rec	Recover, recovering, recovered, recovery, recommend, recorder, recording	resid	Residual, residue
		Res. O.N.	Research Octane Number
recd	Received	ret	Retain (ed) (er) (ing), retard (ed) return
recip	Reciprocate(ing)		
recirc	Recirculate	retd	Returned
recomp	Recomplete (d) (ion)	retr ret	Retrievable retainer
recond	Recondition(ed)	rev	Reverse(d), revise (d) (ing) (ion), revolution (s)
recp	Receptacle		
rect	Rectifier, rectangle, rectangular		
		rev/O	Reversed out
recy	Recycle	RF	Raised face, rig floor
red	Reducing, reducer	RFFE	Raised face flanged end
red bal	Reducing balance	RFWN	Raised face weld neck
redrld	Redrilled	RG	Ring groove
ref	Reference, refine (d) (r) (ry)	rg	Ring
refer	Refrigeration	Rge	Range
referg	Refrigerant	rgh	Rough
refg	Refining	RH	Rat hole, right hand
refgr	Refrigerator	RHD	Right hand door
refl	Reflection	rheo	Rheostat
refl	Reflux	RHM	Rat hole mud
reform	Reformate, reformer, reforming	RHN	Rockwell hardness number
		RI	Royalty interest
refr	Refraction, refractory	Rib	Ribbon sand
reg	Regular, regulator, register	Rier	Rierdon
regen	Regenerator	rig rel	Rig released
reinf	Reinforce (d) (ing)	RIH	Ran in hole
reinf conc	Reinforced concrete	RIL	Red indicating lamp
rej	Reject	riv	Rivet
rej'n	Rejection	RJ	Ring Joint
Rek	Reklaw	RJFE	Ring joing flanged end
rel	Relay, release(d)	rk	Rock
REL	Running electric log	rky	Rocky
reloc	Relocate(d)	RL	Random lengths
rem	Remains, remedial, remove (al) (able)	R&L	Road & location
		R&LC	Road & location complete
		rlf	Relief
Ren	Renault	rlg	Railing
rent	Rental	ris (d) (ing)	Release (d) (ing)
Reo bath	Reophax bathysiphoni	rly	Relay
rep	Repair (ed) (ing) (s), replace (d), report	rm	Room, ream
		rmd	Reamed
reperf	Reperforated	rmg	Reaming
repl	Replace(ment)	rmn	Remains
req	Requisition	RMS	Root mean square
reqd	Required	rmv	Removable
reqmt	Requirement	rnd	Rounded
reqn	Requisition	rng	Running

RO	Reversed out	RUR	Rigging up rotary
Ro	Rosiclare sand	RURT	Rigging rotary tools
R.O.	Red Oak	RUSR	Rigging up service rig
R & O	Rust & oxidation	RUST	Rigging up standard tools
ro	Rose	RUT	Rigging up tools
Rob	Robulus	RVP	Reid vapor pressure
Rod	Rodessa	rvs(d)	Reverse(d)
ROF	Rich oil fractionator	R/W	Right of way
ROI	Return on investment	rwk(d)	Rework(ed)
ROL	Rig on location	RWTP	Returned well to production
ROM	Run of mine	Ry	Railway
RON	Research Octane Number		
ROP	Rate of penetration		
ROR	Rate of return		
rot	Rotary, rotate, rotator		

S

ROW	Right-of-way	S/2	South half
roy	Royalty	S/	Swabbed
RP	Rock pressure	Sab	Sabinetown
rpm	Revolutions per minute	sach	Saccharoidal
rpmn	Repairman	Sad Cr	Saddle Creek
RPP	Retail pump price	sadl	Saddle
rps	Revolutions per second	saf	Safety
RR	Railroad, Red River, rig released	sal	Salary, salaried, salinity
		Sal	Salado
RRC	Railroad Commission (Texas)	Sal Bay	Saline Bayou
RR&T	Ran (running) rods and tubing	salv	Salvage
		samp	Sample
RS	Rig skidded, rising stem (valve)	Sana	Sanastee
		San And	San Andres
RSD	Returnable steel drum	San Ang	San Angelo
RSH	Mercaptan	sani	Sanitary
rsns	Resinous	San Raf	San Rafael
RSU	Released swab unit	sap	Saponification
RT	Rotary tools, rotary table	Sap No.	Sapanification number
R & T	Rods & tubing	Sara	Saratoga
R test	Rotary test	sat	Saturated, saturation
RTG	Running tubing	Saw	Sawatch
rtg	Rating	Sawth	Sawtooth
rthy	Earthy	Say Furol	Saybolt Furol
RTJ	Ring tool joint, ring type joint	SB	Sideboom, sleeve bearing, stuffing box
RTLTM	Rate too low to measure		
rtnr	Retainer	Sb	Sunburst
RTTS	Retrievable test treat squeeze (tool)	sb	Sub
		SBA	Secondary butyl alcohol
RU	Rigging up, rigged up	SBB&M	San Bernardino Base and Meridian
RU	Rotary unit		
rub	Rubber	SBHP	Static bottom-hole pressure
RUCT	Rigging up cable tools	S Bomb	Sulfur by bomb method
RUM	Rigging up machine	SC	Show condensate
RUP	Rigging up pump	sc	Scales
rupt	Rupture	SCF	Standard cubic foot

SCFD	Standard cubic feet per day	SE NA	Screw end American National Acme thread
SCFH	Standard cubic feet per hour	SE NC	Screw end American National Coarse thread
SCFM	Standard cubic feet per minute	SE NF	Screw end American National Fine thread
sch	Schedule	SE No.	Steam Emulsion Number
schem	Schematic	SE NPT	Screw End American National Taper Pipe thread
scly	Securaloy		
scolc	Scolescodonts	sep	Separator
scr	Scratcher, screw, screen	sept	Septuplicate
scrd	Screwed	seq	Sequence
scrub	Scrubber	ser	Series, serial
sctrd	Scattered	Serp	Serpentine
SD	Shut down	Serr	Serratt
sd	Sand, sandstone	serv	Service(s)
SDA	Shut down to acidize	serv chg	Service change
SD Ck	Side door choke	set	Settling
SDF	Shut down to fracture	sew	Sewer
sdfract	Sandfract	Sex	Sexton
SDL	Shut down to log	sext	Sextuplicate
SDO	Show of dead oil	SF	Sandfrac
SDO	Shut down for orders	S&F	Swab and flow
sdoilfract	Sand oil fract	sfc	Surface
SDON	Shut down overnight	SFL	Starting fluid level
SDPA	Shut down to plug & abandon	SFLU	Slight, weak or poor fluorescence
SDPL	Shut down for pipe line	SFO	Show of free oil
SDR	Shut down for repairs	sft	Soft
Sd SG	Sand showing gas	SG	Show gas, surface geology
sd & sh	Sand and shale	SG&C	Show gas and condensate
Sd SO	Sand showing oil	SGCM	Slightly gas-cut mud
sdtkr	Sidetrack (ed) (ing)	SGCO	Slightly gas-cut oil
SDW	Shut down for weather	SGCW	Slightly gas-cut water
sdwtrfract	Sand water fract	SGCWB	Slightly gas-cut water blanket
SDWO	Shut down awaiting orders	SG&D	Show gas and distillate
sdy	Sandy	sgd	Signed
sdy li	Sandy lime	sgls	Singles
sdy sh	Sandy shale	SG&O	Show of gas and oil
SE	Southeast	SG&W	Show of gas & water
SE/4	Southeast quarter	sh	Shale, sheet
S/E	Screwed end	Shan	Shannon
Sea	Seabreeze	SHDP	Slim hole drill pipe
sec	Secant, second, secondary, secretary, section	Shin	Shinarump
		shls	Shells
SE/C	Southeast corner	shld	Shoulder
sed	Sediment(s)	shly	Shaley
Sedw	Sedwick	shp	Shaft horsepower
seis	Seismograph, seismic	shpg	Shipping
Sel	Selma	shpt	Shipment
sel	Selenite	shr	Shear
Sen	Senora		

SHT	Straight hole test	SOCW	Slightly oil-cut water
shthg	Sheathing	sod gr	Sodium base grease
SI	Shut in	SOE	Screwed on one end
SIBHP	Shut in bottom hole pressure	SOF	Sand oil fracture
SICP	Shut in casing pressure	SO&G	Show oil and gas
sid	Siderite(ic)	SO&GCM	Slightly oil & gas-cut mud
SIGW	Shut in gas well	SOH	Shot open hole
Sil	Silurian	sol	Solenoid, solids
silic	Silica, siliceous	soln	Solution
silt	Siltstone	solv	Solvent
sim	Similar	som	Somastic
Simp	Simpson	somct	Somastic coated
SIOW	Shut in oil well	SOP	Standard operational
SIP	Shut in pressure		procedure
siph. d.	Siphonina davisi	sort	Sorted(ing)
SITP	Shut in tubing pressure	SO&W	Show oil and water
SIWHP	Shut in well head pressure	sow	Socket weld
SIWOP	Shut in—waiting on potential	SP	Self (Spontaneous) potential,
sk	Sacks		set plug, surface pressure,
Sk Crk	Skull Creek		straddle packer, shot point,
skim	Skimmer		slightly porous
Skn	Skinner	Sp	Sparta
sks	Slickensided	sp	Spare, spore
skt	Socket	s&p	Salt & pepper
SL	Section line, state lease,	spcl	Special
	south line	spcr	Spacer
sl	Sleeve, slight(ly)	spd	Spud (ded) (der)
SLC	Steel line correction	spdl	Spindle
sld	Sealed	SPDT	Single pole double throw
Sli	Sligo	SP—DST	Straddle packer drill stem
sli	Slight(ly)		test
sli SO	Slight show of oil	spec	Specification
slky	Silky	speck	Speckled
SLM	Steel line measurement	spf	Spearfish
slnd	Solenoid	sprf	Spirifers
slt	Silt	spg	Sponge, spring
Slt Mt	Salt Mountain	sp gr	Specific gravity
Slty	Salty	sph	Spherules
slur	Slurry	Sphaer	Sphaerodina
SM	Surface measurement	sphal	Sphalerite
sm	Small	sp ht	Specific heat
Smithw	Smithwick	spic	Spicule(ar)
Smk	Smackover	Spiro, b.	Spiroplectammina barrowi
smls	Seamless	spkt	Sprocket
smth	Smooth	spkr	Sprinkler
SN	Seating nipple	splty	Splintery, split
S O	South offset, shake out,	Spletp	Spindletop
	show oil, slip on, side	splty	Specialty
	opening	sply	Supply
S&O	Stain and odor	SPM	Strokes per minute
SOCM	Slightly oil-cut mud	Spra	Spraberry

Sprin	Springer	St. Gen	Saint Genevieve
S Riv	Seven Rivers	stging	Straightening
SPST	Single pole single throw	STH	Sidetracked hole
SPT	Shallower pool (pay) test	stip	Stippled
sptd	Spotted	stir	Stirrup
sptty	Spotty	stk	Stock, stuck, streaks, streaked
sp. vol.	Specific volume		
sq	Square, squeezed	St L	Saint Louis Lime
sq cg	Squirrel cage	stl	Steel
sq cm	Square centimeter	STM	Steel tape measurement
sq ft	Square foot	stm	Steam
sq in	Square inch	stm cyl oil	Steam cylinder oil
sq km	Square kilometer	stm eng oil	Steam engine oil
sq m	Square meter	stn (d) (g)	Stain (ed) (ing)
sq mm	Square millimeter	stn/by	Stand by
sq pkr	Squeeze packer	Stn Crl	Stone Corral
sq yd	Square yard(s)	Stnka	Satanka
sqz	Squeeze (d) (ing)	stnr	Strainer
SR	Short radius	stoip	Stock tank oil in place
SRL	Single random lengths	stor	Storage
srt (d) (g)	Sort (ed) (ing)	STP	Standard temperature and pressure
SS	Stainless steel, service station, single shot, slow set (cement), string shot, subsea, subsurface, small show	stp	Stopper
		stpd	Stopped
		St Ptr	Saint Peter
s & s	Spigot and spigot	S-T-R	Section-township-range
SSG	Slight show of gas	Str	Strawn
SSO	Slight show of oil	strat	Stratigraphic
SSO&G	Slight show of oil & gas	strd	Straddle, strand(ed)
S/SR	Sliding scale royalty	strg	Strong, storage, stringer
SSU	Saybolt Seconds Universal	stri	Striated
SSUW	Salty sulphur water	strom	Stromatoporoid
ST	Short thread	strt	Straight
ST(g)	Sidetrack(ing)	struc	Structure, structural
S/T	Sample tops	STTD	Sidetracked total depth
sta	Station	stv	Stove oil
stab	Stabilized(er)	stwy	Stairway
Stal	Stalnaker	styo	Styolite, styolitic
Stan	Stanley	Sty Mt	Stony Mountain
stat	Stationary, statistical	sub	Subsidiary, substance
State pot	State potential	Sub Clarks	Sub-Clarksville
STB	Stock tank barrels	subd	Subdivision
STB/D	Stock tank barrels per day	substa	Substation
s, t&b	Sides, tops & bottoms	suc	Sucrose, sucrosic
ST&C	Short threads & Coupling	suct	Suction
stcky	Sticky	sug	Sugary
std (s) (g)	Standard, stand (s) (ing)	sul	Sulphur (sulfur)
stdy	Steady	sulph	Sulphated
Stel	Steele	sul wtr	Sulphur water
steno	Stenographer	sum	Summary
Stens	Stensvad	Sum	Summerville

Sunb	Sunburst
Sund	Sundance
Sup	Supai
supl	Supply (ied) (ier) (ing)
supp	Supplement
suppt	Support
suprv	Supervisor
supsd	Superseded
supt	Superintendent
sur	Survey
surf	Surface
surp	Surplus
SUS	Saybolt Universal Seconds
susp	Suspended
svc	Service
svcu	Service unit
SVI	Smoke Volatility Index
Svry	Severy
SW/4	Southwest quarter
SW	Salt wash, salt water, spiral weld, socket weld, southwest
Swas	Swastika
swbd	Switchboard
swb (d) (g)	Swabbed, swabbing
SWC	Sidewall cores
SW/c	Southwest corner
SWD	Salt water disposal
swd	Swaged
SWDS	Salt water disposal system
SWDW	Salt water disposal well
swet	Sweetening
SWF	Sand-water fracture
swg	Swage
swgr	Switchgear
SWI	Salt water injection
SWP	Steam working pressure
SWS	Sidewall samples
SWTS	Salt water to surface
SWU	Swabbing unit
sx	Sacks
sxtu	Sextuple
Syc	Sycamore
Syl	Sylvan
sym	Symbol, symmetrical
syn	Synthetic, synchronous, synchronizing
syn conv	Synchronous converter
sys	System
sz	Size

T

T	Tee
T	Ton (after a number)
T	Township (as T2N)
T/	Top of (a formation)
TA	Temporarily abandoned, turn around
tab	Tabular, tabulating
Tag	Tagliabue
Tal	Tallahatta
Tamp	Tampico
Tan	Tansill
Tann	Tannehill
Tark	Tarkio
Tay	Taylor
TB	Tank battery, thin bedded
T & B	Top and bottom
tb	Tube
TBA	Tertiary butyl alcohol; tires, batteries, and accessories
T&BC	Top & bottom chokes
TBE	Threaded both ends
tbg	Tubing
tbg chk	Tubing choke
tbg press	Tubing pressure
TBP	True boiling point
TC	Temperature controller, tool closed, top choke, tubing choke
T/C	Tank car
T & C	Threaded & coupled, topping and coking
TCC	Thermofor catalytic cracking
TCP	Tricresyl phosphate
TCV	Temperature control valve
TD	Total depth
TDA	Temporary dealer allowance
TDI	Temperature differential indicator
TDR	Temperature differential recorder
tech	Technical, technician
TEFC	Totally enclosed-fan cooled
tel	Telephone, telegraph
TEL	Tetraethyl lead
Tel Cr	Telegraph Creek
Temp	Temperature, temporary (ily)
Tens	Tensleep
Tent	Tentaculites

tent	Tentative	TOE	Threaded one end
Ter	Tertiary	TOF	Top of fish
term	Terminal	TOH	Trip out of hole
termin	Terminate (d) (ing) (ion)	tol	Tolerance
Tex	Texana	TOL	Top of liner
tex	Texture	tolu	Toluene
Text. art.	Textularia articulate	Tonk	Tonkawa
Text. d.	Textularia dibollensis	TOP	Testing on pump
Text. h.	Textularia hockleyenis	topg	Topping
Text. w.	Textularia warreni	topo	Topographic, topography
Tfing	Three Finger	TOPS	Turned over to producing
Tfks	Three Forks		section
tfs	Tuffaceous	Tor	Toronto
T & G	Tongue and groove (joint)	Toro	Toroweap
tgh	Tough	TORT	Tearing out rotary tools
TH	Tight hole	tot	Total
th	Thence	Tow	Towanda
Thay	Thaynes	TP	Travis Peak, tubing pressure,
thd	Thread, threaded		tool pusher
Ther	Thermopolis	T/pay	Top of pay
therm	Thermometer	TPC	Tubing pressure—closed
therst	Thermostat	TPF	Threaded pipe flange, tubing
THF	Tubinghead flange		pressure-flowing
THFP	Top hole flow pressure	tpk	Turnpike
thk	Thick, thickness	Tpka	Topeka
thrling	Throttling	TPSI	Tubing pressure shut in
thrm	Thermal	TR	Temperature recorder
thrm ckr	Thermal cracker	tr	Tract, trace
thru	Through	T&R	Tubing and rods
Thur	Thurman	trans	Transformer
TI	Temperature indicator	trans	Transfer (ed) (ing)
ti	Tight		transmission
TIC	Temperature indicator	transl	Translucent
	controller	transp	Transparent, transportation
TIH	Trip in hole	TRC	Temperature recorder
Tim	Timpas		controller
Timpo	Timpoweap	Tren	Trenton
tk	Tank	Tremp	Teremplealeau
tkg	Tankage	Tri	Triassic
tkr	Tanker(s)	trilo	Trilobite
TLE	Thread large end	Trin	Trinidad
TLH	Top of liner hanger	trip	Tripoli, tripolitic, triplicate,
tl	Tools, tools		tripped (ing)
TML	Tetramethyl lead	trkg	Trackage
tndr	Tender	trk	Truck
TNS	Tight no show	Trn	Trenton
TO	Tool open	trt (d) (g)	Treat (ed) (ing)
TOBE	Thread on both ends	trtr	Treater
TOC	Top of cement	TS	Tensile strength, Tar Springs
TOCP	Top of cement plug		sand
Tod	Todilto	T/S	Top salt

TSD	Temporarily shut down
T/sd	Top of sand
TSE	Thread small end
TSE-WLE	Thread small end, weld large end
TSI	Temporarily shut in
TSITC	Temperature Survey indicated top cement at
tst (d) (g)	Test (ed) (ing)
tste	Taste
TSTM	Too small to measure
tstr	Tester
TT	Tank truck, through tubing
TTF	Test to follow
TTL	Total time lost
TTTT	Turned to test tank
Tuck	Tucker
tuf	Tuffaceous
Tul Cr	Tulip Creek
tung carb	Tungsten carbide
Tus	Tuscaloosa
TV	Television
TVA	Temporary voluntary allowance
TVD	True vertical depth
TVP	True Vapor Process
TW	Tank wagon
T&W	Tarred and wrapped
Tw Cr	Twin Creek
twp	Township
twst	Townsite
twst off	Twisted off
TWTM	Two weak to measure
TWX	Teletype
ty	Type
typ	Typical
tywr	Typewriter

U

U/	Upper (i.e., U/Simpson)
U/C	Under construction
UCH	Use customer's hose
UD	Under digging
UG	Under gauge, underground
UGL	Universal gear lubricant
UHF	Ultra high frequency
U/L	Upper and lower
ult	Ultimate
un	Unit

unbr	Unbranded
unconf	Unconformity
uncons	Unconsolidated
undiff	Undifferentiated
unf	Unfinished
uni	Uniform
univ	University, universal
UR	Under reaming, unsulfonated residue
uv	Union Valley
Uvig lir.	Uvigerina lirettensis
U/W	Used with

V

V	Volume
v	Volt
v.	Very (as very tight)
va	Volt-ampere
vac	Vacuum, vacant, vacation
Vag. reg	Vaginuline regina
Val	Valera
Vang	Vanguard
vap	Vapor
var	Variable, various, volt-ampere reactive
vari	Variegated
v.c.	Very common
VCP	Vitrified clay pipe
vel	Velocity
vent	Ventilator
Ver Cl	Vermillion Cliff
Verd	Verdigris
vert	Vertical
ves	Vesicular
v-f-gr	Very fine-grained
VHF	Very high frequency
v-HOCM	Very heavily oil-cut mud
VI	Viscosity index
Vi	Viola
Virg	Virgelle
vis	Viscosity, visible
vit	Vitreous
Vks	Vicksburg
V/L	Vapor-liquid ratio
VLAC	Very light amber cut
vlv	Valve
VM&P Naphtha	Varnish makers & painters naphtha
v. n.	Very noticeable

Vogts	Vogtsberger
vol	Volume
vol. eff.	Volumetric efficiency
VP	Vapor pressure
V.P.S.	Very poor sample
v. r.	Very rare
vrs	Varas
vrtb	Vertebrate
vrtl	Vertical
vrvd	Varved
V/S	Velocity survey
vs	Versus
VSGCM	Very slight gas-cut mud
v-sli	Very slight
VSP	Very slightly porous
VSSG	Very slight show of gas
VSSO	Very slight show of oil
vug	Vuggy, vugular

W

w	Watt
W	West, wall (if used with pipe)
W/2	West half
w/	With
Wa Sd	Waltersburg sand
WAB	Weak air blow
Wab	Wabaunsee
Wad	Waddell
Wap	Wapanucka
War	Warsaw
Was	Wasatch
Wash	Washita
WB	Water blanket, wet bulb, Woodbine
WBIH	Went back in hole
WC	Wildcat, water cushion (DST), Wolfe City, water cut
WCM	Water-cut mud
WCO	Water-cut oil
W Cr	Wall Creek
WCTS	Water cushion to surface
WD	Water depth
WD	Water disposal well
Wdfd	Woodford
Wd R	Wind River
WE	Weld ends
Web	Weber
Well	Wellington
WF	Waterflood, wide flange

W—F	Washita-Fredericksburg
WFD	Wildcat field discovery
wgt.	Weight
WH	Wellhead
Wh Dol	White Dolomite
whip	Whipstock
Wh Sd	White sand
whse	Warehouse
whsle	Wholesale
wht	White
WI	Washing in, water injection, working interest, wrought iron
Wich.	Wichita
Wich Alb	Wichita Albany
WIH	Water in hole, went in hole
Willb	Willberne
Win	Winona
Winf	Winfield
Wing	Wingate
Winn	Winnipeg
wk	Weak, week
wkd	Worked
wkg	Working
wko	Workover
wkor	Workover rig
WL	West line, wire line, water loss
W/L	Water load
WLC	Wire line coring
wld	Welded, welding
wldr	Welder
WLT	Wireline test
WLTD	Wireline total depth
W'ly	Westerly
WN	Weld neck, welding neck
WNSO	Water not shut-off
WO	Waiting on
WO	Workover, wash over, work order
W/O	West offset, without
WOA	Waiting on allowable, waiting on acid
WOB	Waiting on battery
W.O.B.	Weight on bit
WOC	Waiting on cement
WOCR	Waiting on completion rig
WOCT	Waiting on cable tools, or completion tools
WODP	Without drill pipe
WOG	Water, oil or gas
Wolfc	Wolfcamp

WOO	Waiting on orders
Wood	Woodside
Woodf	Woodford
WOP	Waiting on permit, waiting on pipe, waiting on pump
WOPE	Waiting on production equipment
WOPT	Waiting on potential test
WOPU	Waiting on pumping unit
WOR	Waiting on rig or rotary, water-oil ratio
WORT	Waiting on rotary tools
WOS	Washover string
WOSP	Waiting on state potential
WOST	Waiting on standard tools
WOT	Waiting on test or tools
WOT&C	Waiting on tank & connection
WOW	Waiting on weather
WP	Wash pipe, working pressure
wpr	Wrapper
WR	White River
Wref	Wreford
WS	Whipstock
WSD	Whipstock depth
w shd	Washed
wshg	Washing
WSO	Water shut-off
WSONG	Water shut off no good
WSOOK	Water shut off OK
W/SSO	Water with slight show of oil
W/sulf O	Water with sulphur odor
WSW	Water supply well
WT	Wall thickness (pipe)
wt	Weight
wtg	Waiting
wthd	Weathered
wthr	Weather
wtr(y)	Water, watery
WTS	Water to surface
WW	Wash water, water well
Wx	Wilcox

X

X-bdd(ing)	Crossbedded, crossbedding
X-hvy	Extra heavy
Xing	Crossing
Xlam	Cross-laminated
X-line	Extreme line (casing)
Xln	Crystalline
X-cover	Crossover
X-R	X-Ray
x-stg	Extra strong
xtal	Crystal
Xtree	Christmas tree
XX-Hvy	Double extra heavy

Y

Y	Yates
yd	Yard(s)
YIL	Yellow indicating lamp
yel	Yellow
YMD	Your message of date
YMY	Your message yesterday
Yoak	Yoakum
YP	Yield point
yr	Year
Yz	Yazoo

Z

Z	Zone
zen	Zenith
Zil	Zilpha

A

Abandoned	abd
Abandoned gas well	abd-gw
Abandoned location	abd loc
Abandoned oil & gas well	abdogw
Abandoned oil well	abd-ow
About	abt
Above	ab
Above	abv
Abrasive jet	abrsi jet
Absolute open flow potential (gas well)	AOF
Absorber	asbr
Absorption	absrn
Abstract	abst
Abundant	abun
Account	acct
Accounting	acct
Account of	acct
Accounts receivable	A/R
Accumulative	accum
Acid	ac
Acid-cut mud	ACM
Acid-cut water	ACW
Acid frac	AF
Acid fracture treatment	acfr
Acidize	acd
Acidized	acd
Acidized with	A/
Acidizing	ac
Acidizing	acd
Acid residue	AR
Acid treat (ment)	AT
Acid water	AW
Acoustic cement	A-Cem
Acre	ac
Acreage	ac
Acreage	acrg
Acre feet	ac-ft
Acres	ac
Acrylonitrile butadiene styrene rubber	ABS
Adapter	adpt
Additional	addl
Additive	add
Adjustable	adj
Adjustments and Allowances	A&A
Administration	adm
Administrative	adm
Adomite	ADOM
Adsorption	adspn
Advanced	advan
Affidavit	afft
Affirmed	affd
After acidizing	AA
After fracture	AF
After shot	AS
After top center	ATC
After treatment	AT
Agglomerate	aglm
Aggregate	aggr
Alarm	alm
Albany	Alb
Algae	alg
Alkalinity	alk
Alkylate	alky
Alkylation	akly
Allowable	allow
Allowable not yet available	ANYA
All thread	AT
Alternate	alt
Alternating current	AC
Aluminum conductor steel reinforced	ACSR
Ambient	amb
American Steel & Wire gauge	AS&W ga
American Wire Gauge	AWG
Amorphous	amor
Amortization	amort
Amount	amt
Ampere	amp
Ampere hour	amp hr
Amphipore	amph
Amphistegina	Amph
Analysis	anal
Analytical	anal
And husband	et con
And husband	et vir
And the following	et seq
And others	et al
And wife	et ux
Angle	ang
Angular	ang
Angulogerina	Angul
Anhydrite	anhy
Anhydrite stringer	AS
Anhydritic	anhy
Anhydrous	anhyd
Annular velocity	AV
Apartment	apt

Apparent(ly)	apr
Appearance	app
Appears	app
Appliance	appl
Application	applic
Applied	appl
Approved	appd
Approximate(ly)	approx
Aqueous	aq
Aragonite	arag
Arapahoe	Ara
Arbuckle	Arb
Archeozoic	Archeo
Architectural	arch
Arenaceous	aren
Argillaceous	arg
Argillite	arg
Arkadelphia	Arka
Arkose(ic)	ark
Armature	arm
Aromatics	arom
Around	arnd
As above	AA
Asbestos	asb
Ashern	Ash
Asphalt	asph
Asphaltic	asph
Asphaltic stain	astn
Assembly	assy
Assigned	assgd
Assignment	asgmt
Assistant	asst
Associate(d)(s)	assoc
Association	assn
As soon as possible	ASAP
Atmosphere	atm
Atmospheric	atm
Atoka	At
Atomic	at
Atomic weight	at wt
At rate of	ARO
Attempt(ed)	att
Attorney	atty
Audit Bureau of Circulation	ABC
Auditorium	aud
Austin	Aus
Austin Chalk	AC
Authorization for expenditure	AFE
Authorized	auth
Authorized depth	AD
Automatic	auto

Automatic custody transfer	ACT
Automatic data processing	ADP
Automatic transmission fluid	ATF
Automatic volume control	AVC
Automotive	auto
Automotive gasoline	autogas
Aux Vases sand	AV
Auxiliary	aux
Available	avail
Average	avg
Average flowing pressure	AFP
Average injection rate	AIR
Aviation	av
Aviation gasoline	avgas
Awaiting	awtg
Azeotropic	aztrop
Azimuth	az

B

Back pressure	BP
Back pressure valve	BPV
Back to back	B/B
Backed out (off)	BO
Bailed	bld
Bailed dry	B/dry
Bailed water	BW
Bailer	blr
Bailing	blg
Ball and flange	B&F
Balltown sand	Ball.
Band(ed)	bnd
Barge deck to mean low water	BD-MLW
Barite(ic)	bar
Barker Creek	Bark Crk
Barlow Lime	Bar
Barometer	bar
Barometric	bar
Barrel	BBL
Barrels	BBL
Barrels acid residue	BAR
Barrels acid water	BAW
Barrels Acid Water Per Day	BAWPD
Barrels Acid Water Per Hour	BAWPH
Barrels acid water under load	BAWUL
Barrels condensate per day	BCPD
Barrels condensate per hour	BCPH
Barrels condensate per million	BCPMM
Barrels diesel oil	BDO

Barrels distillate per day	BDPD	Base plate	BSPL
Barrels distillate per hour	BDPH	Basement	bsmt
Barrels fluid	BF	Basement (Granite)	base
Barrels fluid per day	BFPD	Basic sediment	BS
Barrels fluid per hour	BFPH	Basic sediment & water	BS&W
Barrels formation water	BFW	Basket	bskt
Barrels frac oil	BFO	Bateman	Bate
Barrels load	BL	Battery	bat
Barrels load & acid water	BL&AW	Battery	btry
Barrels load oil	BLO	Baume	Be
Barrels load oil recovered	BLOR	Beaded and center beaded	B & CB
Barrels load oil yet to recover	BLOYR	Bearing	brg
Barrels load water	BLW	Bearpaw	BP
Barrels mud	BM	Bear River	Bear R
Barrels new oil	BNO	Becoming	bec
Barrels of acid	BA	Beckwith	Beck
Barrels of condensate	BC	Before acid treatment	BAT
Barrels of distillate	BD	Before top dead center	BTDC
Barrels of pipeline oil	BPLO	Beldon	Bel
Barrels of pipeline oil per day	BPLOPD	Belemnites	Belm
Barrels of water	BW	Bell and bell	B & B
Barrels of water per day	BW/D	Bell & spigot	B & S
Barrels of water per day	BWPD	Belle City	Bel C
Barrels of water per hour	BWPH	Belle Fourche	Bel F
Barrels oil	BO	Bench mark	BM
Barrels oil per calendar day	BOCD, BOPCD	Benoist (Bethel) sand	Ben, BT
Barrels oil per day	BOD	Benton	Ben
Barrels oil per day	BOPD	Bentonite	bent
Barrels oil per hour	BOPH	Bentonitic	bent
Barrels oil per producing day	BOPPD	Benzene	bnz
Barrels per barrel	B/B	Benzene toluene, xylene (Unit)	BTX (unit)
Barrels per calendar day	BPCD	Berea	Be
Barrels per day	B/D	Bethel (Benoist) sand	BT, Ben
Barrels per hour	B/H	Between	btw
Barrels per hour	B/hr	Bevel	bev
Barrels per hour	BPH	Bevel both ends	BBE
Barrels per minute	BPM	Beveled as for welding	bev
Barrels per stream day (refinery)	B/SD	Beveled for welding	BV/WLD
Barrels per well per day	BPWPD	Beveled end	B.E.
Barrels salt water	BSW	Bevel one end	BOE
Barrels salt water per day	BSWPD	Bevel small end	BSE
Barrels salt water per hour	BSWPH	Big Horn	B Hn
Barrels water load	BWL	Big Injun	B. Inj.
Barrels water over load	BWOL	Big Lime	B. Ls
Bartlesville	Bart	Bigenerina	Big.
Basal Oil Creek sand	BOCS	Bigenerina floridana	Big. f.
Base	B/	Bigenerina humblei	Big. h.
Base Blane	B.Bl	Bigenerina nodosaria	Big. nod.
Base of the salt	B slt, B/S	Billion cubic feet	BCF
Base Pennsylvanian	BP	Billion cubic feet per day	BCPD

Bill of lading	B/L	Bottom-hole choke	BHC
Bill of material	B/M	Bottom-hole flowing pressure	BHFP
Bill of sale	B/S	Bottom-hole location	BHL
Biotite	bio	Bottom-hole money	BHM
Birmingham (or Stubbs) iron		Bottom-hole pressure	BHP
wire gauge	BW ga	Bottom-hole pressure, closed	
Bitumen	bit	(See also SIBHP and BHSIP)	BHPC
Bituminous	bit	Bottom-hole pressure, flowing	BHPF
Black	blk	Bottom-hole pressure survey	BHPS
Black Leaf	Blk Lf	Bottom-hole shut-in pressure	BHSIP
Black Lime	Blk Li	Bottom-hole assembly	BHA
Black Magic (mud)	BM	Bottom-hole temperature	BHT
Black malleable iron	BMI	Bottom of given formation	
Black River	B. Riv	(i.e., B/Frio)	B/
Black sulfur water	BSUW	Bottom sediment	BS
Blank liner	blk lnr	Bottom settlings	BS
Blast joint	BL/JT	Boulders	bldrs
Bleeding	bldg	Boundary	bndry
Bleeding gas	bldg	Box(es)	bx
Bleeding oil	bldo	Brachiopod	brach
Blend	blnd	Bracket(s)	brkt(s)
Blended	blnd	Brackish (water)	brksh
Blender	blndr	Brake horsepower	bhp
Blending	blnd	Brake horsepower hour	bhp-hr
Blind (flange)	bld	Brake mean effective	BMEP
Blinebry	Blin	pressure	
Block	blk	Brake specific fuel	
Blossom	Blos	consumption	BSFC
Blow	blo	Break (broke)	brk
Blow-down test	BDT	Breakdown	bkdn
Blow out equipment	BOE	Breakdown acid	BDA
Blowout preventer	BOP	Breakdown pressure	BDP
Blue	bl	Breaker	bkr
Board	bd	Breccia	brec
Board feet	bd ft	Bridged back	BB
Board foot	bd ft	Bridge plug	BP
Bodcaw	Bod	Bridger	Brid
Boiler feed water	BFW	Brinell hardness number	BHN
Boiling point	BP	British thermal unit	BTU
Bois D'Arc	Bd'A	Brittle	brit
Bolibarensis	Bol.	Brittle	brtl
Bolivina a.	Bol. a.	Broke (break) down formation	BDF
Bolivina floridana	Bol. flor.	Broken	brkn
Bolivina perca	Bol. p.	Broken sand	brkn sd
Bone Spring	BS	Bromide	Brom
Bonneterre	Bonne	Brown	bn
Bottom	bot	Brown	brn or br
Bottom	btm	Brown and Sharpe gauge	B&S ga
Bottom choke	BC	Brown lime	Brn Li
Bottom choke	btm chk	Brown oil stain	BOS
Bottomed	btmd	Brown shale	brn sh

Brownish	bnish	Cane River	CR
Bryozoa	bry	Cane River	Cane R
Buckner	Buck	Canvas-lined metal petal	
Buckrange	Buckr	basket	CLMP
Budgeted depth	BD	Canyon	Cany
Buff	bf	Canyon Creek	Cany Crk
Building	bldg.	Capacitor	cap
Building derrick	bldg drk	Capacity	cap
Building rig	BR	Capitan	Cap
Building road	BR	Carbonaceous	carb
Building roads	bldg rds	Carbon copy	CC
Buliminella textularia	Bul. text.	Carbon Residue (Conradson)	CR Con
Bulk plant	BP	Carbon steel	CS
Bullets	blts	Carbon tetrachloride	carb tet
Bull plug	BP	Carburetor air temperature	CAT
Bull waggon	Bul W	Care of	c/o
Bumper	bmpr	Carlile	Car
Burgess	Burg	Carload	CL
Burner	bunr	Carmel	Carm
Bushing	bsg	Carrizo	Cz
Butane	BB	Carton	ctn
Butane and propane mix	BP Mix	Cased Hole	C/H
Butane fraction	BB fraction	Casing	csg
Butt weld	BW	Casing cemented (depth)	CC
Buttress thread	butt	Casing choke	Cck
		Casing collar locator	CCL
		Casing collar perforating	
		record	CCPR
		Casing flange	CF

C

Cable tools	CT	Casing head	csg hd
Caddell	Cadd	Casinghead flange	CHF
Calcareous	calc	Casinghead (gas)	CH
Calceneous	cale	Casinghead gas	CHG
Calcerenite	calc	Casinghead pressure	CHP
Calcite	cal	Casing point	CP
Calcitic	cal	Casing point	csg pt
Calcium	calc	Casing pressure	CP
Calcium base grease	calc gr	Casing pressure	csg press
Calculate(d)	calc	Casing pressure—closed	CPC
Calculated absolute open flow	CAOF	Casing pressure—flowing	CPF
Calculated open flow	COF	Casing pressure shut in	CPSI
Calculated open flow		Casing seat	CS
(potential)	calc OF, COF	Casing set at	CSA
Calendar day	CD	Casper	Casp
Caliche	cal	Cast-iron	CI
Caliper survey	cal	Cast-iron bridge plug	CIBP
Calorie	cal	Cast-steel	CS
Calvin	Calv	Catcracked light gas oil	CCLGO
Cambrian	Camb	Cat Creek	Cat Crk
Camerina	Cam.	Catahoula	Cat
Camp Colorado	Cp Colo	Catalog	cat

Catalyst	cat	Chappel	Chapp
Catalytic	cat	Charge	chg
Catalytic cracker	cat ckr	Charged	chg
Catalytic Cracking Unit	CCU	Charging	chg
Cathodic	cath	Charles	Char
Cattleman	Ctlmn	Chart	cht
Caustic	caus	Chattanooga shale	Chatt
Caving(s)	cvg(s)	Check	ck
Cavity	cav	Checked	chkd
Cedar Mountain	Cdr Mtn	Checkerboard	Chkbd
Cellar	cell	Chemical	chem
Cellar & pits	C&P	Chemically pure	CP
Cellular	cell	Chemically retarded acid	CRA
Cement(ed)	cem	Chemical products	chem prod
Cement (ed) (ing)	cmt (d) (g)	Chemist	chem
Cement friction reducer	CFR	Chemistry	chem
Cement in place	CIP	Cherokee	Cher
Cementer	cmtr	Chert	ch
Cenozoic	Ceno	Chert	cht
Center	ctr	Cherty	chty
Center(ed)	cntr	Chester	Ches
Center (land description)	C	Chicksan	cksn
Center line	C/L	Chimney Hill	Chim H
Center of casinghead flange	CCHF	Chimney Rock	Chim R
Center of gravity	CG	Chinle	Chin
Center of tubing flange	CTHF	Chitin(ous)	chit
Center to center	C to C	Chloride(s)	chl
Center to end	C to E	Chlorine log	chl log
Center to face	C to F	Chloritic	chl
Centigrade temp. scale	C	Choke	ch
Centimeter	cm	Choke	chk
Centimeter-gram-second		Chouteau lime	Chou
system	cgs	Christmas tree	Xtree
Centimeters per second	cm/sec	Chromatograph	chromat
Centipoise	cp	Chrome molybdenum	cr moly
Centistokes	cs	Chromium	chrome
Centralizers	cent	Chugwater	Chug
Centrifugal	centr	Cibicides	Cib
Centrifuge	cntf	Cibicides hazzardi	Cib h
Cephalopod	ceph	Cimarron	Cima
Ceratobulimina eximia	Cert	Circle	cir
Certified public accountant	CPA	Circuit	cir
Cetane number	CN	Circular	cir
Chairman	chrm	Circular mils	cir mils
Chalcedony	chal	Circulate	circ
Chalk	chk	Circulated out	CO
Chalky	chky	Circulating	circ
Change	chng	Circulating & conditioning	C&C
Changed	chng	Circulation	circ
Changed(ing) bits	CB	Cisco	Cis
Changing	chng	Clagget	Clag

Claiborne	Claib	Color, American Standard Test	
Clarksville	Clarks	Method	Col ASTM
Clastic	clas	Colored	COL
Clavalinoides	Clav	Column	COL
Clay filled	CF	Comanche	Com
Claystone	clyst	Comanchean	Cmchn
Clayton	Clay	Comanche Peak	Com Pk
Claytonville	Clay	Comatula	Com
Clean, cleaned, cleaning	cln (d) (g)	Combination	comb
Cleaned out	CO	Combined	comb
Cleaned out to total depth	COTD	Coming out of hole	COH
Cleaning out	CO	Commenced	comm
Cleaning to pits	CTP	Commercial	coml
Clean out	CO	Commission	comm
Clean out & shoot	CO & S	Commission agent	C/A
Clean up	CU	Commissioner	commr
Clear	clr	Common	com
Clearance	clr	Common Business Oriented	
Clearfork	Clfk	Language	COBOL
Clearing	clrg	Communication	comm
Cleveland	Cleve	Communitized	comm
Cleveland open cup	COC	Community	comm
Cliff House	Cliff H	Compact	cmpt
Clockwise	cw	Companion flange one end	CFOE
Closed	clsd	Companion flanges bolted on	CFBO
Closed-in pressure	CIP	Company operated	Co. Op.
Closed cup	CC	Company operated service	
Cloverly	Clov	stations	Co. Op. S.S.
Coarse-grained	c-gr	Compartment	compt
Coarse-grained	cse gr	Complete	comp
Coarse(ly)	c	Completed	compt
Coarse(ly)	crs	Completed natural	comp nat
Coat and wrap (pipe)	C & W	Completed with	C/W
Coated	ctd	Completion	comp
Coated and wrapped	ctw	Components	compnts
Cockfield	Cf	Compressor	compr
Coconino	Coco	Compressor station	compr sta
Codell	Cod	Compression and absorption	
Cody (Wyoming)	Cdy	plant	C&A
Coefficient	coef	Compression-ignition engine	CI engine
Coke oven gas	COG	Compression ratio	CR
Cold rolled	CR	Concentrate	conc
Cold working pressure	CWP	Concentric	cnnc
Cole Junction	Cole J	Concentric	conc
Coleman Junction	Col Jct	Conchoidal	conch
Collar	colr	Conclusion	concl
Collect	coll	Concrete	conc
Collected	coll	Concretion(ary)	conc
Collecting	coll	Condensate	cond
Collection	coll	Condensate-cut mud	CCM
		Condensor	cdsr

Conditioned	cond	Corporation	Corp
Conditioning	cond	Corrected gravity	CG
Conductivity	condt	Correct (ed) (ion), corrosion,	
Conductor (pipe)	condr	corrugated	corr
Confirm	conf	Corrected total depth	CTD
Confirmed	conf	Correlation	correl
Confirming	conf	Correspondence	corres
Conflict	confl	Cost, insurance and freight	CIF
Conglomerate	cglt	Cost per gallon or cents per	
Conglomerate(itic)	cong	gallon	CPG
Conglomeritic	cglt	Cottage Grove	Cott G
Connection	conn	Cotton Valley	CV
Conodonts	cono	Cottonwood	Ctnwd
Conradson carbon residue	CCR	Council Grove	Counc G
Conservation	consv	Counterbalance (pumping	
Conserve	consv	equip)	CB
Consolidated	con	Counterclockwise	ccw
Consolidated	consol	Counter electromotive force	CEMF
Constant	const	County	Cnty
Construction	const	County school lands, center	
Consumer tank car	CTC	section line	CSL
Consumer tank wagon	CTW	Coupling	cplg
Consumer transport truck	CTT	Cow Run	CR
Contact caliper	C-Cal	Cracker	crkr
Container	cntr	Cracking	crkg
Contaminated	contam	Crenulated	cren
Contamination	contam	Cretaceous	Cret
Continue, continued	cont(d)	Crinkled	crnk
Continuous dipmeter survey	CDM	Crinoid(al)	Crin
Continuous weld	CW	Cristellaria	Cris
Contour interval (map)	CI	Critical	crit
Contract depth	CD	Critical compression pressure	CCP
Contractor (i.e., C/John Doe)	C/	Critical compression ratio	CCR
Contractor	contr	Cromwell	Crom
Contractor's responsibility	contr resp	Crossbedded	Crdb
Contribution	contrib	Crossbedded	X-bdd(ing)
Controller	cntr	Crossbedding	X-bdd(ing)
Control(s)	cntl	Crossing	Xing
Control valve	CV	Cross-laminated	Xlam
Converse	conv	Crossover	CX
Conveyor	cnvr	Crossover	X-over
Coquina	coq	Crown block	crn blk
Cook Mountain	Ck Mt	Crude oil	CO
Cooperative	co-op	Crude oil purchasing	COP
Coordinate	coord	Cryptocrystalline	crypto-xln
Coordinating Research		Crystal	xtal
Council, Inc.	CRC	Crystalline	cryst
Core, cored, coring, core hole	cr (d), (g), (h)	Crystalline	Xln
Core barrel	CB	Cubic	cu
Coring	cg	Cubic centimeter	cc
Corner	cor	Cubic feet	CF

Cubic feet gas	CFG
Cubic feet gas per day	CFGPD
Cubic feet of gas per hour	CFGH
Cubic feet per barrel	cu ft/bbl
Cubic feet per minute	CFM
Cubic feet per minute	cu ft/min
Cubic feet per second	CFS
Cubic feet per second	cu ft/sec
Cubic foot	cu ft
Cubic inch	cu in
Cubic meter	cu m
Cubic yard	cu yd
Culvert	culv
Cumulative	cum
Curtis	Cur
Cushion	cush
Cut across grain	CAG
Cutback	cutbk
Cut Bank	Cut B
Cutler	Cutl
Cutting oil	Cut Oil
Cutting oil-active sulphurized—dark	Cut Oil Act
Cutting oil-active sulphurized—transparent	Cut Oil Act Sul-Trans
Cutting oil-inactive-sulphurized	Cut Oil Inact Sul
Cutting oil soluble	Cut Oil Sol
Cutting oil-straight mineral	Cut Oil St Mrl
Cuttings	ctg(s)
Cyclamina	Cyc.
Cyclamina cancellata	Cycl canc
Cycles per minute	CPM
Cycles per second	CPS
Cylinder	cyl
Cylinder Stock	cyl stk
Cypress sand	Cy Sd
Cypridopsis	cyp

D

Daily allowable	DA
Daily average injection, barrels	DAIB
Dakota	Dak
Damper	dmpr
Dantzler	Dan
Dark	dk
Dark brown oil	DBO
Dark brown oil stains	DBOS

Darwin	Dar
Data processing	DP
Date of first production	DFP
Datum	dat
Datum	DM
Datum faulted out	DFO
Days since spudded	DSS
Day to day	D/D
Dead	dd
Dead oil show	DSO
Dead weight tester	DWT
Deadwood	Deadw
Deaerator	deaer
Dealer	dlr
Dealer tank wagon	DTW
Deasphalting	deasph
Debutanizer	debutzr
Decibel	db
Decimal	dec
Decrease (d) (ing)	decr
Deepen	dpn
Deepening	dpg
Deep pool test	DPT
Deethanizer	deethnzr
Deflection	defl
Degonia	deg
Degree day	DD
Degree Fahrenheit	°F.
Deisobutanizer	deisobut
Delaware	Dela
Delayed coker	DC
Deliverability	delv
Delivered	delv
Delivery	delv
Delivery point	delv pt
Del Rio	Del R
Demand meter	DM
Demolition	dml
Demurrage	demur
Dendrite(ic)	dend
Dense	dns
Dense	ds
Density log	D/L
Density log	DENL
Department	dept
Depletion	depl
Depreciation	dep
Depreciation	depec
Depropanizer	deprop
Depth	dpt
Depth recorder	dpt rec

Derrick	drk	Digging slush pits, digging cellar, or digging cellar and slush pits	DCLSP
Derrick floor	DF	Diluted	dilut
Derrick floor elevation	DFE	Dimension	dim
Desalter	desalt	Diminish	dim
Description	desc	Diminishing	dim
Desert Creek	Des Crk	Dinwoody	Din
Design	dsgn	Dip meter	DM
Desk and Derrick	D & D	Direct	dir
Des Moines	Des M	Direct current	DC
Desorbent	desorb	Direction	dir
Destination	dstn	Directional survey	DS
Desulferizer	desulf	Directional survey	dir sur
Detail(s)	det	Director	dir
Detector	det	Discharge	disch
Detergent	deterg	Discorbis	Disc
Detrital	detr	Discorbis gravelli	Disc grav
Detrital	dtr	Discorbis normada	Disc norm
Develop (ed) (ment)	devel	Discorbis yeguaensis	Disc y
Development gas well	DG	Discount	disc
Development oil well	DO	Discovery allowable requested	DAR
Development redrill (sidetrack)	DR	Discover (y) (ed) (ing)	disc
Development well—carbon dioxide	DC	Disseminated	dism
Development well—helium	DH	Dismantle	disman
Development well—sulfur	DSU	Dismantle(ing)	dismtl (g)
Development well workover	DX	Displaced	displ
Deviate	dev	Displacement	displ
Deviation	dev	Distance	dist
Devonian	Dev	Distillate	dist
Dewaxing	dewax	Distillate	dstl
Dew point	DP	Distillate-cut mud	DCM
Dexter	Dext	Distillation	dist
Diagonal	diag	Distribute (d) (ing) (ion)	distr
Diagram	diag	District	dist
Diameter	dia	Ditto	do
Diamond core	DC	Division	div
Diamond core bit	DCB	Division office	D/O
Diaphragm	diaph	Division order	D.O.
Dichloride	dichlor	Dockum	Doc
Dichloro-diphenyl-trichloroethane	DDT	Doctor-treating	doc-tr
Diesel No. 2	D-2	Document	doc
Diesel fuel	DF	Doing business as	d/b/a
Diesel (oil)	dsl	Dolomite(ic)	dolo
Diesel oil cement	DOC	Dolstone	dolst
Diethylene	diethy	Domestic	dom
Difference	diff	Domestic airline	dom AL
Different	diff	Dornick Hills	Dorn H
Differential	diff	Dothan	Doth
Differential valve	DV	Double end	DE
		Double extra heavy	XX-Hvy

Double pole double base (switch)	DPDB	Drill pipe unloaded	DPU
Double pole double throw (switch)	DPDT	Drill stem	DS
		Drill stem test	DST
Double pole single base (switch)	DPSB	Drive	dr
		Dropped	drpd
Double pole single throw (switch)	DPST	Drum	dr
		Druse	dr
Double pole (switch)	DP	Drusy	drsy
Double random lenths	DRL	Dry and abandoned	D&A
Douglas	Doug	Dry Creek	Dr Crk
Down	dn	Dry desiccant dehydrator	DDD
Down	dwn	Dry gas	DG
Dozen	doz	Dry hole contribution	DHC
Draft gage	DG	Dry hole drilled deeper	DHDD
Drain	dr	Dry hole money	DHM
Drainage	drng	Dry hole reentered	DHR
Drawing	dwg	Drying	dry
Drawworks	dwks	Dually complete(d)	DC
Dressed dimension four sides	d-d-4-s	Dual (double) wall packer	DWP
Dressed dimension one side and one edge	d-d-1-s-1-e	Duck Creek	Dk Crk
		Dun & Bradstreet	D & B
Dressed four sides	d-4-s	Duperow	Dup
Dressed one side	d-1-s	Duplex	dx
Dressed two sides	d-2-s	Duplicate	dup
Drier	dry	Dutcher	Dutch
Drill	drl	Dynamic	dyn
Drill and complete	D & C		
Drill collar	DC		
Drill collars	DCS		

E

Drilled	drld	Each	ea
Drill (ed) (ing) out	DO	Eagle	Egl
Drill (ed) (ing) plug	D/P	Eagle Ford	EF
Drilled out cement	DOC	Eagle Mills	EM
Drilled out depth	DOD	Earlsboro	Earls
Drilled out plug	DOP	Earthy	rthy
Driller	drlr	East	E
Driller's tops	D/T	East boundary line	E/BL
Drillers total depth	DTD	East Cimarron Meridian (Oklahoma)	ECM
Drilling	drlg		
Drilling break	DB	East half, quarter, etc.	E/2, E/4
Drilling deeper	DD	East Line	E/L
Drilling mud	DM	East offset	E/O
Drilling suspended indefinitely	DSI	East of Rockies	EOR
Drilling time	DT	East of west line	E of W/L
Drilling with air	DWA	Eau Claire	Eau Clr
Drilling with mud	DWM	Eccentric	ecc
Drilling with oil	DWO	Echinoid	Ech
Drilling with salt water	DWSW	Economics	Econ
Drill pipe	DP	Economizer	Econ
Drill pipe measurement	DPM	Economy	Econ

Ector (county, Tex.)	Ect	Eponides yeguaensis	Ep y
Education	educ	Equal	eq
Edwards	Edw	Equalizer	eq
Edwards lime	Ed lm	Equation	eq
Effective	eff	Equilibrium flash vaporization	EFV
Effective horsepower	EHP	Equipment	equip
Efficiency	eff	Equivalent	equiv
Effluent	effl	Erection	erect
Ejector	eject	Ericson	Eric
Elbert	Elb	Estate	est
Elbow(s)	ell(s)	Estimate (d) (ing)	est
Electric accounting machines	EAM	Estimated time of arrival	ETA
Electric(al)	elec	Estimated yearly consumption	EYC
Electric log tops	El/T	Ethane	eth
Electric resistance weld	ERW	Ethylene	ethyle
Electric weld	EW	Euhedral	euhed
Electromotive force	EMF	Eutaw	Eu
Electronic data processing	EDP	Evaluate	eval
Electron volts	ev	Evaporation	evap
Element(ary)	elem	Evaporite	evap
Elevation	elev	Even-sorted	ev-sort
Elevation ground	el gr	Excavation	exc
Elevator	elev	Excellent	excl
Elgin	Elg	Exchanger	exch
Ellenburger	Ellen	Executor	exr
Ellis-Madison contact	EMS	Executrix	exrx
Elmont	Elm	Exhaust	exh
Embar	Emb	Exhibit	exh
Emergency	emer	Existing	exist
Emergency order	EO	Existing	exst
Employee	empl	Expansion	exp
Empty container	MT	Expendable plug	exp plg
Emulsion	emul	Expense	exp
Enamel	enml	Expiration	expir
Enclosure	encl	Expire	expir
Endicott	End	Expired	expir
End of file	EOF	Expiring	expir
End of line	EOL	Exploration	expl
End of month	EOM	Exploratory	expl
End of quarter	EOQ	Exploratory well	EW
End of year	EOY	Explosive	explos
Endothyra	endo	Extended	ext(n)
End point	EP	Extension	ext(n)
End to end	E/E	Extension manhole	Ext M/H
Engine	eng	Exter	Ex
Engineer(ing)	engr(g)	Exterior	extr
Englewood	Eglwd	External	ext
Enlarged	enl	External upset end	EUE
Entrada	Ent	Extraction	extrac
Eocene	Eoc	Extra heavy	X-hvy
Eponides	Epon	Extra strong	x-stg

Extreme pressure	EP
Extreme line (casing)	X-line

F

Fabricate(d)	fab
Faced and drilled	F&D
Facet(ed)	fac
Face to face	F to F
Failed	fld
Failure	fail
Faint	fnt
Faint air blow	FAB
Fair	fr
Fall River	Fall Riv
Farmington	Farm
Farmout	FO
Fault	flt
Faulted out	FO
Fauna	fau
Favosites	fvst
Federal	fed
Federal Employers Liability Act	FELA
Feed	FD
Feeder	fdr
Feed rate	FR
Feedstock	FS
Feet	ft
Feet per hour	ft/hr
Feet per minute	fpm
Feet per minute	ft/min
Feet per second	fps
Feet per second	ft/sec
Feldspar(rhic)	fld
Female pipe thread	FPT
Female to female angle	FFA
Female to female globe (valve)	FFG
Ferguson	Ferg
Ferruginous	ferr
Ferry Lake anhydrite	FLA
Fertilizer	fert
Fiberglass reinforced plastic	FRP
Fibrous	fib
Field	fld
Field authorized to commence operations	FACO
Field purchase order	FPO
Field wildcat	FWC
Figure	fig

Fillister	fill
Fill up	FU
Filter cage	FC
Filtrate	filt
Final	fin
Final boiling point	FBP
Final bottom-hole pressure flowing	FBHPF
Final bottom-hole pressure shut-in	FBHPSI
Final flowing pressure	FFP
Final hydrostatic pressure	FHP
Final pressure	FP
Final report for rig	FRR
Final report for well	FRQ
Final shut-in pressure	FSIP
Final tubing pressure	FTP
Fine	fn
Fine-grained	f-gr
Finely	fnly
Finely-crystalline	f/xln
Finish	fin
Finish all over	FAO
Finished	fin
Finished drilling	fin drlg
Finish going in hole	FGIH
Finish going in with—	FGIW
Fireproof	fprf
Fire-resistant oil	F-R Oil
Fishing	fish
Fishing	fsg
Fishing for	FF
Fissile	fisl
Fissure	fis
Fittings	ftg
Fixed	fxd
Fixed carbon	FC
Fixture	fix
Flaky	flk
Flanged and dished (heads)	F & D
Flanged and spigot	F & S
Flanged one end, welded one end	FOE—WOE
Flange (d) (s)	flg (d) (s)
Flash Point, Cleveland Open Cup	FI—COC
Flat face	FF
Flathead	Flath
Flattened	flat
Flexible	flex
Flippen	Flip

Float	flt	Formation	fm
Float collar	FC	Formation density	F-D
Floating	fltg	Formation density log	FDL
Float shoe	FS	Formation gas-oil ratio	F/GOR
Floor	FL	Formation interval tester	FIT
Floor drain	FD	Formation water	Fm W
Florence Flint	Flor Fl	Fort Chadbourne	Ft C
Flow	flo	Fort Hayes	Ft H
Flow control valve	FCV	Fort Riley	Ft R
Flow indicator	FI	Fort Worth	Ft W
Flow indicating controller	FIC	Fort Union	Ft U
Flow indicating ratio controller	FIRC	Fortura	Fort
Flow line	FL	Forward	fwd
Flow rate	FR	Foot	ft
Flow recorder	FR	Footage	ftg
Flow recorder control	FRC	Foot-candle	ft-c
Flowed or flowing	fl/	Footing	ftg
Flowed, flowing	F/	Foot-pound	ft lb
Flowed, flowing	flw (d) (g)	Foot-pound-second (system)	fps
Flowed(ing) at rate of	FARO	Foot pounds per hour	ft lbs/hr
Flowerpot	Flwrpt	Fossiliferous	foss
Flowing	flg	Foundation	fdn
Flowing bottom-hole pressure	FBHP	Fountain	Fount
Flowing by heads	FBH	Four-wheel drive	FWD
Flowing casing pressure	FCP	Fox Hills	Fox H
Flowing on test	FOT	Frac finder (Log)	FF
Flowing pressure	Flwg. Pr.	Fractional frosted	fr
Flowing pressure	FP	Fractionation	fract
Flowing surface pressure	FSP	Fractionator	fract
Flowing tubing pressure	FTP	Fracture, fractured, fractures	frac (d) (s)
Flue	flu	Fracture gradient	F.G.
Fluid in hole	FIH	Fragment	frag
Fluid	fl	Framework	frwk
Fluid	flu	Franchise	fran
Fluid catalytic cracking	FCC	Franconia	Franc
Fluid level	FL	Fredericksburg	Fred
Fluid to surface	FTS	Fredonia	Fred
Fluorescence	fluor	Free on board	FOB
Fluorescent	fluor	Free point indicator	FPI
Flush	FL	Freezer	frzr
Flushed	flshd	Freezing point	FP
Flush joint	FJ	Freight	frt
Focused log	FOCL	Frequency	freq
Foliated	fol	Frequency meter	FM
Foraker	forak	Frequency modulation	FM
Foraminifera	foram	Fresh	frs
Foreman	f'man	Fresh break	FB
For example	e.g.	Fresh water	FW
For your information	FYI	Friable	fri
Forged steel	FS	Friction reducing agent	FRA
Forged steel	FST	Froggy	Frgy

From	fr	Gallons condensate per day	GCPD
From east line	FE/L	Gallons condensate per hour	GCPH
From east line	fr E/L	Gallons gelled water	GGW
From north line	FNL	Gallons heavy oil	GHO
From north line	fr N/L	Gallons mud acid	GMA
From northeast line	FNEL	Gallons oil	GO
From northwest line	FNWL	Gallons oil per hour	GOPH
From south and west lines	FS&WLs	Gallons oil per day	GOPD
From southeast line	FSEL	Gallons solution	gal sol
From south line	fr S/L	Gallons water per hour	GWPH
From south line	FSL	Gallons per day	GPD
From southwest line	FSWL	Gallons per hour	GPH
From west line	fr W/L	Gallons per minute	gal/min
From west line	FWL	Gallons per minute	GPM
Front	fr	Gallons per thousand cubic feet	gal/Mcf
Front & side, flange x screwed	F/S	Gallons per thousand cubic feet	gpm
Frontier	Fron	Gallons per second	GPS
Frosted	fros	Gallons regular acid	GRA
Frosted quartz grains	FQG	Gallons salt water	GSW
Fruitland	Fruit	Gallons water	GW
Fuel oil	FO	Galvanized	galv
Fuel oil equivalent	F.O.E.	Gamma ray	GR
Fuels & fractionation	F&F	Gas	G
Fuels & Lubricants	F & L	Gas and mud-cut oil	G&MCO
Fullerton	Full	Gas and oil	G&O
Full hole	FH	Gas and oil-cut mud	G&OCM
Full interest	FI	Gas and water	G&W
Full of fluid	FF	Gas-condensate ratio	GCR
Full open head (grease drum 120 lb)	FOH	Gas cut	GC
Full opening	FO	Gas-cut acid water	GCAW
Funnel viscosity	FV	Gas-cut distillate	GCD
Furfural	furf	Gas-cut load oil	GCLO
Furnace	furn	Gas-cut load water	GCLW
Furnace Fuel Oil	FFO	Gas-cut mud	GCM
Furniture and fixtures	Furn & fix	Gas-cut oil	GCO
Fuson	Fus	Gas-cut salt water	GCSW
Fusselman	Fussel	Gas-cut water	GCW
Future	fut	Gas-distillate ratio	GDR
Fusulinid	Fusul	Gas injection	GI
		Gas-injection well	GIW
		Gasket	gskt
		Gas lift	GL

G

		Gas-liquid ratio	GLR
Gage (d) (ing)	ga	Gas odor	GO
Galena	Glna	Gas odor distillate taste	GODT
Gallatin	Gall	Gas-oil contract	GOC
Gallon	gal	Gas-oil ratio	GOR
Gallons	gal	Gasoline	gaso
Gallons acid	GA	Gasoline plant	GP
Gallons breakdown acid	GBDA		

Gas pay	GP	Government	govt
Bas purchase contract	GPC	Governor	gov
Gas Rock	G Rk	Grade	gr
Gas sales contract	GSC	Grading	grdg
Gas show	GS	Grading location	grd loc
Gas too small to measure	GTSTM	Gradual	grad
Gas to surface (time)	GTS	Gradually	grad
Gastropod	gast	Grain	gr
Gas Unit	GU	Grained (as in fine-grained)	gnd
Gas volume	GV	Grains per gallon	GPG
Gas volume not measured	GVNM	Gram	g
Gas-water contact	GWC	Gram	gm
Gas well	GW	Gram-calorie	g-cal
Gas-well gas	GWG	Gram-calorie	gm-cal
Gas well shut-in	GSI	Gram molecular weight	g mole
Gathering line	G/L	Graneros	Granos
General	genl	Granite	gran
General Land Office (Texas)	GLO	Granite Wash	Gran W
Generator	gen	Grant (of land)	grt
Geological	Geol	Granular	gran
Geologist	Geol	Granular	grnlr
Geology	Geol	Graptolite	grap
Geophysical	Geop	Grating	grtg
Geophysic	Geop	Gravel	gvl
Georgetown	Geo	Gravel packed	GVLPK
Gibson	Gib	Gravitometer	grvt
Gilcrease	Gilc	Gravity	grav
Gilsonite	gil	Gravity	GTY
Glass	gls	Gravity, °API	gr API
Glassy	gl	Gravity meter	GM
Glauconite	glau	Gray	gry
Glauconitic	glau	Grayburg	Grayb
Glen Dean lime	GD	Gray sand	Gr Sd
Glen Rose	GR	Grayson	Gray
Glenwood	Glen	Graywacke	gywk
Globigerina	Glob	Grease	gr
Glorieta	Glor	Greasy	gsy
Glycol	glyc	Green	grn
Gneiss	gns	Greenhorn	GH
Going in hole	GIH	Green River	Grn Riv
Golconda lime	Gol	Green shale	Grn sh
Good	gd	Gritty	grty
Good fluorescence	GFLU	Grooved	grv
Good odor & taste	gd o&t	Grooved ends	GE
Good show of gas	GSG	Gross	grs
Good show of oil	GSO	Gross acre feet	GAF
Goodland	Gdld, Good L	Grross royalty	gr roy
Goodwin	Gdwn	Gross weight	gr wt
Goose egg	g egg	Ground	gr
Gorham	Gor	Ground	grd
Gouldbusk	Gouldb	Ground joint	GJ

Ground level	GL	Heavily water-cut mud	HWCM
Ground measurement		Heavy cycle oil	HCO
(elevation)	GM	Heavy duty	HD
Guard lot	CRDL	Heavy fuel oil	HFO
Gull River	G. Riv	Heavy oil	HO
Gummy	gmy	Heavy steel drum	HSD
Gun barrel	GB	Heebner	Heeb
Gun Perforate	G/P	Height	hgt
Gunsite	Guns	Heirs	hrs
Gypsiferous	gypy	Held by production	HBP
Gypsum	gyp	Hematite	hem
Gypsum Springs	Gyp Sprgs	Herington	Her
Gyroidina	Gyr	Hermosa	Herm
Gyroidina scal	Gyr sc	Hertz (new name for electrical	
		cycles per second)	Hz
		Heterostegina	het

H

		Hexagon(al)	hex
		Hexane	hex
Hackberry	Hackb	Hickory	Hick
Hackly	hky	High detergent	HD
Hand-control valve	HCV	High gas-oil ratio	HGOR
Hand hole	HH	High-level shut-down	HLSD
Handle	hdl	Highly	hily
Hanger	hgr	High pressure	HP
Haragan	Hara	High-pressure gas	HPG
Harbor	hbr	High-resolution dipmeter	HRD
Hard	hd	High temperature	ht
Hardinsburg sand (local)	Hberg	High-temperature shut-down	HTSD
Hard lime	hd li	High tension	ht
Hardness	hdns	High viscosity	HV
Hard sand	hd sd	High viscosity index	HVI
Hardware	hdwe	Highway	hwy
Haskell	Hask	Hilliard	Hill
Haynesville	Haynes	Hockelyensis	hock
Hazardous	haz	Hogshooter	Hog
Head	hd	Holes per foor	HPF
Header	hdr	Hole full of oil	HFO
Headquarters	HQ	Hole full of salt water	HGSW
Heater	htr	Hole full of sulphur water	HF Sul W
Heat exchanger	HEX	Hole full of water	HFW
Heat exchanger	HX	Hollandberg	Holl
Heat treated	ht	Home Creek	Home Cr
Heater treater	ht	Hookwall packer	HWP
Heating oil	HO	Hoover	Hov
Heavy	hvy	Hopper	hop
Heavily	hvly	Horizontal	horiz
Heavily gas-cut mud	HGCM	Horsepower	HP
Heavily gas-cut water	HGCW	Horsepower hour	hp hr
Heavily oil and gas-cut mud	HO&GCM	Hospah	Hosp
Heavily oil-cut mud	HOCM	Hot-rolled steel	HRS
Heavily oil-cut water	HOCW	Hour	hr

Hours	hr	Indicates	indic
Housebrand (regular grade of gasoline)	HB	Indication	indic
		Indistinct	indst
Hoxbar	Hox	Individual	indiv
Humblei	Humb	Induction	ind
Humphreys	Hump	Indurated	indr
Hundred weight	cwt	Inflammable liquid	Inf L
Hunton	Hun	Inflammable solid	Inf S
Hydraulic	HYD	Information	info
Hydraulic horsepower	HHP	Inhibitor	inhib
Hydraulic pump	HP	Initial	init
Hydril	HD	Initial air blow	IAB
Hydril thread	HYD	Initial boiling point	IBP
Hydril Type A joint	HYDA	Initial bottom-hole pressure	IBHP
Hydril Type CA joint	HYDCA	Initial bottom-hole pressure flowing	IBHPF
Hydril Type CS joint	HYDCS		
Hydrocarbon	HC	Iniital bottom-hole pressure shut-in	IBHPSI
Hydrofining	HFG		
Hydrostatic head	HH	Initial flowing pressure	IFP
Hydrostatic pressure	HP	Initial hydrostatic pressure	IHP
Hydrotreater	hydtr	Initial mud weight	IMW
Hygiene	Hyg	Initial potential	IP
		Initial potential flowed	IPF

I

		Initial production flowed (ing)	IPF
		Initial pressure	IP
		Initial production	IP
Identification sign	I.D. Sign	Initial production gas lift	IPG
Idiomorpha	Idio	Initial production on intermitter	IPI
Igneous	Ign	Initial production plunger lift	IPL
Imbedded	imbd	Initial production pumping	IPP
Immediate(ly)	immed	Initial production swabbing	IPS
Impervious	imperv	Initial shut-in pressure (DST)	ISIP
Imperial	Imp	Initial vapor pressure	IVP
Imperial gallon	Imp gal	Injected	inj
Impression block	IB	Injection	inj
Inbedded	inbdd	Injection gas-oil ratio	IGOR
Incandescent	incd	Injection pressure	Inj Pr
Inch(es)	in	Injection rate	IR
Inches mercury	in. Hg	Injection well	IW
Inches per second	in/sec	Inland	inl
Inch-pound	in-lb	Inlet	inl
Incinerator	incin	Inoceramus (paleo)	Inoc
Include	incl	Input/output	I/O
Included	incl	Inside diameter	ID
Including	incl	Inside screw (valve)	IS
Inclusions	incls	Inspect	insp
Incorporated	Inc	Inspected	insp
Increase (d) (ing)	incr	Inspecting	insp
Indicate	indic	Inspection	insp
Indicated horsepower	IHP	Installation(s)	instl
Indicated horsepower hour	IHPHR	Installed (ed) (ing)	inst

Installed pumping equipment	INPE
Installing pumping equipment	INPE
Install(ing) pumping equipment	IPE
Instantaneous	inst
Instantaneous shut-in pressure (frac)	ISIP
Institute	inst
Instrument	instr
Instrumentation	instr
Insulate	ins
Insulate	insul
Insulation	ins
Insurance	ins
Integrator	intgr
Intention to drill	ITD
Interbedded	inbd
Interbedded	interbd
Intercooler	incolr
Inter-crystalline	inter-xln
Interest	int
Integral joint	IJ
Intergranular	ingr
Intergranular	inter-gran
Interior	int
Interlaminated	inlam
Interlaminated	inter-lam
Internal	int
Internal flush	IF
Internal Revenue Service	IRS
Internal upset ends	IUE
Intersect	ints
Interstitial	intl
Interval	int
Interval	intv
Intrusion	intr
Inventory	inven
Invert	inv
Invertebrate	invrtb
Inverted	inv
Invoice	inv
Ireton	Ire
Iridescent	irid
Iron body brass (bronze) mounted (valve)	IBBM
Iron body brass core (valve)	IBBC
Iron body (valve)	IB
Iron case	IC
Iron pipe size	IPS
Iron pipe thread	IPT
Ironstone	Fe-st
Ironstone	irst
Irregular	irreg
Isometric	isom
Isopropyl alcohol	IPA
Isothermal	isoth
Iverson	Ives

J

Jacket	jac
Jackson	Jack
Jackson	Jxn
Jackson sand	Jax
Jammed	jmd
Jasper(oid)	Jasp
Jefferson	Jeff
Jelly-like colloidal suspension	gel
Jet perforated	JP
Jet perforations per foot	JP/ft
Jet propulsion fuel	JP fuel
Jet shots per foot	JSPF
Jobber	jbr
Job complete	JC
Joint interest non-operated (property)	JINO
Joint operating agreement	JOA
Joint operating provisions	JOP
Joint operation	J/O
Joint(s)	jt(s)
Joint venture	JV
Jordan	Jdn
Judith River	Jud R
Junction	jct
Junction box	JB
Junk basket	JB
Junk(ed)	jnk
Junked and abandoned	J&A
Jurassic	Jur
Jurisdiction	juris

K

Kaibab	Kai
Kansas City	KC
Kaolin	kao
Kayenta	Kay
Keener	Ke
Kelly bushing	KB
Kelly bushing measurement	KBM
Kelly drive bushing	KDB
Kelly drive bushing elevation	KDBE

Kelly drive bushing to landing flange	KDB—LDG FLG	Laid (laying) down drill collars	LDDCs
Kelly drill bushing to mean low water	KDB—MLW	Laid (laying) down drill pipe	LDDP
		Lakota	Lak
Kelly drive busing to platform	KDB—Plat	Laminated	lam
Kelvin (temperature scale)	K	Laminations	lam
Keokuk-Burlington	Keo-Bur	La Motte	La Mte
Kerosine	kero	Land(s)	ld(s)
Ketone	ket	Landulina	Land
Keystone	Key	Lansing	Lans
Kiamichi	Kia	Lap joint	LJ
Kibbey	Kib	Lapweld	LW
Kicked off	KO	Laramie	Lar
Kockoff point	KOP	Large	lg
Killed	kld	Large	lrg
Kill(ed) well	KW	Large Discorbis	Lg Disc
Kiln dried	KD	Latitude	lat
Kilocalorie	kcal	Lauders	Laud
Kilocycle	kc	Layton	Layt
Kilogram	kg	Leached	lchd
Kilogram calorie	kg-cal	Leadville	Leadv
Kilogram meter	kg-m	League	Lge
Kilohertz (See Hz-Hertz)	KHz	Leak	lk
Kilometer	km	Lease	lse
Kilovar-hour	kvar hr	Lease automatic custody transfer	LACT
Kilovar; reactive kilovolt ampere	kvar	Lease crude	LC
		Lease use (gas)	LU
Kilovolt	kv	Leavenworth	Lvnwth
Kilovolt-ampere	kva	Le Compton	Le C
Kilovolt-ampere-hour	kvah	Left hand	LH
Kilowatt	kw	Left in hole	LIH
Kilowatt hour	kwh	Legal subdivision (Canada)	LSD
Kilowatt-hourmeter	kwhm	Length	lg
Kincaid lime	Kin, KD	Lennep	Len
Kinderhook	Khk	Lense	lns
Kinematic	Kin	Lenticular	len
Kinematic viscosity	KV	Less-than-carload lot	LCL
Kirtland	Kirt	Less than truck load	LTL
KMA sand	KMA	Letter	ltr
Knock out	KO	Level	lvl
Kootenai	Koot	Level alarm	LA
Krider	Kri	Level controller	LC
		Level control valve	LCV
		Level glass	lg
		Level indicator	LI
		Level indicator controller	LIC
		Level recorder	LR
		Level recorder controller	LRC

L

Labor	Lab	License	lic
Laboratory	Lab	Liebusculla	Lieb
Ladder	lad	Light	lt
Laid down	LD		
Laid down cost	LDC		

Light barrel	LB	Long	lg
Light brown oil stain	LBOS	Long coupling	LC
Lightening avvester	LA	Long handle/round point	LH/RP
Light fuel oil	LFO	Longitude(inal)	long
Lighting	ltg	Long radius	LR
Light iron barrel	LIB	Long threads & coupling	LT&C
Light iron grease barrel	LIGB	Loop	lp
Light steel drum	LSD	Lost circulation	LC
Lignite	lig	Lost circulation material	LCM
Lignitic	lig	Lovell	Lov
Lime	li	Lovington	Lov
Lime	lim	Low pressure	LP
Limestone	li	Low-pressure separator	LP sep
Limestone	ls	Low-temperature extraction unit	LTX unit
Limit	lim	Low-temperature separation unit	LTS unit
Limited	ltd	Low-temperature shut down	LTSD
Limonite	lim	Low viscosity index	LVI
Limy	lmy	Lower	low
Limy shale	Lmy sh	Lower	lwr
Linear	lin	Lower Albany	L/Alb
Linear foot	lin ft	Lower anhydrite stringer	LAS
Line, as in E/L (East line)	/L	Lower, as L/Gallup	L/
Line pipe	L.P.	Lower Cretaceous	L/Cret
Liner	lnr	Lower explosive limit	LEL
Liner	lin	Lower Glen Dean	LGD
Linguloid	lngl	Lower Menard	LMn
Liquefaction	liqftn	Lower Tuscaloosa	L/Tus
Liquefied natural gas	LNG	Lube oil	LO
Liquefied petroleum gas	LP-Gas	Lubricant	lub
Liquid	liq	Lubricate (d) (ing) (ion)	lub
Liquid level controller	LLC	Lueders	Lued
Liquid level gauge	LLG	Lug cover type (5-gallon can)	LC
Liquid volume	LV	Lug cover with pour spout	LCP
Liter	l	Lumber	lbr
Lithographic	litho	Lumpy	lmpy
Little	ltl	Lustre	lstr
Load	ld		
Load acid	LA		
Load oil	LO		
Load water	LW	**M**	
Local purchase order	LPO		
Located	loc		
Location	loc	Macaroni tubing	MT
Location abandoned	loc abnd	Machine	mach
Location graded	loc gr	Mackhank	Mack
Lock	lk	Madison	Mad
Locker	lkr	Magnetic	mag
Lodge pole	LP	Magnetometer	mag
Log mean temperature difference	LMTD	Maintenance	maint
Log total depth	LTD	Major	maj
		Majority	maj

Male and female (joint)	M&F	Maximum efficient rate	MER
Male pipe thread	MPT	Maximum flowing pressure	MFP
Male to female angle	MFA	Maximum operating pressure	MOP
Malleable	mall	Maximum pressure	MP
Malleable iron	MI	Maximum surface pressure	MSP
Management	m'gmt	Maximum top pressure	MTP
Manager	mgr	Maximum tubing pressure	MTP
Manhole	MH	Maximum working pressure	MWP
Manifold	man	Maywood	May
Manifold	MF	McClosky lime	McC
Manitoban	Manit	McCullough	McCul
Manning	Mann	McElroy	McEl
Manual	man	McKee	McK
Manually operated	man op	McLish	McL
Manufactured	mfd	McMillan	McMill
Manufacturing	mfg	Meakin	Meak
Maquoketa	Maq	Mean effective pressure	MEP
Marble Falls	Mbl Fls	Mean low water to platform	MLW-PLAT
Marchand	March	Mean temperature difference	MTD
Marginal	marg	Measure (ed) (ment)	meas
Marginulina	Marg	Measured depth	MD
Marginulina coco	Marg coco	Measured total depth	MTD
Marginulina flat	Marg fl	Measuring and regulating	
Marginulina round	Marg rd	station	M & R Sta.
Marginulina texana	Marg tex	Mechanic (al) (ism)	mech
Marine	mar	Mechanical down time	Mech DT
Marine gasoline	margas	Mechanism	mchsm
Marine rig	MR	Median	med
Marine terminal	M/T	Medicine Bow	Med B
Marine Tuscaloosa	M'Tus	Medina	Med
Marine wholesale distributors	MWD	Medium	med
Market(ing)	mkt	Medium amber cut	MAC
Market demand factor	MDF	Medium fuel oil	med FO
Markham	Mark	Medium-grained	m-gr
Marlstone	mrlst	Medium-grained	med gr
Marly	mly	Medrano	Medr
Marmaton	Marm	Meeteetse	Meet
Maroon	mar	Megacycle	mc
Massilina pratti	Mass pr	Megahertz (megacycles per	
Massive	mass	second)	MHz
Massive Anhydrite	MA	Melting point	MP
Master	mstr	Member (geologic)	mbr
Material	matl	Memorandum	memo
Material	mtl	Menard lime	Men
Mathematics	math	Menefee	Mene
Matrix	mtx	Meramec	Mer
Matter	mat	Mercaptan	mercap
Maximum	max	Mercaptan	RSH
Maximum & final pressure	M&FP	Merchandise	mdse
Maximum daily delivery		Mercury	merc
obligation	MDDO	Meridian	merid

Mesaverde	mvde	Million electron volts	mev
Mesozoic	Meso	Million standard cubic feet per	
Metal petal basket	MPB	day	MMSCFD
Metamorphic	meta	Millivolt	mv
Meter	m	Mineral	mnrl
Meter	mtr	Minerals	min
Meter run	MR	Mineral interest	MI
Methane	meth	Minimum	min
Methane-rich gas	MRG	Minimum pressure	min P
Methanol	methol	Minnekahta	Mkta
Methyl chloride	meth-cl	Minnelusa	Minl
Methyl ethyl ketone	MEK	Minute(s)	min
Methy isobutyl ketone	MIK	Miocene	Mio
Methylene blue	MB, meth-bl	Miscellaneous	misc
Metric	metr	Misener	Mise
Mezzanine	mezz	Mission Canyon	Miss Cany
Mica	mic	Mississipian	Miss
Micaceous	mic	Mixed	mxd
Microcrystalline	mcr-x	Mixer	mix
Microcrystalline	micro-xin	Mobile	mob
Microfarad	mfd	Model	mod
Microfossil(iferous)	micfos	Moderate(ly)	mod
Microwave	MW	Modification	mod
Middle	M/	Modular	modu
Middle	mdl	Moenkopi	Moen
Middle	mdl	Moisture, impurities, and	
Midway	Mid	unsaponifiables (grease	
Midway	Mwy	testing)	MIU
Mile(s)	MI	Molas	mol
Miles per hour	MPH	Mole	mol
Military	mil	Molecular weight	mol wt
Milky	mky	Mollusca	mol
Mill wrapped plain end	MWPE	Monitor	mon
Milled	mld	Montoya	Mont
Milled other end	MOE	Moddy's Branch	MB
Milliampere	ma	Moore County Lime	MC Ls
Millidarcies	md	Mooringsport	Moor
Milligram	mg	More or less	m/l
Millihenry	mh	Morrison	Morr
Milliliter	ml	Morrow	Mor
Millimeters of mercury	mm HG	Mortgage	mtge
Milliliters tetrethyl lead per		Mosby	Mos
gallon	ml TEL	Motor	mot
Millimeter	mm	Motor generator	MG
Milling	millg	Motor medium	MM
Milling	mlg	Motor octane number	MON
Milliolitic	mill	Motor oil	MO
Million	mil	Motor oil units	MOU
Million British thermal units	MMBTU	Motor severe	MS
Million cubic feet	MMCF	Motor vehicle	M/V
Million cubic feet/day	MMCFD	Motor vehicle fuel tax	MVFT

Motor vessel	M/V
Mottled	mott
Mount Selman	Mt. Selm
Mounted	mtd
Mounting	mtg
Moving	mov
Moving in (equipment)	MI
Moving in and rigging up	MIRU
Moving in cable tools	MICT
Moving in completion unit	MICU
Moving (moved) in double drum unit	MIDDU
Moving in materials	MIM
Moving in pulling unit	MIPU
Moving in rig	MIR
Moving in rotary tools	MIRT
Moving in service rig	MISR
Moving in standard tools	MIST
Moving in tools	MIT
Moving out	MO
Moving out cable tools	MOCT
Moving out completion unit	MOCU
Moving out rig	MOR
Moving our rotary tools	MORT
Mowry	Mow
Mud acid	MA
Mud acid wash	MAW
Mud cleanout agent	MCA
Mud cut	MC
Mud filtrate	MF
Mud logging unit	MLU
Mud logger	ML
Mud to Surfce	MTS
Mud weight	md wt
Mud weight	mud wt
Mud-cut acid	MCA
Mud-cut oil	MCO
Mud-cut salt water	MCSW
Mud-cut water	MCW
Muddy	Mdy
Muddy water	MW
Mudstone	mudst
Multi-grade	MG
Multiple service acid	MSA
Multipurpose	MP
Multipurpose grease, lithium base	MPGR-Lith
Multipurpose grease, soap base	MPGR-soap
Muscovite	musc

N

Nacatoch	Nac
Nacreous	nac
Nameplate	NP
Naphtha	nap
National	Nat'l
National coarse thread	NC
National Electric Code	NEC
National Fine (thread)	NF
National pipe thread	NPT
National pipe thread, female	NPTF
National pipe thread, male	NPTM
Natural	nat
Natural flow	NF
Natural gas	NG
Natural gas liquids	NGL
Nautical mile	NMI
Navajo	Nav
Navarro	Navr
Negative	neg
Negligible	neg
Neoprene	npne
Net effective pay	NEP
Net tons	NT
Neutral	neut
Neutralization	neut
Neutralization Number	Neut. No.
New Albany shale	New Alb
New bit	NB
Newburg	Nbg
Newcastle	Newc
New field discovery	NFD
New field wildcat	NFW
New oil	NO
New pool discovery	NPD
New pool exempt (nonprorated)	NPX
New pool wildcat	MPW
New total depth	NTD
Nigara	Nig
Nickel plated	NP
Ninnescah	Nine
Niobrara	Niob
Nipple	nip
Nipple (d) (ing) up blowout preventers	NUBOPs
Nipple (d) (ing) down blowout preventers	NDBOPs
Nippling up	NU

Nitrogen blanket	NB	Northeast corner	NEC
Nitroglycerine	nitro	Northeast line	NEL
No appreciable gas	NAG	Northeast quarter	NE/4
Noble-Olson	NO	Northerly	N'ly
No change	NC	Northern Alberta land	
No core	NC	registration district	NALRD
Nodosaria blanpiedi	Nod Blan	North half	N/2
Nodosaria Mexicana	Nod Mex	North line	NL
Nodular	nod	North offset	N/O
Nodule	nod	North quarter	N/4
No fluorescence	NF	Northwest	NW
No fluid	NF	Northwest corner	NW/C
No fuel	NF	Northwest line	NWL
No gas to surface	NGTS	Northwest quarter	NW/4
No gauge	NG	Northwest Territories	NWT
No increase	No Inc	No show	NS
Nominal	nom	No show gas	NSG
Nominal pipe size	NPS	No show oil	NSO
Non-contiguous tract	NCT	No show oil & gas	NSO&G
Non-destructive testing	NDT	Not drilling	ND
Non-detergent	ND	Not applicable	NA
Nonemulsifying agent	NE	Notary public	NP
Non-emulsion acid	NEA	Not available	NA
Nonflammable compressed		No test	N/tst
gas	nonf G	No time	NT
Nonionella	Non	Not in contact	NIC
Nonionella Cockfieldensis	N. Cock.	Not prorated	NP
Non-leaded gas	NL Gas	Not pumping	NP
Non-operated joint ventures	NOJV	Not reported	NR
Non-operating property	NOP	Not to Scale	NTS
Non-porous	NP	Not yet available	NYA
Non-returnable	NR	No visible porosity	NVP
Non-returnable steel barrel	NRSB	No water	NW
Non-returnable steel drum	NRSD	Nozzle	noz
Non-rising stem (valve)	NRS	Nugget	Nug
Non-standard	nstd	Number	NO
Non-standard service station	N/S S/S	Numerous	num
Non-upset	NU		
Non-upset ends	NUE		
Noodle Creek	Ndl Cr		
No order required	NOR		
No paint on seams	npos		
No production	NP	Oakville	Oakv
No recovery	no rec	Object	obj
No recovery	NR	Obsolete	obsol
No report	NR	Occasional(ly)	occ
No returns	NR	Ocean bottom suspension	OBS
Normal	nor	Octagon	oct
Normally closed	NC	Octagonal	oct
Normally open	NO	Octane	oct
Northeast	NE	Octane number requirement	ONR

O

314

Octane number requirement increase	ONRI	Oil-cut water	OCW
O'Dell	Odel	Oil-powered total energy	OTE
Odor	od	Oil-water contact	OWC
Odor, stain & fluorescence	O, S & F	Oil-well gas	OWG
Odor, taste & stain	O, T & S	Old abandoned well	OAW
Odor, taste, stain & fluorescence	O, T, S & F	Old plug-back depth	OPBD
		Old total depth	OTD
Off bottom	OB	Old well drilled deeper	OWDD
Office	off	Old well plugged back	OWPB
Official	off	Old well worked over	OWWO
Official potential test	OPT	Olefin	ole
Off-shore	off-sh	Oligocene	Olig
O'Hara	O'H	On center	OC
Ohm-centimeter	ohm-cm	One thousand foot-pounds	kip-ft
Ohm-meter	ohm-m	One thousand pounds	kip
Oil	O	Oolicastic	ooc
Oil and gas	O & G	Oolimoldic	oom
Oil and Gas Journal	OGJ	Oolitic	ool
Oil and gas lease	O&GL	Open (ing) (ed)	opn
Oil and gas-cut mud	O&GCM	Open cup	OC
Oil and gas-cut salt water	O&GCSW	Open end	OE
Oil & gas-cut sulphur water	O&GC SULW	Open flow	OF
Oil and gas-cut water	O&GCW	Open flow potential	OFP
Oil and salt water	O&SW	Open hearth	OH
Oil & sulphur water-cut mud	O&SWCM	Open hole	OH
Oil and water	O&W	Open tubing	Ot
Oil base mud	OBM	Operate	oper
Oil change	O/C	Operations	oper
Oil circuit breaker	OCB	Operations commenced	OC
Oil Creek	Oil Cr	Operator	oper
Oil cut	OC	Operculinoides	Operc.
Oil emulsion	OE	Opposite	opp
Oil emulsion mud	OEM	Optimum bit weight and rotary speed	OBW & RS
Oil fluorescence	OFLU	Option to farmout	optn to F/O
Oil in hole	OIH	Ordovician	Ord
Oil in place	OIP	Oread	Or
Oil odor	OO	Orifice	orf
Oil pay	OP	Orifice flange one end	OFOE
Oil payment interest	OPI	Organic	org
Oil sand	O sd	Organization	org
Oil show	OS	Original	orig
Oil soluble acid	OSA	Originally	orig
Oil stain	OSTN	Original stock tank oil in place	OSTOIP
Oil string flange	OSF	Oriskany	Orisk
Oil to surface	OTS	Orthoclase	orth
Oil unit	OU	Osage	Os
Oil well flowing	OWF	Osborne	Osb
Oil well shut-in	OSI	Ostracod	Ost
Oil-cut mud	OCM	Oswego	Osw
Oil-cut salt water	OCSW	Other end beveled	OEB

Ounce	oz	Park City	Park C
Ouray	Our	Part	pt
Out of service	O/S	Partings	prtgs
Out of stock	O/S	Partly	prtly
Outboard motor oil	OBMO	Partly	pt
Outer continental shelf	OCS	Parts per million	ppm
Outlet	otl	Patent(ed)	pat
Outpost	OP	Pattern	patn
Outside diameter	OD	Pavement	pvmnt
Outside screw and yoke		Paving	pav
(valve)	OS&Y	Pawhuska	Paw
Over and short (report)	O/S	Payment	PP
Over produced	OP	Payment	payt
Overall	OA	Pearly	prly
Overall height	OAH	Pebble	pbl
Overall length	OAL	Pebbles	pebs
Overhead	OH	Pebbly	pbly
Overhead	ovhd	Pecan Gap	PG
Overriding royalty	ORR	Pecan Gap Chalk	PGC
Overriding royalty interest	ORRI	Pelecypod	plcy
Overshot	OS	Pelletal	pell
Overtime	OT	Palletoidal	pell
Oxidation	ox	Penalize, penalized, penalizing	
Oxidation	ox	penalty	penal
Oxygen	oxy	Penetration test	pen
		Penetration	pen

P

		Penetration asphalt cement	Pen A.C.
		Penetration index	PI
		Pennsylvanian	Penn
Package	pkg	Pensky Martins	PM
Packaged	pkgd	Per acre bonus	PAB
Packed	pkd	Per acre rental	PAR
Packer	pkr	Per day	PD
Packer set at	PSA	Percent	pct
Packing	pkg	Percolation	perco
Paddock	Padd	Perforate (d) (ing) (or)	perf
Pahasapa	Paha	Perforated casing	perf csg
Paid	PD	Performance Number (Av gas)	PN
Paint Creek	PC	Period	prd
Pair	pr	Permanent	perm
Paleontology	paleo	Permanently shut down	PSD
Paleozoic	Paleo	Permeability	perm
Palo Pinto	Palo P	Permeable	perm
Paluxy	Pal	Permian	Perm
Paluxy	Pxy	Permit	prmt
Panel	pnl	Perpendicular	perp
Panhandle Lime	Pan L	Personal and confidential	P & C
Paradox	Para	Personnel	pers
Paraffins, olefins, naphthenes,		Petrochemical	petrochem
aromatics	PONA	Petroleum	pet
Parish	Ph	Petroleum & natural gas	P & NG

Petroliferous	petrf	Pliocene	Plio
Pettet	Pet	Plug on bottom	POB
Pettit	Pett	Plugged	plgd
Pettus sd	Pet sd	Plugged & abandoned	P&A
Phase	ph	Plugged Back	PB
Phosphoria	Phos	Plugged back depth	PBD
Phrrhotite	po	Plugged back total depth	PBTD
Picked up	PU	Plunger	plngr
Pictured Cliff	Pic Cl	Pneumatic	pneu
Piece	pc	Podbielniak	Pod.
Pieces	pcs	Point	pt
Pilot	plt	Point Lookout	Pt Lkt
Pin Oak	P.O.	Poison	pois
Pine Island	PI	Poker chipped	PC
Pink	pk	Polish(ed)	pol
Pinpoint	PP	Polished rod	PR
Pinpoint	pinpt	Polyethylene	polyel
Pinpoint porosity	PPP	Polymerization	poly
Pint	pt	Polymerized	poly
Pipe, buttweld	PBW	Polymerized gasoline	polygas
Piel, electric weld	PEW	Polypropylene	polypl
Pipe, lapweld	PLW	Polyvinyl chloride	poly cl
Pipe, seamless	PSM	Polyvinyl chloride	PVC
Pipe, spiral weld	PSW	Pontotoc	Pont
Pipe to soil potential	PTS pot.	Pooling agreement	PA
Pipeline	PL	Porcelaneous	porc
Pipeline oil	PLO	Porcion	porc
Pipeline terminal	PLT	Pore volume	P.V.
Pisolites	piso	Porosity	por
Pisolitic	piso	Porosity and permeability	P&P
Pitted	pit	Porous	por
Plagioclase	plag	Porous and permeable	P&P
Plain end	PE	Portable	port
Plain end beveled	PEB	Porter Creek	PC
Plain large end	PLE	Position	pos
Plain one end	POE	Positive	pos
Plain small end	PSE	Positive crankcase ventilation	PCV
Plan	pln	Possible(ly)	poss
Plant	plt	Post Laramie	P Lar
Plant fossils	pl fos	Post Oak	P.O.
Plant volume reduction	PVR	Potential	pot
Planulina harangensis	Plan. harangensis	Potential difference	pot dif
		Potential test	PT
Planulina palmarie	Plan. palm.	Potential test to follow	PTTF
Plastic	plas	Pound	lb
Plastic viscosity	PV	Pound per foot	lb/ft
Platform	platf	Pound-inch	lb-in
Platformer	platfr	Pounds per gallon	ppg
Platy	plty	Pounds per square foot	lb/sq ft
Please note and return	PNR	Pounds per square foot	psf
Pleistocene	Pleist	Pounds per square inch	psi

Pounds per square inch absolute	psia	Produce (d) (ing) (tion)	prod
Pounds per square inch gauge	psig	Producing gas well	PGW
Pour point (ASTM Method)	pour ASTM	Producing oil & gas well	POGW
Power	PWR	Producing oil well	POW
Power factor	PF	Producing oil well flowing	POWF
Power factor meter	PFM	Producing oil well pumping	POWP
Pre-Cambrian	Pre Camb	Product(s)	prod
Precast	prcst	Production	PP
Precipitate	ppt	Production department exploratory test	PDET
Precipitation Number	pptn No	Production payment interest	PPI
Predominant	predom	Productivity index	PI
Prefabricated	prefab	Profit & loss	P & L
Preferred	pfd	Profit sharing interest	PSI
Preheater	prehtr	Progress	prog
Preliminary	prelim	Project (ed) (ion)	proj
Premium	prem	Property line	PL
Prepaid	ppd	Proportional	prop
Preparation	Prep	Propose(d)	prop
Prepare	Prep	Proposed bottom hole location	PBHL
Preparing	Prep	Proposed depth	PD
Preparing to take potential test	PRPT	Prorated	pro
Pressed distillate	PD	Protection	prot
Pressure	press	Proterozoic	Protero
Pressure alarm	PA	Provincial	Prov
Pressure control valve	PCV	Pryoclastic	pyrclas
Pressure differential controller	PDC	Pseudo	pdso
Pressure differential indicator	PDI	Pseudo	ps
Pressure differential indicator controller	PDIC	Public relations	PR
		Public School Land	PSL
Pressure differential recorder	PDR	Pull(ed) rods & tubing	PR&T
Pressure differential recorder controller	PDRC	Pulled	pld
		Pulled big pipe	PBP
Pressure indicator	PI	Pulled out	PO
Pressure indicator controller	PIC	Pulled out of hole	POH
Pressure recorder	PR	Pulled pipe	PP
Pressure recorder control	PRC	Pulled up	PU
Pressure switch	PS	Pulling	plg
Pressure-volume-temperature	PVT	Pulling tubing	PTG
Prestressed	prest	Pulling tubing and rods	PTR
Prevent	prev	Pump, pumped, pumping	pmp (d) (g)
Preventive	prev	Pump and flow	P&F
Primary	pri	Pump jack	PJ
Primary Reference Fuel	PRF	Pump job	PJ
Principal	prin	Pump on beam	POB
Principle lessee(s)	prncpl lss	Pump-in pressure	PIP
Prism(atic)	pris	Pumping equipment	PE
Private branch exchange	PBX	Pumping for test	PFT
Privilege	priv	Pumping load oil	PLO
Probable(ly)	prob	Pumping unit	PU
Process	proc	Pumps off	PO

Purchase order	PO	Random lengths	RL
Purple	purp	Range	R
Putting on pump	POP	Range	Rge
Pyrite, pyritic	pyr	Ranger	Rang
Pyrobitumen	pyrbit	Rankine (temp. scale)	R
Pyrolysis	pyls	Rapid curing	RC
		Rat hole	RH
		Rat hole mud	RHM

Q

		Rate of penetration	ROP
		Rate of return	ROR
Quadrangle	quad	Rate too low to measure	RTLTM
Quadrant	quad	Rating	rtg
Quadruple	quad	Reacidize	reacd
Quality	qual	Reacidized	reacd
Quantity	qty	Reacidizing	reacd
Quantity	quan	Reactor	reac
Quantity discount allowance	QDA	Reagan	Reag
Quarry	qry	Ream	RM
Quart(s)	qt	Reamed	rmd
Quarter	qtr	Reaming	rmg
Quartz	qtz	Reboiler	reblr
Quartzite	qtz	Received	recd
Quartzitic	qtz	Receptacle	recp
Quartzose	qtzose	Reciprocate (ing)	recip
Queen City	Q. City	Recirculate	recirc
Queen Sand	Q. Sd	Recommend	rec
Quench	qnch	Recomplete (d) (ion)	recomp
Questionable	quest	Recondition(ed)	recond
Quick ram change	QRC	Recorder	rec
Quintuplicate	quint	Recording	rec
		Recover	rec
		Recovered	rec

R

		Recovering	rec
		Recovery	rec
Radial	rad	Rectangle	rect
Radian	RAD	Rectangular	rect
Radiation	radtn	Rectifier	rect
Radioactive	RA	Recycle	recy
Radiological	RAD	Red Beds	Rd Bds
Radius	RAD	Red cave	RC
Railing	rlg	Red Fork	Rd Fk
Railroad	RR	Red indicating lamp	RIL
Railroad Commission (Texas)	RRC	Red Oak	R.O.
Railway	Ry	Red Peak	Rd Pk
Rainbow show of oil	RBSO	Red River	RR
Raised faced	RF	Redrilled	redrld
Raised face flanged end	RFFE	Reducer	red
Raised face weld neck	RFWN	Reducing	red
Ramsbottom Carbon Residue	RCR	Reducing balance	red bal
Ran in hole	RIH	Reference	ref
Ran (running) rods and tubing	RR&T	Refine (d) (r) (ry)	ref

319

Refining	refg	Research Octane Number	RON
Reflection	refl	Research Octane Number	Res. O. N.
Reflux	refl	Reservation	res
Reformate	reform	Reserve	res
Reformer	reform	Reservoir	res
Reforming	reform	Residual	resid
Refraction	refr	Residue	resid
Refractory	refr	Resinous	rsns
Refrigerant	referg	Resistance	res
Refrigeration	refer	Resistivity	res
Refrigerator	refgr	Resistor	res
Regenerator	regen	Retail pump price	RPP
Regular acid	R/A	Retain (ed) (er) (ing)	ret
Reid vapor pressure	RVP	Retainer	rtnr
Reinforce (d) (ing)	reinf	Retard (ed)	ret
Reinforced concrete	reinf conc	Retrievable bridge plug	RBP
Reinforcing bar	rebar	Retrievable retainer	retr ret
Reject	rej	Retrievable test treat squeeze	
Rejection	rej'n	(tool)	RTTS
Reklaw	Rek	Return	ret
Register	reg	Return on investment	ROI
Regular	reg	Returnable steel drum	RSD
Regulator	reg	Returned	retd
Relay	rel	Returned well to production	RWTP
Relay	rly	Returning circulation oil	RCO
Release (d)	rel	Reverse(d)	rev
Release (d) (ing)	ris (d) (ing)	Reverse(d)	rvs(d)
Released swab unit	RSU	Reverse circulation	RC
Relief	rlf	Reverse circulation rig	RCR
Relocate(d)	reloc	Reversed out	RO
Remains	rem	Reversed out	rev/O
Remains	rmn	Revise (d) (ing) (ion)	rev
Remedial	rem	Revolution(s)	rev
Remote control	RC	Revolutions per minute	rpm
Removable	rmv	Revolutions per second	rps
Remove (al) (able)	rem	Rework(ed)	rwk(d)
Renault	Ren	Rheostat	rheo
Rental	rent	Ribbon sand	Rib
Reophax bathysiphoni	Reo bath	Rich oil fractionator	ROF
Repair (ed) (ing) (s)	rep	Rierdon	Rier
Repairman	rpmn	Rig floor	RF
Reperforated	reperf	Rig on location	ROL
Replace (d)	rep	Rig released	RR
Replace (ment)	repl	Rig released	rig rel
Report	rep	Rig skidded	RS
Required	reqd	Rigged down	RD
Requirement	reqmt	Rigged down swabbing unit	RDSU
Requisition	req	Rigged up	RU
Requisition	reqn	Rigging down	RD
Research	res	Rigging up	RU
Research and development	R&D	Rigging up cable tools	RUCT

Rigging up machine	RUM
Rigging up pump	RUP
Rigging up rotary	RUR
Rigging up rotary tools	RURT
Rigging up service tools	RUSR
Rigging up standard tools	RUST
Rigging up tools	RUT
Right angle	RA
Right hand	RH
Righ hand door	RHD
Right-of-way	R/W
Right-of-way	ROW
Ring	rg
Ring groove	RG
Ring Joint	RJ
Ring joint flanged end	RJFE
Ring tool joint	RTJ
Ring type joint	RTJ
Rising stem (valve)	RS
Rivet	riv
Road	rd
Road & location	R&L
Road & location complete	R&LC
Roads	rds
Robulus	Rob
Rock	rk
Rock bit	RB
Rock pressure	RP
Rockwell hardness number	RHN
Rocky	rky
Rodessa	Rod
Rods & tubing	R & T
Room	rm
Root mean square	RMS
Rose	ro
Rosiclare sand	Ro
Rotary	rot
Rotary bushing	RB
Rotary bushing measurement	RBM
Rotary drive bushing	RDB
Rotary drive bushing to ground	RDB-GD
Rotary table	RT
Rotary test	R test
Rotary tools	RT
Rotary unit	RU
Rotate	rot
Rotator	rot
Rough	rgh
Round	rd
Round thread	rd thd

Round trip	rd tp
Rounded	rdd
Rounded	rnd
Royalty	roy
Royalty interest	RI
Rubber	rbr
Rubber	rub
Rubber ball sand oil frac	RBSOF
Rubber ball sand water frac	RBSWF
Rubber balls	Rbls
Run of mine	ROM
Running	rng
Running casing	RC
Running electric log	REL
Running radioactive log	RALOG
Running tubing	RTG
Rupture	rupt
Rust & oxidation	R & O

S

Sabinetown	Sab
Saccharoidal	sach
Sacks	sk
Sacks	sx
Saddle	sadl
Saddle Creek	Sad Cr
Safety	saf
Saint Genevieve	St. Gen
Saint Louis Lime	St L
Saint Peter	St Ptr
Salado	Sal
Salaried	sal
Salary	sal
Saline Bayou	Sal Bay
Salinity	sal
Salt & pepper	s&p
Salt Mountain	Slt Mt
Salt wash	SW
Salt water	SW
Salt water disposal	SWD
Salt water disposal system	SWDS
Salt water disposal well	SWDW
Salt water injection	SWI
Salt water to surface	SWTS
Salty	slty
Salty sulphur water	SSUW
Salvage	salv
Sample	samp
Sample tops	S/T

San Adres	San And	Scrubber	scrub
San Angelo	San Ang	Seabreeze	Sea
San Bernardino Base and		Sealed	sld
Meridian	SBB&M	Seamless	smls
San Rafael	San Raf	Seating nipple	SN
Sanastee	Sana	Secant	sec
Sand	sd	Second	sec
Sand and shale	sd & sh	Secondary	sec
Sand oil fract	sdoilfract	Secondary butyl alcohol	SBA
Sand oil fracture	SOF	Secretary	sec
Sand showing gas	Sd SG	Section	sec
Sand showing oil	Sd SO	Section line	SL
Sand water fract	sdwtrfract	Section-township-range	S-T-R
Sandfrac	SF	Securaloy	scly
Sandfract	sdfract	Sediment(s)	sed
Sandstone	sd	Sedwick	Sedw
Sand-water fracture	SWF	Seismic	seis
Sandy	sdy	Seismograph	seis
Sandy lime	sdy li	Selenite	sel
Sandy shale	sdy sh	Self (spontaneous) potential	SP
Sanitary	sani	Selma	Sel
Saponification	sap	Senora	Sen
Saponification number	Sap No	Separator	sep
Saratoga	Sara	Septuplicate	sept
Satanka	Stnka	Sequence	seq
Saturated	sat	Serial	ser
Saturation	sat	Series	ser
Sawatch	Saw	Serpentine	Serp
Sawtooth	Sawth	Serratt	Serr
Saybolt Furol	Say Furol	Service(s)	serv
Saybolt Seconds Universal	SSU	Service charge	serv chg
Saybolt Universal Seconds	SUS	Service station	SS
Scales	sc	Set plug	SP
Scattered	sctrd	Settling	set
Schedule	sch	Seven Rivers	S Riv
Schematic	schem	Severy	Svry
Scolecodonts	scolc	Sewer	sew
Scratcher	scr	Sexton	Sex
Screen	scr	Sextuple	sxtu
Screw	scr	Sextuplicate	sext
Screw end American National		Shaft horsepower	shp
Acme thread	SE NA	Shake out	SO
Screw end American National		Shale	sh
Coarse thread	SE NC	Shaley	shly
Screw end American National		Shallower pool (pay) test	SPT
Fine thread	SE NF	Shannon	Shan
Screw end American National		Shear	shr
Taper Pipe thread	SE NPT	Sheathing	shthg
Screwed	scrd	Sheet	sh
Screwed end	S/E	Shells	shls
Screwed on one end	SOE	Shinarump	Shin

Shipment	shpt	Silica	silic
Shipping	shpg	Siliceous	silic
Short radius	SR	Silky	slky
Short thread	ST	Silt	slt
Short threads & coupling	ST&C	Siltstone	silt
Shot open hole	SOH	Silurian	Sil
Shot point	SP	Similar	sim
Shoulder	shld	Simpson	Simp
Show condensate	SC	Single pole double throw	SPDT
Show gas	SG	Single pole single throw	SPST
Show gas and condensate	SG&C	Single random lengths	SRL
Show gas and distillate	SG&D	Single Shot	SS
Show gas & water	SG&W	Singles	sgls
Show of dead oil	SDO	Siphonina davisi	Siph. d.
Show of free oil	SFO	Size	sz
Show of gas and oil	SG&O	Skimmer	skim
Show oil	SO	Skinner	Skn
Show oil and gas	SO&G	Skull Creek	Sk Crk
Show oil and water	SO&W	Sleeve	sl
Shut down	SD	Sleeve bearing	SB
Shut down awaiting orders	SDWO	Slickensided	sks
Shut down for orders	SDO	Sliding scale royalty	S/SR
Shut down for pipe line	SDPL	Slight(ly)	sl
Shut down for repairs	SDR	Slight(ly)	sli
Shut down for weather	SDW	Slight show of gas	SSG
Shut down overnight	SDON	Slight show of oil	sli SO
Shut down to acidize	SDA	Slight show of oil & gas	SSO&G
Shut down to fracture	SDF	Slight show oil	SSO
Shut down to log	SDL	Slight, weak, or poor	
Shut down to plug & abandon	SDPA	fluorescence	SFLU
Shut in	SI	Slightly gas-cut mud	SGCM
Shut in bottom hole pressure	SIBHP	Slightly gas-cut oil	SGCO
Shut in casing pressure	SICP	Slightly gas-cut water	SGCW
Shut in gas well	SIGW	Slightly gas-cut water blanket	SGCWB
Shut in oil well	SIOW	Slightly oil & gas-cut mud	SO&GCM
Shut in pressure	SIP	Slightly oil-cut mud	SOCM
Shut in tubing pressure	SITP	Slightly oil-cut water	SOCW
Shut in well head pressure	SIWHP	Slightly porous	SP
Shut in—waiting on potential	SIWOP	Sligo	Sli
Side door choke	SD Ck	Slim hole drill pipe	SHDP
Side opening	SO	Slip on	SO
Sideboom	SB	Slow set (cement)	SS
Siderite(ic)	sid	Slurry	slur
Sides, tops & bottoms	s, t, & b	Smackover	Smk
Sidetrack (ing)	ST (g)	Small	sm
Sidetrack (ed) (ing)	sdtkr	Small show	SS
Sidetracked hole	STH	Smithwick	Smithw
Sidetracked total depth	STTD	Smoke Volatility Index	SVI
Sidewall cores	SWC	Smooth	smth
Sidewall samples	SWS	Socket	skt
Signed	sgd	Socket weld	SW

Socket weld	sow	Spring	spg
Sodium base grease	sod gr	Springer	Sprin
Sodium		Sprinkler	spkr
carboxymethylcellulose	CMC	Sprocket	spkt
Soft	sft	Spud (ded) (der)	spd
Solenoid	slnd	Square	sq
Solenoid	sol	Square centimeter	sq cm
Solids	sol	Square foot	sq ft
Solution	soln	Square inch	sq in
Solvent	solv	Square kilometer	sq km
Somastic	som	Square meter	sq m
Somastic coated	somct	Square millimeter	sq mm
Sorted (ing)	sort	Square yard(s)	sq yd
Sort (ed) (ing)	srt (d) (g)	Squeeze (d) (ing)	sqz
South half	S/2	Squeeze packer	sq pkr
South line	SL	Squeezed	sq
South offset	SO	Squirrel cage	sq cg
Southeast	SE	Stabilized(er)	stab
Southeast corner	SE/C	Stain (ed) (ing)	stn (d) (g)
Southeast quarter	SE/4	Stain and odor	S&O
Southwest	SW	Stainless steel	SS
Southwest corner	SW/c	Stairway	stwy
Southwest quarter	SW/4	Stalnaker	Stal
Spacer	spcr	Stand (s) (ing)	std (s) (g)
Spare	sp	Stand by	stn/by
Sparta	Sp	Standard	std
Spearfish	Spf	Standard cubic feet per day	SCFD
Special	spcl	Standard cubic feet per hour	SCFH
Specialty	splty	Standard cubic fet per minute	SCFM
Specific gravity	sp gr	Standard cubic foot	SCF
Specific heat	sp ht	Standard operational	
Specific volume	sp. vol.	procedure	SOP
Specification	spec	Standard temperature and	
Speckled	speck	pressure	STP
Sphaerodina	Sphaer	Stanley	Stan
Sphalerite	sphal	Starting fluid level	SFL
Spherules	sph	State lease	SL
Spicule(ar)	spic	State potential	State pot
Spigot and spigot	s & s	Static bottom-hole pressure	SBHP
Spindle	spdl	Station	sta
Spindletop	Spletp	Stationary	stat
Spiral weld	SW	Statistical	stat
Spirifers	spfr	Steady	stdy
Spiroplectammina barrowi	Spiro, b.	Steam	stm
Splintery	splty	Steam cylinder oil	stm cul oil
Split	splty	Steam Emulsion Number	SE No.
Sponge	spg	Steam engine oil	stm eng oil
Spore	sp	Steam working pressure	SWP
Spotted	sptd	Steel	stl
Spotty	sptty	Steel line correction	SLC
Spraberry	Spra	Steel line measurement	SLM

Steel tape measurement	STM	Subsurface	SS
Steele	Stel	Sucrose	suc
Stenographer	steno	Sucrosic	suc
Stensvad	Stens	Suction	suct
Sticky	stcky	Sugary	sug
Stippled	stip	Sulphated	sulph
Stirrup	stir	Sulphur (sulfur)	sul
Stock	stk	Sulfur by bomb method	S Bomb
Stock tank barrels	STB	Sulphur water	sul wtr
Stock tank barrels per day	STB/D	Summary	sum
Stock tank oil in place	STOIP	Summerville	Sum
Stone Corral	Stn Crl	Sunburst	Sb
Stony Mountain	Sty Mt	Sunburst	Sunb
Stopped	stpd	Sundance	Sund
Stopper	stp	Supai	Sup
Storage	stor	Superintendent	supt
Storage	strg	Superseded	supsd
Stove oil	stv	Supervisor	suprv
Straddle	strd	Supplement	supp
Straddle packer	SP	Supply	sply
Straddle packer drill stem test	SP—DST	Supply (ied) (ier) (ing)	supl
Straight	strt	Support	suppt
Straight hole test	SHT	Surface	sfc
Straightening	stging	Surface	surf
Strainer	stnr	Surface geology	SG
Strand(ed)	Strd	Surface measurement	SM
Stratigraphic	strat	Surface pressure	SP
Strawn	Str	Surplus	surp
Streaked	stk	Survey	sur
Streaks	stk	Suspended	susp
Striated	stri	Swab and flow	S&F
String shot	SS	Swabbed	S/
Stringer	strg	Swabbed, swabbing	swb (d) (g)
Strokes per minute	SPM	Swabbed unit	SWU
Stromatoporoid	strom	Swage	swg
Strong	strg	Swaged	swd
Structural	struc	Swastika	Swas
Structure	struc	Sweetening	swet
Stuck	stk	Switchboard	swbd
Stuffing box	SB	Switchgear	swgr
Styolite	styo	Sycamore	Syc
Styolitic	styo	Sylvan	Syl
Service	svc	Symbol	sym
Service unit	svcu	Symmetrical	sym
Sub	sb	Synchronizing	syn
Sub-Clarksville	Sub Clarks	Synchronous	syn
Subdivision	subd	Synchronous converter	syn conv
Subsea	SS	Synthetic	syn
Subsidiary	sub	System	sys
Substance	sub		
Substation	substa		

T

Tabular	tab
Tabulating	tab
Tagliabue	Tag
Tallahatta	Tal
Tampico	Tamp
Tank	tk
Tank battery	TB
Tank car	T/C
Tank truck	TT
Tank wagon	TW
Tankage	tkg
Tanker(s)	tkr
Tannehill	Tann
Tansill	Tan
Tarkio	Tark
Tarred and wrapped	T&W
Tar Springs sand	TS
Taste	tste
Taylor	Tay
Tearing out rotary tools	TORT
Technical	tech
Technician	tech
Tee	T
Telegraph	tel
Telegraph Creek	Tel Cr
Telephone	tel
Teletype	TWX
Television	TV
Temperature	Temp
Temperature control valve	TCV
Temperature controller	TC
Temperature differential indicator	TDI
Temperature differential recorder	TDR
Temperature indicator	TI
Temperature indicator controller	TIC
Temperature recorder	TR
Temperature recorder controller	TRC
Temperature Survey indicated top cement at	TSITC
Temporarily abandoned	TA
Temporarily shut down	TSD
Temporarily shut in	TSI
Temporary(ily)	Temp
Temporary dealer allowance	TDA

Temporary voluntary allowance	TVA
Tender	tndr
Tensile strength	TS
Tensleep	Tens
Tentaculites	Tent
Tentative	tent
Teremplealeau	Tremp
Terminal	term
Terminate (d) (ing) (ion)	termin
Tertiary	Ter
Tertiary butyl alcohol	TBA
Test (ed) (ing)	tst (d) (g)
Test to follow	TTF
Tester	tstr
Testing on pump	TOP
Tetraethyl lead	TEL
Tetramethyl lead	TML
Texana	Tex
Textularia articulate	Text. art.
Textularia dibollensis	Text. d.
Textularia hockleyensis	Text. h.
Textularia warreni	Text. w.
Texture	tex
Thaynes	Thay
Thence	th
Thermal	thrm
Thermal cracker	thrm ckr
Thermofor catalytic cracking	TCC
Thermometer	therm
Thermopolis	Ther
Thermostat	therst
Thick	thk
Thickness	thk
Thin bedded	TB
Thousand cubic feet	MCF
Thousand cubic feet per day	MCFD
Thousand British thermal units	MBTU
Thread	thd
Thread large end	TLE
Thread on both ends	TOBE
Thread small end	TSE
Thread small end, weld large end	TSE—WLE
Threaded	thd
Threaded & coupled	T & C
Threaded both ends	TBE
Threaded one end	TOE
Treaded pipe flange	TPF
Three Finger	Tfing
Three Forks	Tfks

Throttling	thrling	Township	twp
Through	thru	Townsite	twst
Through tubing	TT	Trace	tr
Thurman	Thur	Trackage	trkg
Tight	ti	Tract	tr
Tight hole	TH	Transfer (ed) (ing)	trans
Tight no show	TNS	Transformer	trans
Timpas	Tim	Translucent	transl
Timpoweap	Timpo	Transmission	trans
Tires, batteries and		Transparent	transp
accessories	TBA	Transportation	transp
Todilto	Tod	Travis Peak	TP
Tolerance	tol	Treat (ed) (ing)	trt (d) (g)
Toluene	tolu	Treater	trtr
Ton (after a number)	T	Trenton	Tren
Tongue and groove (joint)	T&G	Trenton	Trn
Tonkawa	Tonk	Triassic	Tri
Too small to measure	TSTM	Triscresyl phosphate	TCP
Too weak to measure	TWTM	Trilobite	trilo
Tool	tl	Trinidad	Trin
Tool closed	TC	Trip in hole	TIH
Tool open	TO	Trip out of hole	TOH
Tool pusher	TP	Triplicate	trip
Tools	tl	Tripoli	trip
Top and bottom	T&B	Tripolitic	trip
Top & bottom chokes	T&BC	Tripped (ing)	trip
Top choke	TC	Truck	trk
Top hole flow pressure	THFP	True boiling point	TBP
Top of (a formation)	T/	True Vapor Process	TVP
Top of cement	TOC	True vertical depth	TVD
Top of cement plug	TOCP	Tube	tb
Top of fish	TOF	Tubing	tbg
Top of liner	TOL	Tubing and rods	T&R
Top of liner hanger	TLH	Tubing choke	TC
Top of pay	T/pay	Tubing choke	tbg chk
Top salt	T/S	Tubing pressure	TP
Top of sand	T/sd	Tubing pressure	tbg press
Topeka	Tpka	Tubing pressure—closed	TPC
Topographic	topo	Tubing pressure—flowing	TPF
Topography	topo	Tubing pressure shut in	TPSI
Topping	topg	Tubinghead flange	THF
Topping and coking	T & C	Tucker	Tuck
Toronto	Tor	Tuffaceous	tfs
Toroweap	Toro	Tuffaceous	tuf
Total	tot	Tulip Creek	Tul Cr
Total depth	TD	Tungsten carbide	tung carb
Total time lost	TTL	Turn around	TA
Totally enclosed-fan cooled	TEFC	Turned over to producing	
Tough	tgh	section	TOPS
Towanda	Tow	Turned to test tank	TTTT
Township (as T2N)	T	Turnpike	tpk

Tuscaloosa	Tus	Various	var
Twin Creek	Tw Cr	Varnish makers & painters	
Twisted off	twst off	naphtha	VM&P Naphtha
Type	ty	Varved	vrvd
Typewriter	tywr	Velocity	vel
Typical	typ	Velocity survey	V/S
		Ventilator	vent
		Verdigris	Verd

U

		Vermillion Cliff	Ver Cl
		Versus	vs
Ultimate	ult	Vertebrate	vrtb
Ultra high frequency	UHF	Vertical	vert
Unbranded	unbr	Vertical	vrtl
Unconformity	unconf	Very (as very tight)	v.
Unconsolidated	uncons	Very common	v.c.
Under construction	U/C	Very fine-grained	v-f-gr
Under digging	UD	Very heavily oil-cut mud	v-HOCM
Under gauge	UG	Very high frequency	VHF
Underground	UG	Very light amber cut	VLAC
Under reaming	UR	Very noticeable	v.n.
Undifferentiated	undiff	Very poor sample	VPS
Unfinished	unf	Very rare	v.r.
Uniform	uni	Very slight	v-sli
Union Valley	UV	Very slight gas-cut mud	VSGCM
Unit	un	Very slight show of gas	VSSG
Universal	univ	Very slight show of oil	VSSO
Universal gear lubricant	UGL	Very slightly porous	VSP
University	univ	Vesicular	ves
Unsulfonated residue	UR	Vicksburg	Vks
Upper (i.e., U/Simpson)	U/	Viola	Vi
Upper and lower	U/L	Virgelle	Virg
Use customer's hose	UCH	Viscosity	vis
Used with	U/W	Viscosity index	VI
Uvigerina lirettensis	Uvig. lir.	Visible	vis
		Vitreous	vit

V

		Vitrified clay pipe	VCP
		Vogtsberger	Vogts
		Volt	v
Vacant	vac	Volt-ampere	va
Vacation	vac	Volt-ampere reactive	var
Vacuum	vac	Volume	V
Vaginulina regina	Vag. reg	Volume	vol
Valera	Val	Volumetric efficiency	vol. eff.
Valve	vlv	Vuggy	vug
Vanguard	Vang	Vugular	vug
Vapor	vap		

W

Vapor pressure	VP		
Vapor-liquid ratio	V/L		
Varas	vrs	Wabaunsee	Wab
Variable	var	Waddell	Wad
Variegated	vari	Waiting	wtg

Waiting on	WO	Water not shut-off	WNSO
Waiting on acid	WOA	Water, oil or gas	WOG
Waiting on allowable	WOA	Water shut-off	WSO
Waiting on battery	WOB	Water shut-off no good	WSONG
Waiting on cable tools	WOCT	Water shut-off ok	SWOOK
Waiting on cement	WOC	Water supply well	WSW
Waiting on completion rig	WOCR	Water to surface	WTS
Waiting on completion tools	WOCT	Water well	WW
Waiting on orders	WOO	Water with slight show of oil	W/SSO
Waiting on permit	WOP	Water with sulphur odor	W/sulf O
Waiting on pipe	WOP	Water-cut mud	WCM
Waiting on potential test	WOPT	Water-cut oil	WCO
Waiting on production		Waterflood	WF
equipment	WOPE	Water-oil ratio	WOR
Waiting on pump	WOP	Watt	w
Waiting on pumping unit	WOPU	Weak	wk
Waiting on rig or rotary	WOR	Weak air blow	WAB
Waiting on rotary tools	WORT	Weather	wthr
Waiting on standard tools	WOST	Weathered	wthd
Waiting on state potential	WOSP	Weber	Web
Waiting on tank & connection	WOT&C	Week	wk
Waiting on test or tools	WOT	Weight	wgt
Waiting on weather	WOW	Weight	wt
Wall (if used with pipe)	W	Weight on bit	W.O.B
Wall Creek	W Cr	Weld ends	WE
Wall thickness (pipe)	WT	Weld neck	WN
Waltersburg sand	Wa Sd	Welded	wld
Wapanucka	Wap	Welder	wldr
Warehouse	whse	Welding	wld
Warsaw	War	Welding neck	WN
Wasatch	Was	Wellhead	WH
Wash over	WO	Wellington	Well
Wash pipe	WP	Went back in hole	WBIH
Wash water	WW	Went in hole	WIH
Washed	wshd	West	W
Washing	wshg	West half	W/2
Washing in	WI	West line	WL
Washita	Wash	West Offset	W/O
Washita-Fredericksburg	W-F	Westerly	W'ly
Washover string	WOS	Wet bulb	WB
Water, watery	wtr (y)	Whipstock	WS
Water blanket	WB	Whipstock	whip
Water cushion (DST)	WC	Whipstock depth	WSD
Water cushion to surface	WCTS	White	wht
Water cut	WC	White Dolomite	Wh Dol
Water depth	WD	White River	WR
Water disposal well	WD	White sand	Wh Sd
Water in hole	WIH	Wholesale	whsle
Water injection	WI	Wichita	Wich.
Water load	W/L	Wichita Albany	Wich Alb
Water loss	WL	Wide flange	WF

Wilcox	Wx	Wrapper	wpr
Wildcat	WC	Wreford	Wref
Wildcat field discovery	WFD	Wrought iron	WI
Willberne	Willb		
Wind River	Wd R		
Winfield	Winf		

X

Wingate	Wing	X-ray	X-R
Winnipeg	Winn		
Winona	Win		
Wire line	WL		
Wire line coring	WLC		

Y

Wireline test	WLT		
Wireline total depth	WLTD	Yard(s)	yd
With	w/	Yates	Y
Without drill pipe	WODP	Yazoo	Yz
Wolfe City	WC	Year	yr
Wolfcamp	Wolfc	Yellow	yel
Woodbine	WB	Yellow indicating lamp	YIL
Woodford	Wdfd	Yield point	YP
Woodford	Woodf	Yoakum	Yoak
Woodside	Wood	Your message of date	YMD
Work order	WO	Your message yesterday	YMY
Worked	wkd		
Working	wkg		
Working interest	WI		

Z

Working pressure	WP		
Workover	wko	Zenith	zen
Workover	WO	Zilpha	Zil
Workover rig	wkor	Zone	Z

ABBREVIATIONS FOR LOGGING TOOLS AND SERVICES

(The appropriate companies and associations have not yet established standard abbreviations for the logging segment of the oil and gas industry. The following lists, by individual companies, are included for convenience.)

DRESSER ATLAS

Log or service	Abbreviation
Acoustilog	ALC
Acoustilog Caliper Gamma Ray	ALC-GR
Acoustilog Caliper Neutron	ALC-N
Acoustilog Caliper Gamma Ray-Neutron	ALC-GRN
Acoustic Cement Bond	CBL
Acoustic Cement Bond Gamma Ray	CLB-GR
Acoustic Cement Bond Neutron	CBL N
Acoustic Cement Bond G/R Neutron	CBL GRN
Acoustic Parameter - Depth	AC PAR D
Acoustic Parameter - Logging	AC PAR L
Acoustic Parameter - 16 mm Scope	AC PAR 16
Acoustic Signature	AC SIGN
Atlantic Chlorinlog	A CHL
Borehole Compensated	BHC
BHC Acoustilog Caliper	BHC ALC
BHC Acoustilog Caliper Gamma Ray	BHC ALC GR
BHC Acoustilog Caliper Neutron	BHC ALC N
BHC Acoustilog Caliper G/R Neutron	BHC ALC GRN
BHC Acoustilog Caliper (Thru Casing)	BHC AL TC
BHC Acoustilog Caliper Gamma Ray (Thru Casing)	BHC AL GR TC
BHC Acoustilog Caliper G/R Neutron (Thru Casing)	BHC AL GRN TC
Caliper	CL
Casing Potential Profile	CPP
Cemotop	CTL
Channelmaster	CML
Channelmaster—Neutron	CML N
Chlorinlog	CHL
Chlorinlog—Gamma Ray	CHL GR
Compensated Densilog Caliper	C DLC
Compensated Densilog Caliper Gamma Ray	C DLC GR
Compensated Densilog Caliper Neutron	C DLC N
Compensated Densilog Caliper G/R Neutron	C DLC GRN
Compensated Densilog Caliper Minilog	C DLC M
Conductivity Derived Porosity	CDP
Corgun	CG
Densilog Caliper Gamma Ray	DLC GR
Depth Determination	DD
Directional Survey	DS
Dual Induction Focused Log	DIFL
Dual Induction Focused Log Gamma Ray	DIFL-GR
Electrolog	EL
Focused Diplog	F DIP
Formation Tester	FT
4 Arm High Resolution Diplog	R H DIP
Frac Log	FRAC L
Frac Log - Gamma Ray	FRAC-GR
Gamma Ray Cased Hole	GR CH
Gamma Ray/Dual Caliper	GR/D CALIPER
Gamma Ray—Open Hole	G/R OH
Gamma Ray Neutron Cased Hole	GRN CH
Geophone	GEO
Gamma Ray Neutron - Open Hole	GR/N OH
Induction Electrolog	IEL
Induction Electrolog Gamma Ray	IEL-GR
Induction Electrolog Neutron	IEL-N
Induction Electrolog Gamma Ray Neutron	IEL-GRN
Induction Log	IL
Induction Log—Gamma Ray	IL-GR
Induction Log Neutron	IL-N

Laterolog	LL
Laterolog-Gamma Ray	LL-GR
Laterolog-Neutron	LL-N
Laterolog-Gamma Ray-Neutron	LL-GRN
Microlaterolog-Caliper	MLLC
Minilog Caliper	ML-C
Minilog Caliper Gamma Ray	ML-C-GR
Movable Oil Plot	MOP
Nuclear Flolog	NFL
Nuclear Flolog - Gamma Ray	NFL GR
Nuclear Flolog - Neutron	NFL N
Nuclear Flolog - Gamma Ray Neutron	NFL GRN
Nuclear Cement Log	NCL
Neutron Cased Hole	N CH
Neutron Open Hole	N OH
Neutron Lifetime	NLL
Neutron Lifetime Gamma Ray	NLL GR
Neutron Lifetime Neutron	NLL N
Neutron Lifetime G/R - Neutron	NLL GRN
Neutron Lifetime CBL	NLL CBL
Neutron Lifetime - G/R - CBL	NLL GR CBL
Neutron Lifetime - CBL Neutron	NL CB N
Neutron Lifetime - CBL G/R Neutron	NLL CBL GR N
Perforating Control	PFC
PFC Gamma Ray	PFC GR
PFC Neutron	PFC N
Photon	PL
Proximity Minilog	PROX-MLC
Sidewall Neutron	SWN
Sidewall Neutron - Gamma Ray	SWN GR
Temperature - Differential	DTL
Temperature - Gamma Ray - Neutron	T GRN
Temperature Log	TL
Temperature Log - Gamma Ray	T GR
Temperature - Neutron	TN
Total Time Integrator	TTI
Tracer Log	TL
Tracer Log - Neutron	TNL
Tracer Material	TM
Tracer Placement with Dump Bailer	TU DB
Tricore	TCS

GO INTERNATIONAL

Log or Service	Abbreviation
Caliper	CALP
Cement Bond Log	CBL
Differential Temperature	DIF-T
Gamma Ray	GR
Gamma Ray Neutron	GR-N
Neutron	N
Temperature Log	T

SCHLUMBERGER WELL SERVICES

Log or Service	Abbreviation
Amplitude Logging	A-BHC
Bore Hole Compensated	BHC
BHC Sonic Logging	BHC
BHC Sonic-Gamma Ray Logging	BHC-GR
Bridge Plug Service	BP
Borehole Televiewer	TVT
Caliper Logging	CAL
Casing Cutter Service	SCE-CC
Cement Bond Logging	CBL
Cement Bond-Gamma Ray Logging	CBL-GR
Cement Bond-Variable Density Logging	CBL-VD
Cement Dump Bailer Service	DB
Computer Processed Interpretation	MCT
Customer Instrument Service	ICS
Data Transmission	TRD
Depth Determinations	DD
Diamond Core Slicer	SS
Dipmeter-Digital	HDT-D
Directional Service	CDR
Dual Induction-Laterologging	DIL
Electric Logging	ES
Flowmeter	CFM, PFM
Formation Density Logging	FDC
Formation Density-Gamma Ray Logging	FDC-GR
Formation Testing	FT
Gamma Ray Logging	GR
Gamma Ray-Neutron Logging	GRN
Gradiomanometer	GM

High Resolution Thermometer	HRT
Induction-Gamma Ray Logging	I-GR
Induction-Electric Logging	I-ES
Junk Catcher	JB
Magnetic Taping	TPG
Microlog	ML
Neutron Logging	NL
Orienting Perforating Service	OPR
Perforating-Ceramic DPC	SCE
Perforating Depth Control	PDC
Perforating-Expendable Shaped Charge	SCE
Perforating-Hyper-Jet	SCH
Perforating-Hyper Scallop	SPH
Pressure Control	PC
Production Combination Tool Logging	PCT
Production Packer Service	PPS
Proximity-Microlog	ML
Radioactive Tracer Logging	RTP
Rwa Logging	FAL
Salt Dome Profiling	ES-ULS
Seismic Reference Service	SRS
Sidewall Coring	CST
SNP Neutron Logging	SNP
SNP Neutron-Gamma Ray Logging	SNP-GR
Synergetic Log Systems	MCT
Temperature Logging	T
Temperature-Gamma Ray Logging	T-GR
Thermal Decay Logging	TDT
Thru-Tubing Caliper	C-C
Tubing, Cutter Service	SCE-CC
Variable Density Logging	BHC-VD
Variable Density-Gamma Ray Logging	VD-GR

WELEX

Log or Service	Abbreviation
Analog Computer Service	An Cpt. Ser.
Caliper Log	Cal
Compensated Acoustic Velocity Log	Com AVL
Compensated Acoustic Velocity Gamma Ray	Com AVL-G

Compensated Acoustic Velocity Neutron Log	Com AVL-N
Compensated Density Log	Com Den
Compensated Density Gamma Ray Log	Com Den-G
Computer Analyzed Logs	CAL
Contact Caliper Log	Cont
Continuous Drift Log	Con Dr.
Density Log	Den
Density Gamma Ray Log	Den-G
Depth Determination	DeDet
Digital Tape Recording Service	Dgt Tp Rec
Dip Log Digital Recording Service	Dgt Dip Rec.
Drift Log	Dr
Drill Pipe Electric Log	DPL
Electric Log	EL
Electro-Magnetic Corrosion Detector	Cor Det
Fluid Travel Log	FTrL
Formation Tester	FT
FoRxo Caliper Log	FoRxo
Frac-Finder Micro-Seismogram Log	FF-MSG
Frac-Finder Micro-Seismogram Gamma Log	FF-MSG-G
Frac-Finder Micro-Seismogram Neutron Log	FF-MSG-N
Gamma Guard Log	G-Grd
Gamma Ray Log	GR
Gamma Ray Depth Control Log	GRDC
Guard Log	Grd
High Temperature Equipment	HTEq
Induction Electric Gamma Ray Log	IEL-G
Induction Electric Neutron Log	IEL-N
Induction Electric Log	IEL
Induction Gamma Ray Log	Ind-G
Micro-Seismogram Log, Cased Hole	MSG-CBL
Micro-Seismogram Gamma Collar Log, Cased	MSG-CBL-G
Micro-Seismogram Neutron Collar Log, Cased	MSG-CBL-N
Neutron Log	NL
Neutron Depth Control Log	NDC

Precision Temperature Log	Pr Temp
Radiation Guard Log	R/A Grd
Radioactive Tracer Log	R/A Tra
Resistivity Dip Log	Dip
Side Wall Coring	SWC
Sidewall Neutron Log	SWN
Sidewall Neutron Gamma Ray Log	SWN-G
Simultaneous Gamma Ray Neutron	GRN
Special Insturment Service	Sp Inst Ser
True Vertical Depth	TVD

COMPANIES AND ASSOCIATIONS
U.S. and Canada

AAODC	See IADC
AAPG	American Association of Petroleum Geologists
AAPL	American Association of Petroleum Landmen
ACS	American Chemical Society
ADDC	Association of Desk and Derrick Clubs of North America
AEC	Atomic Energy Commission
AECRB	Alberta Energy Conservation Resources Board
AGA	American Gas Association
AGI	American Geological Institute
AGTL	Alberta Gas Trunkline Co., Ltd.
AGU	American Geophysical Union
AIChE	American Institute of Chemical Engineers
AIME	American Institue of Mining, Metallurgical and Petroleum Engineers
AISI	American Iron and Steel Institute
ANSI	American National Standards Institute
AOCS	American Oil Chemists Society
AOPL	Association of Oil Pipelines
AOSC	Association of Oilwell Servicing Contractors
API	American Petroleum Institute
APRA	American Petroleum Refiners Association

APW	Association of Petroleum Writers
ARCO	Atlanta Richfield Co.
ARKLA	Arkansas Louisiana Gas Co.
ASCE	American Society of Civil Engineers
ASHRAE	American Society of Heating, Refrigerating, and Air-Conditioning Engineers, Inc.
ASLE	American Society of Lubricating Engineers
ASME	American Society of Mechanical Engineers
ASPG	American Society of Professional Geologists
ASSE	American Society of Safety Engineers
ASTM	American Society for Testing Materials
AWS	American Welding Society
BLM	Bureau of Land Management
BLS	Bureau of Labor Statistics
BP	British Petroleum
BuMines	Bureau of Mines, U.S. Department of the Interior
CAGC	A combine: Continental Oil Co., Atlantic Richfield Co., Getty Oil Co., and Cities Service Oil Co.
CAODS	Canadian Association of Oilwell Drilling Contractors
CCCOP	Conservation Committee of California Oil Producers
CDS	Canadian Development Corp.
CFR	Coordinating Fuel Research Committee
GA	Canadian Gas Association
CGA	Clean Gulf Association
CGTC	Columbia Gas Transmission Corp.
CONOCO	Continental Oil Co
CORCO	Commonwealth Oil Refining Co., Inc.
CORS	Canadian Operational Research Society
CPA	Canadian Petroleum Association
CRC	Coordinating Research Council, Inc.
DOT	Department of Transportation

EMR	Department of Energy, Mines, and Resources (Canada)	LL&E	Louisiana Land & Exploration Co.
EPA	Environmental Protection Agency	MIOP	Mandatory Oil Import Program
ERCB	Energy Resource Conservation Board (Alberta, Canada)	NACE	National Association of Corrosion Engineers
		NACOPS	National Advisory Committee on Petroleum Statistics (Canada)
FAA	Federal Aviation Agency		
FCC	Federal Communications Commission	NAS	National Academy of Science
		NASA	National Aeronautical and Space Administration
FPC	Federal Power Commission		
FTC	Federal Trade Commission	NEB	National Energy Board (Canada)
GAMA	Gas Appliance Manufacturers Association		
		NEMA	National Electrical Manufacturers Association
GNEC	General Nuclear Engineering Co.	NEPA	National Environmental Policy Act of 1969
IADC	International Association of Drilling Contractors (formerly AAODC)	NGPA	Natural Gas Processors Association
		NGPSA	Natural Gas Processors Suppliers Association
IAE	Institute of Automotive Engineers		
		NLGI	National Lubricating Grease Institute
ICC	Interstate Commerce Commission		
IEEE	Institute of Electrical and Electronics Engineers	NLPGA	National Liquefied Petroleum Gas Association
IGT	Institute of Gas Technology	NLRB	National Labor Relations Board
INGAA	Independent Natural Gas Association of America		
		NOFI	National Oil Fuel Institute
IOCA	Independent Oil Compounders Association	NOIA	National Ocean Industries Association
IOCC	Interstate Oil Compact Commission	NOJC	National Oil Jobbers Council
		NOMADS	National Oil-Equipment Manufacturers and Delegates Society
IOSA	International Oil Scouts Association		
IP	Institute of Petroleum	NPC	National Petroleum Council
IPAA	Independent Petroleum Association of America	NPRA	National Petroleum Refiners Association
IPAC	Independent Petroleum Association of Canada	NSF	National Science Foundation
		OCR	Office of Coal Research
IPE	International Petroleum Exposition	OEP	Office of Emergency Preparedness
IPP/L	Interprovincial Pipe Line Co.	OIA	Oil Import Administration
IRAA	Independent Refiners Association of America	OIAB	Oil Import Appeals Board
		OIC	Oil Information Committee
ISA	Instrument Society of America	OIPA	Oklahoma Independent Petroleum Association
KERMAC	Kerr-McGee Corp	OOC	Offshore Operators Committee
KIOGA	Kansas Independent Oil and Gas Association	OPC	Oil Policy Committee
		OXY	Occidental Petroleum Corp.

PAD	Petroleum Administration for Defense
PESA	Petroleum Equipment Suppliers Association
PETCO	Petroleum Corporation of Texas
PGCOA	Pennsylvania Grade Crude Oil Association
PIEA	Petroleum Industry Electrical Association
Plato	Penzoil Louisiana and Texas Offshore
PLCA	Pipe Line Contractors Association
POGO	Penzoil Offshore Gas Operators
PPI	Plastic Pipe Institute
PPROA	Panhandle Producers and Royalty Owners Association
RMOGA	Rocky Mountain Oil and Gas Association
R-PAT	Regional Petroleum Associations of Texas
SACROC	Scurry Area Canyon Reef Operators Committee
SAE	Society of Automotive Engineers
SEG	Society of Exploration Geophysicists
SEPM	Society of Economic Paleontologists and Mineralogists
SGA	Southern Gas Association
SLAM	A combine: Signal Oil and Gas Co., Louisiana Land & Exploration Co., Amerada Hess Corp., and Marathon Oil Co.
SOCAL	Standard Oil Company of California
SOHIO	Standard Oil Co. of Ohio
SPE	Society of Petroleum Engineers of AIME
SPEE	Society of Petroleum Evaluation Engineers
SPWLA	Society of Professional Well Log Analysts
STATCAN	Statistics Canada ex Dominion Bureau of Statistics (DBS)

TCP	Trans-Canada Pipe Lines Ltd.
TETCO	Texas Eastern Transmission Corp.
TGT	Tennessee Gas Transmission Co.
THUMS	A combine: Texaco, Inc., Humble Oil & Refining Co., Union Oil Co. of California, Mobile Oil Corp., and Shell Oil Co.
TIPRO	Texas Independent Producers and Royalty Owners Association
TRANSCO	Transcontinental Gas Pipe Line Corp.
USGS	United States Geological Survey
WeCTOGA	West Central Texas Oil and Gas Association
WPC	World Petroleum Congress

COMPANIES AND ASSOCIATIONS
Outside the U.S. and Canada

AAOC	American Asiatic Oil Corp. (Philipines)
ABCD	Asfalti Bitumi Cementi Derivati, S.A. (Italy)
ACNA	Aziende Colori Nazionali Affini (Italy)
A.C.P.H.A.	Association Cooperative pour la Recherche et l'Exploration des Hydrocarbures en Algerie (Algeria)
ADCO-HH	African Drilling Co.—H. Hamouda (Libya)
AGIP S.p.A.	Azienda Generale Italiana Petroli S.p.A. (Italy)
A.H.I. BAU	Allegemeine Hoch-und Ingenieurbau AG (Germany)
AIOC	American International Oil Co. (U.S.A.)
AITASA	Aguas Industriales de Tarragona, S.A. (Spain)
AK CHEMI	GmbH & Co. KG - subsidiary of Associated Octel, Ltd., London, Eng. (Germany)

AKU	Algemene Kunstzijde Unie, N.V. (Netherlands)	ATAS	Anadolu Tastiyehanesi A.S. (Turkey)
ALFOR	Societe Algerienne de Forage (Algeria)	AUXERAP	Societe Auxiliare de la Regie Autonome des Petroles (France)
ALGECO	Alliance & Gestion Commerciale (France)	AZOLACQ	Societe Chimique d'Engrais et de Produits de Synthese (France)
A.L.O.R.	Australian Lubricating Oil Refinery Ltd. (Australia)		
AMATEX	Amsterdamsch Tankopslagbedrifj N.V. (Netherlands)	BAPCO	Bahrain Petroleum Co. Ltd. (Bahrain)
AMI	Ausonia Mineraria (Italy)	BASF	Badische Anilin & Soda-Frabrik AB (Germany)
AMIF	Ausonia Miniere Francaise (France)	BASUCOL	Barranquilla Supply & Co. (Columbia)
AMINOIL	American Independent Oil Co. (U.S.A.)	B.I.P.M.	Bataafse Internationale Petroleum Mij. N.V. (Netherlands)
AMOSEAS	American Overseas Petroleum Co., Ltd. (Libya)	BOGOC	Bolivian Gulf Oil Co. (Bolivia)
AMOSPAIN	American Overseas Petroleum Ltd. (Spain)	BORCO	Bahamas Oil Refining Co. (Bahamas)
ANCAP	Administracion Nacional de Combustibles, Alcohol y Portland (Uraguay)	BP	British Petroleum Co., Ltd. (England)
ANIC	Azienda Nazionale Idrogenazione Combustibili S.p.A. (Italy)	BRGG	Bureau de Recherches Geologique et Geophysique (France)
AOC	Aramco Overseas Co. (Switzerland, Netherlands, Japan)	BRGM	Bureau de Recherches Geologiques et Minieres (France)
APC	Azote et Produits Chimiques (France)	BRIGITTA	Gewerkschaft Brigitta (Germany)
APEX	American Petrofina Exploration Co. (Spain)	BRP	Bureau du Recherche de Petrole (France)
API	Anonima Petroli Italiana (Italy)	BRPM	Bureau de Recherches et de Participations Mineres (Morocco)
AQUITAINE	Societe Nationale des Petroles D'Aquitaine (France)	CALSPAIN	California Oil Co. of Spain (Spain)
ARGAS	Arabian Geophysical & Surveying Co. (Saudi Arabia)	CALTEX	Various affiliates of Texaco Inc. and Std. of Calif.
ARAMCO	Arabian American Oil Co.	CALVO SOTELO	Empresa Nacional Calvo Sotelo (Spain)
ASCOP	Association Cooperative pour la Recherche et l'Exploration des Hydrocarbures en Algerie (Algeria)	CAMEL	Campagnie Algerienne du Methane Liquide (France, Algeria)
ASED	Amoniaque Synthetique et Derives (Belgium)	CAMPSA	Compania Arrendataria del Monopolio de Petroleos, S.A. (Spain)
ASESA	Alfaltos Espanoles S.A. (Spain)	CAPAG	Enterprise Moderne de Canalisations Petrolieres, Aquiferes et Gazieres (France)

CARBESA	Carbon Black Espanola, S.A. (Conoco affiliate)
CAREP	Compagnie Algerienne de Recherche et d'Exploitation Petrolieres (Algiers)
CCC	Compania Carbonos Coloidais (Brazil)
C.E.C.A.	Carbonisation et Charbons Actifs S.A. (Spain)
CEICO	Central Espanol Ingenieria y Control S.A. (Spain)
CEL	Central European Pipeline (Germany)
CEOA	Centre Europe de'Exploitation de l'OTAN (France)
CEP	Compagnie D'Exploration Petroliere (France)
CEPSA	Compania Espanola de Petroleos, S.A. (Spain)
CETRA	Compagnie Europeanne de Canalisations et de Travaux (France)
CFEM	Compagnie Francaise d'Enterprises Metalliques (France)
CFM	Compagnie Ferguson Morrison-Knudsen (France)
CFMK	Compagnie Francaise du Petroles (France)
CFP	Compagnie Francaise du Petroles (France)
CFPA	Compagnie Francaise des Petroles (Algeria) (France)
CFPS	Compagnie Francaise de Prospection Sismique (France)
CFR	Compagnie Francaise de Raffinage (France)
CGG	Compagnie Generale de Geophysique (France, Australia, Singapore)
CIAGP	N.V. Chemische Industrie aku-Goodrich (Netherlands)
CIEPSA	Compania de Investigacion y Explotaciones Petroliferas, S.A. (Spain)
CIM	Compagnie Industrielle Maritime (France)
CIMI	Compania Italiana Montaggi Industriali S.p.A. (Italy)
CINSA	Compania Insular del Nitrogena, S.A. (Spain)
CIPAO	Compagnie Industrielle des Petroles de l'A.O. (France)
CIPSA	Compania Iberica de Prospecciones, S.A. (Spain)
CIRES	Compania Industrial de Resinas Sinteticas (Portugal)
CLASA	Carburanti Lubrificanti Affini S.p.A. (Italy)
CMF	Construzioni Metalliche Finsider S.p.A. (Italy)
COCHIME	Compagnie Chimique de la Meterranee (France)
CODI	Colombianos Distribuidores de Combustibles S.A. (Columbia)
COFIREP	Compagnie Financiere de Recherches Petrolieres (France)
COFOR	Compagnie Generale de Forages (France)
COLCITO	Colombia-Cities Service Petroleum Corp. (Columbia)
COLPET	Colombian Petroleum Co. (Colombia)
COMEX	Compagnie Maritime d'Expertises (France)
CONSPAIN	Continental Oil Co. of Spain (Spain) Conoco Espanola S.A. (Spain)
COPAREX	Compagnie de Participations, de Recherches et D'Exploitations Petrolieres (France)
COPE	Compagnie Orientale des Petroles d'Egypte (Egypt)
COPEBRAS	Compania Petroquimica Brasileira (Brazil)
COPEFA	Compagnie des Petroles France-Afrique (France)
COPETAO	Compagnie des Petroles Total (Afrique Quest)(France)
COPETMA	Compagnie les Petroles Total (Madagascar)
COPISA	Compania Petrolifera Iberica, Sociedad Anonima (Spain)
COPOSEP	Compagnie des Petroles du Sud est Parisien (France)

C.O.R.I.	Compania Richerche Idrocarburi S.p.A. (Italy)
COS	Coordinated Oil Services (France)
CPA	Compagnie des Petroles d'Algerie (Algeria)
CPC	Chinese Petroleum Corporation, Taiwan, China
CPTL	Compagnie des Petroles Total (Libye) (France)
CRAN	Compagnie de Raffinage en Afrique du Nord (Algeria)
CREPS	Compagnie de Recherches et d'Exploitation de Petrole au Sahara (Algeria)
CRR	Compagnie Rhenane de Raffinage (France)
CSRPG	Chambre Syndicale de la Recherche et de la Production du Petrole et du Gaz Naturel (France)
CTIP	Compania Tecnica Industrie Petroli S.p.a. (Italy)
CVP	Corporacion Venezolano del Petroleo (Venezuela)
DCEA	Direction Centrale des Essences des Armees (France)
DEA	Deutsche Erdol-Actiengesellschaft (Germany)
DEMINEX	Deutsch Erdolversorgungsgesellschaft mbH (Germany)(Trinidad)
DIAMEX	Diamond Chemicals de Mexico, S.A. de C.V. (Mexico)
DICA	Direction des Carburants (France)
DICA	Distilleria Italiana Carburanti Affini (Italy)
DITTA	Macchia Averardo (Italy)
DUPETCO	Dubai Petroleum Company (Trucial States)
E.A.O.R.	East African Oil Refineries, Ltd. (Kenya)
ECF	Essences et Carbutants de France (France)
ECOPETROL	Empresa Colombiana de Petroleos (Columbia)

EGTA	Enterprises et Grand Travaux de l'Atlantique (France)
ELF-ERAP	Enterprise de Recherches et d'Activites Petrolieres (France)
ELF-U.I.P.	Elf Union Industrielle des Petroles (France)
ELF-SPAFE	Elf des Petroles D'Afrique Equatoriale (France)
ELGI	M. All. Eotvos Lorand Geofizikai Intezet (Hungary)
ENAP	Empresa Nacional del Petroleo (Chile)
ENCAL	Engenheiros Consultores Associados S.A. (Brazil)
ENCASO	Empresa Nacional Calvo Sotelo de Combutibles Liquidos y Lubricantes, S.A. (Spain)
ENGEBRAS	Engenharia Especializada Brasileira S.A. (Brazil) (Venezuela)
ENI	Ente Nazionale Idrocarburi (Italy)
ENPASA	Empresa Nacional de Petroleos de Aragon, S.A. (Spain)
ENPENSA	Empresa Nacional de Petroleos de Navarra, S.A. (Spain)
ERAP	Enterprise de Recherches et d'Activites Petrolieres (France)
ESSAF	Esso Standard Societe Anonyme Francaise (France)
ESSOPETROL	Esso Petroleos Espanoles, S.A. (Spain)
ESSO REP	Societe Esso de Recherches et Exploitation Petrolieres (France)
E.T.P.M.	Societe Entrepose G.T.M. pour les Travaux Petroliers Maritimes (France)
EURAFREP	Societe de Recherches et D'Exploitation de Petrole (France)
FERTIBERIA	Fertilizantes de Iberia, S.A. (Spain)
FFC	Federation Francaise des Carburants (France)

FINAREP	Societe Financiere des Petroles (France)	ICIANZ	Imperial Chemical Industries of Australia & New Zealand Ltd. (Australia, New Zealand)
FOREX	Societe Forex Forages et Exploitation Petrolieres (United Kingdom)	ICIP	Industrie Chimiche Italiane del Petrolio (Italy)
FRANCAREP	Compagnie Franco-Africaine de Recherches Petroliers (France)	IEOC	International Egyptian Oil Co. Inc. (Egypt)
FRAP	Societe de Construction de Feeders, Raffineries, Adductions d'Eau et Pipe-Lines (France)	IFCE	Institut Francais des Combustibles et de l'Energie (France)
FRISIA	Erdolwerke Frisia A.G. (Germany)	IFP	Institute Francaise du Petrole (France)
GARRONE	Garrone (Dott. Edoardo) Raffineria Petroli S.a.S. (Italy)	IGSA	Investigaciones Geologicas, S.A. (Spain)
GBAG	Gelsenberg Benzin (Germany)	IIAPCO	Independent Indonesian American Petroleum Co. (Indonesia)
GESCO	General Engineering Services (Columbia)	I.L.S.E.A.	Industria Leganti Stradali et Affini (United Kingdom)
GHAIP	Ghanian Italian Petroleum Co., Ltd. (Ghana)	I.M.E.	Industrias Matarazzo de Energia (Brazil)
GPC	The General Petroleum Co. (Egypt)	IMEG	International Management & Engineering Group of Britain Ltd. (United Kingdom)
G.T.M.	Les Grands Travaux de Marseille (France)		
HELIECUADOR	Helicopteros Nacionales S.A. (Ecuador)	IMEG	Iranian Management & Engineering Group Ltd. (Iran)
HDC	Hoechst Dyes & Chemicals Ltd. (India)	IMINOCO	Iranian Marine Internation Oil Co. (Iran)
HIDECA	Hidrocarburos y Derivados C.A. (Brazil, Uraguay & Venezuela)	IMS	Industria Metalurgica de Salvador, S/Z (Brazil)
HIP	Hemijska Industrija Pancevo (Yugoslavia)	I.N.C.I.S.A.	Impresa Nazionale Condotte Industriali Strade Affini (United Kingdom)
H.I.S.A.	Herramientas Interamericanas, S.A. de C.V. (Mexico)	INDEIN	Ingenieria Y Desarrolio Industrial S.A. (Spain)
HISPANOIL	Hispanica de Petroleos, S.A. (Spain)	INI	Instituto Nacional de Industria (Spain)
HOC	Hindustan Organic Chemicals Ltd. (India)	INOC	Iraq National Oil Co (Iraq)
		INTERCOL	International Petroleum (Columbia) Ltd. (Columbia)
HYLSA	Hojalata Y Lamina, S.A. (Mexico)	IODRIC	International Oceanic Development Research Information Center (Japan)
IAP	Institut Algerien du Petrole (Algeria)	IOE & PC	Iranian Oil Exploration & Producing Co. (Iran)
ICI	Imperial Chemical Industries, Ltd. (England)	IORC	Iranian Oil Refining Co. N.V. (United Kingdom)

IPAC	Iran Pan American Oil Co. (Iran)	MENEG	Mene Grande Oil Co. (Venezuela)
I.P.L.O.M.	Industria Piemontese Lavorazione Oil Minerali (United Kingdom)	METG	Mittelrheinische Ergastransport GmbH (Germany)
IPLAS	Industrija Plastike (Yugoslavia)	M.I.T.I.	Ministry of International Trade and Industry (Japan)
IPRAS	Istanbul Petrol Rafinerisi A.S. (Turkey)	MODEC	Mitsui Ocean Development & Engineering Co. Ltd. (Japan)
IRANOP	Iranian Oil Participants Limited (England)	MPL	Murco Petroleum Limited (England)
IROM	Industria Raffinazione Oil Minerali (Italy)	NAKI	Nagynyomasu Kiserleti Intezet (Hungary)
IROPCO	Iranian Offshore Petroleum Company (United Kingdom)	NAM	N.V. Nederlandse Aardolie Mij. (Netherlands)
IROS	Iranian Oil Services, Ltd. (England)	NAPM	N.V. Nederlands Amerikaanse Pijpleiding Maatschappij (Netherlands)
IVP	Instituto Venezolano do Petroquimica (Venezuela)	NCM	Nederlandse Constructiebedrijven en Machinefrabriken N.V. (Netherlands)
JAPEX	Japan Petroleum Trading Co. Ltd. (Japan)		
KIZ	Kemijska Industrijska Zajednica (Yugoslavia)	NDSM	Nederlandsche Dok en Scheepsbouw Maatschappij (Netherlands)
KNPC	Kuwait National Petroleum Co. (Arabia)	NED.	North Sea Diving Services, N.V. (Netherlands)
KSEPL	Kon./Shell Exploration and Production Laboratory (Netherlands)	NEPTUNE	Soc. de Forages en Mer Neptune (France)
KSPC	Kuwait Spanish Petroleum Co. (Kuwait)	NETG	Nordrheinische Erdgastransport Gesselschaft mbH (Germany)
KUOCO	Kuwait Oil Co., Ltd. (England)		
LAPCO	Lavan Petroleum Co. (Iran)	NEVIKI	Nehezvegyipari Kutato Intezet (Hungary)
LEMIGAS	Lembaga Minjak Dan Gas Bumi (Indonesia)	NIOC	National Iranian Oil Co. (Iran)
LINOCO	Libyan National Oil Corp. (Libya)	NORDIVE	North Sea Diving Services Ltd. (United Kingdom)
L.M.B.H.	Lemgage Kebajoran & Gas Bumi (Libya)	NOSODECO	North Sumatra Oil Development Cooperation Co. Ltd. (Indonesia)
MABANAFT	Marquard & Bahls B.m.b.H. (Germany)		
MATEP	Materials, Tecnicos de Petroleo S.A. (Brazil)	NPC	Nederlandse Pijpleiding Constructie Combinatie (Netherlands)
MAWAG	Mineraloel Aktien Gesellschaft ag (Germany)	NPCI	National Petroleum Co. of Iran (Iran)
MEDRECO	Mediterranean Refining Co. (Lebanon)		
MEKOG	N.B. Maatschappij Tot Exploitatie van Kooksovengassen (Netherlands)	N.V.A.I.G.B.	N.V. Algemene Internationale Gasleidingen Bouw (Netherlands)

N.V.G.	Nordsee Versorgungsschiffahrt GmbH (Germany)	PREPA	Societe de Prospection et Exploitations Petrolieres en Alsace (France)
NWO	Nord-West Oelleitung GmbH (Germany)	PRODESA	Productos de Estireno S.A. de C.V. (Mexico)
OCCR	Office Central de Chauffe Rationnelle (France)	PROTEXA	Construcciones Protexa, S.A. de C.V. (Mexico)
OEA	Operaciones Especiales Argentinas (Argentina)	PYDESA	Petroleos y Derivados, S.A. (Spain)
OKI	Organsko Kenijska Industrija (Yugoslavia)	QUIMAR	Quimica del Mar, S.A. (Mexico)
OMNIREX	Omnium de Recherches et Exploitations Petrolieres (France)	RAP	Regie Autonome des Petroles (France)
OMV	Oesterreichische Mineraloelverwaltung A.G. (Austria)	RASIOM	Raffinerie Siciliane Olii Minerali (Esso Standard Italiana S.p.A.) (Italy)
OPEC	Organization of Petroleum Exporting Countries	RDM	De Rotterdamsche Droogdok Mij. N.V. (Netherlands)
OTP	Omnium Techniques des Transports par Pipelines (France)	RDO	Rhein-Donau-Oelleitung GmbH (Germany)
PCRB	Compagnie des Produits Chimiques et Raffineries de Berrre (France)	REDCO	Rehabilitation, Engineering and Development Co. (Indonesia)
PEMEX	Petroleos Mexicanos (Mexico)	REPESA	Refineria de Petroleos de Escombreras, S.A. (Spain)
PERMAGO	Perforaciones Marinas del Golfo S.A. (Mexico)	REPGA	Recherche et Exploitation de Petrole et de Gaz (France)
PETRANGOL	Companhia de Petroleos de Angola (Angola)	RIOGULF	Rio Gulf de Petroleos, S.A. (Spain)
PETRESA	Petroquimica Espanola, S.A. (Spain)	SACOR	Sociedade Anonima Concessionaria da Refinacao de Petroleos em Portugal (Portugal)
PETROBRAS	Petroleo Brasileiro S.A. (Brazil)		
PETROLIBER	Compania Iberica Refinadora de Petroleos, S.A. (Spain)	SAEL	Sociedad Anonima Espanola de Lubricantes (Spain)
PETROMIN	General Petroleum and Mineral Organization (Saudi Arabia)	S.A.F.C.O.	Saudi Arabian Refinery Co. (Saudi Arabia)
		SAFREP	Societe Anonyme Francaise de Recherches et D'Exploitation de Petrole (France)
PETRONOR	Refineria de Petroleos del Norte, S.A. (Spain)	SAIC	Sociedad Anonima Industrial y Commerical (Argentina)
PETROPAR	Societe de Participations Petrolieres (France)	SAM	Societe d'Approvis de Material Patrolier (France)
PETROREP	Societe Petroliere de Recherches Dans La Region Parisienne (France)	SAP	Societe Africaine des Petroles (France)
POLICOLSA	Poliolefinas Colombianas S.A. (Columbia)	SAPPRO	Societe Anonyme de Pipeline a Produits Petroliers sur Territoire Genevois (Switzerland)

SAR	Societe Africaine de Raffinage (Dakar)	
SARAS	S.p.a. Raffinerie Sarde (Italy)	
SARL	Chimie Development International (Germany)	
SAROC	Saudi Arabia Refinery Co. (Saudi Arabia)	
SAROM	Societa Azionaria Raffinazione Olii Minerali (Italy)	
SARPOM	Societa per Azioni Raffineria Padana Olii Minerali (Italy)	
SASOL	South African Coal, Oil and Gas Corp. Ltd. (South Africa)	
S.A.V.A.	Societa Alluminio Veneto per Azioni (Italy)	
SCC	Societe Chimiques des Charbonnages (France)	
SCI	Societe Chimie Industrielle (France)	
SCP	Societe Cherifienne des Petroles (Morocco)	
SECA	Societe Europreeme des Carburants (Belgium)	
SEHR	Societe d'Exploitation des Hydrocarbures d'Hassi R'Mel (France)	
SEPE	Sociedad de Exploracion de Petroleos Espanoles, S.A. (Spain)	
SER	Societe Equatoriale de Raffinage (Gabon)	
SERCOP	Societe Egyptienne pour le Raffinage et le Commerce du Petrole (Egypt)	
SEREPT	Societe de Recherches et D'Exploitation des Petroles en Tunisia (Tunisia)	
SER VIPETROL	Transportes Y Servicios Petroleros (Ecuador)	
SETRAPEM	Societe Equatoriale de Travaux Petroliers Maritimes (France, Germany)	
SEPLJ	Societe Francaise de Pipe Line du Jura (France)	
SHELLREX	Societe Shell de Recherches et D'Exploitations (France)	
S.I.B.P.	Societe Industrielle Belge des Petroles (Belgium)	

SIF	Societe Tunisienne de Sondages, Injections, Forages (Tunisia)
SINCAT	Societa Industriale Cantese S.p.a. (Italy)
SIPSA	Sociedad Investigadora Petrolifera S.A. (Spain)
SIR	Societa Italiana Resine (Italy)
SIREP	Societe Independante de Recherches et d'Exploitation du Petrole (France)
SIRIP	Societe Irano-Italienne des Petroles (Iran)
SITEP	Societe Italo-Tunisienne d'Exploitation (Italy, Tunisia)
SMF	Societe de Fabrication de Material de Forage (France)
SMP	Svenska Murco Petroleum Aktiebolag (Sweden)
SMR	Societe Malagache de Raffinage (Malagasy)
SNGSO	Societe Nationale des Gas de Sud-Ouest (France)
SN MAREP	Societe Nationale de Material pour la Recherche et l'Exploitation du Petrole (France)
SNPA	Societe Nationale des Petroles d'Aquitaine (France)
SN REPAL	Societe Nationale de Recherches et d'Exploitation des Petroles en Algerie (France)
SOCABU	Societe du Caoutchouc Butyl (France)
SOCEA	Societe Eau et Assainissement (France)
SOCIR	Societe Congo-Italienne de Raffinage (Congo Republic)
SOFEI	Societe Francaise d'Enterprises Industrielles (France)
SOGARES	Societe Gabonaise de Realisation de Structures (France)
SOMALGAZ	Societe Mixte Algerienne de Gaz (Algeria)
SOMASER	Societe Maritime de Service (France)

SONAP	Sociedade Nacional de Petroleos S.A.R.I. (Portugal)	STIR	Societe Tuniso-Italienne de Raffinage (Tunisia)
SONAREP	Sociedade Nacional de Refinacao de Petroleos S.A.R.L. (Mozambique)	TAL	Deutsche Transalpine Oelleitung GmbH (Germany)
SONATRACH	Societe Nationale de Transport et de Commercialisation des Hydrocarbures (Algeria, France)	TAMSA	Tubos de Acero de Mexico, S.A. (Mexico)
		TATSA	Tanques de Acero Trinity, S.A. (Mexico)
		TECHINT	Compania Technica Internacional (Brazil)
SONPETROL	Sondeos Petroliferos S.A. (Spain)	TECHNIP	Compagnie Francaise d'Etudes et de Construction Technip (France)
SOPEFAL	Societe Petroliere Francaise en Algeria (Algeria)	TEXSPAIN	Texaco (Spain) Inc. (Spain)
SOPEG	Societe Petroliere de Gerance (France)	TORC	Thai Oil Refinery Co. (Thailand)
SOREX	Societe de Recherches et d'Exploitations Petrolieres (France)	T.P.A.O.	Turkiye Petrolleri A.O. (United Kingdom)
SOTEI	Societe Tunisienne de Enterprises Industrielles (Tunisia)	TRAPIL	Societe des Transports Petroliers Par Pipeline (France)
SOTHRA	Societe de Transport du Gaz Naturel D'Hassi-er-r'mel a Arzew (Algeria)	TRAPSA	Compagnie des Transports par Pipe-Line au Sahara (Algeria)
SPAFE	Societe des Petroles d'Afrique Equatoriale (France)	UCSIP	Union des Chambres Syndicales de l'Industrie du Petrole (France)
SPANGOC	Spanish Gulf Oil Co. (Spain)	UGP	Union Generale des Petroles (France)
SPEICHIM	Societe Pour l/Equipment des Industries, Chimiques (France)	UIE	Union Industrielle et d'Enterprise (France)
SPG	Societe des Petroles de la Garrone (France)	UNIAO	Refinaria e Exploracao de Petroleo "UNIAO" S.A. (Brazil)
S.P.I.	Societa Petrolifera Italiana (Italy)	URAG	Unterweser Reederei GmbH (Germany)
SPIC	Southern Petrochemical Industries Corporation Ltd.	URG	Societe pour l'Utilisation Rationnelle des Gaz (France)
SPLSE	Societe du Pipe-Line Sud Europeen (France)		
SPM	Societe des Petroles de Madagascar (France)	WEPCO	Western Desert Operating Petroleum Company (Egypt)
SPV	Societe des Petroles de Valence (France)	YPE	Yacimientos Petroliferos Fiscales (Argentina)
SSRP	Societe Saharienne de Recherches Petrolieres (France)	YPFB	Yacimientos Petroliferos Fiscales Bolivianos (Bolivia)
STEG	Societe Tunisienne d'Electricite et de Gaz (Tunisia)		

API STANDARD OIL-MAPPING SYMBOLS

Location .... ⭕

Abandoned location .... erase symbol

Dry hole .... ✛

Oil well .... ⚫

Abandoned oil well .... ⚫✛

Gas well .... ☼

Abandoned gas well .... ☼

Distillate well .... ◐

Abandoned distillate well .... ◐

Dual completion—oil .... ◉

Dual completion—gas .... ⊕

Drilled water-input well .... ⌀w

Converted water-input well .... ⬤ w

Drilled gas-input well .... ⌀G

Converted gas-input well .... ⬤G

Bottom-hole location .... ⭕-----x
    (x indicates bottom of hole. Changes in well
      status should be indicated as in symbols
      above.)

Salt-water disposal well .... ⊕ SWD

Courtesy American Petroleum Institute, Division of Production.

MATHEMATICAL SYMBOLS AND SIGNS

$+$	plus	$\therefore$	therefore
$-$	minus	$\because$	because
$\pm$	plus or minus	$:$	is to; divided by
$\times$	multiplied by	$::$	as; equals
$\cdot$	multiplied by	$:\!:$	geometrical proportion
$\div$	divided by	$\propto$	varies as
$/$	divided by	$\doteq$	approaches a limit
$=$	equal to	∞	infinity
$\neq$	not equal to	$\int$	integral
$\approx$	nearly equal to	d	differential
$\cong$	congruent to	∂	partial differential
$\equiv$	identical with	Σ	summation of
$\not\equiv$	not identical with	$!$	factorial product
$\Leftrightarrow$	equivalent to	π	pi (3.1416)
$>$	greater than	e	epsilon (2.7183)
$\not>$	not greater than	$\circ$	degree
$<$	less than	$'$	minute; prime
$\not<$	not less than	$''$	second
$\geqq$	greater than or equal to	$\angle$	angle
$\leqq$	less than or equal to	$\llcorner$	right angle
$\sim$	difference between	$\perp$	perpendicular
$\simeq$	difference between	$\bigcirc$	circle
$-\!:$	difference between	$\frown$	arc
$\sqrt{}$	square root	$\triangle$	triangle
$\sqrt[3]{}$	cube root	$\square$	square
$\sqrt[n]{}$	nth root	$\square$	rectangle

GREEK ALPHABET

A	α	Alpha	N	ν	Nu
B	β	Beta	Ξ	ξ	Xi
Γ	γ	Gamma	O	o	Omicron
Δ	δ	Delta	Π	π	Pi
E	ϵ	Epsilon	P	ρ	Rho
Z	ζ	Zeta	Σ	σ	Sigma
H	η	Eta	T	τ	Tau
Θ	θ	Theta	Y	υ	Upsilon
I	ι	Iota	Φ	ϕ	Phi
K	κ	Kappa	X	χ	Chi
Λ	λ	Lambda	Ψ	ψ	Psi
M	μ	Mu	Ω	ω	Omega

Universal Conversion Factors

compiled and edited by
STEVEN GEROLDE

ACRE: =

0.0015625	square miles or sections
0.004046875	square kilometers
0.1	square furlongs
0.4046875	square hektometers
10	square chains
40.46875	square dekameters
160	square rods
4,046.875	square meters
4,840	square yards
5,645.41213	square varas (Texas)
43,560	square feet
77,440	square spans
100,000	square links
400,000	square hands
404,687.5	square decimeters
40,468.750	square centimeters
6,272,640	square inches
4,046,875,000	square millimeters
0.4046875	hectares
40.46875	ares
4,046.875	centares (centiares)

ACRE FOOT: =

1,233.48766	kiloliters
1,233.48766	cubic meters
1,633.3333	cubic yards
43,560	cubic feet
7,758.34	barrels
12,334.876	hektoliters
35,000	bushels—U.S. (dry)
33,933.16195	bushels—Imperial (dry)
123,348.766	dekaliters
140,000	pecks
325,850.28	gallons—U.S. (liquid)
280,092.5925	gallons—U.S. (dry)
271,325.745265	gallons—Imperial
1,303,401.12	quarts (liquid)
1,120,370.370	quarts (dry)
1,233,487.66	liters
1,233,487.660	cubic decimeters
2,606,802.24	pints
10,427,208.96	gills
12,334,876.6	deciliters
75,271,680	cubic inches
123,348,766	centiliters
1,233,487,660	milliliters
1,233,487,660	cubic centimeters
1,233,487,660,000	cubic millimeters

ACRE FOOT PER DAY: =

7,758.34	barrels per day
323.264167	barrels per hour
5.387736	barrels per minute
0.0897956	barrels per second
325,850.28	gallons (U.S.) per day
13,577.09400	gallons (U.S.) per hour
226.284900	gallons (U.S.) per minute
3.771415	gallons (U.S.) per second
271,325.745265	gallons (Imperial) per day
11,305.238400	gallons (Imperial) per hour
188.42066	gallons (Imperial) per minute
3.140344	gallons (Imperial) per second
1,233.48766	cubic meters per day
51.395319	cubic meters per hour
0.856589	cubic meters per minute
0.0142765	cubic meters per second
1,633.33333	cubic yards per day
68.0555552	cubic yards per hour
1.134259	cubic yards per minute
0.0189043	cubic yards per second
43,560	cubic feet per day
1,815.0	cubic feet per hour
30.25000	cubic feet per minute
0.504167	cubic feet per second
75,271,680.00	cubic inches per day
3,136,320.00	cubic inches per hour
52,272.00	cubic inches per minute
871.200	cubic inches per second

ATMOSPHERE: =

0.103327	hektometers of water @ 60°F.
1.03327	dekameters of water @ 60°F.
10.3327	meters of water @ 60°F.
33.9007	feet of water @ 60°F.
406.8084	inches of water @ 60°F.
103.327	decimeters of water @ 60°F.
1,033.27	centimeters of water @ 60°F.
10,332.7	millimeters of water @ 60°F.
0.00760	hektometers of mercury @ 32°F.
0.0760	dekameters of mercury @ 32°F.
0.760	meters of mercury @ 32°F.
2.49343	feet of mercury @ 32°F.
29.9212	inches of mercury @ 32°F.
7.6	decimeters of mercury @ 32°F.
76	centimeters of mercury @ 32°F.
760	millimeters of mercury @ 32°F.
113,893.88	tons per square hektometer
1,138.9388	tons per square dekameter

ATMOSPHERE: (cont'd)

11.389388	tons per square meter
1.0581	tons per square foot
0.00734792	tons per square inch
0.11389388	tons per square decimeter
0.0011389388	tons per square centimeter
0.000011389388	tons per square millimeter
103,327,000	kilograms per square hektometer
1,033,270	kilograms per square dekameter
10,332.7	kilograms per square meter
959.931252	kilograms per square foot
6.666189	kilograms per square inch
103.327	kilograms per square decimeter
1.03327	kilograms per square centimeter
0.0103327	kilograms per square millimeter
227,774,851.2	pounds per square hektometer
2,277,748.512	pounds per square dekameter
22,777.48512	pounds per square meter
2,116.080	pounds per square foot
14.696	pounds per square inch
227.7748512	pounds per square decimeter
2.277748512	pounds per square centimeter
0.02277748512	pounds per square millimeter
1,033,270,000	hektograms per square hektometer
10,332,700	hektograms per square dekameter
103,327	hektograms per square meter
9,599,31252	hektograms per square foot
66.66189	hektograms per square inch
1,033.27	hektograms per square decimeter
10.3327	hektograms per square centimeter
0.103327	hektograms per square millimeter
10,332,700,000	dekagrams per square hektometer
103,327,000	dekagrams per square dekameter
1,033,270	dekagrams per square meter
95,993.1252	dekagrams per square foot
666.6189	dekagrams per square inch
10,332.7	dekagrams per square decimeter
103.327	dekagrams per square centimeter
1.03327	dekagrams per square millimeter
3,644,397,619.2	ounces per square hektometer
36,443,976.192	ounces per square dekameter
364,439.76192	ounces per square meter
33,857.28	ounces per square foot
235.136	ounces per square inch
3,644.39762	ounces per square decimeter
36.44398	ounces per square centimeter
0.36444	ounces per square millimeter
103,327,000,000	grams per square hektometer
1,033,270,000	grams per square dekameter
10,332,700	grams per square meter
959,931.252	grams per square foot
6,666,189	grams per square inch

103,327	grams per square decimeter
1,033.27	grams per square centimeter
10.3327	grams per square millimeter
1,033,270,000,000	decigrams per square hektometer
10,332,700,000	decigrams per square dekameter
103,327,000	decigrams per square meter
9,599,312.52	decigrams per square foot
66,661.89	decigrams per square inch
1,033,270	decigrams per square decimeter
10,332.7	decigrams per square centimeter
103.327	decigrams per square millimeter
10,332,700,000,000	centigrams per square hektometer
103,327,000,000	centigrams per square dekameter
1,033,270,000	centigrams per square meter
95,933,125.2	centigrams per square foot
666,618.9	centigrams per square inch
10,332,700	centigrams per square decimeter
103,327	centigrams per square centimeter
1,033.270	centigrams per square millimeter
103,327,000,000,000	milligrams per square hektometer
1,033,270,000,000	milligrams per square dekameter
10,332,700,000	milligrams per square meter
959,931,252	milligrams per square foot
6,666,189	milligrams per square inch
103,327,000	milligrams per square decimeter
1,033,270	milligrams per square centimeter
10,332.7	milligrams per square millimeter
101,325,000,000,000	dynes per square hektometer
1,013,250,000,000	dynes per square dekameter
10,132,500,000	dynes per square meter
941,343,587	dynes per square foot
6,537,096	dynes per square inch
101,325,000	dynes per square decimeter
1,013,250	dynes per square centimeter
10,132.5	dynes per square millimeter
1.01325	bars

BARREL: =

0.158987	kiloliters
0.158987	cubic meters
0.20794	cubic yards
1.58987	hectoliters
4.511274	bushels—U.S. (dry)
4.373766	bushels—Imperial (dry)
5.6146	cubic feet
15.89871	dekaliters
18.045097	pecks
42	gallons—U.S. (liquid)
36.09798	gallons—U.S. (dry)
34.99089	gallons—Imperial

BARREL: (cont'd)

168	quarts (liquid)
144.408516	quarts (dry)
158.987146	liters
158.987146	cubic decimeters
336	pints
1,344	gills
1,589.87146	deciliters
9,702.0288	cubic inches
15,898.71459456	centiliters
158,987.1459456	milliliters
158,987.1459456	cubic centimeters
158,987,145.9456	cubic millimeters
0.174993	tons (short) of water @ 62°F.
0.1562438	tons (long) of water @ 62°F.
0.1587512	tons (metric) of water @ 62°F.
158.7512	kilograms of water @ 62°F.
349.986	pounds of water @ 62°F.
15.87512	hektograms of water @ 62°F.
1.587512	dekagrams of water @ 62°F.
5,599.776	ounces of water @ 62°F.
0.1587512	grams of water @ 62°F.
0.01587512	decigrams of water @ 62°F.
0.001587512	centigrams of water @ 62°F.
0.0001587512	milligrams of water @ 62°F.
5.1042	sacks of cement
2,449,902	grains
404.25	pounds of salt walter @ 60°F. of 1.155 specific gravity

BARREL OF CEMENT: =

0.158987	kiloliters
0.158987	cubic meters
0.20794	cubic yards
1.58987	hectoliters
4.511274	bushels—U.S. (dry)
4.373766	bushels—Imperial (dry)
5.6146	cubic feet
15.89871	dekaliters
18.045097	pecks
42	gallons—U.S. (liquid)
36.10213	gallons—U.S. (dry)
34.99089	gallons—Imperial
168	quarts (liquid)
144.408516	quarts (dry)
158.987146	liters
158.987146	cubic decimeters
336	pints
1,344	gills
1,589.87146	deciliters
9,702.0288	cubic inches

BARREL OF CEMENT: (cont'd)

15,898.71459456	centiliters
158,987.1459456	milliliters
158,987.1459456	cubic centimeters
158,987,145.9456	cubic millimeters
0.188	tons (short)
0.16796	tons (long)
0.170551	tons (metric)
170.55097	kilograms
376	pounds
1705.5097	hektograms
17,055.097	dekagrams
6,016	ounces
170,550.97	grams
1,705,509.7	decigrams
17,055,097	centigrams
170,550,970	milligrams

BARREL PER DAY: =

0.041667	barrels per hour
0.00069444	barrels per minute
0.000011574	barrels per second
0.1589871	kiloliters per day
0.0066245	kiloliters per hour
0.00011041	kiloliters per minute
0.000001840	kiloliters per second
0.1589871	cubic meters per day
0.0066245	cubic meters per hour
0.00011041	cubic meters per minute
0.000001840	cubic meters per second
0.20794	cubic yards per day
0.0086642	cubic yards per hour
0.0001444	cubic yards per minute
0.0000024067	cubic yards per second
1.589871	hektoliters per day
0.066245	hektoliters per hour
0.0011041	hektoliters per minute
0.00001840	hektoliters per second
5.6146	cubic feet per day
0.233942	cubic feet per hour
0.00389903	cubic feet per minute
0.0000649838	cubic feet per second
15.89871	dekaliters per day
0.66245	dekaliters per hour
0.011041	dekaliters per minute
0.0001840	dekaliters per second
42	gallons (U.S.) per day
1.71875	gallons (U.S.) per hour
0.029167	gallons (U.S.) per minute
0.0004861	gallons (U.S.) per second
34.99089	gallons (Imperial) per day

BARREL PER DAY: (cont'd)

1.45795	gallons (Imperial) per hour
0.024299	gallons (Imperial) per minute
0.00040499	gallons (Imperial) per second
168	quarts (U.S.) per day
6.875	quarts (U.S.) per hour
0.11668	quarts (U.S.) per minute
0.0019444	quarts (U.S.) per second
158.98714	liters per day
6.6245	liters per hour
0.11041	liters per minute
0.001840	liters per second
158.98714	cubic decimeters per day
6.6245	cubic decimeters per hour
0.11041	cubic decimeters per minute
0.001840	cubic decimeters per second
336	pints per day
13.75	pints per hour
0.23336	pints per minute
0.0038888	pints per second
1,344	gills per day
55	gills per hour
0.933344	gills per minute
0.0155552	gills per second
1,589.87146	deciliters per day
66.245	deciliters per hour
1.1041	deciliters per minute
0.01840	deciliters per second
9,702.0288	cubic inches per day
404.2	cubic inches per hour
6.7375	cubic inches per minute
0.112292	cubic inches per second
15,898.7146	centiliters per day
662.45	centiliters per hour
11.041	centiliters per minute
0.1840	centiliters per second
158,987.145946	milliliters per day
6,624.5	milliliters per hour
110.41	milliliters per minute
1.84	milliliters per second
158,987.145946	cubic centimeters per day
6,624.5	cubic centimeters per hour
110.41	cubic centimeters per minute
1.840	cubic centimeters per second
158,987,145.946	cubic millimeters per day
6,624,500	cubic millimeters per hour
110,410	cubic millimeters per minute
1,840	cubic millimeters per second
24	barrels per day
0.016667	barrels per minute
0.000277778	barrels per second
3.81567	kiloliters per day

BARREL PER DAY: (cont'd)

0.1589871	kiloliters per hour
0.0026498	kiloliters per minute
0.000044163	kiloliters per second
3.81567	cubic meters per day
0.1589871	cubic meters per hour
0.0026498	cubic meters per minute
0.000044163	cubic meters per second
4.99056	cubic yards per day
0.20794	cubic yards per hour
0.0034657	cubic yards per minute
0.000057761	cubic yards per second
38.1567	hektoliters per day
1.589871	hektoliters per hour
0.026498	hektoliters per minute
0.00044163	hektoliters per second
134.7504	cubic feet per day
5.6146	cubic feet per hour
0.093577	cubic feet per minute
0.15596	cubic feet per second
381.567	dekaliters per day
15.89871	dekaliters per hour
0.26498	dekaliters per minute
0.0044163	dekaliters per second
1,008	gallons (U.S.) per day
42	gallons (U.S.) per hour
0.7	gallons (U.S.) per minute
0.11667	gallons (U.S.) per second
839.78136	gallons (Imperial) per day
34.99089	gallons (Imperial) per hour
0.58318	gallons (Imperial) per minute
0.0097197	gallons (Imperial) per second
4,032	quarts (U.S.) per day
168	quarts (U.S.) per hour
2.80	quarts (U.S.) per minute
0.046667	quarts (U.S.) per second
3,815.6904	liters per day
158.98714	liters per hour
2.64979	liters per minute
0.044163	liters per second
3,815.6904	cubic decimeters per day
158.98714	cubic decimeters per hour
2.64979	cubic decimeters per minute
0.044163	cubic decimeters per second
8,064	pints per day
336	pints per hour
5.60	pints per minute
0.93333	pints per second
32,256	gills per day
1,344	gills per hour
22.40	gills per minute
0.37333	gills per second

BARREL PER DAY: (cont'd)

38,156.904	deciliters per day
1,589.87146	deciliters per hour
26.49786	deciliters per minute
0.44163	deciliters per second
232,848	cubic inches per day
9,702.0288	cubic inches per hour
161.7014	cubic inches per minute
2.695	cubic inches per second
381,569.04	centiliters per day
15,898.7146	centiliters per hour
264.97858	centiliters per minute
4.4163	centiliters per second
3,815,690.4	milliliters per day
158,987.145946	milliliters per hour
2,649.78576	milliliters per minute
44.163	milliliters per second
3,815,690.4	cubic centimeters per day
158,987.145946	cubic centimeters per hour
2,649.78576	cubic centimeters per minute
44.163	cubic centimeters per second
3,815,690,400.0	cubic millimeters per day
158,987,145,946	cubic millimeters per hour
2,649,785.76	cubic millimeters per minute
44,163.096	cubic millimeters per second

BARREL PER MINUTE: =

1,440	barrels per day
60	barrels per hour
0.016667	barrels per second
228.94272	kiloliters per day
9.53928	kiloliters per hour
0.158987	kiloliters per minute
0.0026498	kiloliters per second
228.94272	cubic meters per day
9.53928	cubic meters per hour
0.158987	cubic meters per minute
0.0026498	cubic meters per second
299.43648	cubic yards per day
12.47652	cubic yards per hour
0.20794	cubic yards per minute
0.0034657	cubic yards per second
2,289.4272	hektoliters per day
95.3928	hektoliters per hour
1.58987	hektoliters per minute
0.026498	hektoliters per second
8,085.05280	cubic feet per day
336.8772	cubic feet per hour
5.6146	cubic feet per minute
0.093577	cubic feet per second
22,894.272	dekaliters per day

BARREL PER MINUTE: (cont'd)

953.928	dekaliters per hour
15.8987	dekaliters per minute
0.26498	dekaliters per second
60,480	gallons (U.S.) per day
2,520	gallons (U.S.) per hour
42	gallons (U.S.) per minute
0.7	gallons (U.S.) per second
50,386.7520	gallons (Imperial) per day
2,099.4480	gallons (Imperial) per hour
34.99089	gallons (Imperial) per minute
0.58318	gallons (Imperial) per second
241,920	quarts per day (U.S.)
10,080	quarts per hour (U.S.)
168	quarts per minute (U.S.)
2.80	quarts per second (U.S.)
228,941.48966	liters per day
9,539.22874	liters per hour
158.987146	liters per minute
2.64979	liters per second
228,941.48966	cubic decimeters per day
9,539.22874	cubic decimeters per hour
158.987146	cubic decimeters per minute
2.64979	cubic decimeters per second
483,840	pints per day
20,160	pints per hour
336	pints per minute
5.60	pints per second
1,935,360	gills per day
80,640	gills per hour
1,344	gills per minute
22.40	gills per second
2,289,414.89664	deciliters per day
95,392.28736	deciliters per hour
1,589.87146	deciliters per minute
26.49786	deciliters per second
13,970,921.472	cubic inches per day
582,121.7280	cubic inches per hour
9,702.0288	cubic inches per minute
161.70048	cubic inches per second
22,894,148.9664	centiliters per day
953,922.8736	centiliters per hour
15,898.71459	centiliters per minute
264.97858	centiliters per second
228,941,489.664	milliliters per day
9,539,228.736	milliliters per hour
158,987.14595	milliliters per minute
2,649.78576	milliliters per second
228,941,489.664	cubic centimeters per day
9,539,228.736	cubic centimeters per hour
158,987.14595	cubic centimeters per minute
2,649.78576	cubic centimeters per second
228,941,489,664	cubic millimeters per day

BARREL PER MINUTE: (cont'd)

9,539,228,736	cubic millimeters per hour
158,987,145.946	cubic millimeters per minute
2,649,785.76	cubic millimeters per second

BARREL PER SECOND: =

86,400	barrels per day
3,600	barrels per hour
60	barrels per minute
13,736.47680	kiloliters per day
572.35320	kiloliters per hour
9.53922	kiloliters per minute
0.158987	kiloliters per second
13,736.47680	cubic meters per day
572.35320	cubic meters per hour
9.53922	cubic meters per minute
0.158987	cubic meters per second
17,966.0160	cubic yards per day
748.5840	cubic yards per hour
12.47640	cubic yards per minute
0.20794	cubic yards per second
137,364.7680	hektoliters per day
5,723.5320	hektoliters per hour
95.3922	hektoliters per minute
1.58987	hektoliters per second
485,101.44	cubic feet per day
20,212.560	cubic feet per hour
336.8760	cubic feet per minute
5.6146	cubic feet per second
1,373,648.5440	dekaliters per day
57,235,356	dekaliters per hour
953.92260	dekaliters per minute
15.89871	dekaliters per second
3,628,800	gallons (U.S.) per day
151,200	gallons (U.S.) per hour
2,520	gallons (U.S.) per minute
42	gallons (U.S.) per second
3,023,212.8960	gallons (Imperial) per day
125,967.2040	gallons (Imperial) per hour
2,099.45340	gallons (Imperial) per minute
34.99089	gallons (Imperial) per second
14,515,200	quarts (U.S.) per day
604,800	quarts (U.S.) per hour
10,080	quarts (U.S.) per minute
168	quarts (U.S.) per second
13,736,489.4144	liters per day
572,353.7256	liters per hour
9,539.22876	liters per minute
158.987146	liters per second
13,736,489.4144	cubic decimeters per day
572,353.7256	cubic decimeters per hour

BARREL PER SECOND: (cont'd)

9,539.22876	cubic decimeters per minute
158.987146	cubic decimeters per second
29,030,400	pints per day
1,209,600	pints per hour
20,160	pints per minute
336	pints per second
116,121,600	gills per day
4,838,400	gills per hour
80,640	gills per minute
1,344	gills per second
137,364,894.144	deciliters per day
5,723,537.256	deciliters per hour
95,392.28754	deciliters per minute
1,589.87146	deciliters per second
838,255,288.320	cubic inches per day
34,927,303.6800	cubic inches per hour
582,131.7280	cubic inches per minute
9,702.0288	cubic inches per second
1,373,648,941.440	centiliters per day
57,235,372.56	centiliters per hour
953,922.8754	centiliters per minute
15,898.71459	centiliters per second
13,736,489,414.40	milliliters per day
572,353,725.6	milliliters per hour
9,539,228.760	milliliters per minute
158,987.14594	milliliters per second
13,736,489,414.4	cubic centimeters per day
572,353,725.6	cubic centimeters per hour
9,539,228.760	cubic centimeters per minute
158,987.14595	cubic centimeters per second
13,736,489,414,400	cubic millimeters per day
572,353,725,600	cubic millimeters per hour
9,539,228,754	cubic millimeters per minute
158,987,145.9456	cubic millimeters per second

BTU (60°F.): =

25,030	foot poundals
300,360	inch poundals
777.97265	foot pounds
9,335.67120	inch pounds
0.00027776	ton (short) calories
0.25198	kilogram calories
0.55552	pound calories
2.5198	hektogram calories
25.198	dekagram calories
8.88832	ounce calories
251.98	gram calories
2,519.8	decigram calories
25,198	centigram calories
251.980	milligram calories

BTU (60°F.): (cont'd)

0.000012201	kilowatt days
0.00029283	kilowatt hours
0.01757	kilowatt minutes
1.0546	kilowatt seconds
0.012201	watt days
0.29283	watt hours
17.57	watt minutes
1,054.6	watt seconds
0.11856	ton meters
107.56	kilogram meters
237.12678	pound meters
1,075.6	hektogram meters
10,756	dekagram meters
3,794.02848	ounce meters
107,560	gram meters
1,075,600	decigram meters
10,756,000	centigram meters
107,560,000	milligram meters
0.0011856	ton hektometers
1.0756	kilogram hektometers
2.37127	pound hektometers
10.756	hektogram hektometers
107.56	dekagram hektometers
37.94028	ounce hektometers
1,075.6	gram hektometers
10,756	decigram hektometers
107,560	centigram hektometers
1,075,600	milligram hektometers
0.011856	ton dekameters
10.756	kilogram dekameters
23.7127	pound dekameters
107.56	hektogram dekameters
1,075.6	dekagram dekameters
379.40285	ounce dekameters
10,756	gram dekameters
107,560	decigram dekameters
1,075,600	centigram dekameters
10,756,000	milligram dekameters
0.388977	ton feet
352.887473	kilogram feet
777.97265	pound feet
3,528.874731	hektogram feet
35,288.747308	dekagram feet
12,447.611780	ounce feet
352,887.473080	gram feet
3,528.875	decigram feet
35,288,747	centigram feet
352,887,473	milligram feet
4.667724	ton inches
4,234.649677	kilogram inches
9,335.671800	pound inches

BTU (60°F.): (cont'd)

42,346.496772	hektogram inches
423,464.96772	dekagram inches
149,371.34136	ounce inches
4,234,650	gram inches
42,346,497	decigram inches
423,464,968	centigram inches
4,234,649,677	milligram inches
1.1856	ton decimeters
1.0756	kilogram decimeters
2,371.2678	pound decimeters
10,756	hektogram decimeters
107,560	dekagram decimeters
37,940.2848	ounce decimeters
1,075,600	gram decimeters
10,756,000	decigram decimeters
107,560,000	centigram decimeters
1,075,600,000	milligram decimeters
11.856	ton centimeters
10,756	kilogram centimeters
23,712.678	pound centimeters
107,560	hektogram centimeters
1,075,600	dekagram centimeters
379,402.848	ounce centimeters
10,756,000	gram centimeters
107,560,000	decigram centimeters
1,075,600,000	centigram centimeters
10,756,000,000	milligram centimeters
118.56	ton millimeters
107,560	kilogram millimeters
237,126.780	pound millimeters
1,075,600	hektogram millimeters
10,756,000	dekagram millimeters
3,794,028.48	ounce millimeters
107,560,000	gram millimeters
1,075,600,000	decigram millimeters
10,756,000,000	centigram millimeters
107,560,000,000	milligram millimeters
0.0104028	kiloliter-atmospheres
0.104028	hektoliter-atmospheres
1.040277	dekaliter-atmospheres
10.40277	liter-atmospheres
104.0277	deciliter-atmospheres
1,040.277	centiliter-atmospheres
10,402.77	milliliter-atmospheres
0.0000000000104104	cubic kilometer-atmospheres
0.0000000104104	cubic hektometer-atmospheres
0.000010410	cubic dekameter-atmospheres
0.0104104	cubic meter-atmospheres
0.3676637	cubic feet-atmospheres
635.277597	cubic inch-atmospheres
10.410432	cubic decimeter-atmospheres

BTU (60°F.): (cont'd)

10,410.4320	cubic centimeter-atmospheres
10,410,432	cubic millimeter-atmospheres
1,054.198	joules
0.0003982	Cheval-vapeur hours
0.000016372	horsepower days
0.00039292	horsepower hours
0.0235757	horsepower minutes
1.41451	horsepower seconds
0.0000685	pounds of carbon oxidized with perfect efficiency
0.001030	pounds of water evaporated from and at 212°F.

BTU (60°F.) PER DAY: =

25,030	foot poundals per day
1,042.92	foot poundals per hour
17.3820	foot poundals per minute
0.2897	foot poundals per second
777.97265	foot pounds per day
32.41553	foot pounds per hour
0.54026	foot pounds per minute
0.0090043	foot pounds per second
0.25198	kilogram calories per day
0.010499	kilogram calories per hour
0.00017498	kilogram calories per minute
0.0000029164	kilogram calories per second
8.88832	ounce calories per day
0.37035	ounce calories per hour
0.0061724	ounce calories per minute
0.00010287	ounce calories per second
107.56	kilogram meters per day
4.48164	kilogram meters per hour
0.074694	kilogram meters per minute
0.0012449	kilogram meters per second
10.40277	liter-atmospheres per day
0.43344	liter-atmospheres per hour
0.007224	liter-atmospheres per minute
0.00012040	liter-atmospheres per second
0.3676637	cubic foot-atmospheres per day
0.0153166	cubic foot-atmospheres per hour
0.000255276	cubic foot-atmospheres per minute
0.0000042546	cubic foot-atmospheres per second
0.000016148	Cheval-vapeurs
0.000016372	horsepower
0.000012201	kilowatts
0.012201	watts
1,054.198	joules per day
43.9236	joules per hour
0.73206	joules per minute
0.012201	joules per second

BTU (60°F.) PER DAY: (cont'd)

0.0000685	pounds of carbon oxidized with perfect efficiency per day
0.00000285415	pounds of carbon oxidized with perfect efficiency per hour
0.000000047569	pounds of carbon oxidized with perfect efficiency per minute
0.00000000079282	pounds of carbon oxidized with perfect efficiency per second
0.001030	pounds of water evaporated from and at 212°F. per day
0.000042916	pounds of water evaporated from and at 212°F. per hour
0.00000071526	pounds of water evaporated from and at 212°F. per minute
0.000000011921	pounds of water evaporated from and at 212°F. per second
1.0	BTU per day
0.0416667	BTU per hour
0.00069444	BTU per minute
0.000011574	BTU per second

BTU PER HOUR: =

600,720.1920	foot poundals per day
25,030	foot poundals per hour
417.16680	foot poundals per minute
6.95278	foot poundals per second
18,671.34359	foot pounds per day
777.97265	foot pounds per hour
12.96621	foot pounds per minute
0.21610	foot pounds per second
6.04748	kilogram calories per day
0.25198	kilogram calories per hour
0.0041996	kilogram calories per minute
0.000069994	kilogram calories per second
213.31968	ounce calories per day
8.88832	ounce calories per hour
0.14814	ounce calories per minute
0.0024690	ounce calories per second
2,581.4592	kilogram meters per day
107.56	kilogram meters per hour
1.79268	kilogram meters per minute
0.029878	kilogram meters per second
249.67008	liter-atmospheres per day
10.40277	liter-atmospheres per hour
0.173382	liter-atmospheres per minute
0.0028897	liter-atmospheres per second
8.822304	cubic foot-atmospheres per day
0.3676637	cubic foot-atmospheres per hour
0.00612660	cubic foot-atmospheres per minute
0.00010211	cubic foot-atmospheres per second

BTU PER HOUR: (cont'd)

0.00038754	Cheval-vapeurs
0.00039292	horsepower
0.00029283	kilowatts
0.29283	watts
25,300.5120	joules per day
1,054.198	joules per hour
17.569967	joules per minute
0.292833	joules per second
0.00164402	pounds of carbon oxidized with perfect efficiency per day
0.0000685	pounds of carbon oxidized with perfect efficiency per hour
0.0000011417	pounds of carbon oxidized with perfect efficiency per minute
0.000000019028	pounds of carbon oxidized with perfect efficiency per second
0.02472	pounds of water evaporated from and at 212°F. per day
0.001030	pounds of water evaporated from and at 212°F. per hour
0.000017167	pounds of water evaporated from and at 212°F. per minute
0.00000028611	pounds of water evaporated from and at 212°F. per second
24	BTU per day
1.0	BTU per hour
0.016667	BTU per minute
0.00027778	BTU per second

BTU PER MINUTE: =

36,043,200	foot poundals per day
1,501,800	foot poundals per hour
25,030	foot poundals per minute
417.16667	foot poundals per second
1,120,281	foot pounds per day
46,678.35899	foot pounds per hour
777.97265	foot pounds per minute
12.96621	foot pounds per second
362.854	kilogram calories per day
15.11892	kilogram calories per hour
0.25198	kilogram calories per minute
0.0041997	kilogram calories per second
12,799.180803	ounce calories per day
533.299200	ounce calories per hour
8.88832	ounce calories per minute
0.148139	ounce calories per second
154,886.688000	kilogram meters per day
6,453.6120	kilogram meters per hour
107.56	kilogram meters per minute
1.79267	kilogram meters per second

BTU PER MINUTE: (cont'd)

14,980.0320	liter-atmospheres per day
624.168	liter-atmospheres per hour
10.40277	liter-atmospheres per minute
0.17338	liter-atmospheres per second
529.435728	cubic foot-atmospheres per day
22.059822	cubic foot-atmospheres per hour
0.3676637	cubic foot-atmospheres per minute
0.00612773	cubic foot-atmospheres per second
0.023252	Cheval-vapeur hours
0.023575	horsepower
0.01757	kilowatts
17.57	watts
1,518,048	joules per day
63,252	joules per hour
1,054.198	joules per minute
17.569967	joules per second
0.098643	pounds of carbon oxidized with perfect efficiency per day
0.0041101	pounds of carbon oxidized with perfect efficiency per hour
0.0000685	pounds of carbon oxidized with perfect efficiency per minute
0.0000011417	pounds of carbon oxidized with perfect efficiency per second
1.48320	pounds of water evaporated from and at 212°F. per day
0.06180	pounds of water evaporated from and at 212°F. per hour
0.001030	pounds of water evaporated from and at 212°F. per minute
0.000017167	pounds of water evaporated from and at 212°F. per second
1,440	BTU per day
60	BTU per hour
1.0	BTU per minute
0.016667	BTU per second

BTU PER SECOND: =

2,162,592,000	foot poundals per day
90,108,000	foot poundals per hour
1,501,800	foot poundals per minute
25,030	foot poundals per second
67,216,836.960	foot pounds per day
2,800,701.540	foot pounds per hour
46,678.3590	foot pounds per minute
777.97265	foot pounds per second
21,771.0720	kilogram calories per day
907.1280	kilogram calories per hour
15.1188	kilogram calories per minute
0.25198	kilogram calories per second

BTU PER SECOND: (cont'd)

767,950.8480	ounce calories per day
31,997.9520	ounce calories per hour
533.29920	ounce calories per minute
8.88832	ounce calories per second
9,293,184.0	kilogram meters per day
387,216.0	kilogram meters per hour
6,453.60	kilogram meters per minute
107.56	kilogram meters per second
898,799.3280	liter-atmospheres per day
37,449.9720	liter-atmospheres per hour
624.16620	liter-atmospheres per minute
10.40277	liter-atmospheres per second
31,766.143680	cubic foot-atmospheres per day
1,323.589320	cubic foot-atmospheres per hour
22.0598220	cubic foot-atmospheres per minute
0.3676637	cubic foot-atmospheres per second
1.39519	Cheval-vapeurs
1.41454	horsepower
1.055	kilowatts
1,055	watts
91,082,702.2	joules per day
3,795,112.8	joules per hour
63,251.880	joules per minute
1,054.198	joules per second
5.9184	pounds of carbon oxidized with perfect efficiency per day
0.2466	pounds of carbon oxidized with perfect efficiency per hour
0.00411	pounds of carbon oxidized with perfect efficiency per minute
0.0000685	pounds of carbon oxidized with perfect efficiency per second
88.992000	pounds of water evaporated from and at 212°F. per day
3.70800	pounds of water evaporated from and at 212°F. per hour
0.061800	pounds of water evaporated from and at 212°F. per minute
0.001030	pounds of water evaporated from and at 212°F. per second
86,400	BTU per day
3,600	BTU per hour
60	BTU per minute
1	BTU per second

BTU PER SQUARE FOOT PER DAY: =

23,505.12	kilowatts per square hektometer
235.0512	kilowatts per square dekameter
2.350512	kilowatts per square meter
25.30051	kilowatts per square foot

BTU PER SQUARE FOOT PER DAY: (cont'd)

0.17569 . . . . kilowatts per square inch
0.023505 . . . . kilowatts per square decimeter
0.00023505 . . . . kilowatts per square centimeter
0.0000023505 . . . . kilowatts per square millimeter
23,505,120 . . . . watts per square hektometer
235,051.2 . . . . watts per square dekameter
2,350.512 . . . . watts per square meter
25,300.512 . . . . watts per square foot
175.6944 . . . . watts per square inch
23.50512 . . . . watts per square decimeter
0.23505 . . . . watts per square centimeter
0.0023502 . . . . watts per square millimeter
31,537.728 . . . . horsepower per square hektometer
315.37728 . . . . horsepower per square dekameter
3.15377 . . . . horsepower per square meter
33.94829 . . . . horsepower per square foot
0.23576 . . . . horsepower per square inch
0.031538 . . . . horsepower per square decimeter
0.00031538 . . . . horsepower per square centimeter
0.0000031538 . . . . horsepower per square millimeter

BTU PER SQUARE FOOT PER HOUR: =

979.380 . . . . kilowatts per square hektometer
9.7938 . . . . kilowatts per square dekameter
0.097938 . . . . kilowatts per square meter
1.05419 . . . . kilowatts per square foot
0.0073206 . . . . kilowatts per square inch
0.000979380 . . . . kilowatts per square decimeter
0.00000979380 . . . . kilowatts per square centimeter
0.0000000979380 . . . . kilowatts per square millimeter
97,938 . . . . watts per square hektometer
9,793.800 . . . . watts per square dekameter
97.93800 . . . . watts per square meter
1,054.188 . . . . watts per square foot
7.3206 . . . . watts per square inch
0.979380 . . . . watts per square decimeter
0.0097938 . . . . watts per square centimeter
0.000097938 . . . . watts per square millimeter
1,314.0720 . . . . horsepower per square hektometer
13.14072 . . . . horsepower per square dekameter
0.13141 . . . . horsepower per square meter
1.41451 . . . . horsepower per square foot
0.00982332 . . . . horsepower per square inch
0.00131407 . . . . horsepower per square decimeter
0.000013141 . . . . horsepower per square centimeter
0.00000013141 . . . . horsepower per square millimeter

BTU PER SQUARE FOOT PER MINUTE: =

16.323 . . . . kilowatts per square hektometer
0.16323 . . . . kilowatts per square dekameter

BTU PER SQUARE FOOT PER MINUTE: (cont'd)

0.0016323	kilowatts per square meter
0.01757	kilowatts per square foot
0.00012201	kilowatts per square inch
0.000016323	kilowatts per square decimeter
0.00000016323	kilowatts per square centimeter
0.0000000016323	kilowatts per square millimeter
16,323	watts per square hektometer
163.23	watts per square dekameter
1.6323	watts per square meter
17.57	watts per square foot
0.12201	watts per square inch
0.016323	watts per square decimeter
0.00016323	watts per square centimeter
0.0000016323	watts per square millimeter
21.90118	horsepower per square hektometer
0.21901	horsepower per square dekameter
0.0021901	horsepower per square meter
0.023575	horsepower per square foot
0.00016372	horsepower per square inch
0.000021901	horsepower per square decimeter
0.00000021901	horsepower per square centimeter
0.0000000021901	horsepower per square millimeter

BTU PER SQUARE FOOT PER SECOND: =

0.27205	kilowatts per square hektometer
0.0027205	kilowatts per square dekameter
0.000027205	kilowatts per square meter
0.00029283	kilowatts per square foot
0.0000020335	kilowatts per square inch
0.00000027205	kilowatts per square decimeter
0.0000000027205	kilowatts per square centimeter
0.000000000027205	kilowatts per square millimeter
272.05	watts per square hektometer
2.7205	watts per square dekameter
0.027205	watts per square meter
0.29283	watts per square foot
0.0020335	watts per square inch
0.00027205	watts per square decimeter
0.0000027205	watts per square centimeter
0.000000027205	watts per square millimeter
0.36502	horsepower per square hektometer
0.0036502	horsepower per square dekameter
0.000036502	horsepower per square meter
0.00039292	horsepower per square foot
0.0000027287	horsepower per square inch
0.00000036502	horsepower per square decimeter
0.0000000036502	horsepower per square centimeter
0.000000000036502	horsepower per square millimeter

BUSHEL—U.S. (DRY): =

0.035238	kiloliters
0.035238	cubic meters
0.04609	cubic yards
0.304785	barrels—U.S.
0.35238	hectoliters
0.96945	bushels—Imp. (dry)
1.24446	cubic feet
3.5238	dekaliters
4	pecks
9.3088	gallons —U.S. (liquid)
8	gallons—U.S. (dry)
7.81457	gallons—Imp.
37.2353	quarts (liquid)
32	quarts (dry)
35.238	liters
35.238	cubic decimeters
64	pints (dry)
74.8706	pints (liquid)
299.4824	gills (liquid)
352.38	deciliters
2,150.42	cubic inches
3,523.8	centiliters
35,238	millimeters
35,238	cubic centimeters
35,238,000	cubic millimeters
0.053335	tons (short)
0.047621	tons (long)
0.048385	tons (metric)
48.38492	kilograms
106.67048	pounds
483.84924	hektograms
4,838.4924	dekagrams
7,741.58787	ounces
48,384.924	grams
483,849.24	decigrams
4,838,492.4	centigrams
48,384,924	milligrams

BUSHEL—IMPERIAL: =

0.036348	kiloliters
0.036348	cubic meters
0.047542	cubic yards
0.31439	barrels
0.36348	hectoliters
1.03151	bushels—U.S.
1.2843	cubic feet
3.63484	dekaliters
4.12604	pecks

BUSHEL—IMPERIAL: (cont'd)

9.60212	gallons—U.S. (liquid)
8.25208	gallons—U.S. (dry)
8	gallons—Imp.
38.40858	quarts (liquid)
33.00832	quarts (dry)
36.34835	liters
36.34835	cubic decimeters
66.01664	pints (dry)
76.81716	pints (liquid)
307.26856	gills (liquid)
363.4835	deciliters
2,219.3	cubic inches
3,634.835	centiliters
36,348.35	milliliters
36,348.35	cubic centimeters
36,348,350	cubic millimeters
0.055016	tons (short)
0.049122	tons (long)
0.049910	tons (metric)
49.90953	kilograms
110.031667	pounds
499.095330	hektograms
4,990.95330	dekagrams
7,985.52530	ounces
49,909.53296	grams
499,095.32955	decigrams
4,990,953.29552	centigrams
49,909,532.95524	milligrams

CENTARE (CENTIARE): =

0.0000003831	square miles or sections
0.0000001111	square kilometers
0.000024710	square furlongs
0.00024710	acres
0.0001111	square hektometers
0.00247104	square chains
0.01111	square dekameters
0.039537	square rods
1	square meters
1.19598	square yards
1.39498	square varas (Texas)
10.7639	square feet
19.13580	square spans
27.7104	square links
96.8750	square hands
100	square decimeters
10,000	square centimeters
1,550	square inches
1,000,000	square millimeters
0.0001	hectares
0.01	ares

CENTIGRAM: =

0.000000011231	tons (short)
0.00000000984206	tons (long)
0.00000001	tons (metric)
0.00001	kilograms
0.0000267923	pounds (Troy)
0.000022406	pounds (Avoir)
0.0001	hektograms
0.001	dekagrams
0.000321507	ounces (Troy)
0.000352739	ounces (Avoir)
0.01	grams
0.1	decigrams
10	milligrams
0.1543236	grains
0.00257206	drachmas (fluid)
0.00257206	drams (Troy)
0.0056438	drams (Avoir)
0.006430149	pennyweight
0.00771618	scruples
0.05	carats (metric)

CENTILITERS: =

0.000001	kiloliters
0.00001	cubic meters
0.00001308	cubic yards
0.000062897	barrels
0.0001	hectoliters
0.00028377	bushels—U.S. (dry)
0.00027510	bushels—Imp. (dry)
0.00035314	cubic feet
0.001	dekaliters
0.0026417	gallons—U.S. (liquid)
0.0022707	gallons—U.S. (dry)
0.0021997	gallons—Imp.
0.010567	quarts (liquid)
0.0090828	quarts (dry)
0.01	liters
0.01	cubic decimeters
0.018161	pints
0.072663	gills
0.1	deciliters
0.61025	cubic inches
10	milliliters
10	cubic centimeters
10,000	cubic millimeters
0.33815	ounces (fluid)
2.70518	drams (fluid)
0.0000062137	miles

CENTILITERS: (cont'd)

0.00001	kilometers
0.000049709	furlongs
0.0001	hektometers
0.00049709	chains
0.001	dekameters
0.0019884	rods
0.01	meters
0.010936	yards
0.011811	varas (Texas)
0.032808	feet
0.043744	spans
0.049709	links
0.098424	hands
0.1	decimeters
0.3937	inches
1.00	centimeters
10	millimeters
393.70	mils
10,000	microns
10,000,000	milli-microns
10,000,000	micro-millimeters
100,000,000	Angstrom units
15,531.6	wave lengths of red line of cadmium

CENTIMETERS PER DAY: =

0.000006214	miles per day
0.00000025892	miles per hour
0.0000000043153	miles per minute
0.000000000071921	miles per second
0.00001	kilometers per day
0.00000041667	kilometers per hour
0.0000000069444	kilometers per minute
0.00000000011574	kilometers per second
0.000049709	furlongs per day
0.0000020712	furlongs per hour
0.000000034520	furlongs per minute
0.00000000057534	furlongs per second
0.0001	hektometers per day
0.0000041667	hektometers per hour
0.000000069444	hektometers per minute
0.0000000011574	hektometers per second
0.00049709	chains per day
0.000020712	chains per hour
0.00000034520	chains per minute
0.0000000057534	chains per second
0.001	dekameters per day
0.000041667	dekameters per hour
0.00000069444	dekameters per minute
0.000000011574	dekameters per second
0.0019884	rods per day

CENTIMETERS PER DAY: (cont'd)

0.000082850	rods per hour
0.0000013808	rods per minute
0.000000023014	rods per second
0.01	meters per day
0.00041667	meters per hour
0.0000069444	meters per minute
0.00000011574	meters per second
0.010936	yards per day
0.00045667	yards per hour
0.0000075944	yards per minute
0.00000012657	yards per second
0.011811	varas (Texas) per day
0.00049212	varas (Texas) per hour
0.000008202	varas (Texas) per minute
0.00000013670	varas (Texas) per second
0.032808	feet per day
0.0013670	feet per hour
0.000022783	feet per minute
0.00000037972	feet per second
0.043744	spans per day
0.0018227	spans per hour
0.000030378	spans per minute
0.00000050630	spans per second
0.049709	links per day
0.0020712	links per hour
0.000034520	links per minute
0.00000057534	links per second
0.098424	hands per day
0.0041010	hands per hour
0.000068350	hands per minute
0.0000011392	hands per second
0.1	decimeters per day
0.0041667	decimeters per hour
0.000069444	decimeters per minute
0.0000011574	decimeters per second
1	centimeters per day
0.041667	centimeters per hour
0.00069444	centimeters per minute
0.000011574	centimeters per second
0.3937	inches per day
0.016404	inches per hour
0.00027340	inches per minute
0.0000045567	inches per second
10	millimeters per day
0.41667	millimeters per hour
0.0069444	millimeters per minute
0.00011574	millimeters per second
393.70	mils per day
16.40417	mils per hour
0.27340	mils per minute
0.0045567	mils per second

CENTIMETERS PER DAY: (cont'd)

10,000	microns per day
416.66667	microns per hour
6.9444	microns per minute
0.11574	microns per second

CENTIMETERS PER HOUR: =

0.00014914	miles per day
0.000006214	miles per hour
0.00000010357	miles per minute
0.0000000017261	miles per second
0.00024000	kilometers per day
0.00001	kilometers per hour
0.00000016667	kilometers per minute
0.0000000027778	kilometers per second
0.0011930	furlongs per day
0.000049709	furlongs per hour
0.00000082848	furlongs per minute
0.000000013808	furlongs per second
0.0024000	hektometers per day
0.0001	hektometers per hour
0.0000016667	hektometers per minute
0.000000027778	hektometers per second
0.011930	chains per day
0.00049709	chains per hour
0.0000082848	chains per minute
0.00000013808	chains per second
0.024000	dekameters per day
0.001	dekameters per hour
0.000016667	dekameters per minute
0.00000027778	dekameters per second
0.047722	rods per day
0.0019884	rods per hour
0.000033140	rods per minute
0.00000055233	rods per second
0.24000	meters per day
0.01	meters per hour
0.00016667	meters per minute
0.0000027778	meters per second
0.262464	yards per day
0.010936	yards per hour
0.00018227	yards per minute
0.0000030378	yards per second
0.28346	varas (Texas) per day
0.011811	varas (Texas) per hour
0.00019685	varas (Texas) per minute
0.0000032808	varas (Texas) per second
0.78739	feet per day
0.032808	feet per hour
0.00054680	feet per minute
0.0000091133	feet per second

CENTIMETERS PER HOUR: (cont'd)

1.049856	spans per day
0.043744	spans per hour
0.00072907	spans per minute
0.000012151	spans per second
1.19302	links per day
0.049709	links per hour
0.00082848	links per minute
0.000013808	links per second
2.36218	hands per day
0.098424	hands per hour
0.0016404	hands per minute
0.000027340	hands per second
2.40000	decimeters per day
0.1	decimeters per hour
0.0016667	decimeters per minute
0.000027778	decimeters per second
24.00000	centimeters per day
1	centimeters per hour
0.016667	centimeters per minute
0.00027778	centimeters per second
9.44880	inches per day
0.3937	inches per hour
0.0065617	inches per minute
0.00010936	inches per second
240.00000	millimeters per day
10	millimeters per hour
0.16667	millimeters per minute
0.0027778	millimeters per second
9,448.8	mils per day
393.70	mils per hour
6.56167	mils per minute
0.109361	mils per second
240,000	microns per day
10,000	microns per hour
166.66667	microns per minute
2.77778	microns per second

CENTIMETER PER MINUTE: =

0.0089482	miles per day
0.00037284	miles per hour
0.000006214	miles per minute
0.00000010357	miles per second
0.014400	kilometers per day
0.00060000	kilometers per hour
0.00001	kilometers per minute
0.00000016667	kilometers per second
0.071581	furlongs per day
0.0029825	furlongs per hour
0.000049709	furlongs per minute
0.00000082848	furlongs per second

0.14400	hektometers per day
0.0060000	hektometers per hour
0.0001	hektometers per minute
0.0000016667	hektometers per second
0.71581	chains per day
0.029825	chains per hour
0.00049709	chains per minute
0.0000082848	chains per second
1.44000	dekameters per day
0.060000	dekameters per hour
0.001	dekameters per minute
0.000016667	dekameters per second
2.86330	rods per day
0.11930	rods per hour
0.0019884	rods per minute
0.000033140	rods per second
14.40000	meters per day
0.60000	meters per hour
0.01	meters per minute
0.00016667	meters per second
15.74784	yards per day
0.656160	yards per hour
0.010936	yards per minute
0.00018227	yards per second
17.00784	varas (Texas) per day
0.70866	varas (Texas) per hour
0.011811	varas (Texas) per minute
0.00019685	varas (Texas) per second
47.24352	feet per day
1.96848	feet per hour
0.032808	feet per minute
0.00054680	feet per second
62.99136	spans per day
2.62464	spans per hour
0.043744	spans per minute
0.00072907	spans per second
71.58096	links per day
2.98254	links per hour
0.049709	links per minute
0.00083848	links per second
141.73056	hands per day
5.90544	hands per hour
0.098424	hands per minute
0.0016404	hands per second
144.00	decimeters per day
6.00	decimeters per hour
0.1	decimeters per minute
0.0016667	decimeters per second
1,440.00	centimeters per day
60.00	centimeters per hour
1.0	centimeters per minute

CENTIMETER PER MINUTE: (cont'd)

0.016667	centimeters per second
566.92800	inches per day
23.62200	inches per hour
0.3937	inches per minute
0.0065617	inches per second
14,440.00	millimeters per day
600.00	millimeters per hour
10	millimeters per minute
0.16667	millimeters per second
566,928	mils per day
23,622	mils per hour
393.70	mils per minute
6.56167	mils per second
14,440,000	microns per day
600,000	microns per hour
10,000	microns per minute
166.66667	microns per second

CENTIMETER PER SECOND: =

0.53689	miles per day
0.022370	miles per hour
0.00037284	miles per minute
0.000006214	miles per second
0.8640	kilometers per day
0.0360	kilometers per hour
0.0006	kilometers per minute
0.00001	kilometers per second
4.29486	furlongs per day
0.17895	furlongs per hour
0.0029825	furlongs per minute
0.000049709	furlongs per second
8.640	hektometers per day
0.360	hektometers per hour
0.006	hektometers per minute
0.0001	hektometers per second
42.94858	chains per day
1.78952	chains per hour
0.029825	chains per minute
0.00049709	chains per second
86.400	dekameters per day
3.600	dekameters per hour
0.060	dekameters per minute
0.001	dekameters per second
171.79776	rods per day
7.15824	rods per hour
0.11930	rods per minute
0.0019884	rods per second
864.0	meters per day
36.0	meters per hour
0.60	meters per minute

CENTIMETER PER SECOND: (cont'd)

0.01	meters per second
944.8704	yards per day
39.36960	yards per hour
0.656160	yards per minute
0.010936	yards per second
1,020.4704	varas (Texas) per day
42.51960	varas (Texas) per hour
0.70866	varas (Texas) per minute
0.01181	varas (Texas) per second
2,834.6112	feet per day
118.1088	feet per hour
1.96848	feet per minute
0.032808	feet per second
3,779.4816	spans per day
157.4784	spans per hour
2.62464	spans per minute
0.043744	spans per second
4,294.8576	links per day
178.9524	links per hour
2.98254	links per minute
0.049709	links per second
8,503.8336	hands per day
354.3264	hands per hour
5.90544	hands per minute
0.098424	hands per second
8,640.0	decimeters per day
360.0	decimeters per hour
6.0	decimeters per minute
0.1	decimeters per second
86,400	centimeters per day
3,600.0	centimeters per hour
60.0	centimeters per minute
1	centimeters per second
34,015.68	inches per day
1,417.32	inches per hour
23.6220	inches per minute
0.3937	inches per second
864,000	millimeters per day
36,000	millimeters per hour
600	millimeters per minute
10	millimeters per second
34,015,680	mils per day
1,417,320	mils per hour
23,622	mils per minute
393.70	mils per second
864,000,000	microns per day
36,000,000	microns per hour
600,000	microns per minute
10,000	microns per second

CENTIMETER OF MERCURY (0°C.): =

0.0013595	hektometers of water@ 60°F.
0.013595	dekameters of water@ 60°F.
0.13595	meters of water@ 60°F.
0.44604	feet of water@ 60°F.
0.44604	ounces of water@ 60°F.
5.35248	inches of water@ 60°F.
1.35952	decimeters of water@ 60°F.
13.595299	centimeters of water@ 60°F.
135.95299	millimeters of water@ 60°F.
0.0001	hektometers of mercury@ 32°F.
0.001	dekameters of mercury@ 32°F.
0.01	meters of mercury@ 32°F.
0.032808	feet of mercury@ 32°F.
0.032808	ounces of mercury@ 32°F.
0.3937	inches of mercury@ 32°F.
0.1	decimeters of mercury@ 32°F.
1.0	centimeters of mercury@ 32°F.
10.0	millimeters of mercury@ 32°F.
1,498.62505	tons per square hektometer
14.98625	tons per square dekameter
0.14986	tons per square meter
0.013923	tons per square foot
0.000096685	tons per square inch
0.0014986	tons per square decimeter
0.000014986	tons per square centimeter
0.00000014986	tons per square millimeter
1,359,529.9	kilograms per square hektometer
13,595.299	kilograms per square dekameter
135.95299	kilograms per square meter
12.63034	kilograms per square foot
0.087711	kilograms per square inch
1.35953	kilograms per square decimeter
0.013595	kilograms per square centimeter
0.00013595	kilograms per square millimeter
2,997,249.93506	pounds per square hektometer
29,972.49935	pounds per square dekameter
299.7250	pounds per square meter
27.845	pounds per square foot
0.19337	pounds per square inch
2.99725	pounds per square decimeter
0.029973	pounds per square centimeter
0.00029973	pounds per square millimeter
13,595,299	hektograms per square hektometer
135,952.99	hektograms per square dekameter
1,359.5299	hektograms per square meter
125.3034	hektograms per square foot
0.87711	hektograms per square inch
13.5953	hektograms per square decimeter
0.13595	hektograms per square centimeter

0.0013595 . . . . hektograms per square millimeter
135,952,990 . . . . dekagrams per square hektometer
1,59,529.9 . . . . dekagrams per square dekameter
13,595.299 . . . . dekagrams per square meter
1,263.034 . . . . dekagrams per square foot
8.7711 . . . . dekagrams per square inch
135.95299 . . . . dekagrams per square decimeter
1.35953 . . . . dekagrams per square centimeter
0.013595 . . . . dekagrams per square millimeter
47,955,999 . . . . ounces per square hektometer
479,560 . . . . ounces per square dekameter
4,795.6 . . . . ounces per square meter
445.520 . . . . ounces per square foot
3.09392 . . . . ounces per square inch
47.9560 . . . . ounces per square decimeter
0.479560 . . . . ounces per square centimeter
0.00479560 . . . . ounces per square millimeter
1,359,529,900 . . . . grams per square hektometer
13,595,299 . . . . grams per square dekameter
135,952.99 . . . . grams per square meter
12,630.34 . . . . grams per square foot
87.71111 . . . . grams per square inch
1,359.5299 . . . . grams per square decimeter
13.595299 . . . . grams per square centimeter
0.13595 . . . . grams per square millimeter
13,595,299,000 . . . . decigrams per square hektometer
135,952,990 . . . . decigrams per square dekameter
1,359,529.9 . . . . decigrams per square meter
126,303.4 . . . . decigrams per square foot
877.11111 . . . . decigrams per square inch
13,595.299 . . . . decigrams per square decimeter
135.95299 . . . . decigrams per square centimeter
1.35953 . . . . decigrams per square millimeter
135,952,990,000 . . . . centigrams per square hektometer
1,359,529,900 . . . . centigrams per square dekameter
13,595,299 . . . . centigrams per square meter
1,263,034 . . . . centigrams per square foot
8,771.1111 . . . . centigrams per square inch
135,952.99 . . . . centigrams per square decimeter
1,359.5299 . . . . centigrams per square centimeter
13.595299 . . . . centigrams per square millimeter
1,359,529,900,000 . . . . milligrams per square hektometer
13,595,299,000 . . . . milligrams per square dekameter
135,952,990 . . . . milligrams per square meter
12,630,340 . . . . milligrams per square foot
87,711.1111 . . . . milligrams per square inch
1,359,529.9 . . . . milligrams per square decimeter
13,595.299 . . . . milligrams per square centimeter
135.95299 . . . . milligrams per square millimeter
0.013333 . . . . bars
0.013158 . . . . atmospheres

CENTIMETER OF MERCURY (0°C.): (cont'd)

1,322,220,000,000	dynes per square hektometer
13,332,200,000	dynes per square dekameter
133,322,000	dynes per square meter
12,385,916.01	dynes per square foot
86,013.3056	dynes per square inch
1,333,220	dynes per square decimeter
13,332.20	dynes per square centimeter
133.3220	dynes per square millimeter

CENTIMETER PER SECOND PER SECOND: =

0.000006214	miles per second per second
0.00001	kilometers per second per second
0.000049709	furlongs per second per second
0.0001	hektometers per second per second
0.00049709	chains per second per second
0.001	dekameters per second per second
0.0019884	rods per second per second
0.01	meters per second per second
0.010936	yards per second per second
0.011811	varas (Texas) per second per second
0.032808	feet per second per second
0.043744	spans per second per second
0.049709	links per second per second
0.098424	hands per second per second
0.10	decimeters per second per second
0.3937	inches per second per second
1.00	centimeters per second per second
10.00	millimeters per second per second
393.70	mils per second per second
10,000	microns per second per second

CHAIN (SURVEYOR'S OR GUNTER'S): =

0.0125	miles
0.020117	kilometers
0.1	furlongs
0.20117	hektometers
2.0117	dekameters
4	rods
20.117	meters
22	yards
23.76	varas (Texas)
66	feet
88	spans
100	links
198	hands
201.17	decimeters
2,011.7	centimeters
792	inches

CHAIN (SURVEYOR'S OR GUNTER'S): (cont'd)

20,117	millimeters
792,000	mils
21,117,000	microns
21,117,000,000	millimicrons
21,117,000,000	micromillimeters
31,244,672	wave lengths of red line of cadmium
$21,117 \times 10^9$	Angstrom Units

CIRCULAR MIL: =

0.00000000000000019564	square miles or sections
0.0000000000000050671	square kilometers
0.00000000000012521	square furlongs
0.0000000000050671	square hektometers
0.0000000000012521	square chains
0.0000000000050671	square dekameters
0.000000000020034	square rods
0.00000000050671	square meters
0.00000000060602	square yards
0.00000000070687	square varas (Texas)
0.0000000054542	square feet
0.0000000096964	square spans
0.000000012521	square links
0.000000049186	square hands
0.000000050671	square decimeters
0.0000050671	square centimeters
0.0000007854	square inches
0.00050671	square millimeters
0.000000000000050671	hectares
0.0000000000050671	ares
0.00000000050671	centares (centiares)
0.7854	square mils
0.000001	circular inches
0.001	inches
0.000645143	circular millimeters

CHEVAL-VAPEUR (METRIC HORSEPOWER): =

62,832,926.34	foot poundals
753,995,116.08	inch poundals
1,952,910	foot pounds
23,434,920	inch pounds
0.69727	ton (short) calories
632.551	kilogram calories
1,394.53604	pound calories
6,325.51	hektogram calories
63,255.1	dekagram calories
22,312.57664	ounce calories
632,551	gram calories
6,325,510	decigram calories

CHEVAL-VAPEUR (METRIC HORSEPOWER): (cont'd)

63,255,100	centigram calories
632,551,000	milligram calories
17.6448	kilowatt days
0.7352	kilowatt hours
0.012533	kilowatt minutes
0.00020422	kilowatt seconds
30.633333	watt days
735.2	watt hours
44,120	watt minutes
2,646,720	watt seconds
297.62401	ton meters
270,000	kilogram meters
595,248.02	pound meters
2,700,000	hektogram meters
27,000,000	dekagram meters
9,523,968.32	ounce meters
270,000,000	gram meters
2,700,000,000	decigram meters
27,000,000,000	centigram meters
270,000,000,000	milligram meters
2.97624	ton hektometers
2,700	kilogram hektometers
5,952.4802	pound hektometers
27,000	hektogram hektometers
270,000	dekagram hektometers
95,239.6832	ounce hektometers
2,700,000	gram hektometers
27,000,000	decigram hektometers
270,000,000	centigram hektometers
2,700,000,000	milligram hektometers
29.7624	ton dekameters
27,000	kilogram dekameters
59,524.802	pound dekameters
270,000	hektogram dekameters
2,700,000	dekagram dekameters
952,396.832	ounce dekameters
27,000,000	gram dekameters
270,000,000	decigram dekameters
2,700,000,000	centigram dekameters
27,000,000,000	milligram dekameters
9.07158	ton feet
82,296	kilogram feet
181,431.5965	pound feet
822,960	hektogram feet
8,229,600	dekagram feet
2,902,905.54394	ounce feet
82,296,000	gram feet
822,960,000	decigram feet
8,229,600,000	centigram feet
82,296,000,000	milligram feet
0.75596	ton inches

CHEVAL-VAPEUR (METRIC HORSEPOWER): (cont'd)

```
6,858 ......................................... kilogram inches
15,119.29971 ..................................... pound inches
68,580 ....................................... hektogram inches
685,800 ....................................... dekagram inches
241,908.79536 .................................... ounce inches
6,858,000 ......................................... gram inches
68,580,000 ................................... decigram inches
685,800,000 ................................. centigram inches
6,858,000,000 ............................... milligram inches
2,976.24 ...................................... ton decimeters
2,700,000 .............................. kilogram decimeters
5,952,480.2 .............................. pound decimeters
27,000,000 ............................. hektogram decimeters
270,000,000 ............................ dekagram decimeters
95,239,683.2 .............................. ounce decimeters
2,700,000,000 ............................... gram decimeters
27,000,000,000 ........................... decigram decimeters
270,000,000,000 ......................... centigram decimeters
2,700,000,000,000 ........................ milligram decimeters
29,762.4 ...................................... ton centimeters
27,000,000 .............................. kilogram centimeters
59,524,802 ............................... pound centimeters
270,000,000 ............................ hektogram centimeters
2,700,000,000 ........................... dekagram centimeters
952,396,832 .............................. ounce centimeters
27,000,000,000 .............................. gram centimeters
270,000,000,000 .......................... decigram centimeters
2,700,000,000,000 ......................... milligram centimeters
297,624 ........................................ ton millimeters
270,000,000 .............................. kilogram millimeters
595,248,020 ................................ pound millimeters
2,700,000,000 ........................... hektogram millimeters
27,000,000,000 ........................... dekagram millimeters
9,523,968,320 .............................. ounce millimeters
270,000,000,000 .............................. gram millimeters
2,700,000,000,000 ......................... decigram millimeters
27,000,000,000,000 ......................... milligram millimeters
26.1298 .................................. kiloliter atmospheres
261.298 ................................. hektoliter atmospheres
2,612.98 ................................ dekaliter atmospheres
26,129.8 .................................... liter-atmospheres
261,298 ................................. deciliter-atmospheres
2,612,980 ............................... centiliter-atmospheres
26,129,800 .............................. milliliter-atmospheres
0.00000002613 ......................... cubic kilometer-atmospheres
0.00002613 .......................... cubic hektometer-atmospheres
0.02613 ............................. cubic dekameter-atmospheres
26.1298 .............................. cubic meter-atmospheres
922.74776 ............................. cubic feet-atmospheres
26,129.8 ............................. cubic decimeters-atmospheres
26,129,800 .......................... cubic centimeter-atmospheres
```

CHEVAL-VAPEUR (METRIC HORSEPOWER): (cont'd)

26,129,800,000 . . . . cubic millimeter-atmospheres
2,648,700 . . . . joules
0.0410967 . . . . horsepower days
0.98632 . . . . horsepower hours
59.17920 . . . . horsepower minutes
3,550.752 . . . . horsepower seconds
0.171945 . . . . pounds of carbon oxidized with
100% efficiency
2.5877 . . . . pounds of water evaporated from
and at 212°F.
2,510.152 . . . . B.T.U.

CUBIC CENTIMETERS: =

0.00000000000000062137 . . . . cubic miles
0.000000000000001 . . . . cubic kilometers
0.0000000000000122833 . . . . cubic furlongs
0.000000000001 . . . . cubic hektometers
0.000000000122833 . . . . cubic chains
0.000000001 . . . . cubic dekameters
0.0000000078613 . . . . cubic rods
0.000001 . . . . cubic meters
0.000001 . . . . kiloliters
0.0000013079 . . . . cubic yards
0.0000016479 . . . . cubic varas (Texas)
0.0000062897 . . . . barrels
0.00001 . . . . hectoliters
0.000028378 . . . . bushels—U.S. (dry)
0.000027496 . . . . bushels—Imperial (dry)
0.000035314 . . . . cubic feet
0.000083707 . . . . cubic spans
0.0001 . . . . dekaliter
0.000113512 . . . . pecks
0.000122833 . . . . cubic links
0.00026417 . . . . gallons—U.S. (liquid)
0.00022705 . . . . gallons—U.S. (dry)
0.00021997 . . . . gallons—Imperial
0.00095635 . . . . cubic hands
0.0010567 . . . . quarts (liquid)
0.00090808 . . . . quarts (dry)
0.001 . . . . liters
0.001 . . . . cubic decimeters
0.0021134 . . . . pints (liquid)
0.0018162 . . . . pints (dry)
0.0084536 . . . . gills (liquid)
0.01 . . . . deciliters
0.061023 . . . . cubic inches
0.1 . . . . centiliters
1 . . . . milliliters
1 . . . . cubic centimeters
1000 . . . . cubic millimeters

CUBIC CENTIMETERS: (cont'd)

0.033814	ounces (fluid)
0.27051	drams (fluid)

CUBIC FOOT: =

0.000000000017596	cubic miles
0.000000000028317	cubic kilometers
0.0000000034783	cubic furlongs
0.000000028317	cubic hektometers
0.0000034783	cubic chains
0.000028317	cubic dekameters
0.00022261	cubic rods
0.028317	cubic meters
0.028317	kiloliters
0.037036	cubic yards
0.046656	cubic varas (Texas)
0.17811	barrels
0.28317	hectoliters
0.80358	bushels—U.S. (dry)
0.77860	bushels—Imperial (dry)
2.37033	cubic spans
2.8317	dekaliters
3.2143	pecks
3.48327	cubic links
7.48050	gallons—U.S. (liquid)
6.42937	gallons—U.S. (dry)
6.22889	gallons—Imperial
27.08096	cubic hands
29.92257	quarts (liquid)
25.71410	quarts (dry)
28.317	liters
28.317	cubic decimeters
59.84515	pints (liquid)
51.4934	pints (dry)
239.38060	gills (liquid)
283.17	deciliters
1,727.98829	cubic inches
2,831.7	centiliters
28,317	milliliters
28,317	cubic centimeters
28,317,000	cubic millimeters
957.51104	ounces (fluid)
7,660.03167	drams (fluid)
0.9091	sacks of cement (set)
62.35	pounds of water @ 60°F.
64.3	pounds of salt water
72.0	pounds of salt water @ 60°F. at 1.155 specific gravity
489.542	pounds of steel of 7.851 specific gravity

CUBIC FOOT PER DAY: =

0.17811	barrels per day
0.0074214	barrels per hour
0.00012369	barrels per minute
0.0000020615	barrels per second
0.028317	kiloliters per day
0.0011799	kiloliters per hour
0.000019664	kiloliters per minute
0.00000032774	kiloliters per second
0.028317	cubic meters per day
0.0011799	cubic meters per hour
0.000019664	cubic meters per minute
0.00000032774	cubic meters per second
0.037036	cubic yards per day
0.0015432	cubic yards per hour
0.00002572	cubic yards per minute
0.00000042866	cubic yards per second
0.28317	hektoliters per day
0.011799	hektoliters per hour
0.00019664	hektoliters per minute
0.0000032774	hektoliters per second
1	cubic feet per day
0.041666	cubic feet per hour
0.00069444	cubic feet per minute
0.000011574	cubic feet per second
2.8317	dekaliters per day
0.11799	dekaliters per hour
0.0019664	dekaliters per minute
0.000032774	dekaliters per second
7.48050	gallons per day
0.31169	gallons per hour
0.0051948	gallons per minute
0.000086580	gallons per second
6.22889	gallons (Imperial) per day
0.25954	gallons (Imperial) per hour
0.0043256	gallons (Imperial) per minute
0.000072094	gallons (Imperial) per second
29.92257	quarts per day
1.24679	quarts per hour
0.020780	quarts per minute
0.00034633	quarts per second
28.317	liters per day
1.17988	liters per hour
0.019664	liters per minute
0.00032774	liters per second
28.317	cubic decimeters per day
1.17988	cubic decimeters per hour
0.019664	cubic decimeters per minute
0.00032774	cubic decimeters per second
59.84515	pints per day

CUBIC FOOT PER DAY: (cont'd)

2.49354	pints per hour
0.041559	pints per minute
0.00069265	pints per second
239.38060	gills per day
9.97416	gills per hour
0.166236	gills per minute
0.0027706	gills per second
283.17	deciliters per day
11.79875	deciliters per hour
0.19664	deciliters per minute
0.0032774	deciliters per second
1,727.98829	cubic inches per day
72.0	cubic inches per hour
1.20	cubic inches per minute
0.020	cubic inches per second
2,831.7	centiliters per day
117.98750	centiliters per hour
1.9664	centiliters per minute
0.032774	centiliters per second
28,317	milliliters per day
1,179.8750	milliliters per hour
19.664	milliliters per minute
0.32774	milliliters per second
28,317	cubic centimeters per day
1,179.8750	cubic centimeters per hour
19.664	cubic centimeters per minute
0.32774	cubic centimeters per second
28,317,000	cubic millimeters per day
1,179,875	cubic millimeters per hour
19,664	cubic millimeters per minute
327.74	cubic millimeters per second

CUBIC FOOT PER HOUR: =

4.27464	barrels per day
0.17811	barrels per hour
0.0029685	barrels per minute
0.000049475	barrels per second
0.67961	kiloliters per day
0.028317	kiloliters per hour
0.00047195	kiloliters per minute
0.0000078658	kiloliters per second
0.67961	cubic meters per day
0.028317	cubic meters per hour
0.00047195	cubic meters per minute
0.0000078658	cubic meters per second
0.88888	cubic yards per day
0.037036	cubic yards per hour
0.00061728	cubic yards per minute
0.000010288	cubic yards per second
6.79605	hectoliters per day

CUBIC FOOT PER HOUR: (cont'd)

0.28317	hectoliters per hour
0.0047195	hectoliters per minute
0.000078658	hectoliters per second
24	cubic feet per day
1	cubic feet per hour
0.016667	cubic feet per minute
0.00027778	cubic feet per second
67.96051	dekaliters per day
2.8317	dekaliters per hour
0.047195	dekaliters per minute
0.00078658	dekaliters per second
179.53056	gallons per day
7.48050	gallons per hour
0.12467	gallons per minute
0.0020779	gallons per second
149.48928	gallons (Imperial) per day
6.22889	gallons (Imperial) per hour
0.10381	gallons (Imperial) per minute
0.0017302	gallons (Imperial) per second
718.13952	quarts per day
29.92257	quarts per hour
0.49871	quarts per minute
0.0083118	quarts per second
679.60512	liters per day
28.317	liters per hour
0.47195	liters per minute
0.0078658	liters per second
679.60512	cubic decimeters per day
28.317	cubic decimeters per hour
0.47195	cubic decimeters per minute
0.0078658	cubic decimeters per second
1,436.31360	pints per day
59.84515	pints per hour
0.99744	pints per minute
0.016624	pints per second
5,745.25440	gills per day
239.38060	gills per hour
3.98976	gills per minute
0.066496	gills per second
6,796.05120	deciliters per day
283.17	deciliters per hour
4.71948	deciliters per minute
0.078658	deciliters per second
41,472.0	cubic inches per day
1,727.98829	cubic inches per hour
28.80	cubic inches per minute
0.480	cubic inches per second
67,960.512	centiliters per day
2,831.7	centiliters per hour
47.19480	centiliters per minute
0.78658	centiliters per second

CUBIC FOOT PER HOUR: (cont'd)

679,605.120	milliliters per day
28,317	milliliters per hour
471.94980	milliliters per minute
7.86583	milliliters per second
679,605.120	cubic centimeters per day
28,317	cubic centimeters per hour
471.94980	cubic centimeters per minute
7.86583	cubic centimeters per second
679,605,120	cubic millimeters per day
28,317,000	cubic millimeters per hour
471,949.80	cubic millimeters per minute
7,865.83	cubic millimeters per second

CUBIC FOOT PER MINUTE: =

256.47840	barrels per day
10.68660	barrels per hour
0.17811	barrels per minute
0.0029685	barrels per second
40.77648	kiloliters per day
1.69902	kiloliters per hour
0.028317	kiloliters per minute
0.00047195	kiloliters per second
40.77648	cubic meters per day
1.69902	cubic meters per hour
0.028317	cubic meters per minute
0.00047195	cubic meters per second
53.33213	cubic yards per day
2.22217	cubic yards per hour
0.037036	cubic yards per minute
0.00061727	cubic yards per second
407.76480	hectoliters per day
16.99020	hectoliters per hour
0.28317	hectoliters per minute
0.0047195	hectoliters per second
1,440	cubic feet per day
60	cubic feet per hour
1	cubic feet per minute
0.016667	cubic feet per second
4,077.6480	dekaliters per day
169.9020	dekaliters per hour
2.8317	dekaliters per minute
0.047195	dekaliters per second
10,771.920	gallons per day
448.830	gallons per hour
7.48050	gallons per minute
0.12468	gallons per second
8,969.184	gallons (Imperial) per day
373.7160	gallons (Imperial) per hour
6.22889	gallons (Imperial) per minute
0.10381	gallons (Imperial) per second

CUBIC FOOT PER MINUTE: (cont'd)

43,088.5440	quarts per day
1,795.3560	quarts per hour
29.92257	quarts per minute
0.49871	quarts per second
40,776.480	liters per day
1,699.020	liters per hour
28.3170	liters per minute
0.47195	liters per second
40,776.480	cubic decimeters per day
1,699.020	cubic decimeters per hour
28.3170	cubic decimeters per minute
0.47195	cubic decimeters per second
86,177.0880	pints per day
3,590.712	pints per hour
59.84515	pints per minute
0.99742	pints per second
344,708.3520	gills per day
14,362.848	gills per hour
239.38060	gills per minute
3.98968	gills per second
407,764.80	deciliters per day
16,990.20	deciliters per hour
283.17	deciliters per minute
4.7195	deciliters per second
2,488,303.584	cubic inches per day
103,769.3160	cubic inches per hour
1,727.98829	cubic inches per minute
28.79981	cubic inches per second
4,077,648.0	centiliters per day
169,902.0	centiliters per hour
2,831.7	centiliters per minute
47.195	centiliters per second
40,776,480.0	milliliters per day
1,699,020.0	milliliters per hour
28,317	milliliters per minute
471.95	milliliters per second
40,776,480.0	cubic centimeters per day
1,699,020.0	cubic centimeters per hour
28,317	cubic centimeters per minute
471.95	cubic centimeters per second
40,776,480,000	cubic millimeters per day
1,699,020,000	cubic millimeters per hour
28,317,000	cubic millimeters per minute
471,950	cubic millimeters per second

CUBIC FOOT PER SECOND: =

15,388.70400	barrels per day
641.19600	barrels per hour
10.68660	barrels per minute
0.17811	barrels per second

2,446.58880 . . . . kiloliters per day
101.94120 . . . . kiloliters per hour
1.69902 . . . . kiloliters per minute
0.028317 . . . . kiloliters per second
2,446.5880 . . . . cubic meters per day
101.94120 . . . . cubic meters per hour
1.69902 . . . . cubic meters per minute
0.028317 . . . . cubic meters per second
3,199.91040 . . . . cubic yards per day
133.32960 . . . . cubic yards per hour
2.22216 . . . . cubic yards per minute
0.037036 . . . . cubic yards per second
24,465.8880 . . . . hectoliters per day
1,019.41200 . . . . hectoliters per hour
16.99020 . . . . hectoliters per minute
0.28317 . . . . hectoliters per second
86,400 . . . . cubic feet per day
3,600 . . . . cubic feet per hour
60 . . . . cubic feet per minute
1 . . . . cubic feet per second
244,658.880 . . . . dekaliters per day
10,194.120 . . . . dekaliters per hour
169.9020 . . . . dekaliters per minute
2.8317 . . . . dekaliters per second
646,315.20 . . . . gallons per day
26,929.80 . . . . gallons per hour
448.830 . . . . gallons per minute
7.48050 . . . . gallons per second
538,176.096 . . . . gallons (Imperial) per day
22,424.004 . . . . gallons (Imperial) per hour
373.73340 . . . . gallons (Imperial) per minute
6.22889 . . . . gallons (Imperial) per second
2,585,310.0480 . . . . quarts per day
107,721.2520 . . . . quarts per hour
1,795.35420 . . . . quarts per minute
29.92257 . . . . quarts per second
2,446,588.80 . . . . liters per day
101,941.20 . . . . liters per hour
1,699.020 . . . . liters per minute
28.317 . . . . liters per second
2,446,588.80 . . . . cubic decimeters per day
101,941.20 . . . . cubic decimeters per hour
1,699.020 . . . . cubic decimeters per minute
28.317 . . . . cubic decimeters per second
5,170,620.960 . . . . pints per day
215,442.540 . . . . pints per hour
3,590.7090 . . . . pints per minute
59.84515 . . . . pints per second
20,682,483.84 . . . . gills per day
861,770.16 . . . . gills per hour
14,362.8360 . . . . gills per minute

CUBIC FOOT PER SECOND: (cont'd)

239.38060 . . . . gills per second
24,465,880 . . . . deciliters per day
1,019,412.0 . . . . deciliters per hour
16,990.20 . . . . deciliters per minute
283.17 . . . . deciliters per second
149,298,188.2560 . . . . cubic inches per day
6,220,757.8440 . . . . cubic inches per hour
103,679.29740 . . . . cubic inches per minute
1,727.98829 . . . . cubic inches per second
244,658,880.0 . . . . centiliters per day
10,194,120.0 . . . . centiliters per hour
169,902.0 . . . . centiliters per minute
2,831.7 . . . . centiliters per second
2,446,588,800.0 . . . . milliliters per day
101,941,200.0 . . . . milliliters per hour
1,699,020.0 . . . . milliliters per minute
28,317 . . . . milliliters per second
2,446,588,800.0 . . . . cubic centimeters per day
101,941,200.0 . . . . cubic centimeters per hour
1,699,020.0 . . . . cubic centimeters per minute
28,317 . . . . cubic centimeters per second
2.446,588,800,000 . . . . cubic millimeters per day
101,941,200,000 . . . . cubic millimeters per hour
1,699,020,000 . . . . cubic millimeters per minute
28,317,000 . . . . cubic millimeters per second

CUBIC INCH: =

0.000000000000010183 . . . . cubic miles
0.000000000000016387 . . . . cubic kilometers
0.0000000000020129 . . . . cubic furlongs
0.000000000016387 . . . . cubic hektometers
0.0000000020129 . . . . cubic chains
0.000000016387 . . . . cubic dekameters
0.00000012883 . . . . cubic rods
0.000016387 . . . . cubic meters
0.000016387 . . . . kiloliters
0.000021434 . . . . cubic yards
0.000027000 . . . . cubic varas (Texas)
0.00010307 . . . . barrels
0.00016387 . . . . hectoliters
0.00046503 . . . . bushels—U.S. (dry)
0.00045058 . . . . bushels—Imperial (dry)
0.0005787 . . . . cubic feet
0.0013717 . . . . cubic spans
0.0016387 . . . . dekaliters
0.0018601 . . . . pecks
0.0020129 . . . . cubic links
0.0043290 . . . . gallons—U.S. (liquid)
0.003721 . . . . gallons —U.S. (dry)
0.003607 . . . . gallons (Imperial)

CUBIC INCH: (cont'd)

0.015672	cubic hands
0.017316	quarts (liquid)
0.014881	quarts (dry)
0.016387	liters
0.016387	cubic decimeters
0.034632	pints (liquid)
0.029762	pints (dry)
0.13853	gills (liquid)
0.16387	deciliters
1.63871	centiliters
16.38716	milliliters
16.38716	cubic centimeters
16,387.16	cubic millimeters
4.4329	drams (fluid)
0.2833	pounds of steel (specific gravity—7.851)
0.03607	pounds of water @ 60°F.
0.5541	ounces of fluid
0.041667	pounds of salt water @ 60°F. and 1.155 specific gravity
0.0005261	sacks of cement (set)
0.000016387	kiloliters per day
0.00000068278	kiloliters per hour
0.000000011380	kiloliters per minute
0.00000000018966	kiloliters per second
0.000016387	cubic meters per day
0.00000068278	cubic meters per hour
0.000000011380	cubic meters per minute
0.00000000018966	cubic meters per second
0.000021434	cubic yards per day
0.00000089309	cubic yards per hour
0.000000014885	cubic yards per minute
0.00000000024808	cubic yards per second
0.00010307	barrels per day
0.0000042944	barrels per hour
0.000000071574	barrels per minute
0.0000000011929	barrels per second
0.00016387	hectoliters per day
0.0000068278	hectoliters per hour
0.00000011380	hectoliters per minute
0.0000000018966	hectoliters per second
0.0005787	cubic feet per day
0.000024112	cubic feet per hour
0.00000040187	cubic feet per minute
0.0000000066979	cubic feet per second
0.0016387	dekaliters per day
0.000068278	dekaliters per hour
0.0000011380	dekaliters per minute
0.000000018966	dekaliters per second
0.0043290	gallons per day
0.00018037	gallons per hour

CUBIC INCH: (cont'd)

0.0000030062	gallons per minute
0.000000050104	gallons per second
0.003607	gallons (Imperial) per day
0.00015029	gallons (Imperial) per hour
0.0000025049	gallons (Imperial) per minute
0.000000041748	gallons (Imperial) per second
0.016387	liters per day
0.00068278	liters per hour
0.000011380	liters per minute
0.00000018966	liters per second
0.016387	cubic decimeters per day
0.00068278	cubic decimeters per hour
0.000011380	cubic decimeters per minute
0.00000018966	cubic decimeters per second
0.017316	quarts per day
0.00072151	quarts per hour
0.000012025	quarts per minute
0.00000020042	quarts per second
0.034632	pints per day
0.0014430	pints per hour
0.000024050	pints per minute
0.00000040083	pints per second
0.13853	gills per day
0.0057722	gills per hour
0.000096024	gills per minute
0.0000016034	gills per second
0.16387	deciliters per day
0.0068278	deciliters per hour
0.00011380	deciliters per minute
0.0000018966	deciliters per second
1.0	cubic inches per day
0.041666	cubic inches per hour
0.00069444	cubic inches per minute
0.000011574	cubic inches per second
1.63871	centiliters per day
0.068278	centiliters per hour
0.0011380	centiliters per minute
0.000018966	centiliters per second
16.38716	milliliters per day
0.68278	milliliters per hour
0.011380	milliliters per minute
0.00018966	milliliters per second
16.38716	cubic centimeters per day
0.68278	cubic centimeters per hour
0.011380	cubic centimeters per minute
0.00018966	cubic centimeters per second
16,387.16	cubic millimeters per day
682.7760	cubic millimeters per hour
11.3796	cubic millimeters per minute
0.18966	cubic millimeters per second

CUBIC INCH PER HOUR: =

0.00039328	kiloliters per day
0.000016387	kiloliters per hour
0.00000027311	kiloliters per minute
0.0000000045519	kiloliters per second
0.00039328	cubic meters per day
0.000016387	cubic meters per hour
0.00000027311	cubic meters per minute
0.0000000045519	cubic meters per second
0.00051442	cubic yards per day
0.000021434	cubic yards per hour
0.00000035723	cubic yards per minute
0.0000000059539	cubic yards per second
0.0024737	barrels per day
0.00010307	barrels per hour
0.0000017179	barrels per minute
0.000000028631	barrels per second
0.0039328	hectoliters per day
0.00016387	hectoliters per hour
0.0000027311	hectoliters per minute
0.000000045519	hectoliters per second
0.013889	cubic feet per day
0.0005787	cubic feet per hour
0.000009645	cubic feet per minute
0.00000016075	cubic feet per second
0.039328	dekaliters per day
0.0016387	dekaliters per hour
0.000027311	dekaliters per minute
0.00000045519	dekaliters per second
0.10390	gallons per day
0.0043290	gallons per hour
0.000072150	gallons per minute
0.0000012025	gallons per second
0.086564	gallons (Imperial) per day
0.003607	gallons (Imperial) per hour
0.000060114	gallons (Imperial) per minute
0.0000010019	gallons (Imperial) per second
0.39328	liters per day
0.016387	liters per hour
0.00027311	liters per minute
0.0000045519	liters per second
0.39328	cubic decimeters per day
0.016387	cubic decimeters per hour
0.00027311	cubic decimeters per minute
0.0000045519	cubic decimeters per second
0.41558	quarts per day
0.017316	quarts per hour
0.0002886	quarts per minute
0.000004810	quarts per second
0.83117	pints per day

CUBIC INCH PER HOUR: (cont'd)

0.034632	pints per hour
0.0005772	pints per minute
0.000009620	pints per second
3.32476	gills per day
0.13853	gills per hour
0.0023089	gills per minute
0.000038481	gills per second
3.93284	deciliters per day
0.16387	deciliters per hour
0.0027311	deciliters per minute
0.000045519	deciliters per second
24	cubic inches per day
1.0	cubic inches per hour
0.016667	cubic inches per minute
0.00027778	cubic inches per second
39.32842	centiliters per day
1.63871	centiliters per hour
0.027311	centiliters per minute
0.00045519	centiliters per second
393.28416	milliliters per day
16.38716	milliliters per hour
0.27311	milliliters per minute
0.0045519	milliliters per second
393.28416	cubic centimeters per day
16.38716	cubic centimeters per hour
0.27311	cubic centimeters per minute
0.0045519	cubic centimeters per second
393,284.16	cubic millimeters per day
16,387.16	cubic millimeters per hour
273.114	cubic millimeters per minute
4.5519	cubic millimeters per second

CUBIC INCH PER MINUTE: =

0.023598	kiloliters per day
0.00098323	kiloliters per hour
0.000016387	kiloliters per minute
0.00000027312	kiloliters per second
0.023598	cubic meters per day
0.00098323	cubic meters per hour
0.000016387	cubic meters per minute
0.00000027312	cubic meters per second
0.030865	cubic yards per day
0.0012860	cubic yards per hour
0.000021434	cubic yards per minute
0.00000035723	cubic yards per second
0.14842	barrels per day
0.0061841	barrels per hour
0.00010307	barrels per minute
0.0000017178	barrels per second
0.23598	hectoliters per day

CUBIC INCH PER MINUTE: (cont'd)

0.0098323	hectoliters per hour
0.00016387	hectoliters per minute
0.0000027312	hectoliters per second
0.83333	cubic feet per day
0.034722	cubic feet per hour
0.0005787	cubic feet per minute
0.0000096450	cubic feet per second
2.35976	dekaliters per day
0.098323	dekaliters per hour
0.0016387	dekaliters per minute
0.000027312	dekaliters per second
6.23376	gallons per day
0.25974	gallons per hour
0.0043290	gallons per minute
0.000072150	gallons per second
5.19411	gallons (Imperial) per day
0.21642	gallons (Imperial) per hour
0.003607	gallons (Imperial) per minute
0.000060117	gallons (Imperial) per second
23.59757	liters per day
0.98323	liters per hour
0.016387	liters per minute
0.00027312	liters per second
23.59757	cubic decimeters per day
0.98323	cubic decimeters per hour
0.016387	cubic decimeters per minute
0.00027312	cubic decimeters per second
24.93504	quarts per day
1.038960	quarts per hour
0.017316	quarts per minute
0.0002886	quarts per second
49.87008	pints per day
2.07792	pints per hour
0.034632	pints per minute
0.0005772	pints per second
199.48032	gills per day
8.31168	gills per hour
0.13853	gills per minute
0.0023088	gills per second
235.97568	deciliters per day
9.83232	deciliters per hour
0.16387	deciliters per minute
0.0027312	deciliters per second
1,440	cubic inches per day
60	cubic inches per hour
1	cubic inches per minute
0.016667	cubic inches per second
2,359.7568	centiliters per day
98.3232	centiliters per hour
1.63871	centiliters per minute
0.027312	centiliters per second

CUBIC INCH PER MINUTE: (cont'd)

23,597.568	milliliters per day
983.232	milliliters per hour
16.38716	milliliters per minute
0.27312	milliliters per second
23,597.568	cubic centimeters per day
983.232	cubic centimeters per hour
16.38716	cubic centimeters per minute
0.27312	cubic centimeters per second
23,597,568	cubic millimeters per day
983,232	cubic millimeters per hour
16,387.16	cubic millimeters per minute
273.11933	cubic millimeters per second

CUBIC INCH PER SECOND: =

1.41584	kiloliters per day
0.058993	kiloliters per hour
0.00098322	kiloliters per minute
0.000016387	kiloliters per second
1.41584	cubic meters per day
0.058993	cubic meters per hour
0.00098322	cubic meters per minute
0.000016387	cubic meters per second
1.85190	cubic yards per day
0.077162	cubic yards per hour
0.0012860	cubic yards per minute
0.000021434	cubic yards per second
8.90525	barrels per day
0.37105	barrels per hour
0.0061842	barrels per minute
0.00010307	barrels per second
14.15837	hectoliters per day
0.58993	hectoliters per hour
0.0098322	hectoliters per minute
0.00016387	hectoliters per second
49.99968	cubic feet per day
2.08332	cubic feet per hour
0.034722	cubic feet per minute
0.0005787	cubic feet per second
141.58368	dekaliters per day
5.89932	dekaliters per hour
0.098332	dekaliters per minute
0.0016387	dekaliters per second
374.02560	gallons per day
15.58440	gallons per hour
0.25974	gallons per minute
0.0043290	gallons per second
311.64480	gallons (Imperial) per day
12.98520	gallons (Imperial) per hour
0.21642	gallons (Imperial) per minute
0.003607	gallons (Imperial) per second

CUBIC INCH PER SECOND: (cont'd)

1,415.83680	liters per day
58.99320	liters per hour
0.98322	liters per minute
0.016387	liters per second
1,415.83680	cubic decimeters per day
58.99320	cubic decimeters per hour
0.98322	cubic decimeters per minute
0.016387	cubic decimeters per second
1,496.10240	quarts per day
62.33760	quarts per hour
1.038960	quarts per minute
0.017316	quarts per second
2,992.20480	pints per day
124.6752	pints per hour
2.077920	pints per minute
0.034632	pints per second
11,968.9920	gills per day
498.70800	gills per hour
8.31180	gills per minute
0.13853	gills per second
14,158.36800	deciliters per day
589.93200	deciliters per hour
9.83326	deciliters per minute
0.16387	deciliters per second
86,400	cubic inches per day
3,600	cubic inches per hour
60	cubic inches per minute
1.0	cubic inches per second
141,584.5440	centiliters per day
5,899.3560	centiliters per hour
98.32260	centiliters per minute
1.63871	centiliters per second
1,415,850.6240	milliliters per day
58,993.7760	milliliters per hour
983.22960	milliliters per minute
16.38716	milliliters per second
1,415,850.6240	cubic centimeters per day
58,993.7760	cubic centimeters per hour
983.22960	cubic centimeters per minute
16.38716	cubic centimeters per second
1,415,850,624	cubic millimeters per day
58,993,776	cubic millimeters per hour
983,229.6	cubic millimeters per minute
16,387.16	cubic millimeters per second

CUBIC METER: =

0.00000000062139	cubic miles
0.000000001	cubic kilometers
0.00000012283	cubic furlongs
0.000001	cubic hektometers

CUBIC METER: (cont'd)

0.00012283	cubic chains
0.001	cubic dekameters
0.0078613	cubic rods
1	kiloliters
1.307943	cubic yards
1.64763	cubic varas (Texas)
6.28994	barrels
10	hectoliters
28.37798	bushels (U.S.) dry
27.49582	bushels (Imperial) dry
35.314445	cubic feet
83.70688	cubic spans
100	dekaliters
113.51120	pecks
122.83316	cubic links
264.17762	gallons (U.S.) liquid
227.026407	gallons (U.S.) dry
219.97542	gallons (Imperial)
956.34894	cubic hands
1,000	liters
1,000	cubic decimeters
1,056.71088	quarts (liquid)
908.10299	quarts (dry)
2,113.42176	pints (liquid)
1,816.19834	pints (dry)
8,453.68704	gills (liquid)
10,000	deciliters
61,022.93879	cubic inches
100,000	centiliters
1,000,000	milliliters
1,000,000	cubic centimeters
1,000,000,000	cubic millimeters
2,204.62	pounds of water @ 39°F.
2,201.82790	pounds of water @ 60°F.
2,542.608	pounds of salt water @ 60°F. and 1.155 specific gravity
32.10396	sacks of cement (set)
33,813.54487	ounces (fluid)
270,506.35839	drams (fluid)

CUBIC METERS PER DAY: =

1	kiloliters per day
0.041667	kiloliters per hour
0.00069444	kiloliters per minute
0.000011574	kiloliters per second
1	cubic meters per day
0.041667	cubic meters per hour
0.00069444	cubic meters per minute
0.000011574	cubic meters per second
1.30794	cubic yards per day

0.054497	cubic yards per hour
0.00090828	cubic yards per minute
0.000015138	cubic yards per second
6.289943	barrels per day
0.26208	barrels per hour
0.0043680	barrels per minute
0.000072800	barrels per second
10	hectoliters per day
0.41667	hectoliters per hour
0.0069444	hectoliters per minute
0.00011574	hectoliters per second
35.31444	cubic feet per day
1.47143	cubic feet per hour
0.024524	cubic feet per minute
0.00040873	cubic feet per second
100	dekaliters per day
4.1666	dekaliters per hour
0.069444	dekaliters per minute
0.0011574	dekaliters per second
264.17762	gallons per day
11.0074008	gallons per hour
0.18346	gallons per minute
0.0030576	gallons per second
219.97542	gallons (Imperial) per day
9.16564	gallons (Imperial) per hour
0.15276	gallons (Imperial) per minute
0.0025460	gallons (Imperial) per second
1,000	liters per day
41.66640	liters per hour
0.69444	liters per minute
0.011574	liters per second
1,000	cubic decimeters per day
41.66640	cubic decimeters per hour
0.69444	cubic decimeters per minute
0.011574	cubic decimeters per second
1,056.71088	quarts per day
44.02962	quarts per hour
0.73383	quarts per minute
0.012230	quarts per second
2,113.42176	pints per day
88.059240	pints per hour
1.46765	pints per minute
0.024461	pints per second
8,453.68704	gills per day
352.23696	gills per hour
5.87062	gills per minute
0.097844	gills per second
10,000	deciliters per day
416.66400	deciliters per hour
6.94440	deciliters per minute
0.11574	deciliters per second

CUBIC METERS PER DAY: (cont'd)

61,022.93879	cubic inches per day
2,542.6080	cubic inches per hour
42.3768	cubic inches per minute
0.70628	cubic inches per second
100,000	centiliters per day
4,166.6400	centiliters per hour
69.4440	centiliters per minute
1.15740	centiliters per second
1,000,000	milliliters per day
41,666.40	milliliters per hour
694.440	milliliters per minute
11.57407	milliliters per second
1,000,000	cubic centimeters per day
4,166.64000	cubic centimeters per hour
69.4440	cubic centimeters per minute
1.15740	cubic centimeters per second
1,000,000,000	cubic millimeters per day
4,166,640	cubic millimeters per hour
69,444	cubic millimeters per minute
1,157.4	cubic millimeters per second

CUBIC METERS PER HOUR: =

24	kiloliters per day
1	kiloliters per hour
0.016667	kiloliters per minute
0.00027778	kiloliters per second
24	cubic meters per day
1	cubic meters per hour
0.01667	cubic meters per minute
0.00027778	cubic meters per second
31.39085	cubic yards per day
1.30794	cubic yards per hour
0.021799	cubic yards per minute
0.00036332	cubic yards per second
150.95863	barrels per day
6.289943	barrels per hour
0.10483	barrels per minute
0.0017472	barrels per second
240	hectoliters per day
10	hectoliters per hour
0.16667	hectoliters per minute
0.0027778	hectoliters per second
847.54944	cubic feet per day
35.31444	cubic feet per hour
0.58858	cubic feet per minute
0.0098096	cubic feet per second
2,400	dekaliters per day
100	dekaliters per hour
1.66668	dekaliters per minute
0.027778	dekaliters per second

CUBIC METERS PER HOUR: (cont'd)

6,340.26288	gallons per day
264.17762	gallons per hour
4.40296	gallons per minute
0.073383	gallons per second
5,279.41008	gallons (Imperial) per day
219.97542	gallons (Imperial) per hour
3.66626	gallons (Imperial) per minute
0.06114	gallons (Imperial) per second
24,000	liters per day
1,000	liters per hour
16.66680	liters per minute
0.27778	liters per second
24,000	cubic decimeters per day
1,000	cubic decimeters per hour
16.66680	cubic decimeters per minute
0.27778	cubic decimeters per second
25,360.9920	quarts per day
1,056.71088	quarts per hour
17.61180	quarts per minute
0.29353	quarts per second
50,722.12224	pints per day
2,113.42176	pints per hour
35.22370	pints per minute
0.58706	pints per second
202,888.48896	gills per day
8,453.68704	gills per hour
140.89478	gills per minute
2.34825	gills per second
240,000	deciliters per day
10,000	deciliters per hour
166.6680	deciliters per minute
2.77778	deciliters per second
1,464,550.8480	cubic inches per day
61,022.93879	cubic inches per hour
1,017.04920	cubic inches per minute
16.95082	cubic inches per second
2,400,000	centiliters per day
100,000	centiliters per hour
1,666.680	centiliters per minute
27.77778	centiliters per second
24,000,000	milliliters per day
1,000,000	milliliters per hour
16,666.80	milliliters per minute
277.77778	milliliters per second
24,000,000	cubic centimeters per day
1,000,000	cubic centimeters per hour
16,666.666680	cubic centimeters per minute
277.777778	cubic centimeters per second
24,000,000,000	cubic millimeters per day
1,000,000,000	cubic millimeters per hour

CUBIC METERS PER HOUR: (cont'd)

16,666,800 .... cubic millimeters per minute
277,777.777778 .... cubic millimeters per second

CUBIC METERS PER MINUTE: =

1,440 .... kiloliters per day
60 .... kiloliters per hour
1 .... kiloliters per minute
0.016667 .... kiloliters per second
1,440 .... cubic meters per day
60 .... cubic meters per hour
1 .... cubic meters per minute
0.016667 .... cubic meters per second
1,883.43360 .... cubic yards per day
78.47640 .... cubic yards per hour
1.30794 .... cubic yards per minute
0.021799 .... cubic yards per second
9,057.51792 .... barrels per day
377.39658 .... barrels per hour
6.289943 .... barrels per minute
0.10483 .... barrels per second
14,400 .... hectoliters per day
600 .... hectoliters per hour
10 .... hectoliters per minute
0.16667 .... hectoliters per second
50,852.4480 .... cubic feet per day
2,118.8520 .... cubic feet per hour
35.31444 .... cubic feet per minute
0.58857 .... cubic feet per second
144,000 .... dekaliters per day
6,000 .... dekaliters per hour
100 .... dekaliters per minute
1.66667 .... dekaliters per second
380,415.7728 .... gallons per day
15,850.65720 .... gallons per hour
264.17762 .... gallons per minute
4.40296 .... gallons per second
316,764.60480 .... gallons (Imperial) per day
13,198.5252 .... gallons (Imperial) per hour
219.97542 .... gallons (Imperial) per minute
3.66626 .... gallons (Imperial) per second
1,440,000 .... liters per day
60,000 .... liters per hour
1,000 .... liters per minute
16.66667 .... liters per second
1,440,000 .... cubic decimeters per day
60,000 .... cubic decimeters per hour
1,000 .... cubic decimeters per minute
16.66667 .... cubic decimeters per second
1,521,663.6672 .... quarts per day
63,402.65280 .... quarts per hour

CUBIC METERS PER MINUTE: (cont'd)

1,056.71088	quarts per minute
17.61185	quarts per second
3,043,327.3344	pints per day
126,805.30560	pints per hour
2,113.42176	pints per minute
35.22370	pints per second
12,173,309.8560	gills per day
507,221.2440	gills per hour
8,453.68704	gills per minute
140.89479	gills per second
14,400,000	deciliters per day
600,000	deciliters per hour
10,000	deciliters per minute
166.66667	deciliters per second
87,873,031.8720	cubic inches per day
3,661,376.3280	cubic inches per hour
61,022.93879	cubic inches per minute
1,017.04898	cubic inches per second
144,000,000	centiliters per day
6,000,000	centiliters per hour
100,000	centiliters per minute
1,666.66667	centiliters per second
1,440,000,000	milliliters per day
60,000,000	milliliters per hour
1,000,000	milliliters per minute
16,666.66667	milliliters per second
1,440,000,000	cubic centimeters per day
60,000,000	cubic centimeters per hour
1,000,000	cubic centimeters per minute
16,666.66667	cubic centimeters per second
1,440,000,000,000	cubic millimeters per day
60,000,000,000	cubic millimeters per hour
1,000.000.000	cubic millimeters per minute
16,666,666.66667	cubic millimeters per second

CUBIC METERS PER SECOND: =

86,400	kiloliters per day
3,600	kiloliters per hour
60	kiloliters per minute
1	kiloliters per second
86,400	cubic meters per day
3,600	cubic meters per hour
60	cubic meters per minute
1	cubic meters per second
113,006.0160	cubic yards per day
4,708.5840	cubic yards per hour
78.47640	cubic yards per minute
1.30794	cubic yards per second
543,451.07520	barrels per day
22,643.7948	barrels per hour

CUBIC METERS PER SECOND: (cont'd)

377.39658	barrels per minute
6.28994	barrels per second
864,000	hektoliters per day
36,000	hektoliters per hour
600	hektoliters per minute
10	hektoliters per second
3,051,167.616	cubic feet per day
127,131.9840	cubic feet per hour
2,118.86640	cubic feet per minute
35.31444	cubic feet per second
8,640,000	dekaliters per day
360,000	dekaliters per hour
6,000	dekaliters per minute
100	dekaliters per second
22,824,946.3680	gallons per day
951,039.4320	gallons per hour
15,850.65720	gallons per minute
264.17762	gallons per second
19,005,798.5280	gallons (Imperial) per day
791,908.2720	gallons (Imperial) per hour
13,198.47120	gallons (Imperial) per minute
219.97542	gallons (Imperial) per second
86,400,000	liters per day
3,600,000	liters per hour
60,000	liters per minute
1,000	liters per second
86,400,000	cubic decimeters per day
3,600,000	cubic decimeters per hour
60,000	cubic decimeters per minute
1,000	cubic decimeters per second
91,299,820.0320	quarts per day
3,804,159.16800	quarts per hour
63,402.65280	quarts per minute
1,056.71088	quarts per second
182,599,640.0640	pints per day
7,608,318.3360	pints per hour
126,805.30560	pints per minute
2,113.42176	pints per second
730,398,560.2560	gills per day
30,433,273.3340	gills per hour
507,221.22240	gills per minute
8,453.68704	gills per second
864,000,000	deciliters per day
36,000,000	deciliters per hour
600,000	deciliters per minute
10,000	deciliters per second
5,272,381,911.4560	cubic inches per day
219,682,579.6440	cubic inches per hour
3,661,376.32740	cubic inches per minute
61,022.93879	cubic inches per second
8,640,000,000	centiliters per day

360,000,000	centiliters per hour
6,000,000	centiliters per minute
100,000	centiliters per second
86,400,000,000	milliliters per day
3,600,000,000	milliliters per hour
60,000,000	milliliters per minute
1,000,000	milliliters per second
86,400,000,000	cubic centimeters per day
3,600,000,000	cubic centimeters per hour
60,000,000	cubic centimeters per minute
1,000,000	cubic centimeters per second
86,400,000,000,000	cubic millimeters per day
3,600,000,000,000	cubic millimeters per hour
60,000,000,000	cubic millimeters per minute
1,000,000,000	cubic millimeters per second

CUBIC POISE CENTIMETER PER GRAM: =

0.0000000001	square kilometers per second
0.00000001	square hektometers per second
0.000001	square dekameters per second
0.0001	square meters per second
0.0010764	square feet per second
0.1550	square inches per second
0.01	square decimeters per second
1.0	square centimeters per second
100	square millimeters per second

CUBIC POISE FOOT PER POUND: =

0.000000006243	square kilometers per second
0.0000006243	square hektometers per second
0.00006243	square dekameters per second
0.006243	square meters per second
0.067200	square feet per second
9.67680	square inches per second
0.6243	square decimeters per second
62.43	square centimeters per second
6,243	square millimeters per second

CUBIC POISE INCH PER GRAM: =

0.0000000016387	square kilometers per second
0.00000016387	square hektometers per second
0.000016387	square dekameters per second
0.0016387	square meters per second
0.017639	square feet per second
2.540	square inches per second
0.16387	square decimeters per second
16.387	square centimeters per second
1,638.7	square millimeters per second

CUBIC YARD: =

0.00000000047509	cubic miles
0.00000000076456	cubic kilometers
0.000000093914	cubic furlongs
0.00000076456	cubic hektometers
0.000093914	cubic chains
0.00076456	cubic dekameters
0.0060105	cubic rods
0.76456	kiloliters
0.76456	cubic meters
1.25971	cubic varas (Texas)
4.80897	barrels
7.64559	hektoliters
21.69666	bushels (U.S.) dry
21.0220	bushels (Imperial) dry
27	cubic feet
63.99891	cubic spans
76.45595	dekaliters
86.78610	pecks
93.91329	cubic links
201.97350	gallons (U.S.) liquid
173.59299	gallons (U.S.) dry
168.18003	gallons (Imperial)
731.18592	cubic hands
764.55945	liters
764.55945	cubic decimeters
807.89400	quarts (liquid)
694.28070	quarts (dry)
1,615.78800	pints (liquid)
1,388.59218	pints (dry)
6,463.15200	gills (liquid)
7,645.5945	deciliters
46,656	cubic inches
76,455.945	centiliters
764,559.45	milliliters
764,559.45	cubic centimeters
764,559,450	cubic millimeters
25,852.79808	ounces (fluid)
206,820.85509	drams (fluid)
24.5457	sacks of cement (set)
1,683.45	pounds of water @ 60°F.
1,736.10	pounds of salt water
1,944	pounds of salt water @ 60°F. and 1.155 specific gravity
13,217.634	pounds of steel of 7.851 specific gravity

CUBIC YARD PER DAY: =

0.76456	kiloliters per day
0.031857	kiloliters per hour

CUBIC YARD PER DAY: (cont'd)

0.00053095	kiloliters per minute
0.0000088491	kiloliters per second
0.76456	cubic meters per day
0.031857	cubic meters per hour
0.00053095	cubic meters per minute
0.0000088491	cubic meters per second
1.0	cubic yards per day
0.04166	cubic yards per hour
0.00069444	cubic yards per minute
0.000011574	cubic yards per second
4.80897	barrels per day
0.20037	barrels per hour
0.0033395	barrels per minute
0.000055659	barrels per second
7.64559	hektoliters per day
0.31857	hektoliters per hour
0.0053095	hektoliters per minute
0.00008849	hektoliters per second
27	cubic feet per day
1.1250	cubic feet per hour
0.018750	cubic feet per minute
0.00031250	cubic feet per second
76.45595	dekaliters per day
3.18568	dekaliters per hour
0.053095	dekaliters per minute
0.00088491	dekaliters per second
201.97350	gallons per day
8.41572	gallons per hour
0.14026	gallons per minute
0.0023377	gallons per second
168.18003	gallons (Imperial) per day
7.00740	gallons (Imperial) per hour
0.11679	gallons (Imperial) per minute
0.0019465	gallons (Imperial) per second
764.55945	liters per day
31.85676	liters per hour
0.53095	liters per minute
0.0088491	liters per second
764.55945	cubic decimeters per day
31.85676	cubic decimeters per hour
0.53095	cubic decimeters per minute
0.0088491	cubic decimeters per second
807.8940	quarts per day
33.66216	quarts per hour
0.56104	quarts per minute
0.0093506	quarts per second
1,615.7880	pints per day
67.32360	pints per hour
1.12206	pints per minute
0.018701	pints per second
6,463.152	gills per day

CUBIC YARD PER DAY: (cont'd)

269.2980	gills per hour
4.48830	gills per minute
0.074805	gills per second
7,645.5945	deciliters per day
318.56760	deciliters per hour
5.30946	deciliters per minute
0.088491	deciliters per second
46,656	cubic inches per day
1,944.0	cubic inches per hour
32.40	cubic inches per minute
0.5400	cubic inches per second
76,455.945	centiliters per day
3,185.6760	centiliters per hour
53.09460	centiliters per minute
0.88491	centiliters per second
764,559.45	milliliters per day
31,856.760	milliliters per hour
530.946	milliliters per minute
8.84907	milliliters per second
764,559.45	cubic centimeters per day
31,856.760	cubic centimeters per hour
530.946	cubic centimeters per minute
8.84907	cubic centimeters per second
764,559,450	cubic millimeters per day
31,856,760	cubic millimeters per hour
530,946	cubic millimeters per minute
8,849.07	cubic millimeters per second

CUBIC YARD PER HOUR: =

18.34963	kiloliters per day
0.76456	kiloliters per hour
0.012743	kiloliters per minute
0.00021238	kiloliters per second
18.34963	cubic meters per day
0.76456	cubic meters per hour
0.012743	cubic meters per minute
0.00021238	cubic meters per second
24.0	cubic yards per day
1.0	cubic yards per hour
0.016667	cubic yards per minute
0.00027778	cubic yards per second
115.41312	barrels per day
4.80897	barrels per hour
0.080148	barrels per minute
0.0013358	barrels per second
183.49632	hektoliters per day
7.64559	hektoliters per hour
0.12743	hektoliters per minute
0.0021238	hektoliters per second
648.0	cubic feet per day

CUBIC YARD PER HOUR: (cont'd)

27.0	cubic feet per hour
0.450	cubic feet per minute
0.00750	cubic feet per second
1,834.9632	dekaliters per day
76.45595	dekaliters per hour
1.27428	dekaliters per minute
0.021238	dekaliters per second
4,847.3856	gallons per day
201.97350	gallons per hour
3.36624	gallons per minute
0.056104	gallons per second
4,036.34880	gallons (Imperial) per day
168.18003	gallons (Imperial) per hour
2.80302	gallons (Imperial) per minute
0.046717	gallons (Imperial) per second
18,349.632	liters per day
764.55945	liters per hour
12.7428	liters per minute
0.21238	liters per second
18,349.632	cubic decimeters per day
764.55945	cubic decimeters per hour
12.7428	cubic decimeters per minute
0.21238	cubic decimeters per second
19,389.8880	quarts per day
807.8940	quarts per hour
13.4652	quarts per minute
0.22442	quarts per second
38,778.9120	pints per day
1,615.7780	pints per hour
26.92980	pints per minute
0.44883	pints per second
155,115.648	gills per day
6,463.152	gills per hour
107.7192	gills per minute
1.79532	gills per second
183,496.32	deciliters per day
7,645.5945	deciliters per hour
127.428	deciliters per minute
2.12378	deciliters per second
1,119,744	cubic inches per day
46,656	cubic inches per hour
777.6	cubic inches per minute
12.960	cubic inches per second
1,834,963.2	centiliters per day
76,455.945	centiliters per hour
1,274.28	centiliters per minute
21.23776	centiliters per second
18,349.632	milliliters per day
764,559.45	milliliters per hour
12,742.8	milliliters per minute
212.37763	milliliters per second

CUBIC YARD PER HOUR: (cont'd)

18,349,632	cubic centimeters per day
764,559.45	cubic centimeters per hour
12,742.8	cubic centimeters per minute
212.37763	cubic centimeters per second
18,349,632,000	cubic millimeters per day
764,559,450	cubic millimeters per hour
12,742,800	cubic millimeters per minute
212,377.63	cubic millimeters per second

CUBIC YARD PER MINUTE: =

1,100.96640	kiloliters per day
45.87360	kiloliters per hour
0.76456	kiloliters per minute
0.012743	kiloliters per second
1,100.96640	cubic meters per day
45.87360	cubic meters per hour
0.76456	cubic meters per minute
0.012743	cubic meters per second
1,440	cubic yards per day
60	cubic yards per hour
1.0	cubic yards per minute
0.016667	cubic yards per second
6,924.960	barrels per day
288.540	barrels per hour
4.80897	barrels per minute
0.080150	barrels per second
11,009.66399	hektoliters per day
458.73560	hektoliters per hour
7.64559	hektoliters per minute
0.12743	hektoliters per second
38,880	cubic feet per day
1,620	cubic feet per hour
27.0	cubic feet per minute
0.450	cubic feet per second
110,096.633994	dekaliters per day
4,587.3600	dekaliters per hour
76.45595	dekaliters per minute
1.27427	dekaliters per second
290,842.2720	gallons per day
12,118.4280	gallons per hour
201.97350	gallons per minute
3.36623	gallons per second
242,179.20	gallons (Imperial) per day
10,090.80	gallons (Imperial) per hour
168.18003	gallons (Imperial) per minute
2.80300	gallons (Imperial) per second
1,100,966.39942	liters per day
45,873.59998	liters per hour
764.55945	liters per minute
12.74267	liters per second

CUBIC YARD PER MINUTE: (cont'd)

1,100,966.39942	cubic decimeters per day
45,873.59998	cubic decimeters per hour
764.55945	cubic decimeters per minute
12.74267	cubic decimeters per second
1,163,367.36	quarts per day
48,473.64	quarts per hour
807.8940	quarts per minute
13.4649	quarts per second
2,326,734.72	pints per day
96,947.28	pints per hour
1,615.7880	pints per minute
26.9298	pints per second
9,306,938.88	gills per day
387,789.12	gills per hour
6,463.152	gills per minute
107.71920	gills per second
11,009,663.99424	deciliters per day
458,735.99976	deciliters per hour
7,645.5945	deciliters per minute
127.42667	deciliters per second
67,184,640	cubic inches per day
2,799,360.0	cubic inches per hour
46,656.0	cubic inches per minute
777.6	cubic inches per second
110,096,639.9424	centiliters per day
4,587,359.9976	centiliters per hour
76,455.945	centiliters per minute
1,274.26667	centiliters per second
1,100,966,399.424	milliliters per day
45,873,599.976	milliliters per hour
764,559.45	milliliters per minute
12,742.66667	milliliters per second
1,100,966,399.424	cubic centimeters per day
45,873,599.976	cubic centimeters per hour
764,559.45	cubic centimeters per minute
12,742.66667	cubic centimeters per second
1,100,966,399,424	cubic millimeters per day
45,873,499,976	cubic millimeters per hour
764,559,450	cubic millimeters per minute
12,742,666.67	cubic millimeters per second

CUBIC YARD PER SECOND: =

66,057.93648	kiloliters per day
2,752.41402	kiloliters per hour
45.87357	kiloliters per minute
0.76456	kiloliters per second
66,057.93648	cubic meters per day
2,752.41402	cubic meters per hour
45.87357	cubic meters per minute
0.76456	cubic meters per second

CUBIC YARD PER SECOND: (cont'd)

86,400	cubic yards per day
3,600	cubic yards per hour
60.0	cubic yards per minute
1.0	cubic yards per second
415,495.008	barrels per day
17,312.2920	barrels per hour
288.53820	barrels per minute
4.80897	barrels per second
660,579.3648	hectoliters per day
27,524.1402	hectoliters per hour
458.73567	hectoliters per minute
7.64559	hectoliters per second
2,332,800	cubic feet per day
97,200	cubic feet per hour
1,620	cubic feet per minute
27	cubic feet per second
6,605,793.648	dekaliters per day
275,241.402	dekaliters per hour
4,587.3567	dekaliters per minute
76.45595	dekaliters per second
17,450,510.40	gallons per day
727,104.60	gallons per hour
12,118.410	gallons per minute
201.97350	gallons per second
14,530,754.5920	gallons (Imperial) per day
605,448.1080	gallons (Imperial) per hour
10,090.80180	gallons (Imperial) per minute
168.18003	gallons (Imperial) per second
66,057,936.48	liters per day
2,752,414.02	liters per hour
45,873.567	liters per minute
764.55945	liters per second
66,057,936.48	cubic decimeters per day
2,752,414.02	cubic decimeters per hour
45,873.567	cubic decimeters per minute
764.55945	cubic decimeters per second
69,802,041.6	quarts per day
2,908,418.40	quarts per hour
48,473.640	quarts per minute
807.8940	quarts per second
139,604,083.2	pints per day
5,816,836.8	pints per hour
96,947.280	pints per minute
1,615.7880	pints per second
558,416,332.8	gills per day
23,267,347.20	gills per hour
387,789.120	gills per minute
6,463.152	gills per second
660,579,364.8	deciliters per day
27,524,140.2	deciliters per hour
458,735.67	deciliters per minute

CUBIC YARD PER SECOND: (cont'd)

7,645.5945	deciliters per second
4,031,078,400	cubic inches per day
167,961,600	cubic inches per hour
2,799,360	cubic inches per minute
46,656	cubic inches per second
6,605,793,648	centiliters per day
275,241,402	centiliters per hour
4,587,356.7	centiliters per minute
76,445.945	centiliters per second
66,057,936.480	milliliters per day
2,752,414,020	milliliters per hour
45,873,567	milliliters per minute
764,559.45	milliliters per second
66,057,936,480	cubic centimeters per day
2,752,414,020	cubic centimeters per hour
45,873,565	cubic centimeters per minute
764,559.45	cubic centimeters per second
66,057,936,480,000	cubic millimeters per day
2,752,414,020,000	cubic millimeters per hour
45,873,565,000	cubic millimeters per minute
764,559,450	cubic millimeters per second

DECIGRAM: =

0.0000001102311150	tons (short)
0.0000000984206383	tons (long)
0.0000001	tons (metric)
0.0001	kilograms
0.000267922895	pounds (Troy)
0.00022046223	pounds (Avoir.)
0.001	hektograms
0.01	dekagrams
0.00321507	ounces (Troy)
0.00352739568	ounces (Avoir.)
0.1	grams
10	centigrams
100	milligrams
1.543236	grains
0.0257205	drams (Troy)
0.05643833088	drams (Avoir.)
0.5	carats (metric)

DECILITER: =

0.0001	kiloliters
0.0001	cubic meters
0.0001308	cubic yards
0.00062900	barrels
0.001	hektoliters
0.0028380	bushels (U.S.)

DECILITER: (cont'd)

0.0027513	bushels (Imperial) dry
0.0035316	cubic feet
0.01	dekaliters
0.011352	pecks
0.026418	gallons (U.S.) liquid
0.022706	gallons (U.S.) dry
0.021997	gallons (Imperial)
0.10567	quarts (liquid)
0.090816	quarts (dry)
0.1	liters
0.1	cubic decimeters
0.18163	pints (dry)
0.21134	pints (liquid)
0.84536	gills (liquid)
6.1025	cubic inches
1.0	centiliters
100	milliliters
100	cubic centimeters
100,000	cubic millimeters
3.38147	ounces (fluid)
27.05179	drams (fluid)

DECIMETER: =

0.000062137	miles
0.0001	kilometers
0.00049710	furlongs
0.001	hektometers
0.0049710	chains
0.01	dekameters
0.019884	rods
0.1	meters
0.10935	yards
0.11811	varas (Texas)
0.3280833	feet
0.43744	spans
0.49710	links
0.98425	hands
1	decimeter
10	centimeter
3.93700	inches
100	millimeters
3,937	mils
100,000	microns
100,000,000	millimicrons
100,000,000	micromillimeters
155,316	wave lengths of red line of cadmium
100,000,000,000	Angstrom units

DEGREE (ANGLE): =

0.0027778	circumferences
0.0027778	revolutions
0.01111	quadrants
0.017453	radians
1	hours
60	minutes
3,600	seconds

DEGREE PER DAY: =

0.0027778	revolutions per day
0.00011574	revolutions per hour
0.0000019290	revolutions per minute
0.000000032150	revolutions per second
0.01111	quadrants per day
0.00046292	quadrants per hour
0.0000077154	quadrants per minute
0.00000012859	quadrants per second
0.017453	radians per day
0.00072720	radians per hour
0.000012120	radians per minute
0.00000020200	radians per second
1.0	degrees per day
0.041666	degrees per hour
0.00069444	degrees per minute
0.000011574	degrees per second
1.0	hours per day
0.041666	hours per hour
0.00069444	hours per minute
0.000011574	hours per second
60	minutes per day
2.5	minutes per hour
0.041666	minutes per minute
0.00069444	minutes per second
3,600	seconds per day
150	seconds per hour
2.5	seconds per minute
0.041667	seconds per second

DEGREE PER HOUR: =

0.066667	revolutions per day
0.0027778	revolutions per hour
0.000046297	revolutions per minute
0.00000077161	revolutions per second
0.26664	quadrants per day
0.01111	quadrants per hour
0.00018517	quadrants per minute
0.0000030861	quadrants per second

DEGREE PER HOUR: (cont'd)

0.41888 . . . . radians per day
0.017453 . . . . radians per hour
0.00029088 . . . . radians per minute
0.0000048481 . . . . radians per second
24 . . . . degrees per day
1.0 . . . . degrees per hour
0.016667 . . . . degrees per minute
0.00027778 . . . . degrees per second
24 . . . . hours per day
1.0 . . . . hours per hour
0.016667 . . . . hours per minute
0.00027778 . . . . hours per second
1,440 . . . . minutes per day
60 . . . . minutes per hour
1 . . . . minutes per minute
0.016667 . . . . minutes per second
86,400 . . . . seconds per day
3,600 . . . . seconds per hour
60 . . . . seconds per minute
1 . . . . seconds per second

DEGREE PER MINUTE: =

4.0 . . . . revolutions per day
0.16667 . . . . revolutions per hour
0.0027778 . . . . revolutions per minute
0.000046297 . . . . revolutions per second
15.99869 . . . . quadrants per day
0.66661 . . . . quadrants per hour
0.01111 . . . . quadrants per minute
0.00018517 . . . . quadrants per second
25.13203 . . . . radians per day
1.04717 . . . . radians per hour
0.017453 . . . . radians per minute
0.00029088 . . . . radians per second
1,440 . . . . degrees per day
60 . . . . degrees per hour
1.0 . . . . degrees per minute
0.016667 . . . . degrees per second
1,440 . . . . hours per day
60 . . . . hours per hour
1.0 . . . . hours per minute
0.016667 . . . . hours per second
86,400 . . . . minutes per day
3,600 . . . . minutes per hour
60 . . . . minutes per minute
1 . . . . minutes per second
5,184,000 . . . . seconds per day
216,000 . . . . seconds per hour
3,600 . . . . seconds per minute
60 . . . . seconds per second

DEGREE PER MINUTE: (cont'd)

240	revolutions per day
10	revolutions per hour
0.16667	revolutions per minute
0.0027778	revolutions per second
959.904	quadrants per day
39.9960	quadrants per hour
0.6666	quadrants per minute
0.01111	quadrants per second
1,507.9392	radians per day
62.83080	radians per hour
1.047180	radians per minute
0.017453	radians per second
86,400	degrees per day
3,600	degrees per hour
60	degrees per minute
1.0	degrees per second
86,400	hours per day
3,600	hours per hour
60	hours per minute
1.0	hours per second
5,184,000	minutes per day
216,000	minutes per hour
3,600	minutes per minute
60	minutes per second
311,040,000	seconds per day
12,960,000	seconds per hour
216,000	seconds per minute
3,600	seconds per second

DEKAGRAM: =

0.00000984206383	tons (long)
0.00001	tons (metric)
0.0000110231115	tons (short)
0.01	kilograms
0.022046223	pounds (Avoir.)
0.0267922895	pounds (Troy)
0.1	hektograms
0.321507	ounces (Troy)
0.352739568	ounces (Avoir.)
2.572053	drams (Troy)
5.6438330880	drams (Avoir.)
6.430149	pennyweights
7.71618	scruples
10	grams
100	decigrams
154.3234765625	grains
1,000	centigrams
10,000	milligrams
50	carats (metric)

DEKALITER: =

0.01	kiloliters
0.01	cubic meters
0.01308	cubic yards
0.06290	barrels
0.1	hektoliters
0.28380	bushels (U.S.)
0.27513	bushels (Imperial) dry
0.35316	cubic feet
1.1352	pecks
2.6418	gallons (U.S.) liquid
2.2706	gallons (U.S.) dry
2.1997	gallons (Imperial)
10.567	quarts (liquid)
9.0816	quarts (dry)
10	liters
10	cubic decimeters
18.163	pints (dry)
21.134	pints (liquid)
84.536	gills (liquid)
100	deciliters
610.25	cubic inches
1,000	centiliters
10,000	milliliters
10,000	cubic centimeters
100,000,000	cubic millimeters
338.147	ounces (fluid)
2,705.179	drams (fluid)

DEKAMETER: =

0.0062137	miles
0.01	kilometers
0.049710	furlongs
0.1	hektometers
0.49710	chains
1.0	dekameters
1.98838	rods
10	meters
10.935	yards
11.811	varas (Texas)
32.80830	feet
43.744	spans
49.710	links
98.425	hands
100	decimeters
1000	centimeters
393.70	inches
10,000	millimeters
393,700	mils

DEKAMETER: (cont'd)

10,000,000	microns
10,000,000,000	millimicrons
10,000,000,000	micromillimeters
15,531,595	wave lengths of red line of cadmium
10,000,000,000 x 10³	Angstrom Units

DRAM (AVOIRDUPOIS): =

0.0000017439	tons (long)
0.0000017718	tons (metric)
0.0000019531	tons (short)
0.001771845	kilograms
0.00390625	pounds (Avoir.)
0.0047471788	pounds (Troy)
0.01771845	hektograms
0.1771845	dekagrams
0.056966146	ounces (Troy)
0.0625	ounces (Avoir.)
0.4557292	drams (Troy)
1.139323	pennyweights
1.3671875	scruples
1.771845	grams
17.71845	decigrams
27.34375	grains
177.1845	centigrams
1,771.845	milligrams
8.85923	carats (metric)

DRAM (FLUID): =

0.0000036966	kiloliters
0.0000036966	cubic meters
0.0000048352	cubic yards
0.000023252	barrels
0.000036966	hektoliters
0.00013055	cubic feet
0.00036966	dekaliters
0.00097658	gallons (U.S.) liquid
0.00081318	gallons (Imperial) liquid
0.00390625	quarts (liquid)
0.0036966	liters
0.0036966	cubic decimeters
0.0078125	pints (liquid)
0.03125	gills (liquid)
0.036966	deciliters
0.225586	cubic inches
0.36966	centiliters
3.69661	milliliters
3.69661	cubic centimeters

DRAM (FLUID): (cont'd)

3,696.61	cubic millimeters
0.125	ounces (fluid)
60	minims

DRAM (TROY OR APOTHECARY): =

0.0000038265308	tons (long)
0.0000038879351	tons (metric)
0.0000042857145	tons (short)
0.0038879351	kilograms
0.008571429	pounds (Avoir.)
0.010416667	pounds (Troy)
0.038879351	hektograms
0.38879351	dekagrams
0.12500	ounces (Troy)
0.1371429	ounces (Avoir.)
2.194286	drams (Avoir.)
2.50	pennyweight
3.0	scruples
3.8879351	grams
38.879351	decigrams
60.0	grains
388.79351	centigrams
3,887.9351	milligrams
19.43968	carats (metric)

FOOT (OR ENGINEER'S LINK): =

0.0001893939	miles
0.0003048006	kilometers
0.00151515	furlongs
0.003048006	hektometers
0.0151515	chains
0.03048006	dekameters
0.0606061	rods
0.3048006	meters
0.33333	yards
0.3600	varas (Texas)
1.33333	spans
1.515152	links
3.00	hands
3.048006	decimeters
30.48006	centimeters
12	inches
304.8006	millimeters
12,000	mils
304,801	microns
304,801,200	millimicrons
304,801,200	micromillimeters
473,404	wave lengths of red line of cadmium
3,048,012,000	Angstrom Units

FOOT PER DAY: =

0.0001893939	miles per day
0.0000078914	miles per hour
0.00000013152	miles per minute
0.0000000021921	miles per second
0.0003048006	kilometers per day
0.0000127	kilometers per hour
0.00000021167	kilometers per minute
0.0000000035278	kilometers per second
0.00151515	furlongs per day
0.000063131	furlongs per hour
0.0000010522	furlongs per minute
0.000000017536	furlongs per second
0.003048006	hektometers per day
0.000127	hektometers per hour
0.0000021167	hektometers per minute
0.000000035278	hektometers per second
0.0151515	chains per day
0.00063131	chains per hour
0.000010522	chains per minute
0.00000017536	chains per second
0.03048006	dekameters per day
0.00127	dekameters per hour
0.000021167	dekameters per minute
0.00000035278	dekameters per second
0.0606061	rods per day
0.0025253	rods per hour
0.000042088	rods per minute
0.00000070146	rods per second
0.3048006	meters per day
0.0127	meters per hour
0.00021167	meters per minute
0.0000035278	meters per second
0.33333	yards per day
0.013889	yards per hour
0.00023148	yards per minute
0.0000038580	yards per second
0.3600	varas (Texas) per day
0.015	varas (Texas) per hour
0.00025	varas (Texas) per minute
0.0000041667	varas (Texas) per second
1.0	feet per day
0.041667	feet per hour
0.00069444	feet per minute
0.000011574	feet per second
1.33333	spans per day
0.055556	spans per hour
0.00092593	spans per minute
0.000015432	spans per second
1.515152	links per day
0.0163131	links per hour

FOOT PER DAY: (cont'd)

0.0010522	links per minute
0.000017536	links per second
3.00	hands per day
0.125	hands per hour
0.0020833	hands per minute
0.000034722	hands per second
3.048006	decimeters per day
0.127	decimeters per hour
0.0021167	decimeters per minute
0.000035278	decimeters per second
30.48006	centimeters per day
1.270	centimeters per hour
0.021167	centimeters per minute
0.00035278	centimeters per second
12.0	inches per day
0.50	inches per hour
0.0083333	inches per minute
0.00013889	inches per second
304.8006	millimeters per day
12.70004	millimeters per hour
0.21167	millimeters per minute
0.0035278	millimeters per second

FOOT PER HOUR: =

0.0045455	miles per day
0.0001893939	miles per hour
0.0000031566	miles per minute
0.000000052609	miles per second
0.0073152	kilometers per day
0.0003048006	kilometers per hour
0.00000508	kilometers per minute
0.000000084667	kilometers per second
0.036364	furlongs per day
0.00151515	furlongs per hour
0.000025253	furlongs per minute
0.00000042088	furlongs per second
0.073152	hektometers per day
0.003048006	hektometers per hour
0.0000508	hektometers per minute
0.00000084667	hektometers per second
0.36364	chains per day
0.0151515	chains per hour
0.00025253	chains per minute
0.0000042088	chains per second
0.73152	dekameters per day
0.03048006	dekameters per hour
0.000508	dekameters per minute
0.0000084667	dekameters per second
1.45456	rods per day
0.0606061	rods per hour

FOOT PER HOUR: (cont'd)

0.00101010 . . . . rods per minute
0.000016835 . . . . rods per second
7.31521 . . . . meters per day
0.3048006 . . . . meters per hour
0.00508 . . . . meters per minute
0.000084667 . . . . meters per second
8.0000 . . . . yards per day
0.33333 . . . . yards per hour
0.0055556 . . . . yards per minute
0.000092593 . . . . yards per second
8.640 . . . . varas (Texas) per day
0.3600 . . . . varas (Texas) per hour
0.006 . . . . varas (Texas) per minute
0.0001 . . . . varas (Texas) per second
24.0 . . . . feet per day
1.0 . . . . feet per hour
0.016667 . . . . feet per minute
0.00027778 . . . . feet per second
32.000 . . . . spans per day
1.33333 . . . . spans per hour
0.022222 . . . . spans per minute
0.00037037 . . . . spans per second
36.36364 . . . . links per day
1.51515 . . . . links per hour
0.025253 . . . . links per minute
0.00042088 . . . . links per second
72.0 . . . . hands per day
3.0 . . . . hands per hour
0.05000 . . . . hands per minute
0.00083333 . . . . hands per second
73.15214 . . . . decimeters per day
3.048006 . . . . decimeters per hour
0.0508 . . . . decimeters per minute
0.00084667 . . . . decimeters per second
731.52144 . . . . centimeters per day
30.48006 . . . . centimeters per hour
0.5080 . . . . centimeters per minute
0.0084667 . . . . centimeters per second
288.0 . . . . inches per day
12.0 . . . . inches per hour
0.200 . . . . inches per minute
0.0033333 . . . . inches per second
7,315.21440 . . . . millimeters per day
304.8006 . . . . millimeters per hour
5.080 . . . . millimeters per minute
0.084667 . . . . millimeters per second

FOOT PER MINUTE: =

0.27273 . . . . miles per day
0.011364 . . . . miles per hour

FOOT PER MINUTE: (cont'd)

0.0001893939	miles per minute
0.0000031566	miles per second
0.43891	kilometers per day
0.018288	kilometers per hour
0.0003048006	kilometers per minute
0.00000508	kilometers per second
2.18182	furlongs per day
0.090909	furlongs per hour
0.0015151	furlongs per minute
0.000025253	furlongs per second
4.38913	hektometers per day
0.18288	hektometers per hour
0.003048006	hektometers per minute
0.0000508	hektometers per second
21.81818	chains per day
0.90909	chains per hour
0.015151	chains per minute
0.00025253	chains per second
43.89129	dekameters per day
1.82880	dekameters per hour
0.03048006	dekameters per minute
0.000508	dekameters per second
87.27273	rods per day
3.63636	rods per hour
0.0606061	rods per minute
0.0010101	rods per second
438.91286	meters per day
18.28804	meters per hour
0.3048006	meters per minute
0.00508	meters per second
480.0	yards per day
20.0	yards per hour
0.33333	yards per minute
0.0055556	yards per second
518.400	varas (Texas) per day
21.60	varas (Texas) per hour
0.3600	varas (Texas) per minute
0.0060000	varas (Texas) per second
1,440.0	feet per day
60.0	feet per hour
1.0	feet per minute
0.016667	feet per second
1,920	spans per day
80.0	spans per hour
1.33333	spans per minute
0.022222	spans per second
2,181.81818	links per day
90.90909	links per hour
1.51515	links per minute
0.025253	links per second
4,320	hands per day

FOOT PER MINUTE: (cont'd)

180.0	hands per hour
3.0	hands per minute
0.050000	hands per second
4,389.12864	decimeters per day
182.88036	decimeters per hour
3.048006	decimeters per minute
0.050800	decimeters per second
43,891.2864	centimeters per day
1,828.80360	centimeters per hour
30.48006	centimeters per minute
0.50800	centimeters per second
17,280	inches per day
720	inches per hour
12.0	inches per minute
0.20000	inches per second
438,912.864	millimeters per day
18,288.0360	millimeters per hour
304.8006	millimeters per minute
5.08001	millimeters per second

FOOT PER SECOND: =

16.36363	miles per day
0.68182	miles per hour
0.011364	miles per minute
0.0001893939	miles per second
26.33477	kilometers per day
1.097282	kilometers per hour
0.018288	kilometers per minute
0.0003048006	kilometers per second
130.90909	furlongs per day
5.45455	furlongs per hour
0.090909	furlongs per minute
0.0015151	furlongs per second
263.34772	hektometers per day
10.97282	hektometers per hour
0.18288	hektometers per minute
0.003048006	hektometers per second
1,309.09090	chains per day
54.54545	chains per hour
0.90909	chains per minute
0.015151	chains per second
2,633.47718	dekameters per day
109.72823	dekameters per hour
1.82880	dekameters per minute
0.03048006	dekameters per second
5,236.36364	rods per day
218.18182	rods per hour
3.63636	rods per minute
0.0606061	rods per second
26,334.77184	meters per day

FOOT PER SECOND: (cont'd)

1,097.28216	meters per hour
18.28804	meters per minute
0.3048006	meters per second
28,800	yards per day
1,200	yards per hour
20.0	yards per minute
0.33333	yards per second
31,104	varas (Texas) per day
1,296.0	varas (Texas) per hour
21.60	varas (Texas) per minute
0.3600	varas (Texas) per second
86,400	feet per day
3,600	feet per hour
60	feet per minute
1	feet per second
115,200	spans per day
4,800	spans per hour
80.0	spans per minute
1.33333	spans per second
130,909.090909	links per day
5,454.54545	links per hour
90.90909	links per minute
1.51515	links per second
259,200	hands per day
10,800	hands per hour
180.0	hands per minute
3.0	hands per second
263,347.7184	decimeters per day
10,972.8216	decimeters per hour
182.88036	decimeters per minute
3.048006	decimeters per second
2,633,477.1840	centimeters per day
109,728.2160	centimeters per hour
1,828.8036	centimeters per minute
30.48006	centimeters per second
1,036,800	inches per day
43,200	inches per hour
720	inches per minute
12	inches per second
26,334,771.8400	millimeters per day
1,097,282.1600	millimeters per hour
18,288.0360	millimeters per minute
304.8006	millimeters per second

FOOT POUNDS: =

32.174	foot poundals
386.088	inch poundals
1.0	foot pounds
12	inch pounds
0.00000035703	ton (short) calories

FOOT POUNDS: (cont'd)

0.00032389	kilogram calories
0.00071406	pound calories
0.0032389	hektogram calories
0.032389	dekagram calories
0.011425	ounce calories
0.32389	gram calories
3.2389	decigram calories
32.389	centigram calories
323.89	milligram calories
0.0000000156925	kilowatt days
0.00000037662	kilowatt hours
0.0000225972	kilowatt minutes
0.00135583	kilowatt seconds
0.0000156925	watt days
0.00037662	watt hours
0.0225972	watt minutes
1.35583	watt seconds
0.0015240	ton meters (short)
0.138255	kilogram meters
0.30480	pound meters
1.38255	hektogram meters
13.8255	dekagram meters
4.8768	ounce meters
138.255	gram meters
1,382.55	decigram meters
13,825.5	centigram meters
138,255	milligram meters
0.000015240	ton hektometers
0.00138255	kilogram hektometers
0.0030480	pound hektometers
0.0138255	hektogram hektometers
0.138255	dekagram hektometers
0.048768	ounce hektometers
1.38255	gram hektometers
13.8255	decigram hektometers
138.255	centigram hektometers
1,382.55	milligram hektometers
0.00015240	ton dekameters
0.0138255	kilogram dekameters
0.030480	pound dekameters
0.138255	hektogram dekameters
1.38255	dekagram dekameters
0.48768	ounce dekameters
13.8255	gram dekameters
138.255	decigram dekameters
1,382.55	centigram dekameters
13,825.5	milligram dekameters
0.0050000	ton feet (short)
0.45359	kilogram feet
1.0	pound feet
4.53592	hektogram feet

FOOT POUNDS: (cont'd)

45.35916	dekagram feet
16	ounce feet
453.59157	gram feet
4,535.91566	decigram feet
45,359.15664	centigram feet
453,591.56642	milligram feet
0.060000	ton inches
5.44308	kilogram inches
12.0	pound inches
54.43104	hektogram inches
544.30992	dekagram inches
192	ounch inches
5,443.09984	gram inches
54,430.98792	decigram inches
544,309.87968	centigram inches
5,443,098.7968	milligram inches
0.015240	ton decimeters
1.38255	kilogram decimeters
3.04800	pound decimeters
13.8255	hektogram decimeters
138.255	dekagram decimeters
48.760	ounce decimeters
1,382.55	gram decimeters
13,825.5	decigram decimeters
138,255	centigram decimeters
1,382,550	milligram decimeters
0.15240	ton centimeters
13.8255	kilogram centimeters
30.480	pound centimeters
138.255	hektogram centimeters
1,382.55	dekagram centimeters
487.60	ounce centimeters
13,825.5	gram centimeters
138,255	decigram centimeters
1,382,550	centigram centimeters
13,825,500	milligram centimeters
1.5240	ton millimeters
138.255	kilogram millimeters
304.80	pound millimeters
1,382.55	hektogram millimeters
13,825.5	dekagram millimeters
4,876.0	ounce millimeters
138,255	gram millimeters
1,382,550	decigram millimeters
13,825,500	centigram millimeters
138,255,000	milligram millimeters
0.000013381	kiloliter-atmospheres
0.00013381	hektoliter-atmospheres
0.0013381	dekaliter-atmospheres
0.013381	liter-atmospheres
0.13381	deciliter-atmospheres

FOOT POUNDS: (cont'd)

1.3381	centiliter-atmospheres
13.381	milliliter-atmospheres
0.000000000000013381	cubic kilometer-atmospheres
0.000000000013381	cubic hektometer-atmospheres
0.000000013381	cubic dekameter-atmospheres
0.000013381	cubic meter-atmospheres
0.00047253	cubic feet-atmospheres
0.013381	cubic decimeter-atmospheres
13.381	cubic centimeter-atmospheres
13,381	cubic millimeter-atmospheres
1.35582	joules (absolute)
0.00000049814	Cheval-vapeur hours
0.0000000210438	horsepower day
0.00000050505	horsepower hours
0.0000303030	horsepower minutes
0.00181818	horsepower seconds
0.00000008808	pounds of carbon oxidized with 100% efficiency
0.0000013256	pounds of water evaporated from and at 212°F.
0.0012854	B.T.U.
13,558,200	ergs
13,558,200	centimeters dynes

FOOT POUND PER DAY: =

32.174	foot poundals per day
1.34058	foot poundals per hour
0.022343	foot poundals per minute
0.00037238	foot poundals per second
1.0	foot pounds per day
0.041667	foot pounds per hour
0.00069444	foot pounds per minute
0.000011574	foot pounds per second
0.00032389	kilogram calories per day
0.000013495	kilogram calories per hour
0.00000022492	kilogram calories per minute
0.0000000037487	kilogram calories per second
0.011425	ounce calories per day
0.00047604	ounce calories per hour
0.0000079340	ounce calories per minute
0.00000013223	ounce calories per second
0.138255	kilogram meters per day
0.0057606	kilogram meters per hour
0.00009601	kilogram meters per minute
0.0000016002	kilogram meters per second
0.013381	liter-atmospheres per day
0.00055754	liter-atmospheres per hour
0.0000092924	liter-atmospheres per minute
0.00000015487	liter-atmospheres per second
0.00047253	cubic foot atmospheres per day

FOOT POUND PER DAY: (cont'd)

0.000019689	cubic foot atmospheres per hour
0.00000032815	cubic foot atmospheres per minute
0.0000000054691	cubic foot atmospheres per second
0.000000021333	Cheval-vapeurs
0.000000021044	horsepower
0.000000015692	kilowatts
0.000015692	watts
1.35582	joules per day
0.056492	joules per hour
0.00094154	joules per minute
0.000015692	joules per second
0.00000008808	pounds of carbon oxidized with perfect efficiency per day
0.0000000036700	pounds of carbon oxidized with perfect efficiency per hour
0.0000000000611667	pounds of carbon oxidized with perfect efficiency per minute
0.0000000000010194	pounds of carbon oxidized with perfect efficiency per second
0.0000013256	pounds of water evaporated from and at 212°F. per day
0.000000055233	pounds of water evaporated from and at 212°F. per hour
0.00000000092056	pounds of water evaporated from and at 212°F. per minute
0.000000000015343	pounds of water evaporated from and at 212°F. per second
0.0012854	BTU per day
0.000053558	BTU per hour
0.00000089264	BTU per minute
0.000000014877	BTU per second

FOOT POUND PER HOUR: =

722.1760	foot poundals per day
32.174	foot poundals per hour
0.53623	foot poundals per minute
0.0089372	foot poundals per second
24.0	foot pounds per day
1.0	foot pounds per hour
0.016667	foot pounds per minute
0.00027778	foot pounds per second
0.0077734	kilogram calories per day
0.00032389	kilogram calories per hour
0.0000053982	kilogram calories per minute
0.000000089969	kilogram calories per second
0.2742	ounce calories per day
0.011425	ounce calories per hour
0.00019042	ounce calories per minute
0.0000031736	ounce calories per second
3.31812	kilogram meters per day

0.138255	kilogram meters per hour
0.0023042	kilogram meters per minute
0.000038404	kilogram meters per second
0.32114	liter-atmospheres per day
0.013381	liter-atmospheres per hour
0.00022302	liter-atmospheres per minute
0.0000037169	liter-atmospheres per second
0.011341	cubic foot atmospheres per day
0.00047253	cubic foot atmospheres per hour
0.0000078755	cubic foot atmospheres per minute
0.00000013126	cubic foot atmospheres per second
0.0000005120	Cheval-vapeurs
0.00000050505	horsepower
0.00000037661	kilowatts
0.00037661	watts
32.53968	joules per day
1.35582	joules per hour
0.022597	joules per minute
0.00037662	joules per second
0.0000021139	pounds of carbon oxidized with perfect efficiency per day
0.00000008808	pounds of carbon oxidized with perfect efficiency per hour
0.0000000014680	pounds of carbon oxidized with perfect efficiency per minute
0.000000000024467	pounds of carbon oxidized with perfect efficiency per second
0.000031814	pounds of water evaporated from and at 212°F. per day
0.0000013256	pounds of water evaporated from and at 212°F. per hour
0.000000022093	pounds of water evaporated from and at 212°F. per minute
0.00000000036822	pounds of water evaporated from and at 212°F. per second
0.030850	BTU per day
0.0012854	BTU per hour
0.000021423	BTU per minute
0.0000035706	BTU per second

FOOT POUND PER MINUTE: =

46,330.56	foot poundals per day
1,930.44	foot poundals per hour
32.174	foot poundals per minute
0.53623	foot poundals per second
1,440	foot pounds per day
60	foot pounds per hour
1.0	foot pounds per minute
0.016667	foot pounds per second
0.46640	kilogram calories per day

FOOT POUND PER MINUTE: (cont'd)

0.019433	kilogram calories per hour
0.00032389	kilogram calories per minute
0.0000053982	kilogram calories per second
16.452	ounce calories per day
0.6855	ounce calories per hour
0.011425	ounce calories per minute
0.00019042	ounce calories per second
199.0872	kilogram meters per day
8.29530	kilogram meters per hour
0.138255	kilogram meters per minute
0.0023043	kilogram meters per second
19.26864	liter-atmospheres per day
0.80286	liter-atmospheres per hour
0.013381	liter-atmospheres per minute
0.00022302	liter-atmospheres per second
0.68044	cubic foot atmospheres per day
0.028352	cubic foot atmospheres per hour
0.00047253	cubic foot atmospheres per minute
0.0000078755	cubic foot atmospheres per second
0.000030719	Cheval-vapeurs
0.000030303	horsepower
0.000022597	kilowatts
0.022597	watts
1,952.3808	joules per day
81.34920	joules per hour
1.35582	joules per minute
0.022597	joules per second
0.00012684	pounds of carbon oxidized with perfect efficiency per day
0.0000052848	pounds of carbon oxidized with perfect efficiency per hour
0.00000008808	pounds of carbon oxidized with perfect efficiency per minute
0.000000001468	pounds of carbon oxidized with perfect efficiency per second
0.0019089	pounds of water evaporated from and at 212°F. per day
0.000079536	pounds of water evaporated from and at 212°F. per hour
0.0000013256	pounds of water evaporated from and at 212°F. per minute
0.000000022093	pounds of water evaporated from and at 212°F. per second
1.85098	BTU per day
0.077124	BTU per hour
0.0012854	BTU per minute
0.000021423	BTU per second

FOOT POUND PER SECOND: =

2,779,833.6	foot poundals per day
115,826.40	foot poundals per hour

FOOT POUND PER SECOND: (cont'd)

1,930.440	foot poundals per minute
32.174	foot poundals per second
86,400	foot pounds per day
3,600	foot pounds per hour
60	foot pounds per minute
1.0	foot pounds per second
27.9841	kilogram calories per day
1.1660	kilogram calories per hour
0.019433	kilogram calories per minute
0.00032389	kilogram calories per second
987.120	ounce calories per day
41.130	ounce calories per hour
0.68550	ounce calories per minute
0.011425	ounce calories per second
11,945.2320	kilogram meters per day
497.7180	kilogram meters per hour
8.29530	kilogram meters per minute
0.138255	kilogram meters per second
1,156.11840	liter-atmospheres per day
48.17160	liter-atmospheres per hour
0.80286	liter-atmospheres per minute
0.013381	liter-atmospheres per second
40.82659	cubic foot atmospheres per day
1.70111	cubic foot atmospheres per hour
0.028352	cubic foot atmospheres per minute
0.00047253	cubic foot atmospheres per second
0.0018432	Cheval-vapeurs
0.0018182	horsepower
0.0013558	kilowatts
1.35582	watts
117,142.8480	joules per day
4,880.9520	joules per hour
81.34920	joules per minute
1.35582	joules per second
0.0076101	pounds of carbon oxidized with perfect efficiency per day
0.00031709	pounds of carbon oxidized with perfect efficiency per hour
0.0000052848	pounds of carbon oxidized with perfect efficiency per minute
0.00000008808	pounds of carbon oxidized with perfect efficiency per second
0.11453	pounds of water evaporated from and at 212°F. per day
0.0047722	pounds of water evaporated from and at 212°F. per hour
0.000079536	pounds of water evaporated from and at 212°F. per minute
0.0000013256	pounds of water evaporated from and at 212°F. per second
111.05856	BTU per day

FOOT POUND PER SECOND: (cont'd)

4.62744	BTU per hour
0.077124	BTU per minute
0.0012854	BTU per second

FOOT OF WATER AT 60°F: =

0.00304801	hektometers of water at 60°F.
0.0304801	dekameters of water at 60°F.
0.304801	meters of water at 60°F.
1.0	feet of water at 60°F.
12	inches of water at 60°F.
3.04801	decimeters of water at 60°F.
30.4801	centimeters of water at 60°F.
304.801	millimetrs of water at 60°F.
0.002240	hektometers of mercury at 32°F.
0.02240	dekameters of mercury at 32°F.
0.2240	meters of mercury at 32°F.
0.73491	feet of mercury at 32°F.
8.81888	ounces of mercury at 32°F.
0.882612	inches of mercury at 32°F.
2.240	decimeters of mercury at 32°F.
22.40	centimeters of mercury at 32°F.
224.0	millimeters of mercury at 32°F.
3,359.85556	tons per square hektometer
33.59856	tons per square dekameter
0.33599	tons per square meter
0.031214	tons per square foot
0.00021677	tons per square inch
0.0033599	tons per square decimeter
0.000033599	tons per square centimeter
0.00000033599	tons per square millimeter
3,048,010	kilograms per square hektometer
30,480.1	kilograms per square dekameter
304.801	kilograms per square meter
28.31705	kilograms per square foot
0.19665	kilograms per square inch
3.04801	kilograms per square decimeter
0.0304801	kilograms per square centimeter
0.000304801	kilograms per square millimeter
6,719,721.9	pounds per square hektometer
67,197.2	pounds per square dekameter
671.97219	pounds per square meter
62.42832	pounds per square foot
0.433530	pounds per square inch
6.71972	pounds per square decimeter
0.067197	pounds per square centimeter
0.00067197	pounds per square millimeter
30,480,100	hektograms per square hektometer
304,801	hektograms per square dekameter
3,048.01	hektograms per square meter
283.1705	hektograms per square foot

FOOT OF WATER AT 60°F: (cont'd)

1.9665	hektograms per square inch
30.4801	hektograms per square decimeter
0.304801	hektograms per square centimeter
0.00304801	hektograms per square millimeter
304,801,000	dekagrams per square hektometer
3,048,010	dekagrams per square dekameter
30,480.1	dekagrams per square meter
2,831.705	dekagrams per square foot
19.655	dekagrams per square inch
304.801	dekagrams per square decimeter
3.04801	dekagrams per square centimeter
0.0304801	dekagrams per square millimeter
107,515,550.4	ounces per square hektometer
1,075,155.504	ounces per square dekameter
10,751.55504	ounces per square meter
998.85312	ounces per square foot
6.93648	ounces per square inch
107.51556	ounces per square decimeter
1.075156	ounces per square centimeter
0.0107516	ounces per square millimeter
3,048,010,000	grams per square hektometer
30,480,100	grams per square dekameter
304,801	grams per square meter
28,317.05	grams per square foot
196.55	grams per square inch
3,048.01	grams per square decimeter
30.4801	grams per square centimeter
0.304801	grams per square millimeter
30,480,100,000	decigrams per square hektometer
304,801,000	decigrams per square dekameter
3,048,010	decigrams per square meter
283,170.5	decigrams per square foot
1,966.5	decigrams per square inch
30,480.1	decigrams per square decimeter
304.801	decigrams per square centimeter
3.04801	decigrams per square millimeter
304,801,000,000	centigrams per square hektometer
3,048,010,000	centigrams per square dekameter
30,480,100	centigrams per square meter
2,831,705	centigrams per square foot
19,665	centigrams per square inch
304,801	centigrams per square decimeter
3,048.01	centigrams per square centimeter
30.4801	centigrams per square millimeter
3,048,010,000,000	milligrams per square hektometer
30,480,100,000	milligrams per square dekameter
304,801,000	milligrams per square meter
28,317,050	milligrams per square foot
196,650	milligrams per square inch
3,048,010	milligrams per square decimeter
30,480.1	milligrams per square centimeter

FOOT OF WATER AT 60°F: (cont'd)

304.801	milligrams per square millimeter
0.029889	bars
0.0294979	atmospheres
821.2	feet of air @ 62°F. and 29.92 barometer pressure
2,988,874,717,500	dynes per square hektometer
29,888,747,175	dynes per square dekameter
298,887,471.75	dynes per square meter
27,767,658.99497	dynes per square foot
192,830.60410	dynes per square inch
2,988,874.7175	dynes per square decimeter
29,888.74718	dynes per square centimeter
298.88747	dynes per square millimeter

GALLON (DRY): =

0.00044040	kiloliters
0.00044040	cubic meters
0.0057601	cubic yards
0.027701	barrels
0.0044040	hektoliters
0.12497	bushels—U.S. (dry)
0.12116	bushels—Imperial (dry)
0.15553	cubic feet
0.44040	dekaliters
1.16342	gallons—U.S. (liquid)
1	gallons—U.S. (dry)
0.96874	gallons—Imperial
4.65368	quarts (liquid)
4	quarts (dry)
4.4040	liters
4.4040	cubic decimeters
9.30736	pints (fluid)
37.22943	gills (fluid)
44.04010	deciliters
268.75	cubic inches
440.40097	centiliters
4,404.00974	milliliters
4,404.00974	cubic centimeters
4,404,009.74	cubic millimeters
0.14139	sacks of cement
71,481	minims
148.91775	ounces (fluid)
1,191.34199	drams (fluid)
0.0043301	tons (long) of water @ 62°F.
0.0043996	tons (metric) of water @ 62°F.
0.0048498	tons (short) of water @ 62°F.
4.40005	kilograms of water @ 62°F.
9.69943	pounds (Avoir) of water @ 62°F.
11.78750	pounds (Troy) of water @ 62°F.
44.00054111	hektograms

UNIVERSAL CONVERSION FACTORS
GALLON (DRY): (cont'd)

440.0054111	dekagrams
141.45004	ounces (Troy) of water @ 62°F.
155.19091	ounces (Avoir.) of water @ 62°F.
1,131.60029	drams (Troy) of water @ 62°F.
2,483.05454	drams (Avoir.) of water @ 62°F.
2,829.00071	pennyweights of water @ 62°F.
3,394.80086	scruples (Avoir.) of water @ 62°F.
4,400.054111	grams of water @ 62°F.
44,000.54111	decigrams of water @ 62°F.
67,896.022703	grains of water @ 62°F.
440,005.41110	centigrams of water @ 62°F.
4,400,054.1197	milligrams of water @ 62°F.

GALLON (IMPERIAL): =

0.00045460	kiloliters
0.00045460	cubic meters
0.0059459	cubic yards
0.028594	barrels
0.045460	hektoliters
0.12900	bushels—U.S. (dry)
0.125066	bushels—Imperial (dry)
0.16054	cubic feet
0.45460	dekaliters
1.20094	gallons—U.S. (liquid)
1.032184	gallons—U.S. (dry)
1	gallon—Imperial
4.80376	quarts (liquid)
4.12820	quarts (dry)
4.54596	liters
4.54596	cubic decimeters
9.60752	pints (liquid)
38.43008	gills (liquid)
45.4596	deciliters
277.41714	cubic inches
454.596	centiliters
4,545.96	milliliters
4,545.96	cubic centimeters
4,545,960	cubic millimeters
0.14595	sacks of cement
73,785.7536	minims
153.72032	ounces (fluid)
1,299.76256	drams (fluid)
0.0044698	tons (long) of water @ 62°F.
0.0045415	tons (metric) of water @ 62°F.
0.0050061	tons (short) of water @ 62°F.
4.54196	kilograms of water @ 62°F.
10.012237	pounds (Avoir.) of water @ 62°F.
12.16765	pounds (Troy) of water @ 62°F.
45.41955	hektograms of water @ 62°F.
454.19551	dekagrams of water @ 62°F.

GALLON (IMPERIAL): (cont'd)

146.011774	ounces (Troy) of water @ 62°F.
160.19579	ounces (Avoir.) of water @ 62°F.
1,168.094195	drams (Troy) of water @ 62°F.
2,563.13262	drams (Avoir.) of water @ 62°F.
2,902.23549	pennyweights of water @ 62°F.
3,504.28258	scruples of water @ 62°F.
4,541.95508	grams of water @ 62°F.
45,419.5508	decigrams of water @ 62°F.
70,085.65746	grains of water @ 62°F.
454,195.508	centigrams of water @ 62°F.
4,541,955.08	milligrams of water @ 62°F.

GALLON (LIQUID): =

0.0037854	kiloliters
0.0037854	cubic meters
0.004951	cubic yards
0.0238095	barrels
0.037854	hektoliters
0.10742	bushels—U.S. (dry)
0.10414	bushels—Imperial (dry)
0.133681	cubic feet
0.37854	dekaliters
1	gallons—U.S. (liquid)
0.85948	gallons—U.S. (Dry)
0.83268	gallons—Imperial
4	quarts (liquid)
3.43747	quarts (dry)
3.78544	liters
3.78544	cubic decimeters
8	pints (liquid)
32	gills (liquid)
37.8544	deciliters
231	cubic inches
378.544	centiliters
3,785.44	milliliters
3,785.44	cubic centimeters
3,785,440	cubic millimeters
0.12153	sacks of cement
61,440	minims
128	ounces (fluid)
1,024	drams (fluid)
0.0037219	tons (long) of water @ 62°F.
0.0037816	tons (metric) of water @ 62°F.
0.0041685	tons (short) of water @ 62°F.
3.7820	kilograms of water @ 62°F.
8.337	pounds (Avoir.) of water @ 62°F.
10.13177	pounds (Troy) of water @ 62°F.
37.820	hektograms of water @ 62°F.
378.20	dekagrams of water @ 62°F.
121.58124	ounces (Troy) of water @ 62°F.

GALLON (LIQUID): (cont'd)

133.392	ounces (Avoir.) of water @ 62°F.
972.64992	drams (Troy) of water @ 62°F.
2,134.272	drams (Avoir.) of water @ 62°F.
2,431.6284	pennyweights of water @ 62°F.
2,917.94976	scruples (Avoir.) of water @ 62°F.
3,782	grams of water @ 62°F.
37,820	decigrams of water @ 62°F.
58,359	grains of water @ 62°F.
378,200	centigrams of water @ 62°F.
3,782,000	milligrams of water @ 62°F.

GALLON (U.S.) PER DAY: =

0.00037854	kiloliters per day
0.000015773	kiloliters per hour
0.00000026288	kiloliters per minute
0.0000000043813	kiloliters per second
0.00037854	cubic meters per day
0.000015773	cubic meters per hour
0.00000026288	cubic meters per minute
0.0000000043813	cubic meters per second
0.004951	cubic yards per day
0.00020629	cubic yards per hour
0.0000034382	cubic yards per minute
0.000000057303	cubic yards per second
0.02381	barrels per day
0.00099208	barrels per hour
0.000016535	barrels per minute
0.00000027558	barrels per second
0.0037854	hektoliters per day
0.00015773	hektoliters per hour
0.0000026288	hektoliters per minute
0.000000043813	hektoliters per second
0.10742	bushels—U.S. (dry) per day
0.0044758	bushels—U.S. (dry) per hour
0.000074597	bushels—U.S. (dry) per minute
0.0000012433	bushels—U.S. (dry) per second
0.10414	bushels (Imperial) per day
0.00433917	bushels (Imperial) per hour
0.0000723194	bushels (Imperial) per minute
0.00000120532	bushels (Imperial) per second
0.133681	cubic feet per day
0.00557002	cubic feet per hour
0.0000928337	cubic feet per minute
0.00000154723	cubic feet per second
0.378544	dekaliters per day
0.0157725	dekaliters per hour
0.000262875	dekaliters per minute
0.00000438128	dekaliters per second
1.0	gallons—U.S. (liquid) per day
0.0416667	gallons—U.S. (liquid) per hour

GALLON (U.S.) PER DAY: (cont'd)

0.000694444	gallons—U.S. (liquid) per minute
0.0000115741	gallons—U.S. (liquid) per second
0.83268	gallons (Imperial) per day
0.0346950	gallons (Imperial) per hour
0.000578250	gallons (Imperial) per minute
0.00000963750	gallons (Imperial) per second
3.78544	liters per day
0.157726	liters per hour
0.00262877	liters per minute
0.0000438128	liters per second
3.78543	cubic decimeters per day
0.157726	cubic decimeters per hour
0.00262877	cubic decimeters per minute
0.0000438128	cubic decimeters per second
4.0	quarts per day
0.166667	quarts per hour
0.00277778	quarts per minute
0.0000462963	quarts per second
8.0	pints per day
0.333333	pints per hour
0.00555556	pints per minute
0.0000925925	pints per second
32.0	gills per day
1.333333	gills per hour
0.0222222	gills per minute
0.000370370	gills per second
37.8544	deciliters per day
1.57726	deciliters per hour
0.0262877	deciliters per minute
0.000438128	deciliters per second
231.0	cubic inches per day
9.625	cubic inches per hour
0.160417	cubic inches per minute
0.00267361	cubic inches per second
378.544	centiliters per day
15.772608	centiliters per hour
0.262877	centiliters per minute
0.00438128	centiliters per second
3,785.44	milliliters per day
157.726080	milliliters per hour
2.628768	milliliters per minute
0.0438128	milliliters per second
3,785.44	cubic centimeters per day
157.726080	cubic centimeters per hour
2.628768	cubic centimeters per minute
0.0438128	cubic centimeters per second
3,785,440	cubic millimeters per day
157,726.249920	cubic millimeters per hour
2,628.770832	cubic millimeters per minute
43.8128	cubic millimeters per second

GALLON PER HOUR: =

0.0908506	kiloliters per day
0.00378544	kiloliters per hour
0.0000630907	kiloliters per minute
0.00000105151	kiloliters per second
0.0908506	cubic meters per day
0.00378544	cubic meters per hour
0.0000630907	cubic meters per minute
0.00000105151	cubic meters per second
0.11882	cubic yards per day
0.004951	cubic yards per hour
0.000082517	cubic yards per minute
0.0000013753	cubic yards per second
0.57144	barrels per day
0.0238095	barrels per hour
0.00039683	barrels per minute
0.0000066139	barrels per second
0.908506	hektoliters per day
0.0378544	hektoliters per hour
0.000630907	hektoliters per minute
0.0000105151	hektoliters per second
2.57808	bushels—U.S. (dry) per day
0.10742	bushels—U.S. (dry) per hour
0.0017903	bushels—U.S. (dry) per minute
0.000029839	bushels—U.S. (dry) per second
2.49936	bushels (Imperial) per day
0.10414	bushels (Imperial) per hour
0.0017357	bushels (Imperial) per minute
0.000028928	bushels (Imperial) per second
3.20834	cubic feet per day
0.133681	cubic feet per hour
0.0022280	cubic feet per minute
0.000037134	cubic feet per second
9.0850560	dekaliters per day
0.378544	dekaliters per hour
0.00630907	dekaliters per minute
0.000105151	dekaliters per second
24	gallons—U.S. (liquid) per day
1.0	gallons—U.S. (liquid) per hour
0.016667	gallons—U.S. (liquid) per minute
0.00027778	gallons—U.S. (liquid) per second
19.98432	gallons (Imperial) per day
0.83268	gallons (Imperial) per hour
0.013878	gallons (Imperial) per minute
0.0002313	gallons (Imperial) per second
90.850560	liters per day
3.78544	liters per hour
0.0630907	liters per minute
0.00105151	liters per second
90.850560	cubic decimeters per day

GALLON PER HOUR: (cont'd)

3.785440	cubic decimeters per hour
0.0630907	cubic decimeters per minute
0.00105151	cubic decimeters per second
96	quarts per day
4.0	quarts per hour
0.066667	quarts per minute
0.0011111	quarts per second
192	pints per day
8.0	pints per hour
0.13333	pints per minute
0.0022222	pints per second
768	gills per day
32.0	gills per hour
0.53333	gills per minute
0.0088889	gills per second
908.505600	deciliters per day
37.854400	deciliters per hour
0.630907	deciliters per minute
0.0105151	deciliters per second
5,544	cubic inches per day
231	cubic inches per hour
3.85	cubic inches per minute
0.064167	cubic inches per second
9,085.055999	centiliters per day
378.544000	centiliters per hour
6.309067	centiliters per minute
0.105151	centiliters per second
90,850.559990	milliliters per day
3,785.440	milliliters per hour
63.090667	milliliters per minute
1.051511	milliliters per second
90,850.559990	cubic centimeters per day
3,785.440	cubic centimeters per hour
63.090667	cubic centimeters per minute
1.051511	cubic centimeters per second
90,850,560	cubic millimeters per day
3,785,440	cubic millimeters per hour
63,090.666660	cubic millimeters per minute
1,051.511111	cubic millimeters per second

GALLON PER MINUTE: =

5.451034	kiloliters per day
0.227126	kiloliters per hour
0.00378544	kiloliters per minute
0.0000630907	kiloliters per second
5.451034	cubic meters per day
0.227126	cubic meters per hour
0.00378544	cubic meters per minute
0.0000630907	cubic meters per second
7.129440	cubic yards per day

GALLON PER MINUTE: (cont'd)

0.297060	cubic yards per hour
0.004951	cubic yards per minute
0.0000825167	cubic yards per second
34.285680	barrels per day
1.428570	barrels per hour
0.0238095	barrels per minute
0.000396825	barrels per second
54.510336	hektoliters per day
2.271264	hektoliters per hour
0.0378544	hektoliters per minute
0.000630907	hektoliters per second
154.6848	bushels—U.S. (dry) per day
6.4452	bushels—U.S. (dry) per hour
0.10742	bushels—U.S. (dry) per minute
0.00179033	bushels—U.S. (dry) per second
149.9616	bushels (Imperial) per day
6.2484	bushels (Imperial) per hour
0.10414	bushels (Imperial) per minute
0.00173567	bushels (Imperial) per second
192.500640	cubic feet per day
8.020860	cubic feet per hour
0.133681	cubic feet per minute
0.00222802	cubic feet per second
545.103359	dekaliters per day
22.712640	dekaliters per hour
0.378544	dekaliters per minute
0.00630907	dekaliters per second
1,440	gallons—U.S. (liquid) per day
60	gallons—U.S. (liquid) per hour
1.0	gallons—U.S. (liquid) per minute
0.0166667	gallons—U.S. (liquid) per second
1,199.05920	gallons (Imperial) per day
49.960800	gallons (Imperial) per hour
0.83268	gallons (Imperial) per minute
0.0138780	gallons (Imperial) per second
5,451.033594	liters per day
227.126400	liters per hour
3.78544	liters per minute
0.0630907	liters per second
5,451.033594	cubic decimeters per day
227.126400	cubic decimeters per hour
3.78544	cubic decimeters per minute
0.0630907	cubic decimeters per second
5,760	quarts per day
240	quarts per hour
4.0	quarts per minute
0.0666667	quarts per second
11,520	pints per day
480	pints per hour
8.0	pints per minute
0.133333	pints per second

GALLON PER MINUTE: (cont'd)

46,080 . . . . gills per day
1,920 . . . . gills per hour
32.0 . . . . gills per minute
0.533333 . . . . gills per second
54,510.335942 . . . . deciliters per day
2,271.263998 . . . . deciliters per hour
37.854400 . . . . deciliters per minute
0.630907 . . . . deciliters per second
332,640 . . . . cubic inches per day
13,860 . . . . cubic inches per hour
231.0 . . . . cubic inches per minute
3.850 . . . . cubic inches per second
545,103.359424 . . . . centiliters per day
22,712.639976 . . . . centiliters per hour
378.544000 . . . . centiliters per minute
6.309067 . . . . centiliters per second
5,451,034 . . . . milliliters per day
227,126.399760 . . . . milliliters per hour
3,785.440 . . . . milliliters per minute
63.090667 . . . . milliliters per second
5,451,034 . . . . cubic centimeters per day
227,126.399760 . . . . cubic centimeters per hour
3,785.440 . . . . cubic centimeters per minute
63.090667 . . . . cubic centimeters per second
5,451,033,594 . . . . cubic millimeters per day
227,126,400 . . . . cubic millimeters per hour
3,785,440 . . . . cubic millimeters per minute
63,090.6666 . . . . cubic millimeters per second

GALLON PER SECOND: =

327.062016 . . . . kiloliters per day
13.627584 . . . . kiloliters per hour
0.227126 . . . . kiloliters per minute
0.00378544 . . . . kiloliters per second
327.062016 . . . . cubic meters per day
13.627584 . . . . cubic meters per hour
0.227126 . . . . cubic meters per minute
0.00378544 . . . . cubic meters per second
427.766400 . . . . cubic yards per day
17.823600 . . . . cubic yards per hour
0.297060 . . . . cubic yards per minute
0.004951 . . . . cubic yards per second
2,057.140800 . . . . barrels per day
85.714200 . . . . barrels per hour
1.428570 . . . . barrels per minute
0.0238095 . . . . barrels per second
3,270.620160 . . . . hektoliters per day
136.275840 . . . . hektoliters per hour
2.271264 . . . . hektoliters per minute
0.0378544 . . . . hektoliters per second

GALLON PER SECOND: (cont'd)

9,281.08800	bushels—U.S. (dry) per day
386.712000	bushels—U.S. (dry) per hour
6.44520	bushels—U.S. (dry) per minute
0.10742	bushels—U.S. (dry) per second
8,997.696000	bushels (Imperial) per day
374.90400	bushels (Imperial) per hour
6.248400	bushels (Imperial) per minute
0.10414	bushels (Imperial) per second
11,550.038400	cubic feet per day
481.251600	cubic feet per hour
8.020860	cubic feet per minute
0.133681	cubic feet per second
32,706.201600	dekaliters per day
1,362.758400	dekaliters per hour
22.712640	dekaliters per minute
0.378544	dekaliters per second
86,400	gallons—U.S. (liquid) per day
3,600	gallons—U.S. (liquid) per hour
60	gallons—U.S. (liquid) per minute
1.0	gallons—U.S. (liquid) per second
71,943.55200	gallons (Imperial) per day
2,997.64800	gallons (Imperial) per hour
49.96080	gallons (Imperial) per minute
0.83268	gallons (Imperial) per second
327,062.01600	liters per day
13,627.58400	liters per hour
227.12640	liters per minute
3.78544	liters per second
327,062.01600	cubic decimeters per day
13,627.58400	cubic decimeters per hour
227.12640	cubic decimeters per minute
3.78544	cubic decimeters per second
345,600	quarts per day
14,400	quarts per hour
240	quarts per minute
4.0	quarts per second
691,200	pints per day
28,800	pints per hour
480	pints per minute
8.0	pints per second
2,764,800	gills per day
115,200	gills per hour
1,920	gills per minute
32.0	gills per second
3,270,620	deciliters per day
136,275.8400	deciliters per hour
2,271.2640	deciliters per minute
37.8544	deciliters per second
19,958,400	cubic inches per day
831,600	cubic inches per hour
13,860	cubic inches per minute

GALLON PER SECOND: (cont'd)

231	cubic inches per second
32,706,202	centiliters per day
1,362,758	centiliters per hour
22,712.640	centiliters per minute
278.544	centiliters per second
327,062,016	milliliters per day
13,627,584	milliliters per hour
227,126.40	milliliters per minute
3,785.44	milliliters per second
327,062,016	cubic centimeters per day
13,627,584	cubic centimeters per hour
227,126.40	cubic centimeters per minute
3,785.44	cubic centimeters per second
327,061,016,000	cubic millimeters per day
13,627,584,000	cubic millimeters per hour
227,126,400	cubic millimeters per minute
3,785,440	cubic millimeters per second

GRAIN (AVOIRDUPOIS): =

0.000000637755089	tons (long)
0.000000647989857	tons (metric)
0.0000007142857	tons (short)
0.000064798918	kilograms
0.00014285714	pounds (Avoir.)
0.00017361111	pounds (Troy)
0.00064798918	hektograms
0.0064798918	dekagrams
0.00208333	ounces (Troy)
0.0022857	ounces (Avoir.)
0.03657143	drams (Avoir.)
0.0166667	drams (Troy)
0.0416667	pennyweights (Troy)
0.05000	scruples (Troy)
0.064798918	grams
0.64798918	decigrams
6.4798918	centigrams
64.798918	milligrams
0.3240	carats (metric)

GRAIN (AVOIR.) PER BARREL: =

718.956	parts per million
0.000793548	tons (net) per cubic meter
0.71988	kilograms per cubic meter
1.587054	pounds (Avoir.) per cubic meter
25.39320	ounces (Avoir.) per cubic meter
406.29120	drams (Avoir.) per cubic meter
719.88	grams per cubic meter
0.000126161	tons (net) per barrel

GRAIN (AVOIR.) PER BARREL: (cont'd)

0.114449	kilograms per barrel
0.252316	pounds (Avoir.) per barrel
4.037124	ounces (Avoir.) per barrel
64.59390	drams (Avoir.) per barrel
114.44916	grams per barrel
0.00000300384	tons (net) per gallon—U.S.
0.00272498	kilograms per gallon—U.S.
0.00600753	pounds (Avoir.) per gallon—U.S.
0.096122	ounces (Avoir.) per gallon—U.S.
1.53795	drams (Avoir.) per gallon—U.S.
2.724982	grams per gallon—U.S.
0.000000793548	tons (net) per liter
0.00071988	kilograms per liter
0.00158705	pounds (Avoir.) per liter
0.0253932	ounces (Avoir.) per liter
0.406291	drams (Avoir.) per liter
0.71988	grams per liter
0.000000000793548	tons (net) per cubic centimeter
0.00000071980	kilograms per cubic centimeter
0.00000158705	pounds (Avoir.) per cubic centimeter
0.0000253932	ounces (Avoir.) per cubic centimeter
0.000406291	drams (Avoir.) per cubic centimeter
0.00071988	grams per cubic centimeter

GRAIN (AVOIR.) PER CUBIC CENTIMETER: =

0.00452219	parts per million
0.00000000499137	tons (net) per cubic meter
0.00000452800	kilograms per cubic meter
0.00000998248	pounds (Avoir.) per cubic meter
0.000159722	ounces (Avoir.) per cubic meter
0.00255555	drams (Avoir.) per cubic meter
0.00452800	grams
0.000000000793545	tons (net) per barrel
0.000000719878	kilograms per barrel
0.00000158705	pounds (Avoir.) per barrel
0.0000253933	ounces (Avoir.) per barrel
0.000406292	drams (Avoir.) per barrel
0.000719879	grams per barrel
0.0000000000188940	tons (net) per gallon—U.S.
0.00000001714	kilograms per gallon—U.S.
0.0000000377873	pounds (Avoir.) per gallon—U.S.
0.000000604602	ounces (Avoir.) per gallon—U.S.
0.00000967362	drams (Avoir.) per gallon—U.S.
0.00001714	grams per gallon—U.S.
0.00000000000499137	tons (net) per liter
0.00000000452800	kilograms per liter
0.00000000998248	pounds (Avoir.) per liter
0.000000159722	ounces (Avoir.) per liter
0.00000255555	drams (Avoir.) per liter
0.00000452800	grams per liter

GRAIN (AVOIR.) PER CUBIC CENTIMETER: (cont'd)

0.0000000000000499137 . . . . tons (net) per cubic centimeter
0.00000000000452800 . . . . kilograms per cubic centimeter
0.00000000000998248 . . . . pounds (Avoir.) per cubic centimeter
0.000000000159722 . . . . ounces (Avoir.) per cubic centimeter
0.00000000255555 . . . . drams (Avoir.) per cubic centimeter
0.00000000452800 . . . . grams per cubic centimeter

GRAIN (AVOIR.) PER CUBIC METER: =

4,522.192260 . . . . parts per million
0.00499137 . . . . tons (net) per cubic meter
4.528004 . . . . kilograms per cubic meter
9.982479 . . . . pounds (Avoir.) per cubic meter
159.721781 . . . . ounces (Avoir.) per cubic meter
2,555.548489 . . . . drams (Avoir.) per cubic meter
4,528.004167 . . . . grams per cubic meter
0.000793545 . . . . tons (net) per barrel
0.71978 . . . . kilograms per barrel
1.587053 . . . . pounds (Avoir.) per barrel
25.393280 . . . . ounces (Avoir.) per barrel
406.291949 . . . . drams (Avoir.) per barrel
719.878693 . . . . grams per barrel
0.0000188940 . . . . tons (net) per gallon—U.S.
0.01714 . . . . kilograms per gallon—U.S.
0.0377873 . . . . pounds (Avoir.) per gallon—U.S.
0.604602 . . . . ounces (Avoir.) per gallon—U.S.
9.673618 . . . . drams (Avoir.) per gallon—U.S.
17.14 . . . . grams per gallon—U.S.
0.00000499137 . . . . tons (net) per liter
0.00452800 . . . . kilograms per liter
0.00998248 . . . . pounds (Avoir.) per liter
0.159722 . . . . ounces (Avoir.) per liter
2.555548 . . . . drams (Avoir.) per liter
4.528004 . . . . grams per liter
0.00000000499137 . . . . tons (net) per cubic centimeter
0.00000452800 . . . . kilograms per cubic centimeter
0.00000998248 . . . . pounds (Avoir.) per cubic centimeter
0.000159722 . . . . ounces (Avoir.) per cubic centimeter
0.00255555 . . . . drams (Avoir.) per cubic centimeter
0.00452800 . . . . grams

GRAIN (AVOIR.) PER GALLON—U.S.: =

17.118 . . . . parts per million
0.000018894 . . . . tons (net) per cubic meter
0.01714 . . . . kilograms per cubic meter
0.037787 . . . . pounds (Avoir.) per cubic meter
0.60460 . . . . ounces (Avoir.) per cubic meter
9.67360 . . . . drams (Avoir.) per cubic meter
17.14 . . . . grams per cubic meter

GRAIN (AVOIR.) PER GALLON—U.S.: (cont'd)

0.00000300384 .... tons (net) per barrel
0.00272498 .... kilograms per barrel
0.00600753 .... pounds (Avoir.) per barrel
0.096122 .... ounces (Avoir.) per barrel
1.53795 .... drams (Avoir.) per barrel
2.72498 .... grams per barrel
0.000000071520 .... tons (net) per gallon—U.S.
0.000064881 .... kilograms per gallon—U.S.
0.00014304 .... pounds (Avoir.) per gallon—U.S.
0.0022886 .... ounces (Avoir.) per gallon—U.S.
0.036617 .... drams (Avoir.) per gallon—U.S.
0.064881 .... grams per gallon—U.S.
0.000000018894 .... tons (net) per liter
0.00001714 .... kilograms per liter
0.000037787 .... pounds (Avoir.) per liter
0.00060460 .... ounces (Avoir.) per liter
0.0096735 .... drams (Avoir.) per liter
0.01714 .... grams per liter
0.000000000018894 .... tons (net) per cubic centimeter
0.00000001714 .... kilograms per cubic centimeter
0.000000037787 .... pounds (Avoir.) per cubic centimeter
0.00000060460 .... ounces (Avoir.) per cubic centimeter
0.0000096735 .... drams (Avoir.) per cubic centimeter
0.00001714 .... grams per cubic centimeter

GRAIN (AVOIR.) PER LITER: =

4.522192 .... parts per million
0.00000499137 .... tons (net) per cubic meter
0.00452800 .... kilograms per cubic meter
0.00998248 .... pounds (Avoir.) per cubic meter
0.159722 .... ounces (Avoir.) per cubic meter
2.555548 .... drams (Avoir.) per cubic meter
4.52804 .... grams per cubic meter
0.00000793545 .... tons (net) per barrel
0.000719878 .... kilograms per barrel
0.00158705 .... pounds (Avoir.) per barrle
0.0253933 .... ounces (Avoir). per barrel
0.406292 .... drams (Avoir.) per barrel
0.719879 .... grams per barrel
0.0000000188940 .... tons (net) per gallon—U.S.
0.00001714 .... kilograms per gallon—U.S.
0.0000377873 .... pounds (Avoir.) per gallon—U.S.
0.00604602 .... ounces (Avoir.) per gallon—U.S.
0.00967362 .... drams (Avoir.) per gallon—U.S.
0.01714 .... grams per gallon—U.S.
0.0000000049937 .... tons (net) per liter
0.00000452800 .... kilograms per liter
0.00000998248 .... pounds (Avoir.) per liter
0.000159722 .... ounces (Avoir.) per liter
0.00255555 .... drams (Avoir.) per liter

GRAIN (AVOIR.) PER LITER: (cont'd)

0.00452800	grams per liter
0.00000000000499137	tons (net) per cubic centimeter
0.00000000452800	kilograms per cubic centimeter
0.00000000998248	pounds (Avoir.) per cubic centimeter
0.000000159722	ounces (Avoir.) per cubic centimeter
0.00000255555	drams (Avoir.) per cubic centimeter
0.00000452800	grams per cubic centimeter

GRAM: =

0.00000098426	tons (long)
0.000001	tons (metric)
0.00000110231	tons (short)
0.001	kilograms
0.00220462	pounds (Avoir.)
0.00267923	pounds (Troy)
0.01	hektograms
0.1	dekagrams
0.0321507	ounces (Troy)
0.03527392	ounces (Avoir.)
0.257206	drams (Troy)
0.564383	drams (Avoir.)
0.6430149	pennyweights
0.771618	scruples
5.0	carats (metric)
10.0	decigrams
15.4324	grains
100	centigrams
1,000	milligrams

GRAM CALORIE (MEAN): =

99.334	foot poundals
1,192.008	inch poundals
3.0874	foot pounds
37.0488	inch pounds
0.0000001	ton (short) calories
0.001	kilogram calories
0.00204622	pound calories
0.01	hektogram calories
0.1	dekagram calories
0.0327395	ounce calories
1.0	gram calorie
10.0	decigram calories
100.0	centigram calories
1,000	milligram calories
0.00000004845	kilowatt days
0.0000011628	kilowatt hours
0.0000697680	kilowatt minutes
0.004186	kilowatt seconds

0.00004845	watt days
0.0011628	watt hours
0.0697680	watt minutes
4.186	watt seconds
0.00042685	ton meters
0.42685	kilogram meters
0.941043	pound meters
4.2685	hektogram meters
42.685	dekagram meters
15.056688	ounce meters
426.85	gram meters
4,268.5	decigram meters
42,685	centigram meters
426,850	milligram meters
0.0000042685	ton hektometers
0.0042685	kilogram hektometers
0.00941043	pound hektometers
0.042685	hektogram hektometers
0.42685	dekagram hektometers
0.150567	ounce hektometers
4.2685	gram hektometers
42.685	decigram hektometers
426.85	centigram hektometers
4,268.5	milligram hektometers
0.000042685	ton dekameters
0.042685	kilogram dekameters
0.0941043	pound dekameters
0.42685	hektogram dekameters
4.2685	dekagram dekameters
1.50567	ounce dekameters
42.685	gram dekameters
426.85	decigram dekameters
4,268.5	centigram dekameters
42,685	milligram dekameters
0.00140042	ton feet
1.400424	kilogram feet
0.0308704	pound feet
14.004236	hektogram feet
140.042357	dekagram feet
0.493985	ounce feet
1,400.423566	gram feet
14,004.235661	decigram feet
140,042.356605	centigram feet
1,400,423.0	milligram feet
0.0168050	ton inches
16.805083	kilogram inches
37.0488557	pound inches
168.050828	hektogram inches
1,680.0508284	dekagram inches
592.78169	ounce inches
16,805.082792	gram inches

GRAM CALORIE (MEAN): (cont'd)

168,050.82792	decigram inches
1,680,508	centigram inches
16,805,082	milligram inches
0.0042685	ton decimeters
4.2685	kilogram decimeters
9.410430	pound decimeters
42.685	hektogram decimeters
426.85	dekagram decimeters
150.566885	ounce decimeters
4,268.5	gram decimeters
42,685	decigram decimeters
426,850	centigram decimeters
4,268,500	milligram decimeters
0.042685	ton centimeters
42.685	kilogram centimeters
94.104303	pound centimeters
426.85	hektogram centimeters
4,268.5	dekagram centimeters
1,505.668846	ounce centimeters
42,685.0	gram centimeters
426,850	decigram centimeters
4,268,500	centigram centimeters
42,685,000	milligram centimeters
0.42685	ton millimeters
426.85	kilogram millimeters
941.043029	pound millimeters
4,268.5	hektogram millimeters
42,685	dekagram millimeters
15,056.688464	ounce millimeters
426,850	gram millimeters
4,268,500	decigram millimeters
42,685,000	centigram millimeters
426,850,000	milligram millimeters
0.000041311	kiloliter-atmospheres
0.00041311	hektoliter-atmospheres
0.0041311	dekaliter-atmospheres
0.041311	liter-atmospheres
0.41311	deciliter-atmospheres
4.1311	centiliter-atmospheres
41.311	milliliter-atmospheres
0.0000000000000415977	cubic kilometer-atmospheres
0.0000000000415977	cubic hektometer-atmospheres
0.0000000415977	cubic dekameter-atmospheres
0.0000415977	cubic meter-atmospheres
0.001469	cubic feet-atmospheres
2.538415	cubic inch-atmospheres
0.0415977	cubic decimeter-atmospheres
41.597673	cubic centimeter-atmospheres
41,597.673	cubic millimeter-atmospheres
4.185829	joules per gram
0.00000158097	Cheval-vapeur hours

GRAM CALORIE (MEAN): (cont'd)

0.0000000649708 .... horsepower days
0.0000015593 .... horsepower hours
0.0000935980 .... horsepower minutes
0.00561588 .... horsepower seconds
0.000000271842 .... pounds of carbon oxidized
with 100% efficiency
0.00000408756 .... pounds of water evaporated from
and at 212°F.
1.8 .... BTU (mean) per pound
0.0039685 .... BTU
41,858,291 .... ergs

GRAM CALORIE: =
(15°C. per square centimeter per second for temperature gradient of 1°C. per centimeter): =

4.185829 .... joules (absolute) per square
centimeter per second for temperature
gradient of 1°C per centimeter
0.80620 .... BTU (mean) per square foot per
second for a temperature gradient of
1°F. per inch

GRAM WEIGHT SECOND PER SQUARE CENTIMETER: =

980.665 .... poises
98,066.5 .... centipoises

GRAM PER CUBIC CENTIMETER: =

100,000 .... kilograms per cubic meter
2,831.7 .... kilograms per cubic foot
16.38776 .... kilograms per cubic inch
0.001 .... kilograms per cubic centimeter
2,204.62 .... pounds per cubic meter
62.42822 .... pounds per cubic foot
0.036127 .... pounds per cubic inch
0.0022046 .... pounds per cubic centimeter
35,273.92 .... ounces per cubic meter
998.85159 .... ounces per cubic foot
0.57804 .... ounces per cubic inch
0.035274 .... ounces per cubic centimeter
1,000,000 .... grams per cubic meter
28,317.0 .... grams per cubic foot
16.38705 .... grams per cubic inch
1.0 .... grams per cubic centimeter
15,432,400 .... grains per cubic meter
436,999.27080 .... grains per cubic inch
252.89148 .... grains per cubic inch
15.4324 .... grains per cubic centimeter

GRAM PER CUBIC FOOT: =

0.035314	kilograms per cubic meter
0.001	kilograms per cubic foot
0.00000057870	kilograms per cubic inch
0.000000035314	kilograms per cubic centimeter
0.077854	pounds per cubic meter
0.0022046	pounds per cubic foot
0.0000012758	pounds per cubic inch
0.000000077854	pounds per cubic centimeter
1.24568	ounces per cubic meter
0.035274	ounces per cubic foot
0.000020413	ounces per cubic inch
0.00000124578	ounces per cubic centimeter
35.31444	grams per cubic meter
1.0	grams per cubic foot
0.00057870	grams per cubic inch
0.0000353144	grams per cubic centimeter
544.98657	grains per cubic meter
15.4324	grains per cubic foot
0.0089307	grains per cubic inch
0.00054499	grains per cubic centimeter

GRAM PER CUBIC INCH: =

61.203	kilograms per cubic meter
1.72799	kilograms per cubic foot
0.001	kilograms per cubic inch
0.000061203	kilograms per cubic centimeter
134.53131	pounds per cubic meter
3.80952	pounds per cubic foot
0.0022046	pounds per cubic inch
0.00013453	pounds per cubic centimeter
2,152.52530	ounces per cubic meter
60.95306	ounces per cubic foot
0.035274	ounces per cubic inch
0.0021525	ounces per cubic centimeter
610,230	grams per cubic meter
17,279.88291	grams per cubic foot
1.0	grams per cubic inch
0.61023	grams per cubic centimeter
941,731.3452	grains per cubic meter
26,667.0065002	grains per cubic foot
15.4324	grains per cubic inch
0.94173	grains per cubic centimeter

GRAM PER CUBIC METER: =

0.001	kilograms per cubic meter
0.00028317	kilograms per cubic foot

GRAM PER CUBIC METER: (cont'd)

0.00000016387 .... kilograms per cubic inch
0.000000001 .... kilograms per cubic centimeter
0.00220462 .... pounds per cubic meter
0.000062428 .... pounds per cubic foot
0.000000036127 .... pounds per cubic inch
0.00000000220462 .... pounds per cubic centimeter
0.035274 .... ounces per cubic meter
0.00099885 .... ounces per cubic foot
0.00000057803 .... ounces per cubic inch
0.000000035274 .... ounces per cubic centimeter
1.0 .... grams per cubic meter
0.028317 .... grams per cubic foot
0.000016387 .... grams per cubic inch
0.000001 .... grams per cubic centimeter
15.4324 .... grains per cubic meter
0.437 .... grains per cubic foot
0.00025289 .... grains per cubic inch
0.000015434 .... grains per cubic centimeter

GRAM PER CENTIMETER: =

0.100 .... kilograms per meter
0.03048006 .... kilograms per foot
0.00254005 .... kilograms per inch
0.001 .... kilograms per centimeter
0.22046 .... pounds per meter
0.067197 .... pounds per foot
0.0054014 .... pounds per inch
0.0022046 .... pounds per centimeter
3.52736 .... ounces per meter
1.075152 .... ounces per foot
0.086422 .... ounces per inch
0.035274 .... ounces per centimeter
100 .... grams per meter
30.48006 .... grams per foot
2.54005 .... grams per inch
1.0 .... grams per centimeter
1,543.24 .... grains per meter
470.38048 .... grains per foot
39.19822 .... grains per inch
15.4324 .... grains per centimeter

GRAM PER FOOT: =

0.0032808 .... kilograms per meter
0.001 .... kilograms per foot
0.000083333 .... kilograms per inch
0.000032808 .... kilograms per centimeter
0.0072329 .... pounds per meter
0.0022046 .... pounds per foot

GRAM PER FOOT: (cont'd)

0.00018372 . . . . pounds per inch
0.000072329 . . . . pounds per centimeter
0.11573 . . . . ounces per meter
0.035274 . . . . ounces per foot
0.0029395 . . . . ounces per inch
0.0011573 . . . . ounces per centimeter
3.28083 . . . . grams per centimeter
1.0 . . . . grams per foot
0.083333 . . . . grams per inch
0.0328083 . . . . grams per centimeter
50.63113 . . . . grains per meter
15.4324 . . . . grains per foot
1.28603 . . . . grains per inch
0.50631 . . . . grains per centimeter

GRAM PER INCH: =

0.03937 . . . . kilograms per meter
0.012 . . . . kilograms per foot
0.001 . . . . kilograms per inch
0.00039370 . . . . kilograms per centimeter
0.086796 . . . . pounds per meter
0.026455 . . . . pounds per foot
0.0022046 . . . . pounds per inch
0.00086796 . . . . pounds per centimeter
1.38874 . . . . ounces per meter
0.42329 . . . . ounces per foot
0.035274 . . . . ounces per inch
0.013887 . . . . ounces per centimeter
39.37000 . . . . grams per meter
12.0 . . . . grams per foot
1.0 . . . . grams per inch
0.39370 . . . . grams per centimeter
607.57359 . . . . grains per meter
185.1888 . . . . grains per foot
15.4324 . . . . grains per inch
6.075736 . . . . grains per centimeter

GRAM PER LITER: =

8.34543 . . . . pounds per 1000 gallons
1,000 . . . . parts per million
0.00110230 . . . . tons (net) per cubic meter
1.0 . . . . kilograms per cubic meter
2.2046099 . . . . pounds (Avoir.) per cubic meter
35.273758 . . . . ounces (Avoir.) per cubic meter
564.379806 . . . . drams (Avoir.) per cubic meter
1,000 . . . . grams per cubic meter
15,432.258039 . . . . grains per cubic meter
0.0000312140 . . . . tons (net) per cubic foot
0.0283169 . . . . kilograms per cubic foot

GRAM PER LITER: (cont'd)

0.0624280 . . . . pounds (Avoir.) per cubic foot
0.998848 . . . . ounces (Avoir.) per cubic foot
15.981559 . . . . drams (Avoir.) per cubic foot
28.316846 . . . . grams per cubic foot
436.995689 . . . . grains per cubic foot
0.000175254 . . . . tons (net) per barrel
0.158988 . . . . kilograms per barrel
0.350508 . . . . pounds (Avoir.) per barrel
5.608134 . . . . ounces (Avoir.) per barrel
89.730060 . . . . drams (Avoir.) per barrel
158.987766 . . . . grams per barrel
2,543.556 . . . . grains per barrel
0.00000417271 . . . . tons (net) per gallon—U.S.
0.00378542 . . . . kilograms per gallon—U.S.
0.00834543 . . . . pounds (Avoir.) per gallon—U.S.
0.133527 . . . . ounces (Avoir.) per gallon—U.S.
2.136430 . . . . drams (Avoir.) per gallon—U.S.
3.785423 . . . . grams per gallon—U.S.
58.418 . . . . grains per gallon—U.S.
0.00000110231 . . . . tons (net) per liter
0.001 . . . . kilograms per liter
0.00220462 . . . . pounds (Avoir.) per liter
0.0352739 . . . . ounces (Avoir.) per liter
0.564383 . . . . drams (Avoir.) per liter
15.4324 . . . . grains per liter
0.00000000110231 . . . . tons (net) per cubic centimeter
0.000001 . . . . kilograms per cubic centimeter
0.00000220462 . . . . pounds (Avoir.) per cubic centimeter
0.0000352739 . . . . ounces (Avoir.) per cubic centimeter
0.000564383 . . . . drams (Avoir.) per cubic centimeter
0.001 . . . . grams per cubic centimeter
0.0154324 . . . . grains per cubic centimeter

GRAM PER METER: =

0.001 . . . . kilograms per meter
0.00030480 . . . . kilograms per foot
0.000025401 . . . . kilograms per inch
0.00001 . . . . kilograms per centimeter
0.0022046 . . . . pounds per meter
0.00067197 . . . . pounds per foot
0.0000560 . . . . pounds per inch
0.000022046 . . . . pounds per centimeter
0.035274 . . . . ounces per meter
0.010752 . . . . ounces per foot
0.0008960 . . . . ounces per inch
0.00035274 . . . . ounces per centimeter
1.0 . . . . grams per meter
0.30480 . . . . grams per foot
0.025401 . . . . grams per inch
0.01 . . . . grams per centimeter

GRAM PER METER: (cont'd)

15.4324	grains per meter
4.70380	grains per foot
0.39198	grains per inch
0.15432	grains per centimeter

GRAM PER SQUARE CENTIMETER: =

10	kilograms per square meter
0.92903	kilograms per square foot
0.0064516	kilograms per square inch
0.001	kilograms per square centimeter
22.046	pounds per square meter
2.048140	pounds per square foot
0.014223	pounds per square inch
0.0022046	pounds per square centimeter
352.7392	ounces per square meter
32.77053	ounces per square foot
0.22756	ounces per square inch
0.035274	ounces per square centimeter
10,000	grams per square meter
929.03	grams per square foot
6.45156	grams per square inch
1.0	grams per square centimeter
154,324	grains per square meter
14,337.16257	grains per square foot
99.56299	grains per square inch
15.4324	grains per square centimeter
0.73556	millimeters of mercury @ 0°C.
0.00096784	atmospheres
980.665	dynes

GRAM PER SQUARE FOOT: =

0.010764	kilograms per square meter
0.001	kilograms per square foot
0.0000069444	kilograms per square inch
0.0000010764	kilograms per square centimeter
0.023730	pounds per square meter
0.00220462	pounds per square foot
0.000015310	pounds per square inch
0.0000023730	pounds per square centimeter
0.37968	ounces per square meter
0.035274	ounces per square foot
0.00024496	ounces per square inch
0.000037968	ounces per square centimeter
10.76387	grams per square meter
1.0	grams per square foot
0.0069444	grams per square inch
0.00107639	grams per square centimeter
166.11235	grains per square meter

GRAM PER SQUARE FOOT: (cont'd)

15.4324	grains per square foot
0.10717	grains per square inch
0.016611	grains per square centimeter

GRAM PER SQUARE INCH: =

1.55	kilograms per square meter
0.1440	kilograms per square foot
0.001	kilograms per square inch
0.000155	kilograms per square centimeter
3.41713	pounds per square meter
0.31746	pounds per square foot
0.0022046	pounds per square inch
0.00034171	pounds per square centimeter
54.67470	ounces per square meter
5.0794444	ounces per square foot
0.035274	ounces per square inch
0.0054675	ounces per square centimeter
1,550	grams per square meter
143.99965	grams per square foot
1.0	grams per square inch
0.155	grams per square centimeter
23,920.099878	grains per square meter
2,222.25834	grains per square foot
15.4324	grains per square inch
2.39201	grains per square centimeter

GRAM PER SQUARE METER: =

0.001	kilograms per square meter
0.000092903	kilograms per square foot
0.00000064516	kilograms per square inch
0.0000001	kilograms per square centimeter
0.0022046	pounds per square meter
0.00020481	pounds per square foot
0.0000014222	pounds per square inch
0.00000022046	pounds per square centimeter
0.035274	ounces per square meter
0.0032771	ounces per square foot
0.000022757	ounces per square inch
0.0000035274	ounces per square centimeter
1.0	grams per square meter
0.092903	grams per square foot
0.00064521	grams per square inch
0.0001	grams per square centimeter
15.4324	grains per square meter
1.43372	grains per square foot
0.0099563	grains per square inch
0.00154324	grains per square centimeter

HECTARE: =

0.003861	square miles or sections
0.010	square kilometers
0.247104	square furlongs
1.0	square hektometers
24.71044	square chains
100	square dekameters
395.367	square rods
10,000	square meters
11,959.888	square yards
13,949.8	square varas (Texas)
107,639	square feet
191,358	square spans
277,104.4	square links
968,750	square hands
1,000,000	square decimeters
100,000,000	square centimeters
15,500,016	square inches
10,000,000,000	square millimeters
100	ares
1,000	centares (centiares)
2.471044	acres

HEKTOLITER: =

0.000000000053961	cubic miles
0.0000000001	cubic kilometers
0.000000000431688	cubic furlongs
0.0000001	cubic hektometers
0.0000122835	cubic chains
0.0001	cubic dekameters
0.000786142	cubic rods
0.1	kiloliters
0.1	cubic meters
0.13080	cubic yards
0.164759	cubic varas (Texas)
0.628976	barrels
1.0	hektoliters
2.8378	bushels—U.S. (dry)
2.7497	bushel—Imperial (dry)
3.53145	cubic feet
8.370844	cubic spans
10.0	dekaliters
11.3513	pecks
12.283475	cubic links
26.417762	gallons—U.S. (liquid)
22.702	gallons—U.S. (dry)
21.998	gallons—Imperial
95.635866	cubic hands
100	liters

HEKTOLITER: (cont'd)

100	cubic decimeters
105.6710	quarts—U.S. (liquid)
90.8102	quarts—U.S. (dry)
211.34	pints—U.S. (liquid)
181.62	pints—U.S. (dry)
845.38	gills (liquid)
1,000	deciliters
61,025	cubic inches
10,000	centiliters
100,000	milliliters
100,000	cubic centimeters
100,000,000	cubic millimeters
220.46223	pounds of water @ 39°F.
3,381.47	ounces (fluid)
27,051.79	drams (fluid)

HEKTOGRAM: =

0.000098426	tons (long)
0.0001	tons (metric)
0.000110231	tons (short)
0.1	kilograms
0.220462	pounds (Troy)
0.267923	pounds (Avoir.)
1.0	hektograms
10	dekagrams
3.21507	ounces (Troy)
3.527392	ounces (Avoir.)
25.7206	drams (Troy)
56.4383	drams (Avoir.)
64.30149	pennyweights (Troy)
77.1618	scruples (Troy)
500	carats (metric)
1,000	decigrams
1,543.24	grains
10,000	centigrams
100,000	milligrams

HORSEPOWER: =

47,520,000	foot pounds per day
1,980,000	foot pounds per hour
33,000	foot pounds per minute
550	foot pounds per second
570,240,000	inch pounds per day
23,760,000	inch pounds per hour
396,000	inch pounds per minute
6,600	inch pounds per second
15,390.720	kilogram calories (mean) per day
641.280	kilogram calories (mean) per hour

HORSEPOWER: (cont'd)

10.688 . . . . kilogram calories (mean) per minute
0.178133 . . . . kilogram calories (mean) per second
33,930.724525 . . . . pounds calories (mean) per day
1,413.780189 . . . . pound calories (mean) per hour
23.563003 . . . . pound calories (mean) per minute
0.392716 . . . . pound calories (mean) per second
542,891.59248 . . . . ounce calories (mean) per day
22,620.483024 . . . . ounce calories (mean) per hour
377.008048 . . . . ounce calories (mean) per minute
6.283456 . . . . ounce calories (mean) per second
15,390,720 . . . . gram calories (mean) per day
641,280 . . . . gram calories (mean) per hour
10,688 . . . . gram calories (mean) per minute
178.133 . . . . gram calories (mean) per second
61,081.344 . . . . BTU (mean) per day
2,545.5600 . . . . BTU (mean) per hour
42.41760 . . . . BTU (mean) per minute
0.70696 . . . . BTU (mean) per second
0.7452 . . . . kilowatts (g=980)
0.74570 . . . . kilowatts (g=980.665)
745.2 . . . . watts (g=980)
745.70 . . . . watts (g=980.665
1.0139 . . . . horsepower (metric)
1.0139 . . . . Cheval-vapeur hours
0.174 . . . . pounds carbon oxidized with 100% efficiency
2.62 . . . . pounds water evaporated from and @ 212°F.
635.769600 . . . . kiloliter-atmospheres per day
24.490400 . . . . kiloliter-atmospheres per hour
0.441507 . . . . kiloliter-atmospheres per minute
0.00735844 . . . . kiloliter-atmospheres per second
635,769.599962 . . . . liter-atmospheres per day
26,490.399998 . . . . liter-atmospheres per hour
441.506667 . . . . liter-atmospheres per minute
7.358844 . . . . liter-atmospheres per second
635,769,599.962 . . . . milliliter-atmospheres per day
26,490,399.998 . . . . milliliter-atmospheres per hour
441,506.666667 . . . . milliliter-atmospheres per minute
7,358.444444 . . . . milliliter-atmospheres per second

INCH: =

0.00001578 . . . . miles
0.00002540 . . . . kilometers
0.000126263 . . . . furlongs
0.0002560 . . . . hektometers
0.00126263 . . . . chains
0.002540 . . . . dekameters
0.00505051 . . . . rods
0.02540 . . . . meters
0.027777 . . . . yards
0.030000 . . . . varas (Texas)

UNIVERSAL CONVERSION FACTORS

INCH: (cont'd)

0.083333	feet
0.111111	spans
0.126263	links
0.25000	hands
0.2540	decimeters
2.5400	centimeters
1	inches
25.40	millimeters
1000	mils
25,400	microns
39,450.33	wave lengths of red line of cadmium
25,400,000	millimicrons
25,400,000	micromillimeters
254,000,000	Angstrom Units

INCH OF MERCURY @ 32°F. =

0.00345349	hektometers of water @ 60°F.
0.0345349	dekameters of water @ 60°F.
0.345349	meters of water @ 60°F.
1.132944	feet of water @ 60°F.
13.595326	inches of water @ 60°F.
3.45349	decimeters of water @ 60°F.
34.5349	centimeters of water @ 60°F.
345.349	millimeters of water @ 60°F.
0.000254	hektometers of mercury @ 32°F.
0.00254	dekameters of mercury @ 32°F.
0.0254	meters of mercury @ 32°F.
0.0833325	feet of mercury @ 32°F.
1	inches of mercury @ 32°F.
0.254	decimeters of mercury @ 32°F.
2.54	centimeters of mercury @ 32°F.
25.4	millimeters of mercury @ 32°F.
3806.515240	tons per square hektometer
38.065152	tons per square dekameter
0.380652	tons per square meter
0.0353645	tons per square foot
0.000245581	tons per square inch
0.00380652	tons per square decimeter
0.0000380652	tons per square centimeter
0.000000380652	tons per square millimeter
3,453,490	kilograms per square hektometer
34,534.9	kilograms per square dekameter
345.349	kilograms per square meter
32.0811278	kilograms per square foot
0.222786	kilograms per square inch
3.45349	kilograms per square decimeter
0.0345349	kilograms per square centimeter
0.000345349	kilograms per square millimeter
7,613,030	pounds per square hektometer
76,130.300609	pounds per square dekameter

INCH OF MERCURY @ 32°F.: (cont'd)

761.303006 .... pounds per square meter
70.726441 .... pounds per square foot
0.491161 .... pounds per square inch
7.613030 .... pounds per square decimeter
0.0761303 .... pounds per square centimeter
0.000761303 .... pounds per square millimeter
34,534,900 .... hektograms per square hektometer
345,349 .... hektograms per square dekameter
3,453.49 .... hektograms per square meter
320.811278 .... hektograms per square foot
2.227864 .... hektograms per square inch
34.5349 .... hektograms per square decimeter
0.345349 .... hektograms per square centimeter
0.00345349 .... hektograms per square millimeter
345,349,000 .... dekagrams per square hektometer
3,453,490 .... dekagrams per square dekameter
34,534.90 .... dekagrams per square meter
3208.112776 .... dekagrams per square foot
22.278639 .... dekagrams per square inch
345.3490 .... dekagrams per square decimeter
3.453490 .... dekagrams per square centimeter
0.03453490 .... dekagrams per square millimeter
121,808.481 .... ounces per square hektometer
1,218,085 .... ounces per square dekameter
12,180.85 .... ounces per square meter
1,131.623063 .... ounces per square foot
7.858573 .... ounces per square inch
121.8085 .... ounces per square decimeter
1.218085 .... ounces per square centimeter
0.0121809 .... ounces per square millimeter
3,453,490,000 .... grams per square hektometer
34,534,900 .... grams per square dekameter
345,349 .... grams per square meter
32,081.127762 .... grams per square foot
222.786665 .... grams per square inch
3,453.49 .... grams per square decimeter
34.5349 .... grams per square centimeter
0.345349 .... grams per square millimeter
34,534,900,000 .... decigrams per square hektometer
345,349,000 .... decigrams per square dekameter
3,453,490 .... decigrams per square meter
320.811278 .... decigrams per square foot
2,227.866675 .... decigrams per square inch
34,534.90 .... decigrams per square decimeter
345.3490 .... decigrams per square centimeter
3.453490 .... decigrams per square millimeter
345,349,000,000 .... centigrams per square hektometer
3,453,490,000 .... centigrams per square dekameter
34,534,900 .... centigrams per square meter
3,208,113 .... centigrams per square foot
22,278.666751 .... centigrams per square inch

INCH OF MERCURY @ 32°F.: (cont'd)

345,349 .... centigrams per square decimeter
3,453.49 .... centigrams per square centimeter
34.5349 .... centigrams per square millimeter
3,453,490,000,000 .... milligrams per square hektometer
34,534,900,000 .... milligrams per square dekameter
345,349,000 .... milligrams per square meter
32,081,128 .... milligrams per square foot
222,786.667776 .... milligrams per square inch
3,453,490 .... milligrams per square decimeter
34,534.90 .... milligrams per square centimeter
345.3490 .... milligrams per square millimeter
0.0338659 .... bars
0.0334214 .... atmospheres
3,383,845,567,678 .... dynes per square hektometer
33,838,455,677 .... dynes per square dekameter
338,384,557 .... dynes per square meter
31,460,290 .... dynes per square foot
218,471.233172 .... dynes per square inch
3,383,846 .... dynes per square decimeter
33,838.455677 .... dynes per square centimeter
338.384557 .... dynes per square millimeter
930.464111 .... feet of water @ 62°F. and 29.92 Barom. Press

INCH OF WATER @ 60°F.: =

0.000254 .... hektometers of water @ 60°F.
0.00254 .... dekameters of water @ 60°F.
0.0254 .... meters of water @ 60°F.
0.0833332 .... feet of water @ 60°F.
1 .... inches of water @ 60°F.
0.254 .... decimeters of water @ 60°F.
2.54 .... centimeters of water @ 60°F.
25.4 .... millimeters of water @ 60°F.
0.0000186820 .... hektometers of mercury @ 32°F.
0.000186820 .... dekameters of mercury @ 32°F.
0.00186820 .... meters of mercury @ 32°F.
0.00612925 .... feet of mercury @ 32°F.
0.0735510 .... inches of mercury @ 32°F.
0.0186820 .... decimeters of mercury @ 32°F.
0.186820 .... centimeters of mercury @ 32°F.
1.868197 .... millimeters of mercury @ 32°F.
279.973171 .... tons per square hektometer
2.799732 .... tons per square dekameter
0.0279973 .... tons per square meter
0.00260110 .... tons per square foot
0.0000180627 .... tons per square inch
0.000279973 .... tons per square decimeter
0.00000279973 .... tons per square centimeter
0.0000000279973 .... tons per square millimeter
254,000 .... kilograms per square hektometer
2,540 .... kilograms per square dekameter

INCH OF WATER @ 60°F.: **(cont'd)**

25.4	kilograms per square meter
2.359600	kilograms per square foot
0.0163861	kilograms per square inch
0.254	kilograms per square decimeter
0.00254	kilograms per square centimeter
0.0000254	kilograms per square millimeter
559,946	pounds per square hectometer
5,599.463120	pounds per square dekameter
55.994631	pounds per square meter
5.202004	pounds per square foot
0.0361250	pounds per square inch
0.559946	pounds per square decimeter
0.00559946	pounds per square centimeter
0.0000559946	pounds per square millimeter
2,540,000	hektograms per square hektometer
25,400	hektograms per square dekameter
254	hektograms per square meter
23.596005	hektograms per square foot
0.163862	hektograms per square inch
2.54	hektograms per square decimeter
0.0254	hektograms per square centimeter
0.000254	hektograms per square millimeter
25,400,000	dekagrams per square hektometer
254,000	dekagrams per square dekameter
2,540	dekagrams per square meter
235.960045	dekagrams per square foot
1.638617	dekagrams per square inch
25.40	dekagrams per square decimeter
0.2540	dekagrams per square centimeter
0.002540	dekagrams per square millimeter
8,959,141	ounces per square hektometer
89,591.41	ounces per square dekameter
895.9141	ounces per square meter
83.232058	ounces per square foot
0.5780	ounces per square inch
8.959141	ounces per square decimeter
0.0895914	ounces per square centimeter
0.000895914	ounces per square millimeter
254,000,000	grams per square hektometer
2,540,000	grams per square dekameter
25,400	grams per square meter
2,359.600439	grams per square foot
16.386192	grams per square inch
254	grams per square decimeter
2.54	grams per square centimeter
0.0254	grams per square millimeter
2,540,000,000	decigrams per square hektometer
25,400,000	decigrams per square dekameter
254,000	decigrams per square meter
23,596,004	decigrams per square foot
163.861920	decigrams per square inch

INCH OF WATER @ 60°F.: (cont'd)

2,540	decigrams per square decimeter
25.40	decigrams per square centimeter
0.2540	decigrams per square millimeter
25,400,000,000	centigrams per square hektometer
254,000,000	centigrams per square dekameter
2,540,000	centigrams per square meter
235,960	centigrams per square foot
1,638.619198	centigrams per square inch
25,400	centigrams per square decimeter
254	centigrams per square centimeter
2.54	centigrams per square millimeter
254,000,000,000	milligrams per square hektometer
2,540,000,000	milligrams per square dekameter
25,400,000	milligrams per square meter
2,359,600	milligrams per square foot
16,396.192	milligrams per square inch
254,000	milligrams per square decimeter
2,540	milligrams per square centimeter
25.40	milligrams per square millimeter
0.00245562	bars
0.00245818	atmospheres
248,885,374,188	dynes per square hektometer
2,488,853,742	dynes per square dekameter
24,888,537	dynes per square meter
2,313,937	dynes per square foot
16,069.0079	dynes per square inch
248,885	dynes per square decimeters
2,488.853742	dynes per square centimeter
24.888537	dynes per square millimeter
68.44	feet of water @ 62°F. and 29.92 Barom. Press.

JOULE (ABSOLUTE): =

23.730	foot poundals
284.760	inch pounds
0.73756	foot pounds
8.85072	inch pounds
0.000000263331	ton (net) calories
0.00023889	kilogram calories (mean)
0.00526661	pound calories
0.00842658	ounce calories
0.23889	gram calories (mean)
238.89	milligram calories
0.0000000115740	kilowatt days
0.0000002778	kilowatt hours
0.0000166667	kilowatt minutes
0.001	kilowatt seconds
0.0000115740	watt days
0.0002778	watt hours
0.0166667	watt minutes
1	watt seconds

JOULE (ABSOLUTE): (cont'd)

0.000112366	ton (net) meters
0.101937	kilogram meters
0.224733	pounds meters
3.595721	ounce meters
101.937	gram meters
101,937	milligram meters
0.000368654	ton (net) feet
0.334438	kilogram feet
0.737311	pound feet
11.796960	ounce feet
334.438274	gram feet
334,438.273531	milligram feet
0.00442385	ton (net) inches
4.013259	kilogram inches
8.847732	pound inches
141.563520	ounce inches
4,013.259288	gram inches
4,013,259	milligram inches
0.0112366	ton (net) centimeters
10.1937	kilogram centimeters
22.4733	pound centimeters
359.5721	ounce centimeters
10,193.7	gram centimeters
10,193,700	milligram centimeters
0.112366	ton (net) millimeters
101.937	kilogram millimeters
224.733	pound millimeters
3,595.721	ounce millimeters
101,937	gram millimeters
101,937,000	milligram millimeters
0.0000098705	kiloliter-atmosphere
0.000098705	hektoliter-atmosphere
0.0003485	cubic foot-atmosphere
0.00098705	dekaliter-atmosphere
0.0098705	liter-atmosphere
0.098705	deciliter-atmosphere
0.98705	centiliter-atmosphere
9.8705	millimeter-atmosphere
1	joules
0.0000003775	Cheval-vapeur hours
0.0000000155208	horsepower days
0.0000003725	horsepower hours
0.0000223500	horsepower minutes
0.00134100	horsepower seconds
0.0000000642	pounds of carbon oxidized with perfect efficiency
0.0000009662	pounds of water evaporated from and at 212°F.
0.0009480	BTU (mean)
100,000,000	ergs

KILOGRAMS: =

0.000984206	tons (long)
0.001	tons (metric)

KILOGRAMS: (cont'd)

0.00110231	tons (short)
1	kilograms
2.679229	pounds (Troy)
2.204622	pounds (Avoir.)
10	hektograms
100	dekagrams
32.150742	ounces (Troy)
35.273957	ounces (Avoir.)
1,000	grams
257.21	drams (Troy)
564.38	drams (Avoir)
643.01	pennyweights (Troy)
771.62	scruples (Troy)
5,000	carats (metric)
10,000	decigrams
15,432.4	grains
100,000	centigrams
1,000,000	milligrams

KILOGRAM CALORIE (MEAN): =

99,334	foor poundals
1,192,008	inch poundals
3,087.4	foot pounds
37,048.8	inch pounds
0.0001	ton (short) calories
1	kilogram calories
2.04622	pound calories
10	hektogram calories
100	dekagram calories
32.7395	ounce calories
1,000	gram calories
10,000	decigram calories
100,000	centigram calories
1,000,000	milligram calories
0.00004845	kilowatt days
0.0011628	kilowatt hours
0.0697680	kilowatt minutes
4.186	kilowatt seconds
0.04845	watt days
1.1628	watt hours
69.7680	watt minutes
4,186	watt seconds
0.42685	ton meters
426.85	kilogram meters
941.043	pound meters
4,268.5	hektogram meters
42,685	dekagram meters
15,056.688	ounce meters
426,850	gram meters
4,268,500	decigram meters
42,685,000	centigram meters

KILOGRAM CALORIE (MEAN): (cont'd)

426,850,000 .... milligram meters
0.0042685 .... ton hektometers
4.2685 .... kilogram hektometers
9.41043 .... pound hektometers
42.685 .... hektogram hektometers
426.85 .... dekagram hektometers
150.567 .... ounce hektometers
4,268.5 .... gram hektometers
42,685 .... decigram hektometers
426,850 .... centigram hektometers
4,268,500 .... milligram hektometers
0.042685 .... ton dekameters
42.685 .... kilogram dekameters
94.1043 .... pound dekameters
426.85 .... hektogram dekameters
4,268.5 .... dekagram dekameters
1,505.67 .... ounce dekameters
42,685 .... gram dekameters
426,850 .... decigram dekameters
4,268,500 .... centigram dekameters
42,685,000 .... milligram dekameters
1.40042 .... ton feet
1,400.424 .... kilogram feet
30.8704 .... pound feet
14,004.236 .... hektogram feet
140,042.357 .... dekagram feet
493,985 .... ounce feet
1,400,424 .... gram feet
14,004,236 .... decigram feet
140,042,357 .... centigram feet
1,400,423,566 .... milligram feet
16.8050 .... ton inches
16,805.083 .... kilogram inches
307.445 .... pound inches
168,050.828 .... hektogram inches
1,680,051 .... dekagram inches
5,927.117 .... ounce inches
16,805,083 .... gram inches
168,050,828 .... decigram inches
1,680,508,279 .... centigram inches
16,805,082,792 .... milligram inches
4.2685 .... ton decimeters
4,268.5 .... kilogram decimeters
9,410.430 .... pound decimeters
42,685 .... hektogram decimeters
426,850 .... dekogram decimeters
150,566.885 .... ounce decimeters
4,268,500 .... gram decimeters
42,685,000 .... decigram decimeters
426,850,000 .... centigram decimeters
4,268,500,000 .... milligram decimeters

42.685 . . . . ton centimeters
42,685 . . . . kilogram centimeters
94,104.303 . . . . pound centimeters
426,850 . . . . hektogram centimeters
4,268,500 . . . . dekagram centimeters
1,505,669 . . . . ounce centimeters
42,685,000 . . . . gram centimeters
426,850,000 . . . . decigram centimeters
4,268,500,000 . . . . centigram centimeters
42,685,000,000 . . . . milligram centimeters
426.85 . . . . ton millimeters
426,850 . . . . kilogram millimeters
941,043.029 . . . . pound millimeters
4,268,500 . . . . hektogram millimeters
42,685,000 . . . . dekagram millimeters
15,056,688 . . . . ounce millimeters
426,850,000 . . . . gram millimeters
4,268,500,000 . . . . decigram millimeters
42,685,000,000 . . . . centigram millimeters
426,850,000,000 . . . . milligram millimeters
0.041311 . . . . kiloliter-atmosphere
0.41311 . . . . hektoliter-atmosphere
4.1311 . . . . dekaliter-atmosphere
41.311 . . . . liter-atmosphere
413.11 . . . . deciliter-atmosphere
4,131.1 . . . . centiliter-atmosphere
41,311 . . . . milliliter-atmosphere
0.0000000000415977 . . . . cubic kilometer-atmosphere
0.0000000415977 . . . . cubic hektometer-atmosphere
0.0000415977 . . . . cubic dekameter-atmosphere
0.0415977 . . . . cubic meter-atmosphere
1.469 . . . . cubic feet-atmosphere
2,538.415 . . . . cubic feet-atmosphere
41.5977 . . . . cubic decimeter-atmosphere
41,597.673 . . . . cubic centimeter-atmosphere
41,597,673 . . . . cubic millimeter-atmosphere
4,185.8291 . . . . joules
0.00158097 . . . . Cheval-vapeur hours
0.0000649708 . . . . horsepower days
0.0015593 . . . . horsepower hours
0.0935980 . . . . horsepower minutes
5.61588 . . . . horsepower seconds
0.00029909 . . . . pounds of carbon oxidized with 100% efficiency
0.004501 . . . . pounds of water evaporated from at 212°F.
1.800 . . . . BTU (mean) per pound
3.9685 . . . . BTU (mean)
41,858,291,000 . . . . ergs

KILOGRAM CALORIE (MEAN) PER MINUTE: =

4,443,725 . . . . foot pound per day
185,155.2 . . . . foot pound per hour

KILOGRAM CALORIE (MEAN) PER MINUTE: (cont'd)

3,085.920	foot pound per minute
51.432	foot pound per second
0.0935980	horsepower
0.069680	kilowatts
69.7680	watts
232.71	foot poundals
2,792.52	inch poundals
7.2330	foot pounds
86.7960	inch pounds
0.0000023427	ton (net) calories
0.0023427	hektogram calories
0.00516477	pound calories
0.023427	hektogram calories
0.23427	dekagram calories
0.0826363	ounce calories
2.3427	gram calories (mean)
23.427	decigram calories
234.27	centigram calories
2,343.7	kilowatt days
0.000000113479	kilowatt days
0.0000027235	kilowatt hours
0.000163410	kilowatt minutes
0.00980460	kilowatt seconds
0.000113479	watt days
0.0027235	watt hours
0.163410	watt minutes
9.80460	watt seconds
0.001	ton meters
1	kilogram meters
2.204622	pound meters
10	hektogram meters
100	dekagram meters
35.273957	ounce meters
1,000	gram meters
10,000	decigram meters
100,000	centigram meters
1,000,000	milligram meters
0.00001	ton hektometers
0.01	kilogram hektometers
0.02204622	pound hektometers
0.1	hektogram hektometers
1	dekagram hektometers
0.352740	ounce hektometers
10	gram hektometers
100	decigram hektometers
1,000	centigram hektometers
10,000	milligram hektometers
0.0001	ton dekameters
0.1	kilogram dekameters
0.2204622	pound dekameters
1	hektogram dekameters

KILOGRAM CALORIE (MEAN) PER MINUTE: (cont'd)

10	dekagram dekameters
3.527396	ounce dekameters
100	gram dekameters
1,000	decigram dekameters
10,000	centigram dekameters
100,000	milligram dekameters
0.00328084	ton feet
3.280843	kilogram feet
7.233020	pound feet
32.80843	hektogram feet
328.0843	dekagram feet
115.728320	ounce feet
3,280.843	gram feet
32,808.43	decigram feet
328,084.3	centigram feet
3,280,843	milligram feet
0.0393701	ton inches
39.370116	kilogram inches
86.796236	pound inches
393.70116	hektogram inches
3,937.0116	dekagram inches
1,388.739776	ounce inches
39,370.116	gram inches
393,701.16	decigram inches
3,937,012	centigram inches
39,370,116	milligram inches
0.01	ton decimeters
10	kilogram decimeters
22.046223	pound decimeters
100	hektogram decimeters
1,000	dekagram decimeters
352.739568	ounce decimeters
10,000	gram decimeters
100,000	decigram decimeters
1,000,000	centigram decimeters
10,000,000	milligram decimeters
0.1	ton centimeters
100	kilogram centimeters
220.46223	pound centimeters
1,000	hektogram centimeters
10,000	dekagram centimeters
3,527.39568	ounce centimeters
100,000	gram centimeters
1,000,000	decigram centimeters
10,000,000	centigram centimeters
100,000,000	milligram centimeters
1	ton millimeters
1,000	kilogram millimeters
2,204.6223	pound millimeters
10,000	hektogram millimeters
100,000	dekagram millimeters

KILOGRAM CALORIE (MEAN) PER MINUTE: (cont'd)

35,273.9568	ounce millimeters
1,000,000	gram millimeters
10,000,000	decigram millimeters
100,000,000	centigram millimeters
1,000,000,000	milligram millimeters
0.000096782	kiloliter-atmospheres
0.00096782	hektoliter-atmospheres
0.0096782	dekaliter-atmospheres
0.096782	liter-atmospheres
0.96782	deciliter-atmospheres
9.6782	centiliter-atmospheres
96.782	milliliter-atmospheres
0.0000000000000967790	cubic kilometer-atmospheres
0.0000000000967790	cubic hektometer-atmospheres
0.0000000967790	cubic dekameter-atmospheres
0.0000967790	cubic meter-atmospheres
0.0034177	cubic foot-atmospheres
5.905746	cubic inch-atmospheres
0.0967790	cubic decimeter-atmospheres
96.779011	cubic centimeter-atmospheres
96,779.011	cubic millimeter-atmospheres
9.80665	joules
0.000000154324	Cheval-vapeur hours
0.000000152208	horsepower days
0.0000036530	horsepower hours
0.000219180	horsepower minutes
0.0131508	horsepower seconds
0.00000063718	pounds of carbon oxidized with 100% efficiency
0.0000095895	pounds of water evaporated form and at 212°F.
4.216948	BTU (mean) per pounds
0.0092972	BTU (mean)
98,066,500	ergs

KILOGRAM METER PER SECOND: =

0.0098046	kilowatts
9.8046	watts
0.0131508	horsepower

KILOGRAM METER PER MINUTE: =

0.000163410	kilowatts
0.163410	watts
0.00021980	horsepowers

KILOGRAM PER METER: =

1.774004	tons (net) per mile
1.102311	tons (net) per kilometer
0.00110231	tons (net) per meter

KILOGRAM PER METER: (cont'd)

0.001007956	tons (net) per yard
0.000335985	tons (net) per foot
0.0000279987	tons (net) per inch
0.0000110231	tons (net) per centimeter
0.00000110231	tons (net) per millimeter
1,609.349954	kilograms per mile
1,000	kilograms per kilometer
1	kilograms per meter
0.914403	kilograms per yard
0.304801	kilograms per foot
0.0254001	kilograms per inch
0.01	kilograms per centimeter
0.001	kilograms per millimeter
3,548.00896449	pounds (Avoir.) per mile
2,204.622341	pounds (Avoir.) per kilometer
2.204622	pounds (Avoir.) per meter
2.0159127	pounds (Avoir.) per yard
0.671971	pounds (Avoir.) per foot
0.0559976	pounds (Avoir.) per inch
0.0220462	pounds (Avoir.) per centimeter
0.00220462	pounds (Avoir.) per millimeter
56,768.143440	ounces (Avoir.) per mile
35,273.957456	ounces (Avoir.) per kilometer
35.273957	ounces (Avoir.) per meter
32.254604	ounces (Avoir.) per yard
10.751535	ounces (Avoir.) per foot
0.895961	ounces (Avoir.) per inch
0.352740	ounces (Avoir.) per centimeter
0.0352740	ounces (Avoir.) per millimeter
1,609,349.954	grams per mile
1,000,000	grams per kilometer
1,000	grams per meter
914.403	grams per yard
304.801127	grams per foot
25.4001	grams per inch
10	grams per centimeter
1	grams per millimeter
24,836,063	grains per mile
15,432,356	grains per kilometer
15,432.356387	grains per meter
14,111.388900	grains per yard
4,703.79630	grains per foot
391.983025	grains per inch
154.323564	grains per centimeter
15.432356	grains per millimeter

KILOGRAM PER SQUARE CENTIMETER: =

10	meters of water @ 60°F.
32.80833	feet of water @ 60°F.
393.69996	inches of water @ 60°F.

KILOGRAM PER SQUARE CENTIMETER: (cont'd)

1,000 . . . . centimeters of water @ 60°F.
10,000 . . . . millimeters of water @ 60°F.
0.735499 . . . . meters of mercury @ 32°F.
2.413053 . . . . feet of mercury @ 32°F.
28.956632 . . . . inches of mercury @ 32°F.
73.54985 . . . . centimeters of mercury @ 32°F.
735.49845 . . . . millimeters of mercury @ 32°F.
10,000 . . . . kilograms per square meter
929.034238 . . . . kilograms per square foot
6.451626 . . . . kilograms per square inch
1.0 . . . . kilograms per square centimeter
0.01 . . . . kilograms per square millimeter
22,046.233 . . . . pounds per square meter
2,048.1696 . . . . pounds per square foot
14.2234 . . . . pounds per square inch
2.2046223 . . . . pounds per square centimeter
0.0220462 . . . . pounds per square millimeter
352,739.568 . . . . ounces per square meter
32,770.7136 . . . . ounces per square foot
227.5744 . . . . ounces per square inch
35.273957 . . . . ounces per square centimeter
0.352740 . . . . ounces per square millimeter
10,000,000 . . . . grams per square meter
929,034.230 . . . . grams per square foot
6,451.626597 . . . . grams per square inch
1,000 . . . . grams per square centimeter
10 . . . . grams per square millimeter
154,324,000 . . . . grains per square meter
14,337,228 . . . . grains per square foot
99,564.0822955 . . . . grains per square inch
15,432.5 . . . . grains per square centimeter
154.324 . . . . grains per square millimeter
0.967778 . . . . atmosphere

KILOGRAM PER SQUARE FOOT: =

0.0107638 . . . . meters of water @ 60°F.
0.0353518 . . . . feet of water @ 60°F.
0.423774 . . . . inches of water @ 60°F.
1.076387 . . . . centimeters of water @ 60°F.
10.76387 . . . . millimeters of water @ 60°F.
0.000791682 . . . . meters of mercury @ 32°F.
0.00259738 . . . . feet of mercury @ 32°F.
0.0311689 . . . . inches of mercury @ 32°F.
0.0791682 . . . . centimeters of mercury @ 32°F.
0.791682 . . . . millimeters of mercury @ 32°F.
10.76387 . . . . kilograms per square meter
1 . . . . kilograms per square foot
0.00694445 . . . . kilograms per square inch
0.00107639 . . . . kilograms per square centimeter
0.0000107639 . . . . kilograms per square millimeter

KILOGRAM PER SQUARE FOOT: (cont'd)

23.730265	pounds per square meter
2.204622	pounds per square foot
0.0153099	pounds per square inch
0.00237302	pounds per square centimeter
0.0000237302	pounds per square millimeter
379.684288	ounces per square meter
35.273957	ounces per square foot
0.244958	ounces per square inch
0.0379685	ounces per square centimeter
0.000379685	ounces per square millimeter
10,763.87	grams per square meter
1,000	grams per square foot
6.94445	grams per square inch
1.076387	grams per square centimeter
0.0107639	grams per square millimeter
166,112.347388	grains per square meter
15,432.4	grains per square foot
107.169481	grains per square inch
16.611235	grains per square centimeter
0.166112	grains per square millimeter
0.00104170	atmospheres

KILOGRAM PER SQUARE INCH: =

1.549987	meters of water @ 60°F.
5.090659	feet of water @ 60°F.
61.023456	inches of water @ 60°F.
154.999728	centimeters of water @ 60°F.
1,549.99728	millimeters of water @ 60°F.
0.114002	meters of mercury @ 32°F.
0.374022	feet of mercury @ 32°F.
4.488322	inches of mercury @ 32°F.
11.40002	centimeters of mercury @ 32°F.
114.00022	millimeters of mercury @ 32°F.
1,549.99728	kilograms per square meter
144	kilograms per square foot
1	kilograms per square inch
0.1549997	kilograms per square centimeter
0.001549997	kilograms per square millimeter
3,417.158160	pounds per square meter
317.4655568	pounds per square foot
2.2046223	pounds per square inch
0.341715	pounds per square centimeter
0.00341715	pounds per square millimeter
54,674.537472	ounces per square meter
5,079.449808	ounces per square foot
35.273957	ounces per square inch
5.467464	ounces per square centimeter
0.0546746	ounces per square millimeter
1,549,997	grams per square meter
144,000	grams per square foot

KILOGRAM PER SQUARE INCH: (cont'd)

1,000 . . . . grams per square inch
154.999728 . . . . grams per square centimeter
1.549997 . . . . grams per square millimeter
23,920,178 . . . . grains per square meter
2,222,266 . . . . grains per square foot
15,432.4 . . . . grains per square inch
2,392.017840 . . . . grains per square centimeter
23.920178 . . . . grains per square millimeter
0.150005 . . . . atmospheres

KILOGRAM PER SQUARE KILOMETER: =

0.000000001 . . . . meters of water @ 60°F.
0.00000032843 . . . . feet of water @ 60°F.
0.0000003937 . . . . inches of water @ 60°F.
0.0000001 . . . . centimeters of water @ 60°F.
0.000001 . . . . millimeters of water @ 60°F.
0.0000000000735499 . . . . meters of mercury @ 32°F.
0.000000000241305 . . . . feet of mercury @ 32°F.
0.00000000289570 . . . . inches of mercury @ 32°F.
0.00000000735499 . . . . centimeters of mercury @ 32°F.
0.0000000735499 . . . . millimeters of mercury @ 32°F.
0.000001 . . . . kilograms per square meter
0.0000000929034 . . . . kilograms per square foot
0.000000000645163 . . . . kilograms per square inch
0.0000000001 . . . . kilograms per square centimeter
0.000000000001 . . . . kilograms per square millimeter
0.00000220462 . . . . pounds per square meter
0.000000204817 . . . . pounds per square foot
0.00000000142234 . . . . pounds per square inch
0.000000000220462 . . . . pounds per square centimeter
0.00000000000220462 . . . . pounds per square millimeter
0.0000352740 . . . . ounces per square meter
0.00000327707 . . . . ounces per square foot
0.0000000227574 . . . . ounces per square inch
0.00000000352740 . . . . ounces per square centimeter
0.0000000000352740 . . . . ounces per square millimeter
0.001 . . . . grams per square meter
0.0000929034 . . . . grams per square foot
0.000000645163 . . . . grams per square inch
0.0000001 . . . . grams per square centimeter
0.000000001 . . . . grams per square millimeter
0.0154324 . . . . grains per square meter
0.00143372 . . . . grains per square foot
0.00000995640 . . . . grains per square inch
0.00000154324 . . . . grains per square centimeter
0.0000000154324 . . . . grains per square millimeter
0.0000000000967778 . . . . atmospheres

KILOGRAM PER SQUARE METER: =

0.001 . . . . meters of water @ 60°F.
0.0032843 . . . . feet of water @ 60°F.

KILOGRAM PER SQUARE METER: (cont'd)

0.03937	inches of water @ 60°F.
0.1	centimeters of water @ 60°F.
1	millimeters of water @ 60°F.
0.0000735499	meters of mercury @ 32°F.
0.000241300	feet of mercury @ 32°F.
0.00289570	inches of mercury @ 32°F.
0.0073549	centimeters of mercury @ 32°F.
0.0735499	millimeters of mercury @ 32°F.
1	kilograms per square meter
0.0929034	kilograms per square foot
0.000645163	kilograms per square inch
0.0001	kilograms per square centimeter
0.000001	kilograms per square millimeter
2.204622	pounds per square meter
0.204817	pounds per square foot
0.00142234	pounds per square inch
0.000220462	pounds per square centimeter
0.00000220462	pounds per square millimeter
35.273957	ounces per square meter
3.277071	ounces per square foot
0.0227574	ounces per square inch
0.00352740	ounces per square centimeter
0.0000352740	ounces per square millimeter
1,000	grams per square meter
92.903423	grams per square foot
0.645163	grams per square inch
0.1	grams per square centimeter
0.001	grams per square millimeter
15,432.4	grains per square meter
1,433.7228	grains per square foot
9.956408	grains per square inch
1.543240	grains per square centimeter
0.0154324	grains per square millimeter
0.0000967778	atmospheres

KILOGRAM PER SQUARE MILLIMETER: =

1,000	meters of water @ 60°F.
3,280.833	feet of water @ 60°F.
39,369.996	inches of water @ 60°F.
100,000	centimeters of water @ 60°F.
1,000,000	millimeters of water @ 60°F.
73,549845	meters of mercury @ 32°F.
241.3053	feet of mercury @ 32°F.
2,895.6632	inches of mercury @ 32°F.
7,354.9845	centimeters of mercury @ 32°F.
73,549.845	millimeters of mercury @ 32°F.
1,000,000	kilograms per square meter
92,903.4238	kilograms per square foot
645.1626	kilograms per square inch
100	kilograms per square centimeter

KILOGRAM PER SQUARE MILLIMETER: (cont'd)

1	kilograms per square millimeter
2,204,622	pounds per square meter
204,816.6	pounds per square foot
1,422.34	pounds per square inch
220,46223	pounds per square centimeter
2,204622	pounds per square millimeter
35,273,957	ounces per square meter
3,277,071	ounces per square foot
22,757.44	ounces per square inch
3,527.3957	ounces per square centimeter
35.273957	ounces per square millimeter
1,000,000,000	grams per square meter
92,903,423	grams per square foot
645,162.6597	grams per square inch
100,000	grams per square centimeter
1,000	grams per square millimeter
15,432,400,000	grains per square meter
1,433,722,800	grains per square foot
9,956,408	grains per square inch
1,543,240	grains per square centimeter
15,432.4	grains per square millimeter
96.7778	atmospheres

KILOMETER: =

0.53961	miles (nautical)
0.62137	miles (statute)
1	kilometers
4.970974	furlongs
10	hektometers
49.709741	chains
100	dekameters
198.838579	rods
1,000	meters
1,093.6	yards
1,181.1	varas (Texas)
3,280.8	feet
4,374.440070	spans
4,970.974310	links
9,842.50	hands
10,000	decimeters
100,000	centimeters
39,370	inches
1,000,000	millimeters
39,370,000	mils
1,000,000,000	microns
1,000,000,000,000	millimicrons
1,000,000,000,000	micromillimeters
546.81	fathoms

KILOMETER PER DAY: =

0.62137	miles per day
0.0258904	miles per hour
0.000431507	miles per minute
0.00000719718	miles per second
1.0	kilometers per day
0.0416667	kilometers per hour
0.000694444	kilometers per minute
0.0000115741	kilometers per second
4.970974	furlongs per day
0.207124	furlongs per hour
0.00345207	furlongs per minute
0.0000575344	furlongs per second
10	hektometers per day
0.416667	hektometers per hour
0.00694444	hektometers per minute
0.00115741	hektometers per second
49.709741	chains per day
2.071239	chains per hour
0.0345207	chains per minute
0.00057534	chains per second
100	dekameters per day
4.166667	dekameters per hour
0.0694444	dekameters per minute
0.00115741	dekameters per second
198.838579	rods per day
8.284941	rods per hour
0.138082	rods per minute
0.00230137	rods per second
1,000	meters per day
41.666667	meters per hour
0.694444	meters per minute
0.0115741	meters per second
1,093.6	yards per day
45.566667	yards per hour
0.759444	yards per minute
0.0126574	yards per second
1,181.1	varas (Texas) per day
49.21250	varas (Texas) per hour
0.820208	varas (Texas) per minute
0.0136701	varas (Texas) per second
3,280.8	feet per day
136.7	feet per hour
2.278333	feet per minute
0.0379722	feet per second
4,374.440070	spans per day
182.268336	spans per hour
3.0378056	spans per minute
0.0506301	spans per second
4,970.974310	links per day

KILOMETER PER DAY: (cont'd)

207.123916	links per hour
3.452065	links per minute
0.0575344	links per second
9,842.50	hands per day
410.104166	hands per hour
6.835069	hands per minute
0.113918	hands per second
10,000	decimeters per day
416.666664	decimeters per hour
6.944444	decimeters per minute
0.115741	decimeters per second
100,000	centimeters per day
4,166.666640	centimeters per hour
69.444444	centimeters per minute
1.157407	centimeters per second
39,370	inches per day
1,640.416666	inches per hour
27.340278	inches per minute
0.455671	inches per second
1,000,000	millimeters per day
41,666.6664	millimeters per hour
694.444444	millimeters per minute
11.574074	millimeters per second

KILOMETER PER HOUR: =

14.912880	miles per day
0.62137	miles per hour
0.0103562	miles per minute
0.000172603	miles per second
24	kilometers per day
1	kilometers per hour
0.0166667	kilometers per minute
0.000277778	kilometers per second
119.303376	furlongs per day
4.970974	furlongs per hour
0.0828496	furlongs per minute
0.00138083	furlongs per second
240	hektometers per day
10	hektometers per hour
0.166667	hektometers per minute
0.00277778	hektometers per second
1,193.0337599	chains per day
49.709741	chains per hour
0.828496	chains per minute
0.0138083	chains per second
2,400	dekameters per day
100	dekameters per hour
1.666667	dekameters per minute
0.0277778	dekameters per second
4,772.125895	rods per day

KILOMETER PER HOUR: (cont'd)

198.838579	rods per hour
3.313976	rods per minute
0.0552329	rods per second
24,000	meters per day
1,000	meters per hour
16.666667	meters per minute
0.277778	meters per second
26,246.4	yards per day
1,093.6	yards per hour
18.226667	yards per minute
0.303777	yards per second
28,346.399997	varas (Texas) per day
1,181.1	varas (Texas) per hour
19.685	varas (Texas) per minute
0.328083	varas (Texas) per second
78,739.199997	feet per day
3,280.8	feet per hour
54.68	feet per minute
0.911333	feet per second
104,986.561622	spans per day
4,374.440070	spans per hour
72.907334	spans per minute
1.215122	spans per second
119,303.375990	links per day
4,970.974310	links per hour
82.849567	links per minute
1.380826	links per second
236,219.999933	hands per day
9,842.5	hands per hour
164.0416666	hands per minute
2.734028	hands per second
240,000	decimeters per day
10,000	decimeters per hour
166,666667	decimeters per minute
2.777778	decimeters per second
2,400,000	centimeters per day
100,000	centimeters per hour
1,666.666667	centimeters per minute
27.777778	centimeters per second
944,879.999904	inches per day
39,370	inches per hour
656.166667	inches per minute
10.936111	inches per second
24,000,000	millimeters per day
1,000,000	millimeters per hour
16,666.666667	millimeters per minute
277.777778	millimeters per second
0.5396	knots

KILOMETER PER MINUTE: =

894.772800	miles per day
37.282200	miles per hour

KILOMETER PER MINUTE: (cont'd)

0.62137	miles per minute
0.0103562	miles per second
1,440	kilometers per day
60	kilometers per hour
1	kilometers per minute
0.0166667	kilometers per second
7,158.202560	furlongs per day
298.258440	furlongs per hour
4.970974	furlongs per minute
0.0828496	furlongs per second
14,400	hektometers per day
600	hektometers per hour
10	hektometers per minute
0.166667	hektometers per second
71,582.0256029	chains per day
2,982.584400	chains per hour
49.709741	chains per minute
0.828496	chains per second
144,000	dekameters per day
6,000	dekameters per hour
100	dekameters per minute
1.666667	dekameters per second
286,327.553184	rods per day
11,930.314716	rods per hour
198.838579	rods per minute
3.313976	rods per second
1,440,000	meters per day
60,000	meters per hour
1,000	meters per minute
16.666667	meters per second
1,574,784	yards per day
65,616.000012	yards per hour
1,093.6	yards per minute
18.226667	yards per second
1,700,784	varas (Texas) per day
70,866.0	varas (Texas) per hour
1,181.1	varas (Texas) per minute
19.685	varas (Texas) per second
4,724,352	feet per day
196,848	feet per hour
3,280.0	feet per minute
54.68	feet per second
6,299,194	spans per day
262,466.404200	spans per hour
4,374.440070	spans per minute
72.907335	spans per second
7,158,203	links per day
298,258.440012	links per hour
4,970.974310	links per minute
82.849567	links per second
14,173,200	hands per day

KILOMETER PER MINUTE: (cont'd)

590,550	hands per hour
9,842.5	hands per minute
164.0416667	hands per second
14,400,000	decimeters per day
600,000	decimeters per hour
10,000	decimeters per minute
166.666667	decimeters per second
144,000,000	centimeters per day
6,000,000	centimeters per hour
100,000	centimeters per minute
1,666.666667	centimeters per second
56,692,800	inches per day
2,362,200	inches per hour
39,370	inches per minute
656.166667	inches per second
1,440,000,000	millimeters per day
60,000,000	millimeters per hour
1,000,000	millimeters per minute
16,666.666667	millimeters per second

KILOMETER PER SECOND: =

53,686.3680	miles per day
2,236.93200	miles per hour
37.28220	miles per minute
0.62137	miles per second
86,400	kilometers per day
3,600	kilometers per hour
60	kilometers per minute
1	kilometers per second
429,492.180384	furlongs per day
17,895.506400	furlongs per hour
298.258440	furlongs per minute
4.970974	furlongs per second
864,000	hektometers per day
36,000	hektometers per hour
600	hektometers per minute
10	hektometers per second
4,294,922	chains per day
178,955.064	chains per hour
2,982.58440	chains per minute
49.709741	chains per second
8,640,000	dekameters per day
360,000	dekameters per hour
6,000	dekameters per minute
100	dekameters per second
17,179,653	rods per day
715,818.88440	rods per hour
11,930.314740	rods per minute
198.838579	rods per second
86,400,000	meters per day

KILOMETER PER SECOND: (cont'd)

3,600,000	meters per hour
60,000	meters per minute
1,000	meters per second
94,487,040	yards per day
3,936,960	yards per hour
65,616	yards per minute
1,093.6	yards per second
102,047,040	varas (Texas) per day
4,251,960	varas (Texas) per hour
70,866	varas (Texas) per minute
1,181.1	varas (Texas) per second
283,461,120	feet per day
11,810,880	feet per hour
196,848	feet per minute
3,280.8	feet per second
377,951,622	spans per day
15,747,984	spans per hour
262,466.404200	spans per minute
4,374.440070	spans per second
429,492,180	links per day
17,895,508	links per hour
298,258.440	links per minute
4,970.974310	links per second
850,392,000	hands per day
35,433,000	hands per hour
590,550	hands per minute
9,842.5	hands per second
864,000,000	decimeters per day
36,000,000	decimeters per hour
600,000	decimeters per minute
10,000	decimeters per second
8,640,000,000	centimeters per day
360,000,000	centimeters per hour
6,000,000	centimeters per minute
100,000	centimeters per second
3,401,568,000	inches per day
141,732,000	inches per hour
2,362,200	inches per minute
39,370	inches per second
86,400,000,000	millimeters per day
3,600,000,000	millimeters per hour
60,000,000	millimeters per minute
1,000,000	millimeters per second

KILOMETER PER HOUR PER SECOND: =

0.000172594	miles per second per second
0.0002778	kilometers per second per second
0.002778	hektometers per second per second
0.02778	dekameters per second per second
0.2778	meters per second per second

KILOMETER PER HOUR PER SECOND: (cont'd)

0.303767	yards per second per second
0.9113	feet per second per second
2.778	decimeters per second per second
27.78	centimeters per second per second
10.9356	inches per second per second
277.8	millimeters per second per second

KILOLITER: =

1	kiloliters
1	cubic meters
1.3080	cubic yards
10	hektoliters
28.378	bushels (U.S.) dry
27.497	bushels (Imperial) dry
35.316	cubic feet
100	dekaliters
11.3513	pecks
264.18	gallons (U.S.) liquid
227.0574264	gallons (U.S.) dry
219.977402	gallons (Imperial)
1,056.72	quarts (liquid)
908.110825	quarts (dry)
1,000	liters
1,000	cubic decimeters
2,113.44	pints (liquid)
8,453.76	gills (liquid)
10,000	deciliters
34,607.58	cubic inches
100,000	centiliters
1,000,000	milliliters
1,000,000	cubic centimeters
1,000,000,000	cubic millimeters
1.101234	tons (short) of water @ 62°F.
0.983252	tons (long) of water @ 62°F.
1.0	tons (metric) of water @ 62°F.
1,000	kilograms of water @ 62°F.
2,202.46866	pounds (Avoir.) of water @ 62°F.
2,676.611	pounds (Troy) of water @ 62°F.
10,000	hektograms of water @ 62°F.
100,000	dekagrams of water @ 62°F.
32,119.331983	ounces (Troy) of water @ 62°F.
35,239.49856	ounces (Avoir.) of water @ 62°F.
256,954.655866	drams (Troy) of water @ 62°F.
563,831.97696	drams (Avoir.) of water @ 62°F.
642,386.639664	pennyweights of water @ 62°F.
770,863.967597	scruples (Avoir.)
1,000,000	grams of water @ 62°F.
10,000,000	decigrams of water @ 62°F.
15,417,281	grains of water @ 62°F.
100,000,000	centigrams of water @ 62°F.

KILOLITER: (cont'd)

1,000,000,000	milligrams of water @ 62°F.
33,815.04	ounces (fluid)
270,520.32	drams (fluid)
32.105795	sacks of cement

KILOWATT: =

63,725,184	foot pounds per day
2,655,216	foot pounds per hour
44,253.60	foot pounds per minute
737.56	foot pounds per second
764,702,208	inch pounds per day
31,862,592	inch pounds per hour
531,043.20	inch pounds per minute
8,850.72	inch pounds per second
20,640.09600	kilogram calories (mean) per day
860.004	kilogram calories (mean) per hour
14.33340	kilogram calories (mean) per minute
0.23889	kilogram calories (mean) per second
45,503.615916	pound calories (mean) per day
1,895.983996	pound calories (mean) per hour
31.599733	pound calories (mean) per minute
0.526662	pound calories (mean) per second
728,057.85472	ounce calories (mean) per day
30,335.743936	ounce calories (mean) per hour
505.595728	ounce calories (mean) per minute
8.426592	ounce calories (mean) per second
20,640,096	gram calories (mean) per day
860,004	gram calories (mean) per hour
14,333.40	gram calories (mean) per minute
238.89	gram calories (mean) per second
81,930.52800	BTU (mean) per day
3,413.77200	BTU (mean) per hour
56.89620	BTU (mean) per minute
0.94827	BTU (mean) per second
1	kilowatts
1,000	watts
3,600,000	joules
1.341	horsepower
1.3597	horsepower (metric)
1.3597	Cheval-vapeur hours
0.234	pounds carbon oxidized with 100% efficiency
3.52	pounds water evaporated from and at 212°F.
852.647040	kiloliter-atmospheres per day
35.52695	kiloliter-atmospheres per hour
0.592116	kiloliter-atmospheres per minute
0.0098686	kiloliter-atmospheres per second
852,647	liter-atmospheres per day
35,526.95	liter-atmospheres per hour
592.116	liter-atmospheres per minute
9,8686	liter-atmospheres per second

KILOWATT: (cont'd)

8,808,000 .... kilogram meters per day
367,000 .... kilogram meters per hour
6,116.666667 .... kilogram meters per minute
101.944444 .... kilogram meters per second

LINK (SURVEYORS): =

0.0001250 .... miles
0.000201168 .... kilometers
0.001 .... furlongs
0.00201168 .... hektometers
0.01 .... chains
0.0201168 .... dekameters
0.04 .... rods
0.201168 .... meters
0.22 .... yards
0.23760 .... varas (Texas)
0.66 .... feet
0.879998 .... spans
1.98 .... hands
2.011684 .... decimeters
20.11684 .... centimeters
7.92 .... inches
201.1684 .... millimeters
7,920 .... mils
201,168 .... microns
201,168,400 .... millimicrons
201,168,400 .... micromillimeters
312,447 .... wave lengths of red line of cadmium
201,168,400,000 .... Angstrom Units

LITER: =

0.001 .... kiloliters
0.001 .... cubic meters
0.0013080 .... cubic yards
0.00628995 .... barrels
0.01 .... hektoliters
0.028378 .... bushels (U.S.) dry
0.027497 .... bushels (Imperial) dry
0.0353144 .... cubic feet
0.1 .... dekaliters
0.113512 .... pecks (U.S.) dry
0.264178 .... gallons (U.S.) liquid
0.22702 .... gallons (U.S.) dry
0.21998 .... gallons (Imperial)
1.056710 .... quarts (liquid)
0.908102 .... quarts (dry)
1 .... cubic decimeters
1.8162 .... pints (U.S.) dry

LITER: (cont'd)

2.1134	pints (U.S.) liquid
7.0392	gills (Imperial)
8.4538	gills (U.S.)
10	deciliters
61.025	cubic inches
100	centiliters
1,000	milliliters
1,000	cubic centimeters
1,000,000	cubic millimeters
33.8147	ounces (U.S.) fluid
35.196	ounces (Imperial) fluid
270.5179	drams (fluid)
16,231.0740	minims
2.20462	pounds of water at maximum density

LITER ATMOSPHERE: =

2,404.59243	foot poundals
28,855.10916	inch poundals
74.738589	foot pounds
896.863068	inch pounds
0.0000242701	ton calories
0.0242071	kilogram calories
0.0533676	pound calories
0.242071	hektogram calories
2.42071	dekagram calories
0.853881	ounce calories
24.2071	gram calories (mean)
242.071	decigram calories
2,420.71	centigram calories
24,207.1	milligram calories
0.00000117258	kilowatt days
0.0000281419	kilowatt hours
0.00168852	kilowatt minutes
0.101311	kilowatt seconds
0.00117258	watt days
0.0281419	watt hours
1.688516	watt minutes
101.310932	watt seconds
0.010333	ton meters
10.333	kilogram meters
22.780362	pounds meters
103.33	hektogram meters
1,033.3	dekagram meters
364.485798	ounce meters
10,333	gram meters
103,330	decigram meters
1,033,300	centigram meters
10,333,000	milligram meters
0.00010333	ton hektometers
0.10333	kilogram hektometers

LITER ATMOSPHERE: (cont'd)

0.227804	pound hektometers
1.0333	hektogram hektometers
10.333	dekagram hektometers
3.644858	ounce hektometers
103.33	gram hektometers
1,033.3	decigram hektometers
10,333	centigram hektometers
103,330	milligram hektometers
0.0010330	ton dekameters
1.0333	kilogram dekameters
2.278036	pound dekameters
10.333	hektogram dekameters
103.33	dekagram dekameters
36.448583	ounce dekameters
1033.3	gram dekameters
10,333	decigram dekameters
103,330	centigram dekameters
1,033,300	milligram dekameters
0.0339001	ton feet
33.900951	kilogram feet
74.738796	pound feet
339.009507	hektogram feet
3,390.095072	dekagram feet
1,195.820731	ounce feet
33,900.950719	gram feet
339,009.50719	decigram feet
3,390,095	centigram feet
33,900,951	milligram feet
0.406811	ton inches
406.811409	kilogram inches
896.865507	pound inches
4,068.114086	hektogram inches
40,681.140863	dekagram inches
14,349.848105	ounce inches
406,811.408628	gram inches
4,068,114	decigram inches
40,681,141	centigram inches
406,811,409	milligram inches
0.10333	ton decimeters
103.33	kilogram decimeters
227.803622	pound decimeters
1,033.3	hektogram decimeters
10,333	dekagram decimeters
3,644.857956	ounce decimeters
103,330	gram decimeters
1,033,300	decigrams decimeters
10,333,000	centigram decimeters
103,330,000	milligram decimeters
1.0333	ton centimeters
1,033.3	kilogram centimeters
2,278.036223	pound centimeters

LITER ATMOSPHERE: (cont'd)

10,333	hektogram centimeters
103,330	dekagram centimeters
36,448.579561	ounce centimeters
1,033,300	gram centimeters
10,333,000	decigram centimeters
103,330,000	centigram centimeters
1,033,300,000	milligram centimeters
10.333	ton millimeters
10,333	kilogram millimeters
22,780.362226	pound millimeters
103,330	hektogram millimeters
1,033,300	dekagram millimeters
364,485.795614	ounce millimeters
10,333,000	gram millimeters
103,330,000	decigram millimeters
1,033,300,000	centigram millimeters
10,333,000,000	milligram millimeters
0.001	kiloliter-atmospheres
0.01	hektoliter-atmospheres
0.1	dekaliter-atmospheres
1	liter-atmospheres
10	deciliter-atmospheres
100	centiliter-atmospheres
1,000	millimeter-atmospheres
0.000000000001	cubic kilometer-atmospheres
0.000000001	cubic hektometer-atmospheres
0.000001	cubic dekameter-atmospheres
0.001	cubic meter-atmospheres
0.035319	cubic foot-atmospheres
61.025	cubic inch-atmospheres
1	cubic decimeter-atmospheres
1,000	cubic centimeter-atmospheres
1,000,000	cubic millimeter-atmospheres
101.328	joules (absolute)
0.00003827	Cheval-vapeur hours
0.00000157277	horsepower days
0.000037745	horsepower hours
0.00226479	horsepower minutes
0.135887	horsepower seconds
0.00000658398	pounds of carbon oxidized with 100% efficiency
0.00009907	pounds of water evaporated from and at 212°F.
43.573724	BTU (mean) per pound
0.09607	BTU (mean)
1,013,321,145	ergs

LITER PER DAY: =

0.001	kiloliters per day
0.0000416667	kiloliters per hour
0.000000694444	kiloliters per minute
0.0000000115741	kiloliters per second
0.001	cubic meters per day

0.0000416667 . . . . cubic meters per hour
0.000000694444 . . . . cubic meters per minute
0.0000000115741 . . . . cubic meters per second
0.0013080 . . . . cubic yards per day
0.0000545000 . . . . cubic yards per hour
0.000000908333 . . . . cubic yards per minute
0.0000000151389 . . . . cubic yards per second
0.00628996 . . . . barrels per day
0.000262082 . . . . barrels per hour
0.00000436803 . . . . barrels per minute
0.0000000728005 . . . . barrels per second
0.01 . . . . hektoliters per day
0.000416667 . . . . hektoliters per hour
0.00000694444 . . . . hektoliters per minute
0.000000115741 . . . . hektoliters per second
0.028378 . . . . bushels (U.S.—dry) per day
0.00118242 . . . . bushels (U.S.—dry) per hour
0.0000197069 . . . . bushels (U.S.—dry) per minute
0.000000328449 . . . . bushels (U.S.—dry) per second
0.027497 . . . . bushels (Imperial—dry) per day
0.00114571 . . . . bushels (Imperial—dry) per hour
0.0000190951 . . . . bushels (Imperial—dry) per minute
0.000000318252 . . . . bushels (Imperial—dry) per second
0.0353144 . . . . cubic feet per day
0.00147143 . . . . cubic feet per hour
0.000024539 . . . . cubic feet per minute
0.000000408731 . . . . cubic feet per second
0.1 . . . . dekaliters per day
0.00416667 . . . . dekaliters per hour
0.0000694444 . . . . dekaliters per minute
0.00000115741 . . . . dekaliters per second
0.113512 . . . . pecks (U.S.—dry) per day
0.00472967 . . . . pecks (U.S.—dry) per hour
0.0000788278 . . . . pecks (U.S.—dry) per minute
0.000000131380 . . . . pecks (U.S.—dry) per second
0.264178 . . . . gallons (U.S.—liquid) per day
0.0110074 . . . . gallons (U.S.—liquid) per hour
0.000183457 . . . . gallons (U.S.—liquid) per minute
0.00000305762 . . . . gallons (U.S.—liquid) per second
0.22702 . . . . gallons (U.S.—dry) per day
0.00945917 . . . . gallons (U.S.—dry) per hour
0.000157653 . . . . gallons (U.S.—dry) per minute
0.00000262755 . . . . gallons (U.S.—dry) per second
0.21998 . . . . gallons (Imperial) per day
0.00916583 . . . . gallons (Imperial) per hour
0.000152764 . . . . gallons (Imperial) per minute
0.00000254606 . . . . gallons (Imperial) per second
1.056710 . . . . quarts (liquid) per day
0.00440296 . . . . quarts (liquid) per hour
0.000733826 . . . . quarts (liquid) per minute
0.0000122304 . . . . quarts (liquid) per second

LITER PER DAY: (cont'd)

0.908102	quarts (dry) per day
0.0378376	quarts (dry) per hour
0.000630626	quarts (dry) per minute
0.0000105104	quarts (dry) per second
1	liters per day
0.0416667	liters per hour
0.000694444	liters per minute
0.0000115741	liters per second
1	cubic decimeters per day
0.0416667	cubic decimeters per hour
0.00069444	cubic decimeters per minute
0.0000115741	cubic decimeters per second
1.8162	pints (U.S.—dry) per day
0.0756750	pints (U.S.—dry) per hour
0.00126125	pints (U.S.—dry) per minute
0.0000210208	pints (U.S.—dry) per second
2.1134	pints (U.S.—liquid) per day
0.0880583	pints (U.S.—liquid) per hour
0.00146764	pints (U.S.—liquid) per minute
0.0000244606	pints (U.S.—liquid) per second
7.0392	gills (Imperial) per day
0.293300	gills (Imperial) per hour
0.00488833	gills (Imperial) per minute
0.0000814722	gills (Imperial) per second
8.4538	gills (U.S.) per day
0.352242	gills (U.S.) per hour
0.00587069	gills (U.S.) per minute
0.0000978449	gills (U.S.) per second
10	deciliters per day
0.416667	deciliters per hour
0.00694444	deciliters per minute
0.000115741	deciliters per second
61.025	cubic inches per day
2.542708	cubic inches per hour
0.0423785	cubic inches per minute
0.000706308	cubic inches per second
100	centiliters per day
4.166667	centiliters per hour
0.0694444	centiliters per minute
0.00115741	centiliters per second
1,000	milliliters per day
41.666667	milliliters per hour
0.694444	milliliters per minute
0.0115741	milliliters per second
1,000	cubic centimeters per day
41.666667	cubic centimeters per hour
0.694444	cubic centimeters per minute
0.0115741	cubic centimeters per second
1,000,000	cubic millimeters per day
41,666.666400	cubic millimeters per hour
694.444444	cubic millimeters per minute

LITER PER DAY: (cont'd)

11.574074	cubic millimeters per second
33.8147	ounces (U.S.) fluid per day
1.408946	ounces (U.S.) fluid per hour
0.0234824	ounces (U.S.) fluid per minute
0.000391374	ounces (U.S.) fluid per second
35.196	ounces (Imperial—fluid) per day
1.466500	ounces (Imperial—fluid) per hour
0.0244417	ounces (Imperial—fluid) per minute
0.000407361	ounces (Imperial—fluid) per second
270.5179	drams (fluid) per day
11.271579	drams (fluid) per hour
0.187860	drams (fluid) per minute
0.00313099	drams (fluid) per second
16,231.0740	minims per day
676.294747	minims per hour
11.271579	minims per minute
0.187860	minims per second

LITER PER HOUR: =

0.0240	kiloliters per day
0.001	kiloliters per hour
0.0000166667	kiloliters per minute
0.000000277778	kiloliters per second
0.0240	cubic meters per day
0.001	cubic meters per hour
0.0000166667	cubic meters per minute
0.000000277778	cubic meters per second
0.031392	cubic yards per day
0.0013080	cubic yards per hour
0.0000218	cubic yards per minute
0.000000363333	cubic yards per second
0.150959	barrels per day
0.00628995	barrels per hour
0.000104833	barrels per minute
0.00000174721	barrels per second
0.24	hektoliters per day
0.01	hektoliters per hour
0.000166667	hektoliters per minute
0.00000277778	hektoliters per second
0.681072	bushels (U.S.—dry) per day
0.028378	bushels (U.S.—dry) per hour
0.000472967	bushels (U.S.—dry) per minute
0.00000788278	bushels (U.S.—dry) per second
0.659928	bushels (Imperial—dry) per day
0.027497	bushels (Imperial—dry) per hour
0.000458283	bushels (Imperial—dry) per minute
0.00000763806	bushels (Imperial—dry) per second
0.84746	cubic feet per day
0.0353144	cubic feet per hour
0.000588573	cubic feet per minute

LITER PER HOUR: (cont'd)

0.00000980956	cubic feet per second
2.4	dekaliters per day
0.1	dekaliters per hour
0.00166667	dekaliters per minute
0.0000277778	dekaliters per second
2.724288	pecks (U.S.—dry) per day
0.113512	pecks (U.S.—dry) per hour
0.00189187	pecks (U.S.—dry) per minute
0.0000315311	pecks (U.S.—dry) per second
6.340272	gallons (U.S.—liquid) per day
0.264178	gallons (U.S.—liquid) per hour
0.00440297	gallons (U.S.—liquid) per minute
0.0000733828	gallons (U.S.—liquid) per second
5.448480	gallons (U.S.—dry) per day
0.22702	gallons (U.S.—dry) per hour
0.00378367	gallons (U.S.—dry) per minute
0.0000630611	gallons (U.S.—dry) per second
5.279520	gallons (Imperial) per day
0.21998	gallons (Imperial) per hour
0.00366633	gallons (Imperial) per minute
0.0000611056	gallons (Imperial) per second
25.361040	quarts (liquid) per day
1.056710	quarts (liquid) per hour
0.0176118	quarts (liquid) per minute
0.000293531	quarts (liquid) per second
21.794448	quarts (dry) per day
0.908102	quarts (dry) per hour
0.0151350	quarts (dry) per minute
0.000252251	quarts (dry) per second
24	liters per day
1	liters per hour
0.0166667	liters per minute
0.000277778	liters per second
24	cubic decimeters per day
1	cubic decimeters per hour
0.0166667	cubic decimeters per minute
0.000277778	cubic decimeters per second
43.58880	pints (U.S.—dry) per day
1.8162	pints (U.S.—dry) per hour
0.0302700	pints (U.S.—dry) per minute
0.0005045	pints (U.S.—dry) per second
50.721600	pints (U.S.—liquid) per day
2.1134	pints (U.S.—liquid) per hour
0.0352233	pints (U.S.—liquid) per minute
0.000587056	pints (U.S.—liquid) per second
168.940800	gills (Imperial) per day
7.0392	gills (Imperial) per hour
0.117320	gills (Imperial) per minute
0.00195533	gills (Imperial) per second
202.891200	gills (U.S.) per day
8.4538	gills (U.S.) per hour

LITER PER HOUR: (cont'd)

0.140897	gills (U.S.) per minute
0.00234828	gills (U.S.) per second
240	deciliters per day
10	deciliters per hour
0.166667	deciliters per minute
0.00277778	deciliters per second
1,464.6	cubic inches per day
61.025	cubic inches per hour
1.0170833	cubic inches per minute
0.0169514	cubic inches per second
2,400	centiliters per day
100	centiliters per hour
1.666667	centiliters per minute
0.0277778	centiliters per second
24,000	milliliters per day
1,000	milliliters per hour
16.66667	milliliters per minute
0.277778	milliliters per second
24,000	cubic centimeters per day
1,000	cubic centimeters per hour
16.666667	cubic centimeters per minute
0.277778	cubic centimeters per second
24,000,000	cubic millimeters per day
1,000,000	cubic millimeters per hour
16,666.666667	cubic millimeters per minute
277.777778	cubic millimeters per second
811.552800	ounces (U.S.) fluid per day
33.8147	ounces (U.S.) fluid per hour
0.563578	ounces (U.S.) fluid per minute
0.00939297	ounces (U.S.) fluid per second
844.704	ounces (Imperial—fluid) per day
35.196	ounces (Imperial—fluid) per hour
0.586600	ounces (Imperial—fluid) per minute
0.00977667	ounces (Imperial—fluid) per second
6,492.429599	drams (fluid) per day
270.5179	drams (fluid) per hour
4.508632	drams (fluid) per minute
0.0751439	drams (fluid) per second
389,545.775424	minims per day
16,231.0740	minims per hour
270.5179	minims per minute
4.508632	minims per second

LITER PER MINUTE: =

1.440	kiloliters per day
0.0600	kiloliters per hour
0.001	kiloliters per minute
0.0000166667	kiloliters per second
1.440	cubic meters per day
0.0600	cubic meters per hour

LITER PER MINUTE: (cont'd)

0.001	cubic meters per minute
0.0000166667	cubic meters per second
1.883520	cubic yards per day
0.0784800	cubic yards per hour
0.0013080	cubic yards per minute
0.0000218	cubic yards per second
9.0575383	barrels per day
0.377397	barrels per hour
0.00628996	barrels per minute
0.000104833	barrels per second
14.40	hektoliters per day
0.600	hektoliters per hour
0.01	hektoliters per minute
0.000166667	hektoliters per second
40.864320	bushels (U.S.—dry) per day
1.702680	bushels (U.S.—dry) per hour
0.028378	bushels (U.S.—dry) per minute
0.000472967	bushels (U.S.—dry) per second
39.595680	bushels (Imperial—dry) per day
1.649820	bushels (Imperial—dry) per hour
0.027497	bushels (Imperial—dry) per minute
0.000458283	bushels (Imperial—dry) per second
50.852736	cubic feet per day
2.118864	cubic feet per hour
0.0353144	cubic feet per minute
0.000588573	cubic feet per second
144.0	dekaliters per day
6.0	dekaliters per hour
0.1	dekaliters per minute
0.00166667	dekaliters per second
163.457280	pecks (U.S.—dry) per day
6.810720	pecks (U.S.—dry) per hour
0.113512	pecks (U.S.—dry) per minute
0.00189187	pecks (U.S.—dry) per second
380.416320	gallons (U.S.—liquid) per day
15.850680	gallons (U.S.—liquid) per hour
0.264178	gallons (U.S.—liquid) per minute
0.00440297	gallons (U.S.—liquid) per second
326.908800	gallons (U.S.—dry) per day
13.621200	gallons (U.S.—dry) per hour
0.22702	gallons (U.S.—dry) per minute
0.00378367	gallons (U.S.—dry) per second
316.771200	gallons (Imperial) per day
13.198800	gallons (Imperial) per hour
0.21998	gallons (Imperial) per minute
0.00366633	gallons (Imperial) per second
1,521.662400	quarts (liquid) per day
63.402600	quarts (liquid) per hour
1.056710	quarts (liquid) per minute
0.0176118	quarts (liquid) per second
1,307.666880	quarts (dry) per day

LITER PER MINUTE: (cont'd)

54.486120 . . . . quarts (dry) per hour
0.908102 . . . . quarts (dry) per minute
0.0151350 . . . . quarts (dry) per second
1,440 . . . . liters per day
60 . . . . liters per hour
1 . . . . liters per minute
0.0166667 . . . . liters per second
1,440 . . . . cubic decimeters per day
60 . . . . cubic decimeters per hour
1 . . . . cubic decimeters per minute
0.0166667 . . . . cubic decimeters per second
2,615.328 . . . . pints (U.S.—dry) per day
108.972 . . . . pints (U.S.—dry) per hour
1.8162 . . . . pints (U.S.—dry) per minute
0.0302700 . . . . pints (U.S.—dry) per second
3,043.296 . . . . pints (U.S.—liquid) per day
126.804 . . . . pints (U.S.—liquid) per hour
2.1134 . . . . pints (U.S.—liquid) per minute
0.0352233 . . . . pints (U.S.—liquid) per second
10,136.448 . . . . gills (Imperial) per day
422.352 . . . . gills (Imperial) per hour
7.0392 . . . . gills (Imperial) per minute
0.117320 . . . . gills (Imperial) per second
12,173.472003 . . . . gills (U.S.) per day
507.228 . . . . gills (U.S.) per hour
8.4538 . . . . gills (U.S.) per minute
0.140897 . . . . gills (U.S.) per second
14,400 . . . . deciliters per day
600 . . . . deciliters per hour
10 . . . . deciliters per minute
0.166667 . . . . deciliters per second
87,875.999971 . . . . cubic inches per day
3,661.500 . . . . cubic inches per hour
61.025 . . . . cubic inches per minute
1.0170833 . . . . cubic inches per second
144,000 . . . . centiliters per day
6,000 . . . . centiliters per hour
100 . . . . centiliters per minute
1.666667 . . . . centiliters per second
1,440,000 . . . . milliliters per day
60,000 . . . . milliliters per hour
1,000 . . . . milliliters per minute
16.666667 . . . . milliliters per second
1,440,000 . . . . cubic centimeters per day
60,000 . . . . cubic centimeters per hour
1,000 . . . . cubic centimeters per minute
16.666667 . . . . cubic centimeters per second
1,440,000,000 . . . . cubic millimeters per day
60,000,000 . . . . cubic millimeters per hour
1,000,000 . . . . cubic millimeters per minute
16,666.666667 . . . . cubic millimeters per second

LITER PER MINUTE: (cont'd)

48,693.167997 . . . . ounces (U.S.) fluid per day
2,028.882 . . . . ounces (U.S.) fluid per hour
33.8147 . . . . ounces (U.S.) fluid per minute
0.563578 . . . . ounces (U.S.) fluid per second
50,682.240 . . . . ounces (Imperial—fluid) per day
2,111.760 . . . . ounces (Imperial—fluid) per hour
35.196 . . . . ounces (Imperial—fluid) per minute
0.586600 . . . . ounces (Imperial—fluid) per second
389,545.776029 . . . . drams (fluid) per day
16,231.0740 . . . . drams (fluid) per hour
270.5179 . . . . drams (fluid) per minute
4.508632 . . . . drams (fluid) per second
23,372,747 . . . . minims per day
973,864.440 . . . . minims per hour
16,231.0740 . . . . minims per minute
270.5179 . . . . minims per second

LITER PER SECOND: =

86.4 . . . . kiloliters per day
3.60 . . . . kiloliters per hour
0.060 . . . . kiloliters per minute
0.001 . . . . kiloliters per second
86.4 . . . . cubic meters per day
3.60 . . . . cubic meters per hour
0.060 . . . . cubic meters per minute
0.001 . . . . cubic meters per second
113.0112 . . . . cubic yards per day
4.708800 . . . . cubic yards per hour
0.0784800 . . . . cubic yards per minute
0.0013080 . . . . cubic yards per second
543.451886 . . . . barrels per day
22.643829 . . . . barrels per hour
0.377397 . . . . barrels per minute
0.00628995 . . . . barrels per second
864 . . . . hektoliters per day
36 . . . . hektoliters per hour
0.60 . . . . hektoliters per minute
0.01 . . . . hektoliters per second
2,451.859200 . . . . bushels (U.S.—dry) per day
102.160800 . . . . bushels (U.S.—dry) per hour
1.702680 . . . . bushels (U.S.—dry) per minute
0.028378 . . . . bushels (U.S.—dry) per second
2,375.740800 . . . . bushels (Imperial—dry) per day
98.989200 . . . . bushels (Imperial—dry) per hour
1.649820 . . . . bushels (Imperial—dry) per minute
0.27497 . . . . bushels (Imperial—dry) per second
3,051.164160 . . . . cubic feet per day
127.131840 . . . . cubic feet per hour
2.118864 . . . . cubic feet per minute
0.0353144 . . . . cubic feet per second

LITER PER SECOND: (cont'd)

8,640	dekaliters per day
360	dekaliters per hour
6.0	dekaliters per minute
0.1	dekaliters per second
9,807.436800	pecks (U.S.—dry) per day
408.643200	pecks (U.S.—dry) per hour
6.810720	pecks (U.S.—dry) per minute
0.113512	pecks (U.S.—dry) per second
22,824.979200	gallons (U.S.—liquid) per day
951.040800	gallons (U.S.—liquid) per hour
15.850680	gallons (U.S.—liquid) per minute
0.264178	gallons (U.S.—liquid) per second
19,614.52800	gallons (U.S.—dry) per day
817.27200	gallons (U.S.—dry) per hour
13.62120	gallons (U.S.—dry) per minute
0.22702	gallons (U.S.—dry) per second
19,006.27200	gallons (Imperial) per day
791.92800	gallons (Imperial) per hour
13.19880	gallons (Imperial) per minute
0.21998	gallons (Imperial) per second
91,299.744	quarts (liquid) per day
3,804.156	quarts (liquid) per hour
63.402600	quarts (liquid) per minute
1.056710	quarts (liquid) per second
78,460.01280	quarts (dry) per day
3,269.167200	quarts (dry) per hour
54.486120	quarts (dry) per minute
0.908102	quarts (dry) per second
86,400	liters per day
3,600	liters per hour
60	liters per minute
1	liters per second
86,400	cubic decimeters per day
3,600	cubic decimeters per hour
60	cubic decimeters per minute
1	cubic decimeters per second
156,920	pints (U.S.—dry) per day
6,538.320	pints (U.S.—dry) per hour
108.9720	pints (U.S.—dry) per minute
1.8162	pints (U.S.—dry) per second
182,598	pints (U.S.—liquid) per day
7,608.2400	pints (U.S.—liquid) per hour
126.8040	pints (U.S.—liquid) per minute
2.1134	pints (U.S.—liquid) per second
608,187	gills (Imperial) per day
25,341.12	gills (Imperial) per hour
422.3520	gills (Imperial) per minute
7.0392	gills (Imperial) per second
730,408	gills (U.S.) per day
30,433.68	gills (U.S.) per hour
507.2280	gills (U.S.) per minute

LITER PER SECOND: (cont'd)

8.4538	gills (U.S.) per second
864,000	deciliters per day
36,000	deciliters per hour
600	deciliters per minute
10	deciliters per second
5,272,560	cubic inches per day
219,690	cubic inches per hour
3,661.500	cubic inches per minute
61.025	cubic inches per second
8,640,000	centiliters per day
360,000	centiliters per hour
6,000	centiliters per minute
100	centiliters per second
86,400,000	milliliters per day
3,600,000	milliliters per hour
60,000	milliliters per minute
1,000	milliliters per second
86,400,000	cubic centimeters per day
3,600,000	cubic centimeters per hour
60,000	cubic centimeters per minute
1,000	cubic centimeters per second
86,400,000,000	cubic millimeters per day
3,600,000,000	cubic millimeters per hour
60,000,000	cubic millimeters per minute
1,000,000	cubic millimeters per second
2,921,590	ounces (U.S.) fluid per day
121,733	ounces (U.S.) fluid per hour
2,028.8820	ounces (U.S.) fluid per minute
33.8147	ounces (U.S.) fluid per second
3,040,934	ounces (Imperial—fluid) per day
126,706	ounces (Imperial—fluid) per hour
2,111.76	ounces (Imperial—fluid) per minute
35.196	ounces (Imperial—fluid) per second
23,372,747	drams (fluid) per day
973,864	drams (fluid) per hour
16,231.0740	drams (fluid) per minute
270.5179	drams (fluid) per second
1,402,364,794	minims per day
58,431,866	minims per hour
973,864	minims per minute
16,231.0740	minims per second

METER: =

0.00053961	miles (nautical)
0.00062137	miles (statute)
0.001	kilometers
0.00497097	furlongs
0.01	hektometers
0.0497097	chains
0.1	dekameters

METER: (cont'd)

0.198839	rods
1	meters
1.093611	yards
1.811	varas (Texas)
3.280833	feet
4.374440	spans
4.970974	links
9.84250	hands
10	decimeters
100	centimeters
39.370	inches
1,000	millimeters
39,370	mils
1,000,000	microns
1,000,000,000	millimicrons
1,000,000,000	micromillimeters
1,553,164	wave lengths of red line cadmium
1,000,000	Angstrom Units
0.54681	fathoms

METER PER DAY: =

0.00053961	miles (nautical) per day
0.0000224837	miles (nautical) per hour
0.000000374729	miles (nautical) per minute
0.00000000624549	miles (nautical) per second
0.00062137	miles (statute) per day
0.0000258904	miles (statute) per hour
0.000000431507	miles (statute) per minute
0.00000000719178	miles (statute) per second
0.001	kilometers per day
0.0000416667	kilometers per hour
0.000000694444	kilometers per minute
0.0000000115741	kilometers per second
0.00497097	furlongs per day
0.000207124	furlongs per hour
0.00000345206	furlongs per minute
0.0000000575344	furlongs per second
0.01	hektometers per day
0.000416667	hektometers per hour
0.00000694444	hektometers per minute
0.000000115741	hektometers per second
0.0497097	chains per day
0.00207124	chains per hour
0.0000345206	chains per minute
0.000000575344	chains per second
0.1	dekameters per day
0.00416667	dekameters per hour
0.0000694444	dekameters per minute
0.00000115741	dekameters per second

METER PER DAY: (cont'd)

0.198839	rods per day
0.00828496	rods per hour
0.000138083	rods per minute
0.00000230138	rods per second
1	meters per day
0.0416667	meters per hour
0.000694444	meters per minute
0.0000115741	meters per second
1.093611	yards per day
0.0455671	yards per hour
0.000759452	yards per minute
0.0000126575	yards per second
1.1811	varas (Texas) per day
0.0492125	varas (Texas) per hour
0.000820208	varas (Texas) per minute
0.0000136701	varas (Texas) per second
3.280833	feet per day
0.133680	feet per hour
0.00222801	feet per minute
0.0000371334	feet per second
4.374440	spans per day
0.182268	spans per hour
0.00303781	spans per minute
0.0000506301	spans per second
4.970974	links per day
0.207124	links per hour
0.00345206	links per minute
0.0000575344	links per second
10	decimeters per day
0.416667	decimeters per hour
0.00694444	decimeters per minute
0.000115741	decimeters per second
100	centimeters per day
4.166667	centimeters per hour
0.0694444	centimeters per minute
0.00115741	centimeters per second
39.370	inches per day
1.640417	inches per hour
0.0273403	inches per minute
0.000455671	inches per second
1,000	millimeters per day
41.666667	millimeters per hour
0.694444	millimeters per minute
0.0115741	millimeters per second
39,370	mils per day
1,640.416667	mils per hour
27.340278	mils per minute
0.455671	mils per second
1,000,000	microns per day
41,666.666400	microns per hour
694.444444	microns per minute

METER PER DAY: (cont'd)

11.574074	microns per second
1,000,000	Angstrom Units per day
41,666.666400	Angstrom Units per hour
694.444444	Angstrom Units per minute
11.574074	Angstrom Units per second

METER PER HOUR: =

0.0129506	miles (nautical) per day
0.00053961	miles (nautical) per hour
0.00000899350	miles (nautical) per minute
0.000000149892	miles (nautical) per second
0.0149129	miles (statute) per day
0.00062137	miles (statute) per hour
0.0000103562	miles (statute) per minute
0.000000172603	miles (statute) per second
0.0240	kilometers per day
0.001	kilometers per hour
0.0000166667	kilometers per minute
0.000000277778	kilometers per second
0.119303	furlongs per day
0.00497097	furlongs per hour
0.0000828495	furlongs per minute
0.00000138083	furlongs per second
0.240	hektometers per day
0.01	hektometers per hour
0.000166667	hektometers per minute
0.00000277778	hektometers per second
1.193033	chains per day
0.0497097	chains per hour
0.000828495	chains per minute
0.0000138083	chains per second
2.4	dekameters per day
0.1	dekameters per hour
0.00166667	dekameters per minute
0.0000277778	dekameters per second
4.772136	rods per day
0.198839	rods per hour
0.00331398	rods per minute
0.0000552331	rods per second
24	meters per day
1	meters per hour
0.0166667	meters per minute
0.000277778	meters per second
26.246664	yards per day
1.093611	yards per hour
0.0182268	yards per minute
0.000303781	yards per second
28.346400	varas (Texas) per day
1.1811	varas (Texas) per hour
0.0196850	varas (Texas) per minute

METER PER HOUR: (cont'd)

0.000328083	varas (Texas) per second
78.74	feet per day
3.280833	feet per hour
0.0546806	feet per minute
0.000911343	feet per second
104.986560	spans per day
4.374440	spans per hour
0.0729073	spans per minute
0.00121512	spans per second
119.303280	links per day
4.970974	links per hour
0.0828495	links per minute
0.00138083	links per second
240	decimeters per day
10	decimeters per hour
0.166667	decimeters per minute
0.00277778	decimeters per second
2,400	centimeters per day
100	centimeters per hour
1.666667	centimeters per minute
0.0277778	centimeters per second
944.880	inches per day
39.370	inches per hour
0.656167	inches per minute
0.0109361	inches per second
24,000	millimeters per day
1,000	millimeters per hour
16.666667	millimeters per minute
0.277778	millimeters per second
944,879.999004	mils per day
39,370	mils per hour
656.166667	mils per minute
10.936111	mils per second
24,000,000	microns per day
1,000,000	microns per hour
16,666.666667	microns per minute
277.777778	microns per second
24,000,000	Angstrom Units per day
1,000.000	Angstrom Units per hour
16,666.666667	Angstrom Units per minute
277.777778	Angstrom Units per second

METER PER MINUTE: =

0.777038	miles (nautical) per day
0.0323766	miles (nautical) per hour
0.00053961	miles (nautical) per minute
0.00000899350	miles (nautical) per second
0.894773	miles (statute) per day
0.0372822	miles (statute) per hour
0.00062137	miles (statute) per minute

METER PER MINUTE: (cont'd)

0.0000103562	miles (statute) per second
1.440	kilometers per day
0.0600	kilometers per hour
0.001	kilometers per minute
0.0000166667	kilometers per second
7.158197	furlongs per day
0.298258	furlongs per hour
0.00497097	furlongs per minute
0.0000828495	furlongs per second
14.40	hektometers per day
0.600	hektometers per hour
0.01	hektometers per minute
0.000166667	hektometers per second
71.581968	chains per day
2.982582	chains per hour
0.0497097	chains per minute
0.000828495	chains per second
144.0	dekameters per day
6.0	dekameters per hour
0.1	dekameters per minute
0.00166667	dekameters per second
286.328160	rods per day
11.930340	rods per hour
0.198839	rods per minute
0.00331398	rods per second
1,440	meters per day
60	meters per hour
1	meters per minute
0.0166667	meters per second
1,574.799840	yards per day
65.616660	yards per hour
1.093611	yards per minute
0.0182269	yards per second
1,700.784	varas (Texas) per day
70.866	varas (Texas) per hour
1.1811	varas (Texas) per minute
0.0196850	varas (Texas) per second
4,724.400	feet per day
196.85	feet per hour
3.280833	feet per minute
0.0546806	feet per second
6,299.193600	spans per day
262.466400	spans per hour
4.374440	spans per minute
0.0729073	spans per second
7,158.196800	links per day
298.258200	links per hour
4.970974	links per minute
0.0828495	links per second
14,400	decimeters per day
600	decimeters per hour

METER PER MINUTE: (cont'd)

10	decimeters per minute
0.166667	decimeters per second
144,000	centimeters per day
6,000	centimeters per hour
100	centimeters per minute
1.666667	centimeters per second
56,692.800	inches per day
2,362.200	inches per hour
39.370	inches per minute
0.656167	inches per second
1,440,000	millimeters per day
60,000	millimeters per hour
1,000	millimeters per minute
16.666667	millimeters per second
56,692,800	mils per day
2,362,200	mils per hour
39,370	mils per minute
656.166667	mils per second
1,440,000,000	microns per day
60,000,000	microns per hour
1,000,000	microns per minute
16,666.666667	microns per second
1,440,000,000	Angstrom Units per day
60,000,000	Angstrom Units per hour
1,000,000	Angstrom Units per minute
16,666.666667	Angstrom Units per second

METER PER SECOND: =

46.622304	miles (nautical) per day
1.942596	miles (nautical) per hour
0.0323766	miles (nautical) per minute
0.00053961	miles (nautical) per second
53.686368	miles (statute) per day
2.236932	miles (statute) per hour
0.0372822	miles (statute) per minute
0.00062137	miles (statute) per second
86.4	kilometers per day
3.60	kilometers per hour
0.060	kilometers per minute
0.001	kilometers per second
429.491808	furlongs per day
17.895492	furlongs per hour
0.298258	furlongs per minute
0.00497097	furlongs per second
864	hektometers per day
36	hektometers per hour
0.60	hektometers per minute
0.01	hektometers per second
4,294.918080	chains per day
178.954920	chains per hour

METER PER SECOND: (cont'd)

2.982582	chains per minute
0.0497097	chains per second
8,640	dekameters per day
360	dekameters per hour
6.0	dekameters per minute
0.1	dekameters per second
17,179.689600	rods per day
715.820400	rods per hour
11.930340	rods per minute
0.198839	rods per second
86,400	meters per day
3,600	meters per hour
60	meters per minute
1	meters per second
94,487.990400	yards per day
3,936.999600	yards per hour
65.616660	yards per minute
1.093611	yards per second
102,047.0400	varas (Texas) per day
4,251.9600	varas (Texas) per hour
70.8660	varas (Texas) per minute
1.1811	varas (Texas) per second
283,463.971200	feet per day
11,810.998800	feet per hour
196.849980	feet per minute
3.280833	feet per second
377,951.616000	spans per day
15,747.984000	spans per hour
262.466400	spans per minute
4.374440	spans per second
429,492.153600	links per day
17,895.506400	links per hour
298.258440	links per minute
4.970974	links per second
864,000	decimeters per day
36,000	decimeters per hour
600	decimeters per minute
10	decimeters per second
8,640,000	centimeters per day
360,000	centimeters per hour
6,000	centimeters per minute
100	centimeters per second
3,401,568	inches per day
141,732.0	inches per hour
2,362.200	inches per minute
39.370	inches per second
86,400,000	millimeters per day
3,600,000	millimeters per hour
60,000	millimeters per minute
1,000	millimeters per second
3,401,568,000	mils per day

METER PER SECOND: (cont'd)

141,732,000	mils per hour
2,362,200	mils per minute
39,370	mils per second
86,400,000,000	microns per day
3,600,000,000	microns per hour
60,000,000	microns per minute
1,000,000	microns per second
86,400,000,000	Angstrom Units per day
3,600,000,000	Angstrom Units per hour
60,000,000	Angstrom Units per minute
1,000,000	Angstrom Units per second

METER OF MERCURY @ 32°F.:=

0.135964	hektometers of water @ 60°F.
1.359639	dekameters of water @ 60°F.
13.596390	meters of water @ 60°F.
44.604005	feet of water @ 60°F.
535.247985	inches of water @ 60°F.
135.963901	decimeters of water @ 60°F.
1,359.639013	centimeters of water @ 60°F.
13,596.39013	millimeters of water @ 60°F.
0.01	hektometers of mercury @ 32°F.
0.1	dekameters of mercury @ 32°F.
1	meters of mercury @ 32°F.
3.280833	feet of mercury @ 32°F.
39.37	inches of mercury @ 32°F.
10	decimeters of mercury @ 32°F.
100	centimeters of mercury @ 32°F.
1,000	millimeters of mercury @ 32°F.
149,862.505	tons per square hektometer
1,498.62505	tons per square dekameter
14.986251	tons per square meter
1.392300	tons per square foot
0.00966852	tons per square inch
0.149863	tons per square decimeter
0.00149863	tons per square centimeter
0.0000149863	tons per square millimeter
135,963,901	kilograms per square hektometer
1,359,639	kilograms per square dekameter
13,596.39	kilograms per square meter
1,263.034001	kilograms per square foot
8.771085	kilograms per square inch
135.9639	kilograms per square decimeter
1.359639	kilograms per square centimeter
0.0135964	kilograms per square millimeter
299,724,991	pounds per square hektometer
2,997,249	pounds per square dekameter
29,972.49	pounds per square meter
2,784.499982	pounds per square foot
19.337009	pounds per square inch

299.7249 . . . . pounds per square decimeter
2.997249 . . . . pounds per square centimeter
0.0299725 . . . . pounds per square millimeter
13,596,390 . . . . hektograms per square hektometer
135,963.90 . . . . hektograms per square dekameter
1,359.6390 . . . . hektograms per square meter
12,630.340015 . . . . hektograms per square foot
87.711006 . . . . hektograms per square inch
13.596390 . . . . hektograms per square decimeter
0.135964 . . . . hektograms per square centimeter
0.00135964 . . . . hektograms per square millimeter
13,596,390,130 . . . . dekagrams per square hektometer
135,963,901 . . . . dekagrams per square dekameter
1,359,639 . . . . dekagrams per square meter
126,303.4 . . . . dekagrams per square foot
877.110017 . . . . dekagrams per square inch
13,596.39 . . . . dekagrams per square decimeter
135.9639 . . . . dekagrams per square centimeter
1.359639 . . . . dekagrams per square millimeter
4,795,599,897 . . . . ounces per square hektometer
47,955,998 . . . . ounces per square dekameter
479,560 . . . . ounces per square meter
44,552.0 . . . . ounces per square foot
309.392019 . . . . ounces per square inch
4,795.5998 . . . . ounces per square decimeter
47.955998 . . . . ounces per square centimeter
0.479560 . . . . ounces per square millimeter
135,963,901,300 . . . . grams per square hektometer
1,359,639,013 . . . . grams per square dekameter
13,596,390 . . . . grams per square meter
1,263,034 . . . . grams per square foot
8,771.111 . . . . grams per square inch
135,963.90 . . . . grams per square decimeter
1,359.6390 . . . . grams per square centimeter
13.596390 . . . . grams per square millimeter
1,359,639,013 x 10³ . . . . decigrams per square hektometer
13,596,390,130 . . . . decigrams per square dekameter
135,963,900 . . . . decigrams per square meter
12,630,340 . . . . decigrams per square foot
87,711.11 . . . . decigrams per square inch
1,359,639 . . . . decigrams per square decimeter
13,596.390 . . . . decigrams per square centimeter
135.96390 . . . . decigrams per square millimeter
1,359,639,013 x 10⁴ . . . . centigrams per square hektometer
135,963,901,300 . . . . centigrams per square dekameter
1,359,639,013 . . . . centigrams per square meter
126,303.40 . . . . centigrams per square foot
877,111.1 . . . . centigrams per square inch
13,596,390 . . . . centigrams per square decimeter
135,963.9 . . . . centigrams per square centimeter
1,359.6390 . . . . centigrams per square millimeter

METER OF MERCURY @ 32°F.: (cont'd)

1,359,639,013 x 10^5	milligrams per square hektometer
1,359,639,013 x 10^3	milligrams per square dekameter
13,596,390,130	milligrams per square meter
1,263,034	milligrams per square foot
8,771,111	milligrams per square inch
135,963,901	milligrams per square decimeter
1,359,639	milligrams per square centimeter
13,596.390	milligrams per square millimeter
133,222 x 10^9	dynes per square hektometer
133,222 x 10^7	dynes per square dekameter
133,222 x 10^5	dynes per square meter
1,238,591,617	dynes per square foot
8,601,331	dynes per square inch
133,222,000	dynes per square decimeter
1,332,220	dynes per square centimeter
13,322.20	dynes per square millimeter
1.333300	bars
1.315801	atmospheres

METER OF WATER @ 60°F.:=

0.01	hektometers of water @ 60°F.
0.1	dekameters of water @ 60°F.
1.0	meters of water @ 60°F.
3.280833	feet of water @ 60°F.
39.37	inches of water @ 60°F.
10	decimeters of water @ 60°F.
100	centimeters of water @ 60°F.
1,000	millimeters of water @ 60°F.
0.000735510	hektometers of mercury @ 32°F.
0.00735510	dekameters of mercury @ 32°F.
0.0735510	meters of mercury @ 32°F.
2.413086	feet of mercury @ 32°F.
28.957029	inches of mercury @ 32°F.
0.735510	decimeters of mercury @ 32°F.
7.355103	centimeters of mercury @ 32°F.
73.55103	millimeters of mercury @ 32°F.
11,022.543742	tons per square hektometer
110.225437	tons per square dekameter
1.102254	tons per square meter
0.102405	tons per square foot
0.000711128	tons per square inch
0.0110225	tons per square decimeter
0.000110225	tons per square centimeter
0.00000110225	tons per square millimeter
10,000,000	kilograms per square hektometer
100,000	kilograms per square dekameter
1,000	kilograms per square meter
92.897452	kilograms per square foot
0.645121	kilograms per square inch
10	kilograms per square decimeter

METER OF WATER @ 60°F.: (cont'd)

0.10	kilograms per square centimeter
0.001	kilograms per square millimeter
22,045,074	pounds per square hektometer
220,450	pounds per square dekameter
2,204.5074	pounds per square meter
204.802897	pounds per square foot
1.422241	pounds per square inch
22.045074	pounds per square decimeter
0.220451	pounds per square centimeter
0.00220451	pounds per square millimeter
100,000,000	hektograms per square hektometer
1,000,000	hektograms per square dekameter
10,000	hektograms per square meter
928.974717	hektograms per square foot
6.45147	hektograms per square inch
100	hektograms per square decimeter
1	hektograms per square centimeter
0.01	hektograms per square millimeter
1,000,000,000	dekagrams per square hektometer
10,000,000	dekagrams per square dekameter
100,000	dekagrams per square meter
9,289.746972	dekagrams per square foot
64.512351	dekagrams per square inch
1,000	dekagrams per square decimeter
10	dekagrams per square centimeter
0.1	dekagrams per square millimeter
352,721,381	ounces per square hektometer
3,527,213	ounces per square dekameter
35,272.13	ounces per square meter
3,276.846123	ounces per square foot
22.755860	ounces per square inch
352.7213	ounces per square decimeter
3.527213	ounces per square centimeter
0.0352721	ounces per square millimeter
10,000,000,000	grams per square hektometer
100,000,000	grams per square dekameter
1,000,000	grams per square meter
92,897.469283	grams per square foot
645.124379	grams per square inch
10,000	grams per square decimeter
100	grams per square centimeter
1	grams per square millimeter
100,000,000,000	decigrams per square hektometer
1,000,000,000	decigrams per square dekameter
10,000,000	decigrams per square meter
928,975	decigrams per square foot
6,451.243790	decigrams per square inch
100,000	decigrams per square decimeter
1,000	decigrams per square centimeter
10	decigrams per square millimeter
10×10^{11}	centigrams per square hektometer

METER OF WATER @ 60°F.: (cont'd)

10,000,000,000	centigrams per square dekameter
100,000,000	centrigrams per square meter
9,289,747	centigrams per square foot
64,512.437904	centigrams per square inch
1,000,000	centigrams per square decimeter
10,000	centigrams per square centimeter
100	centigrams per square millimeter
10×10^{12}	milligrams per square hektometer
10×10^{10}	milligrams per square dekameter
1,000,000,000	milligrams per square meter
92,897,468	milligrams per square foot
645,124	milligrams per square inch
10,000,000	milligrams per square decimeter
100,000	milligrams per square centimeter
1,000	milligrams per square millimeter
9,798,617,182,254	dynes per square hektometer
97,986,171,823	dynes per square dekameter
979,861,718	dynes per square meter
91,099,700	dynes per square foot
632,637	dynes per square inch
9,798,617	dynes per square decimeter
97,986.171823	dynes per square centimeter
979.861718	dynes per square millimeter
0.0966778	bars
0.0967785	atmospheres

MICRON: =

0.000000000621259	miles (statute)
0.000000001	kilometers
0.00000000497097	furlongs
0.00000001	hektometers
0.0000000497097	chains
0.0000001	dekameters
0.000000198839	rods
0.000001	meters
0.00000109358	yards
0.0000011811	varas (Texas)
0.0000032808	feet
0.00000437440	spans
0.00000497097	links
0.0000098425	hands
0.00001	decimeters
0.0001	centimeters
0.00003937	inches
0.001	millimeters
0.039370	mils
1	microns
1,000	millimicrons
1,000	micromillimeters
1.553159	wave lengths of red line of cadmium
10,000	Angstrom Units

MIL: =

0.00000001578	miles (statute)
0.0000000254	kilometers
0.000000126263	furlongs
0.000000254	hektometers
0.00000126263	chains
0.00000254	dekameters
0.00000505051	rods
0.0000254	meters
0.000027777	yards
0.00003	varas (Texas)
0.000083333	feet
0.000111111	spans
0.000126263	links
0.00025	hands
0.000254	decimeters
0.00254	centimeters
0.001	inches
0.0254	millimeters
1	mils
25.4	microns
25,400	millimicrons
25,400	micromillimeters
39.450445	wave lengths of red line of cadmium
25.4	Angstrom Units

MILE (STATUTE): =

0.86836	miles (nautical)
1	miles (statute)
1.60935	kilometers
8	furlongs
16.0935	hektometers
80	chains
160,935	dekameters
320	rods
1,609.35	meters
1,760	yards
1,900.8	varas (Texas)
5,280	feet
7,040	spans
8,000	links
15,840	hands
16,093.5	decimeters
160,935	centimeters
63,360	inches
1,609,350	millimeters
63,360,000	mils
1,609,344,000	microns
$1,609,344 \times 10^6$	millimicrons

MILE (STATUTE): (cont'd)

1,609,344 x 10^6	micromillimeters
2,499,572,909	wave lengths of red line of cadmium
16,093,440 x 10^6	Angstrom Units

MILE (STATUTE) PER DAY: =

0.86836	miles (nautical) per day
0.0361817	miles (nautical) per hour
0.000603028	miles (nautical) per minute
0.0000100505	miles (nautical) per second
1	miles (statute) per day
0.0416667	miles (statute) per hour
0.000694444	miles (statute) per minute
0.0000115741	miles (statute) per second
1.60935	kilometers per day
0.0670562	kilometers per hour
0.00111760	kilometers per minute
0.0000186267	kilometers per second
8	furlongs per day
0.333333	furlongs per hour
0.00555556	furlongs per minute
0.0000925926	furlongs per second
16.0935	hektometers per day
0.670562	hektometers per hour
0.0111760	hektometers per minute
0.000186267	hektometers per second
80	chains per day
3.333333	chains per hour
0.0555556	chains per minute
0.000925926	chains per second
160.935	dekameters per day
6.705625	dekameters per hour
0.111760	dekameters per minute
0.00186267	dekameters per second
320	rods per day
13.333333	rods per hour
0.222222	rods per minute
0.00370370	rods per second
1,609.35	meters per day
67.056250	meters per hour
1.117604	meters per minute
0.0186267	meters per second
1,760	yards per day
73.333333	yards per hour
1.222222	yards per minute
0.0203703	yards per second
1,900.8	varas (Texas) per day
79.2	varas (Texas) per hour
1.320	varas (Texas) per minute
0.0220	varas (Texas) per second
5,280	feet per day

MILE (STATUTE) PER DAY: (cont'd)

220 . . . . feet per hour
3.666667 . . . . feet per minute
0.06111111 . . . . feet per second
7,040 . . . . spans per day
293.333333 . . . . spans per hour
4.888889 . . . . spans per minute
0.0814814 . . . . spans per second
8,000 . . . . links per day
333.333333 . . . . links per hour
5.555556 . . . . links per minute
0.0925925 . . . . links per second
15,840 . . . . hands per day
660 . . . . hands per hour
11 . . . . hands per minute
0.183333 . . . . hands per second
16,093.5 . . . . decimeters per day
670.56250 . . . . decimeters per hour
11.176042 . . . . decimeters per minute
0.186267 . . . . decimeters per second
160,935 . . . . centimeters per day
6,705.624996 . . . . centimeters per hour
111.760417 . . . . centimeters per minute
1.862674 . . . . centimeters per second
63,360 . . . . inches per day
2,640 . . . . inches per hour
44 . . . . inches per minute
0.733333 . . . . inches per second
1,609,350 . . . . millimeters per day
67,056.249960 . . . . millimeters per hour
1,117.604166 . . . . millimeters per minute
18.626736 . . . . millimeters per second

MILE (STATUTE) PER HOUR: =

20.84064 . . . . miles (nautical) per day
0.86836 . . . . miles (nautical) per hour
0.0144727 . . . . miles (nautical) per minute
0.000241211 . . . . miles (nautical) per second
24 . . . . miles (statute) per day
1 . . . . miles (statute) per hour
0.0166667 . . . . miles (statute) per minute
0.000277778 . . . . miles (statute) per second
38.6244 . . . . kilometers per day
1.60935 . . . . kilometers per hour
0.0268225 . . . . kilometers per minute
0.000447042 . . . . kilometers per second
192 . . . . furlongs per day
8 . . . . furlongs per hour
0.133333 . . . . furlongs per minute
0.00222222 . . . . furlongs per second
386.244 . . . . hektometers per day

MILE (STATUTE) PER HOUR: (cont'd)

16.0935	hektometers per hour
0.268225	hektometers per minute
0.00447042	hektometers per second
1,920	chains per day
80	chains per hour
1.333333	chains per minute
0.0222222	chains per second
3,862.44	dekameters per day
160.935	dekameters per hour
2.682250	dekameters per minute
0.0447042	dekameters per second
7,680	rods per day
320	rods per hour
5.333333	rods per minute
0.0888889	rods per second
38,624.4	meters per day
1,609.35	meters per hour
26.8225	meters per minute
0.447042	meters per second
42,240	yards per day
1,760	yards per hour
29.333333	yards per minute
0.488889	yards per second
45,619.2	varas (Texas) per day
1,900.8	varas (Texas) per hour
31.68	varas (Texas) per minute
0.528	varas (Texas) per second
126,720	feet per day
5,280	feet per hour
88	feet per minute
1.466667	feet per second
168,960	spans per day
7,040	spans per hour
117.333333	spans per minute
1.955556	spans per second
192,000	links per day
8,000	links per hour
133.333333	links per minute
2.222222	links per second
380,160	hands per day
15,840	hands per hour
264	hands per minute
4.40	hands per second
386,244	decimeters per day
16,093.5	decimeters per hour
268.225	decimeters per minute
4.470417	decimeters per second
3,862,440	centimeters per day
160,935	centimeters per hour
2,682.25	centimeters per minute
44.704167	centimeters per second

MILE (STATUTE) PER HOUR: (cont'd)

1,520,640	inches per day
63,360	inches per hour
1,056	inches per minute
17.6	inches per second
38,624,400	millimeters per day
1,609,350	millimeters per hour
26,822.5	millimeters per minute
447.0416667	millimeters per second

MILE (STATUTE) PER MINUTE: =

1,250.438400	miles (nautical) per day
52.101600	miles (nautical) per hour
0.86836	miles (nautical) per minute
0.0144727	miles (nautical) per second
1,440	miles (statute) per day
60	miles (statute) per hour
1	miles (statute) per minute
0.01666667	miles (statute) per second
2,317.464	kilometers per day
96.561	kilometers per hour
1.60935	kilometers per minute
0.0268225	kilometers per second
11,520	furlongs per day
480	furlongs per hour
8	furlongs per minute
0.133333	furlongs per second
23,174.64	hektometers per day
965.61	hektometers per hour
16.0935	hektometers per minute
0.268225	hektometers per second
115,200	chains per day
4,800	chains per hour
80	chains per minute
1.333333	chains per second
231,746	dekameters per day
9,656.1	dekameters per hour
160.935	dekameters per minute
2.682250	dekameters per second
460,800	rods per day
19,200	rods per hour
320	rods per minute
5.333333	rods per second
2,317,464	meters per day
96,561	meters per hour
1,609.35	meters per minute
26.82250	meters per second
2,534,400	yards per day
105,600	yards per hour
1,760	yards per minute
29.333333	yards per second

MILE (STATUTE) PER MINUTE: (cont'd)

2,737,152	varas (Texas) per day
114,048	varas (Texas) per hour
1,900.8	varas (Texas) per minute
31.680	varas (Texas) per second
7,603,200	feet per day
316,800	feet per hour
5,280	feet per minute
88	feet per second
10,137,600	spans per day
422,400	spans per hour
7,040	spans per minute
117.333333	spans per second
11,520,000	links per day
480,000	links per hour
8,000	links per minute
133.333333	links per second
22,809,600	hands per day
950,400	hands per hour
15,840	hands per minute
264.0	hands per second
23,174,640	decimeters per day
965,610	decimeters per hour
16,093.5	decimeters per minute
268.225	decimeters per second
231,746,400	centimeters per day
9,656,100	centimeters per hour
160,935	centimeters per minute
2,682.25	centimeters per second
91,238,400	inches per day
3,801,600	inches per hour
63,360	inches per minute
1,056.0	inches per second
2,317,464,000	millimeters per day
96,561,000	millimeters per hour
1,609,350	millimeters per minute
26,822.5	millimeters per second

MILE (STATUTE) PER SECOND: =

75,026.304	miles (nautical) per day
3,126.096	miles (nautical) per hour
52.10260	miles (nautical) per minute
0.86836	miles (nautical) per second
86,400	miles (statute) per day
3,600	miles (statute) per hour
60	miles (statute) per minute
1	miles (statute) per second
139,048	kilometers per day
5,793.660	kilometers per hour
96.561	kilometers per minute
1.60935	kilometers per second

MILE (STATUTE) PER SECOND: (cont'd)

691,200	furlongs per day
28,800	furlongs per hour
480	furlongs per minute
8	furlongs per second
1,390,480	hektometers per day
57,936.6	hektometers per hour
965.61	hektometers per minute
16.0935	hektometers per second
6,912,000	chains per day
288,000	chains per hour
4,800	chains per minute
80	chains per second
13,904,800	dekameters per day
579,366	dekameters per hour
9,656.1	dekameters per minute
160.935	dekameters per second
27,648,000	rods per day
1,152,000	rods per hour
19,200	rods per minute
320	rods per second
139,048,000	meters per day
5,793,660	meters per hour
96,561.0	meters per minute
1,609.35	meters per second
152,064,000	yards per day
6,336,000	yards per hour
105,600	yards per minute
1,760	yards per second
164,229,120	varas (Texas) per day
6,842,880	varas (Texas) per hour
114,048	varas (Texas) per minute
1,900.8	varas (Texas) per second
456,192,000	feet per day
19,008,000	feet per hour
316,800	feet per minute
5,280	feet per second
608,256,000	spans per day
25,344,000	spans per hour
422,400	spans per minute
7,040	spans per second
691,200,000	links per day
28,800,000	links per hour
480,000	links per minute
8,000	links per second
1,390,478,000	decimeters per day
57,936,600	decimeters per hour
965,610	decimeters per minute
16,093.5	decimeters per second
13,904,780,000	centimeters per day
579,366,000	centimeters per hour
9,656,100	centimeters per minute

MILE (STATUTE) PER SECOND: (cont'd)

160,935	centimeters per second
5,474,304,000	inches per day
228,096,000	inches per hour
3,801,600	inches per minute
63,360	inches per second
139,047,800,000	millimeters per day
5,793,660,000	millimeters per hour
96,561,000	millimeters per minute
1,609,350	millimeters per second

MIL: =

0.0000000137061	miles (nautical)
0.00000001578	miles (statute)
0.0000000254	kilometers
0.000000126263	furlongs
0.000000254	hektometers
0.00000126263	chains
0.00000254	dekameters
0.00000505052	rods
0.0000254	meters
0.0000277778	yards
0.00003	varas (Texas)
0.0000833333	feet
0.000111111	spans
0.000126263	links
0.000250	hands
0.000254	decimeters
0.00254	centimeters
0.001	inches
0.0254	millimeters
1	mils
25.4	microns
25,400	millimicrons
25,400	micromillimeters
39.450445	wave lengths of red line of cadmium
25.4	Angstrom Units

MILLIMETER: =

0.00000053961	miles (nautical)
0.00000062137	miles (statute)
0.000001	kilometers
0.00000497097	furlongs
0.00001	hektometers
0.0000497097	chains
0.0001	dekameters
0.000198839	rods
0.001	meters
0.00109361	yards
0.0011811	varas (Texas)

MILLIMETER: (cont'd)

0.00328083	feet
0.00437444	spans
0.00497097	links
0.00984250	hands
0.01	decimeters
0.1	centimeters
0.039370	inches
1	millimeters
39.370	mils
1,000	microns
1,000,000	millimicrons
1,000,000	micromillimeters
1,553.164	wave lengths of red line of cadmium
1,000	Angstrom Units

MINUTE (ANGLE): =

0.00018519	quadrants
0.000290888	radians
0.0166667	degrees
60	seconds
0.0000462963	circumference or revolutions

MILLIGRAM: =

0.00000000098426	tons (long)
0.000000001	tons (metric)
0.00000000110231	tons (short)
0.000001	kilograms
0.00000267923	pounds (Troy)
0.00000220462	pounds (Avoir.)
0.00001	hektograms
0.0001	dekagrams
0.0000321507	ounces (Troy)
0.0000352739	ounces (Avoir.)
0.001	grams
0.01	decigrams
0.1	centigrams
1	milligrams
0.0154324	grains
0.000257206	drams (Troy)
0.000564383	drams (Avoir.)
0.000643015	pennyweights
0.000771618	scruples
0.005	carats (metric)

MILLIGRAM PER LITER: =

1	parts per million

MILLILITER: =

0.000001	kiloliters
0.000001	cubic meters
0.0000013080	cubic yards
0.00000628995	barrels
0.00001	hektoliters
0.000028378	bushels (U.S.—dry)
0.000027497	bushels (Imperial—dry)
0.000353144	cubic feet
0.0001	dekaliters
0.000113512	pecks (U.S.—dry)
0.000264178	gallons U.S.—liquid)
0.00022702	gallons (U.S.—dry)
0.00021998	gallons (Imperial)
0.00105671	quarts (liquid)
0.000908102	quarts (dry)
0.001	liters
0.001	cubic decimeters
0.0018162	pints (U.S.—dry)
0.0021134	pints (U.S.—liquid)
0.0073092	gills (Imperial)
0.0084538	gills (U.S.)
0.01	deciliters
0.0610234	cubic inches
0.1	centiliters
1	milliliters
1	cubic centimeters
1,000	cubic millimeters
0.0338147	ounces (U.S.—fluid)
0.035196	ounces (Imperial—fluid)
0.270518	drams (fluid)
16.231074	minims
0.00220462	pounds of water @ maximum density

OHM (ABSOLUTE): =

0.00000000000111263	electrostatic cgs unit or statohm
0.000001	megohm (absolute)
0.99948	International ohm
1,000,000	microhms (absolute)
1,000,000,000	electromagnetic cgs or abohms

ohm per kilometer = 0.3048 ohms per 1,000 feet
ohm per 1,000 feet = 3.280833 ohms per kilometer
ohm per 1,000 yards = 1.0936 ohms per kilometer

OUNCE (AVOIRDUPOIS): =

0.0000279018	tons (long)
0.0000283495	tons (metric)

528

OUNCE (AVOIRDUPOIS): (cont'd)

0.00003125	tons (short)
0.0282495	kilograms
0.0759549	pounds (Troy)
0.0625	pounds (Avoir.)
0.283495	hektograms
2.834953	dekagrams
0.9114583	ounces (Troy)
1	ounces (Avoir.)
28.349527	grams
283.49527	decigrams
2,834.9527	centigrams
28,349.527	milligrams
437.5	grains
7.29166	drams (Troy)
16	drams (Avoir.)
18.22917	pennyweights
21.875	scruples
141.75	carats (metric)

OUNCE (FLUID): =

0.0000295729	kiloliters
0.0000295729	cubic meters
0.0000386814	cubic yards
0.000186012	barrels
0.000295729	hektoliters
0.000839221	bushels (U.S.—dry)
0.000813167	bushels (Imperial—dry)
0.00104435	cubic feet
0.00295729	dekaliters
0.00335688	pecks (U.S.—dry)
0.00781252	gallons U.S.—liquid)
0.00671365	gallons (U.S.—dry)
0.00650545	gallons (Imperial)
0.03125	quarts (liquid)
0.0268552	quarts (dry)
0.0295729	liters
0.0295729	cubic decimeters
0.0537104	pints (U.S.—dry)
0.0625	pints (U.S.—liquid)
0.208170	gills (Imperial)
0.25	gills (U.S.)
0.295729	deciliters
1.80469	cubic inches
2.957294	centiliters
29.572937	milliliters
29.572937	cubic centimeters
29,572.9372	cubic millimeters
1	ounces (U.S.—fluid)
1.0408491	ounces (Imperial—fluid)

OUNCE (FLUID): (cont'd)

8	drams (fluid)
480	minims
0.0651972	pounds of water @ maximum density

OUNCE (TROY): =

0.0000306122	tons (long)
0.0000311034	tons (metric)
0.000034285	tons (short)
0.0311035	kilograms
0.0833333	pounds (Troy)
0.0685714	pounds (Avoir.)
0.311035	hektograms
3.110348	dekagrams
1	ounces (Troy)
1.09714	ounces (Avoir.)
31.103481	grams
311.03481	decigrams
3,110,3481	centigrams
31,103.481	milligrams
480	grains
8	drams (Troy)
17.55428	drams (Avoir.)
20	pennyweights
24	scruples
155.52	carats (metric)

OUNCE (WEIGHT) PER SQUARE INCH: =

0.000439419	hektometers of water @ 60°F.
0.00439419	dekameters of water @ 60°F.
0.0439419	meters of water @ 60°F.
0.144174	feet of water @ 60°F.
1.730092	inches of water @ 60°F.
0.439419	decimeters of water @ 60°F.
4.394188	centimeters of water @ 60°F.
43.941875	millimeters of water @ 60°F.
0.0000323219	hektometers of mercury @ 32°F.
0.000323219	dekameters of mercury @ 32°F.
0.00323219	meters of mercury @ 32°F.
0.0106042	feet of mercury @ 32°F.
0.127250	inches of mercury @ 32°F.
0.0323219	decimeters of mercury @ 32°F.
0.323219	centimeters of mercury @ 32°F.
3.232188	millimeters of mercury @ 32°F.
484.379356	tons per square hektometer
4.843794	tons per square dekameter
0.0484379	tons per square meter
0.00450014	tons per square foot
0.0000312500	tons per square inch
0.000484379	tons per square decimeter

0.00000484379 . . . . tons per square centimeter
0.0000000484379 . . . . tons per square millimeter
439,419 . . . . kilograms per square hektometer
4,394.1875 . . . . kilograms per square dekameter
43.941875 . . . . kilograms per square meter
4.0823252 . . . . kilograms per square foot
0.0283494 . . . . kilograms per square inch
0.439419 . . . . kilograms per square decimeter
0.00439419 . . . . kilograms per square centimeter
0.0000439419 . . . . kilograms per square millimeter
968,758 . . . . pounds per square hektometer
9,687.58 . . . . pounds per square dekameter
96.8758 . . . . pounds per square meter
90 . . . . pounds per square foot
0.0625 . . . . pounds per square inch
0.968758 . . . . pounds per square decimeter
0.00968758 . . . . pounds per square centimeter
0.0000968758 . . . . pounds per square millimeter
4,394,190 . . . . hektograms per square hektometer
43,941.875 . . . . hektograms per square dekameter
439.41875 . . . . hektograms per square meter
40.823252 . . . . hektograms per square foot
0.283494 . . . . hektograms per square inch
4.394188 . . . . hektograms per square decimeter
0.043919 . . . . hektograms per square centimeter
0.000439419 . . . . hektograms per square millimeter
43,941,900 . . . . dekagrams per square hektometer
439,419 . . . . dekagrams per square dekameter
4,394.1875 . . . . dekagrams per square meter
408.232519 . . . . dekagrams per square foot
2.834944 . . . . dekagrams per square inch
43.941875 . . . . dekagrams per square decimeter
0.439419 . . . . dekagrams per square centimeter
0.00439419 . . . . dekagrams per square millimeter
15,500,139 . . . . ounces per square hektometer
155,001 . . . . ounces per square dekameter
1,550.0139 . . . . ounces per square meter
144 . . . . ounces per square foot
1 . . . . ounces per square inch
15.500139 . . . . ounces per square decimeter
0.155001 . . . . ounces per square centimeter
0.00155001 . . . . ounces per square millimeter
439,419,000 . . . . grams per square hektometer
4,394,190 . . . . grams per square dekameter
43,941.875 . . . . grams per square meter
4,082.325187 . . . . grams per square foot
28.349438 . . . . grams per square inch
439.4187 . . . . grams per square decimeter
4.394187 . . . . grams per square centimeter
0.0439419 . . . . grams per square millimeter
4,394,190,000 . . . . decigrams per square hektometer

OUNCE (WEIGHT) PER SQUARE INCH: (cont'd)

43,941,900 . . . . decigrams per square dekameter
439,419 . . . . decigrams per square meter
40,823.25187 . . . . decigrams per square foot
283.494375 . . . . decigrams per square inch
4,394.1875 . . . . decigrams per square decimeter
43.94187 . . . . decigrams per square centimeter
0.439419 . . . . decigrams per square millimeter
43,941,900,000 . . . . centigrams per square hektometer
439,419,000 . . . . centigrams per square dekameter
4,394,190 . . . . centigrams per square meter
408,233 . . . . centigrams per square foot
2,834.943750 . . . . centigrams per square inch
43,941.875 . . . . centigrams per square decimeter
439.41875 . . . . centigrams per square centimeter
4.394188 . . . . centigrams per square millimeter
439,419,000,000 . . . . milligrams per square hektometer
4,394,190,000 . . . . milligrams per square dekameter
43,941,900 . . . . milligrams per square meter
4,082,325 . . . . milligrams per square foot
28,349.43750 . . . . milligrams per square inch
439,419 . . . . milligrams per square decimeter
4,394.1875 . . . . milligrams per square centimeter
43.941875 . . . . milligrams per square millimeter
430,920,000,000 . . . . dynes per square hektometer
4,309,200,000 . . . . dynes per square dekameter
43,092,000 . . . . dynes per square meter
4,003,324 . . . . dynes per square foot
28,050.875 . . . . dynes per square inch
430,920 . . . . dynes per square decimeter
4,309.2 . . . . dynes per square centimeter
43.092 . . . . dynes per square millimeter
0.00430919 . . . . bars
0.00425288 . . . . atmosphere

PARTS PER MILLION: =

0.00000110231 . . . . tons (net) per cubic meter
0.001 . . . . kilograms per cubic meter
0.00220462 . . . . pounds (Avoir.) per cubic meter
0.0352739 . . . . ounces (Avoir.) per cubic meter
1.0 . . . . grams per cubic meter
15.4324 . . . . grains per cubic meter
0.000000175250 . . . . tons (net) per barrel
0.000158984 . . . . kilograms per barrel
0.000350499 . . . . pounds (Avoir.) per barrel
0.00560799 . . . . ounces (Avoir.) per barrel
0.158984 . . . . grams per barrel
2.453505 . . . . grains per barrel
0.0000000312133 . . . . tons (net) per cubic foot
0.0000283162 . . . . kilograms per cubic foot
0.0000624264 . . . . pounds (Avoir.) per cubic foot

UNIVERSAL CONVERSION FACTORS

PARTS PER MILLION: (cont'd)

0.000998823 . . . . ounces (Avoir.) per cubic foot
0.0283162 . . . . grams per cubic foot
0.436987 . . . . grains per cubic foot
0.00000000417262 . . . . tons (net) per gallon (U.S.—liquid)
0.00000378524 . . . . kilograms per gallon (U.S.—liquid)
0.00000834522 . . . . pounds (Avoir.) per gallon (U.S.—liquid)
0.000133524 . . . . ounces (Avoir.) per gallon (U.S.—liquid)
0.00378534 . . . . grams per gallon (U.S.—liquid)
0.0584168 . . . . grains per gallons (U.S.—liquid)
0.00000000501107 . . . . grains per gallon (Imperial—liquid)
0.00000454585 . . . . kilograms per gallon (Imperial—liquid)
0.0000100221 . . . . pounds (Avoir.) per gallon (Imp.—liquid)
0.000160355 . . . . ounces (Avoir.) per gallon (Imp.—liquid)
0.00454597 . . . . grams per gallon (Imperial—liquid)
0.0701552 . . . . grains per gallon (Imperial—liquid)
0.00000000110231 . . . . tons (net) per liter
0.000001 . . . . kilograms per liter
0.00000220462 . . . . pounds (Avoir.) per liter
0.0000352739 . . . . ounces (Avoir.) per liter
0.001 . . . . grams per liter
0.0154324 . . . . grains per liter
0.0000000000180663 . . . . tons (net) per cubic inch
0.0000000163867 . . . . kilograms per cubic inch
0.0000000361264 . . . . pounds (Avoir.) per cubic inch
0.000000578023 . . . . ounces (Avoir.) per cubic inch
0.0000163867 . . . . grams per cubic inch
0.000252886 . . . . grains per cubic inch
8.345 . . . . pounds per million gallons

POISE: =

1 . . . . gram per centimeter per second

PENNYWEIGHT: =

0.00000153061 . . . . tons (long)
0.00000155517 . . . . tons (metric)
0.00000171429 . . . . tons (net
0.00155517 . . . . kilograms
0.0041667 . . . . pounds (Troy)
0.00342857 . . . . pounds (Avoir.)
0.0155517 . . . . hektograms
0.155517 . . . . dekagrams
0.05 . . . . ounces (Troy)
0.0548571 . . . . ounces (Avoir.)
1.55517 . . . . grams
15.5517 . . . . decigrams
155.517 . . . . centigrams
1,555.17 . . . . milligrams
24 . . . . grains

PENNYWEIGHT: (cont'd)

0.4	drams (Troy)
0.877714	drams (Avoir.)
1	pennyweights
1.2	scruples
7.776	carats (metric)

POUND (TROY): =

0.000367347	tons (long)
0.000373242	tons (metric)
0.000411429	tons (net
0.373242	kilograms
1	pounds (Troy)
0.822857	pounds (Avoir.)
3.732418	hektograms
37.324176	dekagrams
12	ounces (Troy)
13.165714	ounces (Avoir.)
373.241762	grams
3,732.417621	decigrams
37,324.176213	centigrams
373,242	milligrams
5,760	grains
96	drams (Troy)
210.651425	drams (Avoir.)
240	pennyweights
288	scruples
1,866.239964	carats (metric)

POUND (AVOIRDUPOIS): =

0.000446429	tons (long)
0.000453593	tons (metric)
0.0005	tons (net
0.453592	kilograms
1.215278	pounds (Troy)
1	pounds (Avoir.)
4.535924	hektograms
45.359243	dekagrams
14.5833	ounces (Troy)
16	ounces (Avoir.)
453.592428	grams
4,535.92428	decigrams
45,359.2428	centigrams
453,592	milligrams
7,000	grains
116.666675	drams (Troy)
256	drams (Avoir.)
291.6667	pennyweights
350.1	scruples
2,268	carats (metric)

POUND OF CARBON OXIDIZED WITH 100% EFFICIENCY: =

365,245,567 . . . . foot poundals
4,382,946,806 . . . . inch poundals
11,352,418 . . . . foot pounds
136,229,016 . . . . inch pounds
3.676937 . . . . ton calories
3,676.937455 . . . . kilogram calories
8,106.271602 . . . . pound calories
36,769.374545 . . . . hektogram calories
367,694 . . . . dekagram calories
129,700 . . . . ounce calories
3,676,937 . . . . gram calories (mean)
36,769,375 . . . . decigram calories
367,693,745 . . . . centigram calories
3,676,937,455 . . . . milligram calories
0.178109 . . . . kilowatt days
4.274614 . . . . kilowatt hours
256.477745 . . . . kilowatt minutes
15,388.634345 . . . . kilowatt seconds
178.109039 . . . . watt days
4,274.613901 . . . . watt hours
256,477 . . . . watt minutes
15,388,624 . . . . watt seconds
1,569.531035 . . . . ton meters
1,569,513 . . . . kilogram meters
3,460,223 . . . . pounds meters
15,695,310 . . . . hektogram meters
156,953,104 . . . . dekagram meters
55,363,570 . . . . ounce meters
1,569,531,035 . . . . gram meters
15,695,310,350 . . . . decigram meters
156,953,103,500 . . . . centigram meters
1,569,531,035 x 10^3 . . . . milligram meters
15.695310 . . . . ton hektometers
15,695.31035 . . . . kilogram hektometers
34,602.288580 . . . . pound hektometers
156,953 . . . . hektogram hektometers
1,569,531 . . . . dekagram hektometers
553,636 . . . . ounce hektometers
15,695,310 . . . . gram hektometers
156,953,104 . . . . decigram hektometers
1,569,531,035 . . . . centigram hektometers
15,695,310,350 . . . . milligram hektometers
156.907535 . . . . ton dekameters
156,907 . . . . kilgram dekameters
346,022 . . . . pound dekameters
1,569,075 . . . . hektogram dekameters
15,690.754 . . . . dekagram dekameters
5,536,358 . . . . ounce dekameters
156,907,535 . . . . gram dekameters

POUND OF CARBON OXIDIZED WITH 100% EFFICIENCY: (cont'd)

1,569,075,350	decigram dekameters
15,690,753,500	centigram dekameters
156,907,535 x 10³	milligram dekameters
5,149.255690	ton feet
5,149,256	kilogram feet
11,352,449	pound feet
51,492,557	hektogram feet
514,925,569	dekagram feet
181,639,190	ounce feet
5,149,255,690	gram feet
51,492,556,895	decigram feet
514,925,568,950	centigram feet
51,492,556.895 x 10²	milligram feet
61,792,556845	ton inches
61,792,557	kilogram inches
136,229,386	pound inches
617,925,568	hektogram inches
6,179,255,685	dekagram inches
2,179,670,177	ounce inches
61,792,556,845	gram inches
617,925,568,450	decigram inches
61,792,556,845 x 10²	centigram inches
61,792,556,845 x 10³	milligram inches
15,695.31035	ton decimeters
15,695,310	kilogram decimeters
34,602,231	pound decimeters
156,953,104	hektogram decimeters
1,569,531,035	dekagram decimeters
553,635,699	ounce decimeters
15,695,310,350	gram decimeters
156,953,103,500	decigram decimeters
1,569,531,035 x 10³	centigram decimeters
1,569,531,035 x 10⁴	milligram decimeters
156,953	ton centimeters
156,953,104	kilogram centimeters
346,022,312	pound centimeters
1,569,531,035	hektogram centimeters
15,695,310,350	dekagram centimeters
5,536,356,992	ounce centimeters
156,953,103,500	gram centimeters
1,569,531,035 x 10³	decigram centimeters
1,569,531,035 x 10⁴	centigram centimeters
1,569,531,035 x 10⁵	milligram centimeters
1,569,531	ton millimeters
1,569,531,035	kilogram millimeters
3,460,223,121	pound millimeters
15,695,310,350	hektogram millimeters
156,953,103,500	dekagram millimeters
55,363,569,923	ounces millimeters
1,569,531,035 x 10³	gram millimeters
1,569,531,035 x 10⁴	decigram millimeters

POUND OF CARBON OXIDIZED WITH 100% EFFICIENCY: (cont'd)

1,569,531,035 x 10^5 . . . . centigram millimeters
1,569,531,035 x 10^6 . . . . milligram millimeters
151.895 . . . . kiloliter-atmospheres
1,518.95 . . . . hektoliter-atmospheres
15,189.5 . . . . dekaliter-atmospheres
151,895 . . . . liter-atmospheres
1,518,950 . . . . deciliter-atmospheres
15,189,500 . . . . centiliter-atmospheres
151,895,000 . . . . milliliter-atmospheres
0.000000151895 . . . . cubic kilometer-atmospheres
0.000151895 . . . . cubic hektometer atmospheres
0.151895 . . . . cubic dekameter-atmospheres
151.895 . . . . cubic meter-atmospheres
5,364.779505 . . . . cubic foot-atmospheres
9,269,392 . . . . cubic inch-atmospheres
151,895 . . . . cubic decimeter-atmospheres
151,895,000 . . . . cubic centimeter-atmospheres
151,895 x 10^6 . . . . cubic millimeter-atmospheres
15,391,217 . . . . joules (absolute)
5.813022 . . . . Cheval-vapeur hours
0.238896 . . . . horespower days
5,733277 . . . . horsepower hours
344.0102771 . . . . horsepower minutes
20,640.555865 . . . . horsepower seconds
1 . . . . pounds of carbon oxidized with 100% efficiency
15.0482377 . . . . pounds of water evaporated from and at 212° F.
6,618,631 . . . . BTU (mean) per pound
14,592.55265 . . . . BTU (mean)
153,918,415,320 x 10^3 . . . . ergs

POUND (PRESSURE) PER SQUARE INCH: =

0.0070307 . . . . hektometers of water @ 60°F.
0.070307 . . . . dekameters of water @ 60°F.
0.70307 . . . . meters of water @ 60°F.
2.306787 . . . . feet of water @ 60°F.
27.681473 . . . . inches of water @ 60°F.
7.0307 . . . . decimeters of water @ 60°F.
70,307 . . . . centimeters of water @ 60°F.
703.07 . . . . millimeters of water @ 60°F.
0.00051715 . . . . hektometers of mercury @ 32°F.
0.0051715 . . . . dekameters of mercury @ 32°F.
0.051715 . . . . meters of mercury @ 32°F.
0.169667 . . . . feet of mercury @ 32°F.
2.0360 . . . . inches of mercury @ 32°F.
0.51715 . . . . decimeters of mercury @ 32°F.
5.1715 . . . . centimeters of mercury @ 32°F.
51.715 . . . . millimeters of mercury @ 32°F.
7,750.0696898 . . . . tons per square hektometer
77.500697 . . . . tons per square dekameter
0.775007 . . . . tons per square meter
0.0720023 . . . . tons per quare foot

POUND (PRESSURE) PER SQUARE INCH: (cont'd)

0.0005	tons per square inch
0.00775007	tons per square decimeter
0.0000775007	tons per square centimeter
0.000000775007	tons per square millimeter
7,030,700	kilograms per square hektometer
70,307	kilograms per square hektometer
703.07	kilograms per square meter
65.317203	kilograms per square foot
0.453592	kilograms per square inch
7.0307	kilograms per square decimeter
0.070307	kilograms per square centimeter
0.00070307	kilograms per square millimeter
15,500,130	pounds per square hektometer
155,001	pounds per square dekameter
1,550.0130	pounds per square meter
144	pounds per square foot
1	pounds per square inch
15.500130	pounds per square decimeter
0.155001	pounds per square centimeter
0.00155001	pounds per square millimeter
70,307,000	hektograms per square hektometer
703,070	hektograms per square dekameter
7,030.70	hektograms per square meter
653.172168	hektograms per square foot
4.535933	hektograms per square inch
70.3070	hektograms per square decimeter
0.70307	hektograms per square centimeter
0.0070307	hektograms per square millimeter
703,070,000	dekagrams per square hektometer
7,030,700	dekagrams per square dekameter
70,307	dekagrams per square meter
6,531.721544	dekagrams per square foot
45.359332	dekagrams per square inch
7,030.7	dekagrams per square decimeter
70.307	dekagrams per square centimeter
0.70307	dekagrams per square millimeter
248,002,217	ounces per square hektometer
2,480,022	ounces per square dekameter
24,800.22	ounces per square meter
2,303.985941	ounces per square foot
16	ounces per square inch
248.0022	ounces per square decimeter
2.480022	ounces per square centimeter
0.0248002	ounces per square millimeter
7,030,700,000	grams per square hektometer
70,307,000	grams per square dekameter
703,070	grams per square meter
65,317.215135	grams per square foot
453.593927	grams per square inch
7,030.7	grams per square decimeter
70.307	grams per square centimeter

POUND (PRESSURE) PER SQUARE INCH: (cont'd)

0.70307	grams per square millimeter
70,307,000,000	decigrams per square hektometer
703,070,000	decigrams per square dekameter
7,030,700	decigrams per square meter
653,172	decigrams per square foot
4,535.939265	decigrams per square inch
70,307	decigrams per square decimeter
703.07	decigrams per square centimeter
7.0307	decigrams per square millimeter
703,070,000,000	centigrams per square hektometer
7,030,700,000	centigrams per square dekameter
70,307,000	centigrams per square meter
6,531,720	centigrams per square foot
45,359.392595	centigrams per square inch
703,070	centigrams per square decimeter
7,030.70	centigrams per square centimeter
70.3070	centigrams per square millimeter
$70,307 \times 10^8$	milligrams per square hektometer
$70,307 \times 10^6$	milligrams per square dekameter
$70,307 \times 10^4$	milligrams per square meter
65,317,200	milligrams per square foot
453,594	milligrams per square inch
70,307,000	milligrams per square decimeter
703,070	milligrams per square centimeter
7,030.70	milligrams per square millimeter
$68,947 \times 10^8$	dynes per square hektometer
68,947,000,000	dynes per square dekameters
689,470,000	dynes per square meter
64,053,184	dynes per square foot
448,814	dynes per square inch
6,894,700	dynes per square decimeter
68,947	dynes per square centimeter
689.47	dynes per square millimeter
0.068947	bars
0.068046	atmospheres

POUND (WEIGHT) PER SQUARE FOOT: =

0.0000488243	hektometers of water @ 60°F.
0.000488243	dekameters of water @ 60°F.
0.00488243	meters of water @ 60°F.
0.0160194	feet of water @ 60°F.
0.192232	inches of water @ 60°F.
0.0488243	decimeters of water @ 60°F.
0.488243	centimeters of water @ 60°F.
4.882431	millimeters of water @ 60°F.
0.00000359132	hektometers of mercury @ 32°F.
0.0000359132	dekameters of mercury @ 32°F.
0.000359132	meters of mercury @ 32°F.
0.00117824	feet of mercury @ 32°F.
0.0141389	inches of mercury @ 32°F.

POUND (WEIGHT) PER SQUARE FOOT: (cont'd)

0.00359132	decimeters of mercury @ 32°F.
0.0359132	centimeters of mercury @ 32°F.
0.359132	millimeters of mercury @ 32°F.
53.819930	tons per square hektometer
0.538199	tons per square dekameter
0.00538199	tons per square meter
0.0005000016	tons per square foot
0.00000347222	tons per square inch
0.0000538199	tons per square decimeter
0.000000538199	tons per square centimeter
0.00000000538199	tons per square millimeter
48,824.305552	kilograms per square hektometer
488.243056	kilograms per square dekameter
4.882431	kilograms per square meter
0.453592	kilograms per square foot
0.00314994	kilograms per square inch
0.0488243	kilograms per square decimeter
0.000488243	kilograms per square centimeter
0.00000488243	kilograms per square millimeter
107,640	pounds per square hektometer
1,076.397917	pounds per square dekameter
10.763979	pounds per square meter
1	pounds per square foot
0.00694444	pounds per square inch
0.107640	pounds per square decimeter
0.00107640	pounds per square centimeter
0.0000107640	pounds per square millimeter
488,243	hektograms per square hektometer
4,882.430555	hektograms per square dekameter
48.824306	hektograms per square meter
4.535918	hektograms per square foot
0.0314994	hektograms per square inch
0.488243	hektograms per square decimeter
0.00488243	hektograms per square centimeter
0.0000488243	hektograms per square millimeter
4,882,431	dekagrams per square hektometer
48,824.305552	dekagrams per square dekameter
488.243056	dekagrams per square meter
45.359178	dekagrams per square foot
0.314994	dekagrams per square inch
4.882431	dekagrams per square decimeter
0.0488243	dekagrams per square centimeter
0.000488243	dekagrams per square millimeter
1,722,238	ounces of per square hektometer
17,222.376179	ounces of per square dekameter
172.223762	ounces per square meters
16	ounces per square foot
0.111111	ounces per square inch
1.722238	ounces per square decimeter
0.0172224	ounces per square centimeter
0.000172224	ounces per square millimeter

48,824,306	grams per square hektometer
488,243	grams per square dekameter
4,882.430555	grams per square meter
453.591783	grams per square inch
3.149953	grams per square inch
48.824306	grams per square decimeter
0.488243	grams per square centimeter
0.00488243	grams per square millimeter
488,243,056	decigrams per square hektometer
4,882,431	decigrams per square dekameter
48,824.305552	decigrams per square meter
4,535.917833	decigrams per square foot
3.149953	decigrams per square inch
488.243056	decigrams per square decimeter
4.882431	decigrams per square centimeter
0.0488243	decigrams per square millimeter
4,882,430,555	centigrams per square hektometer
48,824,306	centigrams per square dekameter
488,243	centigrams per square meter
45,359.178330	centigrams per square foot
31.499535	centigrams per square inch
4,882.430555	centigrams per square decimeter
48.824306	centigrams per square centimeter
0.488243	centigrams per square millimeter
48,824,305,552	milligrams per square hektometer
488,243,056	milligrams per square dekameter
4,882.431	milligrams per square meter
453,592	milligrams per square foot
314.995347	milligrams per square inch
48,824.305552	milligrams per square decimeter
488.243056	milligrams per square centimeter
4.882431	milligrams per square millimeter
47,879,860,000	dynes per square hektometer
478,798,600	dynes per square dekameter
4,787,986	dynes per square meter
444,814	dynes per square foot
3,116.763889	dynes per square inch
47,879.861108	dynes per square decimeter
478.798611	dynes per square centimeter
4.787986	dynes per square millimeter
0.00047880	bars
0.00047254	atmospheres

POUND PER INCH: =

19.685057	tons (net) per kilometer
31.680146	tons (net) per mile
0.0196851	tons (net) per meter
0.018	tons (net) per yard
0.006	tons (net) per foot
0.0005	tons (net) per inch

POUND PER INCH: (cont'd)

0.000196851	tons (net) per centimeter
0.0000196851	tons (net) per millimeter
17,857.985915	kilograms per kilometer
28,739.749194	kilograms per mile
17.857985	kilograms per meter
16.329328	kilograms per yard
5.442109	kilograms per foot
0.453592	kilograms per inch
0.178579	kilograms per centimeter
0.0178579	kilograms per millimeter
39,370.11300	pounds (Avoir.) per kilometer
63,360.291357	pounds (Avoir.) per mile
39.370113	pounds (Avoir.) per meter
36	pounds (Avoir.) per yard
12	pounds (Avoir.) per foot
1	pounds (Avoir.) per inch
0.393701	pounds (Avoir.) per centimeter
0.0393701	pounds (Avoir.) per millimeter
629,921.80800	ounces (Avoir.) per kilometer
1,013,764	ounces (Avoir.) per mile
629.9211808	ounces (Avoir.) per meter
576	ounces (Avoir.) per yard
192	ounces (Avoir.) per foot
16	ounces (Avoir.) per inch
6.299218	ounces (Avoir.) per centimeter
0.629921	ounces (Avoir.) per millimeter
17,857,985	grams per kilometer
28,739,794	grams per mile
17,857.9851	grams per meter
16,329.327396	grams per yard
5,443.109132	grams per foot
453.5924277	grams per inch
178.579851	grams per centimeter
17.857985	grams per millimeter
275,590,791	grains per kilometer
443,522,039	grains per mile
275,590.791	grains per meter
252,000	grains per yard
84,000	grains per foot
7,000	grains per inch
2,755.90791	grains per centimeter
275.590791	grains per millimeter

POUND (AVOIR.) PER FOOT: =

1.640239	tons (net) per kilometer
2.639719	tons (net) per mile
0.00164024	tons (net) per meter
0.00149984	tons (net) per yard
0.000499946	tons (net) per foot
0.0000416622	tons (net) per inch

POUND (AVOIR.) PER FOOT: (cont'd)

0.0000164024	tons (net) per centimeter
0.00000164024	tons (net) per millimeter
1,488.161203	kilograms per kilometer
2,394.71280	kilograms per mile
1.488161	kilograms per meter
1.360777	kilograms per yard
0.453592	kilograms per foot
0.0377994	kilograms per inch
0.00148816	kilograms per centimeter
0.00148816	kilograms per millimeter
3,280.8	pounds (Avoir.) per kilometer
5,280	pounds (Avoir.) per mile
3.280833	pounds (Avoir.) per meter
3	pounds (Avoir.) per yard
1	pounds (Avoir.) per foot
0.0833333	pounds (Avoir.) per inch
0.0328083	pounds (Avoir.) per centimeter
0.00328083	pounds (Avoir.) per millimeter
52,493.328	ounces (Avoir.) per kilometer
84,480	ounces (Avoir.) per mile
52.493328	ounces (Avoir.) per meter
48	ounces (Avoir.) per yard
16	ounces (Avoir.) per foot
1.333333	ounces (Avoir.) per inch
0.524933	ounces (Avoir.) per centimeter
0.0524933	ounces (Avoir.) per millimeter
1,488,161	grams per kilometer
2,394,713	grams per mile
1,488.161	grams per meter
1,360.777283	grams per yard
453.5924277	grams per foot
37.799369	grams per inch
14.881612	grams per centimeter
1.488161	grams per millimeter
22,965,899	grains per kilometer
36,960,000	grains per mile
22,965.899250	grains per meter
21,000	grains per yard
7,000	grains per foot
583.333333	grains per inch
229.658993	grains per centimeter
22.965899	grains per millimeter

POUND (WEIGHT) PER CUBIC FOOT: =

17,657.261726	tons (net) per cubic hektometer
17.657262	tons (net) per cubic dekameter
0.01765731	tons (net) per cubic meter
0.0005	tons (net) per cubic foot
0.000000289352	tons (net) per cubic inch
0.0000176573	tons (net) per cubic decimeter

POUND (WEIGHT) PER CUBIC FOOT: (cont'd)

0.0000000176573	tons (net) per cubic centimeter
0.0000000000176573	tons (net) per cubic millimeter
16,018,400	kilograms per cubic hektometer
16,018.4	kilograms per cubic dekameter
16,0184	kilograms per cubic meter
0.453593	kilograms per cubic foot
0.000262496	kilograms per cubic inch
0.0160184	kilograms per cubic decimeter
0.0000160184	kilograms per cubic centimeter
0.0000000160184	kilograms per cubic millimeter
35,314,445	pounds per cubic hektometer
35,314.445	pounds per cubic dekameter
35.314445	pounds per cubic meter
1	pounds per cubic foot
0.000578704	pounds per cubic inch
0.0353144	pounds per cubic decimeter
0.0000353144	pounds per cubic centimeter
0.0000000353144	pounds per cubic millimeter
160,184,000	hektograms per cubic hektometer
160,184	hektograms per cubic dekameter
160.184	hektograms per cubic meter
4.535934	hektograms per cubic foot
0.00262496	hektograms per cubic inch
0.160184	hektograms per cubic decimeter
0.000160184	hektograms per cubic centimeter
0.000000160184	hektograms per cubic millimeter
1,601,840,000	dekagrams per cubic hektometer
1,601,840	dekagrams per cubic dekameter
1,601.840	dekagrams per cubic meter
45.359342	dekagrams per cubic foot
0.0262496	dekagrams per cubic inch
1.60184	dekagrams per cubic decimeter
0.00160184	dekagrams per cubic centimeter
0.00000160184	dekagrams per cubic millimeter
565,031,120	ounces per cubic hektometer
565,031	ounces per cubic dekameter
565.031120	ounces per cubic meter
16	ounces per cubic foot
0.00925926	ounces per cubic inch
0.565031	ounces per cubic decimeter
0.000565031	ounces per cubic centimeter
0.000000565031	ounces per cubic millimeter
16,018,400,000	grams per cubic hektometer
16,018,400	grams per cubic dekameter
16,018.4	grams per cubic meter
453.593422	grams per cubic foot
0.262496	grams per cubic inch
16.0184	grams per cubic decimeter
0.0160184	grams per cubic centimeter
0.0000160184	grams per cubic millimeter
160,184,000,000	decigrams per cubic hektometer

160,184,000	decigrams per cubic dekameter
160,184	decigrams per cubic meter
4,535.934220	decigrams per cubic foot
2.624962	decigrams per cubic inch
160.184	decigrams per cubic decimeter
0.160184	decigrams per cubic centimeter
0.000160184	decigrams per cubic millimeter
$160,184 \times 10^7$	centigrams per cubic hektometer
1,601,840,000	centigrams per cubic dekameter
1,601,840	centigrams per cubic meter
45,359.342205	centigrams per cubic foot
26.249619	centigrams per cubic inch
1,601.840	centigrams per cubic decimeter
1.60184	centigrams per cubic centimeter
0.00160184	centigrams per cubic millimeter
$160,184 \times 10^8$	milligrams per cubic hektometer
16,018,400,000	milligrams per cubic dekameter
16,018,400	milligrams per cubic meter
453,593	milligrams per cubic foot
262.496193	milligrams per cubic inch
16,018.4	milligrams per cubic decimeter
16.0184	milligrams per cubic centimeter
0.0160184	milligrams per cubic millimeter
$157,085,886 \times 10^9$	dynes per cubic hektometer
$157,086 \times 10^6$	dynes per cubic dekameter
157,086,000	dynes per cubic meter
444,820	dynes per cubic foot
257.418981	dynes per cubic inch
157,086	dynes per cubic decimeter
157.0858859	dynes per cubic centimeter
0.157086	dynes per cubic millimeter

POUND PER CUBIC INCH: =

30,511,748	tons (net) per cubic hektometer
30,511.748042	tons (net) per cubic dekameter
30.511748	tons (net) per cubic meter
0.864001	tons (net) per cubic foot
0.0005	tons (net) per cubic inch
0.0305117	tons (net) per cubic decimeter
0.0000305117	tons (net) per cubic centimeter
0.0000000305117	tons (net) per cubic millimeter
27,679,795,200	kilograms per cubic hektometer
27,679,795	kilograms per cubic dekameter
27,679.7952	kilograms per cubic meter
783.809037	kilograms per cubic foot
0.453593	kilograms per cubic inch
27.679795	kilograms per cubic decimeter
0.0276798	kilograms per cubic centimeter
0.0000276798	kilograms per cubic millimeter
61,023,360,960	pounds per cubic hektometer
61,023,361	pounds per cubic dekameter

POUND PER CUBIC INCH: (cont'd)

61,023.360960 . . . . pounds per cubic meter
1,728 . . . . pounds per cubic foot
1 . . . . pounds per cubic inch
61.0233610 . . . . pounds per cubic decimeter
0.0610234 . . . . pounds per cubic centimeter
0.0000610234 . . . . pounds per cubic millimeter
276,797,952,000 . . . . hektograms per cubic hektometer
276,797,952 . . . . hektograms per cubic dekameter
276,798 . . . . hektograms per cubic meter
7,838.0903748 . . . . hektograms per cubic foot
4.535934 . . . . hektograms per cubic inch
276.797952 . . . . hektograms per cubic decimeter
0.276798 . . . . hektograms per cubic centimeter
0.000276798 . . . . hektograms per cubic millimeter
$276,797,952 \times 10^4$ . . . . dekagrams per cubic hektometer
2,767,979,520 . . . . dekagrams per cubic dekameter
2,767,980 . . . . dekagrams per cubic meter
78,380.903748 . . . . dekagrams per cubic foot
45.359342 . . . . dekagrams per cubic inch
2,767.97952 . . . . dekagrams per cubic decimeter
2.767980 . . . . dekagrams per cubic centimeter
0.00276798 . . . . dekagrams per cubic millimeter
976,373,775,360 . . . . ounces per cubic hektometer
975,373,775 . . . . ounces per cubic dekameter
976,374 . . . . ounces per cubic meter
27,648 . . . . ounces per cubic foot
16 . . . . ounces per cubic inch
976.373775 . . . . ounces per cubic decimeter
0.976374 . . . . ounces per cubic centimeter
0.000976374 . . . . ounces per cubic millimeter
$276,797,952 \times 10^5$ . . . . grams per cubic hektometer
27,679,795,200 . . . . grams per cubic dekameter
27,679,795 . . . . grams per cubic meter
783,809 . . . . grams per cubic foot
453.593425 . . . . grams per cubic inch
27,679.7952 . . . . grams per cubic decimeter
27.679795 . . . . grams per cubic centimeter
0.0276798 . . . . grams per cubic millimeter
$276,797,952 \times 10^6$ . . . . decigrams per cubic hektometer
$276,797,952 \times 10^3$ . . . . decigrams per cubic dekameter
276,797,952 . . . . decigrams per cubic meter
7,838,090 . . . . decigrams per cubic foot
4,535.934249 . . . . decigrams per cubic inch
276,798 . . . . decigrams per cubic decimeter
276.797952 . . . . decigrams per cubic centimeter
0.276798 . . . . decigrams per cubic millimeter
$276,797,952 \times 10^7$ . . . . centigrams per cubic hektometer
$276,797,952 \times 10^4$ . . . . centigrams per cubic dekameter
2,767,979,520 . . . . centigrams per cubic meter
78,380,904 . . . . centigrams per cubic foot
45,359.342495 . . . . centigrams per cubic inch

POUND PER CUBIC INCH: (cont'd)

2,767,980	centigrams per cubic decimeter
2,767.979520	centigrams per cubic centimeter
2.767980	centigrams per cubic millimeter
276,797,952 x 10^8	milligrams per cubic hektometer
276,797,952 x 10^5	milligrams per cubic dekameter
276,797,952 x 10^2	milligrams per cubic meter
783,809,040	milligrams per cubic foot
453,593	milligrams per cubic inch
27,679.795	milligrams per cubic decimeter
27,679.7952	milligrams per cubic centimeter
27.679795	milligrams per cubic millimeter
27,144,411 x 10^9	dynes per cubic hektometer
27,144,411 x 10^6	dynes per cubic dekameter
27,144,411 x 10^3	dynes per cubic meter
768,648,960	dynes per cubic foot
444,820	dynes per cubic inch
271,444,110	dynes per cubic decimeter
271,444	dynes per cubic centimeter
271.444411	dynes per cubic millimeter

POUND OF WATER EVAPORATED FROM AND AT 212°F.: =

24,271,651	foot poundals
291,259,807	inch poundals
754,402	foot pounds
9,052,822	inch pounds
0.244343	ton calories
244.343393	kilogram calories
538.685776	pound calories
2,443.433927	hektogram calories
24,434.339270	dekagram calories
8,618.949444	ounce calories
244,343	gram calories (mean)
2,443,434	decigram calories
24,434,339	centigram calories
244,343,393	milligram calories
0.0118359	kilowatt days
0.284061	kilowatt hours
17.0437064	kilowatt minutes
1,022.620366	kilowatt seconds
11.835871	watt days
284.0607707	watt hours
17,043.706381	watt minutes
1,022,620	watt seconds
104.32	tons meters
104,320	kilogram meters
229,942	pound meters
1,043,200	hektogram meters
10,432,000	dekagram meters
3,679,073	ounce meters
104,320,000	gram meters

POUND OF WATER EVAPORATED FROM AND AT 212°F.: (cont'd)

1,043,200,000	decigram meters
10,432,000,000	centigram meters
104,320,000,000	milligram meters
1.0432	ton hektometers
1,043.2	kilogram hektometers
2,299.424641	pounds hektometers
10,432	hektogram hektometers
104,320	dekagram hektometers
36,790.753232	ounce hektometers
1,043,200	gram hektometers
10,432,000	decigram hektometers
104,320,000	centigram hektometers
1,043,200,000	milligram hektometers
10.432	ton dekameters
10,432	kilogram dekameters
22,994.24641	pound dekameters
104,320	hektogram dekameters
1,043,200	dekagram dekameters
367,907	ounce dekameters
10,432,000	gram dekameters
104,320,000	decigram dekameters
1,043,200,000	centigram dekameters
10,432,000,000	milligram dekameters
342.183304	ton feet
342,183	kilogram feet
754,271	pound feet
3,421,830	hektogram feet
34,218,300	dekagram feet
12,070,463	ounce feet
342,183,000	gram feet
3,421,830,000	decigram feet
34,218,300,000	centigram feet
342,183,000,000	milligram feet
4,106.298562	ton inches
4,106,299	kilogram inches
9,052,847	pound inches
41,062,986	hektogram inches
410,629,856	dekagram inches
144,845,540	ounce inches
4,106,298,562	gram inches
41,062,985,620	decigram inches
410,629,856,200	centigram inches
4,106,298,562 x 10^3	milligram inches
1,043.2	ton decimeters
1,043,200	kilogram decimeters
2,299,425	pound decimeters
10,432,000	hektogram decimeters
104,320,000	dekagram decimeters
36,790,700	ounce decimeters
1,043,200,000	gram decimeters
10,432,000,000	decigram decimeters

POUND OF WATER EVAPORATED FROM AND AT 212°F.: (cont'd)

104,320,000,000 .... centigram decimeters
$10,432 \times 10^8$ .... milligram decimeters
10,432 .... ton centimeters
10,432,000 .... kilogram centimeters
22,994250 .... pound centimeters
104,320,000 .... hektogram centimeters
1,043,200,000 .... dekagram centimeters
367,907,000 .... ounce centimeters
10,432,000,000 .... gram centimeters
104,320,000,000 .... decigram centimeters
$10,432 \times 10^8$ .... centigram centimeters
$10,432 \times 10^9$ .... milligram centimeters
104,320 .... ton millimeters
104,320,000 .... kilogram millimeters
229,942,500 .... pound millimeters
1,043,200,000 .... hektogram millimeters
10,432,000,000 .... dekagram millimeters
3,679,070,000 .... ounce millimeters
104,320,000,000 .... gram millimeters
$10,432 \times 10^8$ .... decigram millimeters
$10,432 \times 10^9$ .... centigram millimeters
$10,432 \times 10^{10}$ .... milligram millimeters
10.0938730 .... kiloliter-atmospheres
100.938730 .... hektoliter-atmospheres
1,009.387298 .... dekaliter-atmospheres
10,093.872982 .... liter-atmospheres
100,939 .... deciliter-atmospheres
1,009,387 .... centiliter-atmospheres
10,093,873 .... milliliter-atmospheres
0.0000000100939 .... cubic kilometer-atmospheres
0.0000100939 .... cubic hektometer-atmospheres
0.0100939 .... cubic dekameter-atmospheres
10.0938730 .... cubic meter-atmospheres
356.505500 .... cubic foot-atmospheres
615,0979 .... cubic inch-atmospheres
10,093.872982 .... cubic decimeter-atmospheres
10,093,873 .... cubic centimeter-atmospheres
10,093,872,982 .... cubic millimeter-atmospheres
1,022,792 .... joules (absolute)
0.386293 .... Cheval-vapeur hours
0.0158753 .... horsepower days
0.380993 .... horsepower hours
22.860503 .... horsepower minutes
1,371.626098 .... horsepower seconds
0.0664530 .... pounds of carbon oxidized with 100% efficiency
pounds of water evaporated from and at 212°F.
439,828 .... BTU (mean) per pound
969.718377 .... BTU (mean)

POUND OF WATER AT 60°F.: =

0.0000454248	kiloliters
0.0000454248	cubic meters
0.00059412	cubic yards
0.0028572	barrels
0.00454248	hektoliters
0.0128904	bushels (U.S.—dry)
0.0124968	bushels (Imperial—dry)
0.0160417	cubic feet
0.0454248	dekaliters
0.12	gallons (U.S.—liquid)
0.103138	gallons (U.S.—dry)
0.0999216	gallons (Imperial)
0.48	quarts (liquid)
0.412496	quarts (dry)
0.454248	liters
0.454248	cubic decimeters
0.96	pints
3.84	gills
4.54248	deciliters
27.72	cubic inches
45.4248	centiliters
454.248	milliliters
454.248	cubic centimeters
454,248	cubic millimeters
7,372.80	minims
15.36	ounces (fluid)
122.88	drams (fluid)
0.000446628	tons (long)
0.000453792	tons (metric)
0.000500220	tons (short)
0.453792	kilograms
1	pounds (Avoir.)
1.215812	pounds (Troy)
4.53792	hektograms
45.3792	dekagrams
14.589749	ounces (Troy)
16	ounces (Avoir.)
116.717990	drams (Troy)
256	drams (Avoir.)
35.0153971	scruples
453,792	grams
4,537.92	decigrams
7,000	grains
45,379.2	centigrams
453,792	milligrams
27.6798	inches of water

POUNDS PER MILLION GALLONS: =

0.000000132090	tons (net) per cubic meter
0.00011983	kilograms per cubic meter

POUNDS PER MILLION GALLONS: (cont'd)

0.000264180 . . . . pounds (Avoir.) per cubic meter
0.00422687 . . . . ounces (Avoir.) per cubic meter
0.11983 . . . . grams per cubic meter
1.849264 . . . . grains per cubic meter
0.0000000210002 . . . . tons (net) per barrel
0.0000190510 . . . . kilograms per barrel
0.0000420003 . . . . pounds (Avoir.) per barrel
0.000672005 . . . . ounces (Avoir.) per barrel
0.0190510 . . . . grams per barrel
0.294004 . . . . grains per barrel
0.00000000370029 . . . . tons (net) per cubic foot
0.00000339313 . . . . kilograms per cubic foot
0.00000748056 . . . . pounds (Avoir.) per cubic foot
0.00119689 . . . . ounces (Avoir.) per cubic foot
0.00339313 . . . . grams per cubic foot
0.0523642 . . . . grains per cubic foot
0.0000000005 . . . . tons (net) per gallon (U.S.—liquid)
0.000000453585 . . . . kilograms per gallon (U.S.—liquid)
0.000001 . . . . pounds (Avoir.) per gallon (U.S.—liquid)
0.000016 . . . . ounces (Avoir.) per gallon (U.S.—liquid)
0.000453585 . . . . grams per gallon (U.S.—liquid)
0.007 . . . . grains per gallon (U.S.—liquid)
0.000000000600477 . . . . tons (net) per gallon (Imperial—liquid)
0.000000544729 . . . . kilograms per gallon (Imperial—liquid)
0.00000120095 . . . . pounds (Avoir.) per gallon (Imp.—liquid)
0.0000192153 . . . . ounces (Avoir.) per gallon (Imp.—liquid)
0.000544729 . . . . grams per gallon (Imperial—liquid)
0.00840670 . . . . grains per gallon (Imperial—liquid)
0.000000000132090 . . . . tons (net) per liter
0.00000011983 . . . . kilograms per liter
0.000000264180 . . . . pounds (Avoir.) per liter
0.00000422687 . . . . ounces (Avoir.) per liter
0.00011983 . . . . grams per liter
0.00184926 . . . . grains per liter
0.0000000000216453 . . . . tons (net) per cubic inch
0.00000000196362 . . . . kilograms per cubic inch
0.00000000432903 . . . . pounds (Avoir.) per cubic inch
0.0000000692645 . . . . ounces (Avoir.) per cubic inch
0.00000196362 . . . . grams per cubic inch
0.0000303033 . . . . grains per cubic inch

POUND WEIGHT SECOND PER SQUARE FOOT: =

478.8 . . . . poises

POUND WEIGHT PER SECOND PER SQUARE INCH: =

68,950 . . . . poises

QUADRANT (ANGLE): =

324,000	seconds
5,400	minutes
90	degrees
1.57080	radians
0.25	circumference of revolution
0.7854	pi (π)

QUART (U.S.—DRY): =

0.000110089	kiloliters
0.000110089	cubic meters
0.00143986	cubic yards
0.00692448	barrels
0.0110089	hektoliters
0.0312402	bushels (U.S.—dry)
0.0302863	bushels (Imperial—dry)
0.0388775	cubic feet
0.110089	dekaliters
0.290823	gallons (U.S.—liquid)
0.249956	gallons (U.S.—dry)
0.242162	gallons (Imperial)
1.163290	quarts (liquid)
1	quarts (dry)
1.100889	liters
1.100889	cubic decimeters
2	pints (U.S.—dry)
2.326580	pints (U.S.—liquid)
7.749187	gills (Imperial)
9.306320	gills (U.S.)
11.00888839	deciliters
67.18	cubic inches
110.0888839	centiliters
1,100.888839	millimeters
1,100.888839	cubic centimeters
1,100,889	cubic millimeters (fluid)
17,868.135060	minims (fluid)
37.225281	ounces (fluid)
297.802251	drams (fluid)
0.00108241	tons (long)
0.00109977	tons (metric)
0.00121230	tons (short)
1.0997744	kilograms
2.424587	pounds (Avoir.)
2.946547	pounds (Troy)
10.997744	hektograms
109.977441	dekagrams
35.358561	ounces (Troy)
38.793396	ounces (Avoir.)
282.868492	drams (Troy)

QUART (U.S.—DRY): (cont'd)

620.694342	drams (Avoir.)
707.171230	pennyweights
848.605475	scruples
1,099.774407	grams
10,997.744067	decigrams
16,972.110905	grains
109,977	centigrams
1,099,774	milligrams

QUART (U.S.—LIQUID): =

0.0000946358	kiloliters
0.0000946358	cubic meters
0.00123775	cubic yards
0.0059525	barrels
0.00946358	hektoliters
0.026855	bushels (U.S.—dry)
0.026035	bushels (Imperial—dry)
0.0334203	cubic feet
0.0946358	dekaliter
0.25	gallons (U.S.—liquid)
0.21487	gallons (U.S.—dry)
0.20817	gallons (Imperial)
1	quarts (liquid)
0.859368	quarts (dry)
0.946358	liters
0.946358	cubic decimeters
1.718733	pints (U.S.—dry)
2	pints (U.S.—liquid)
6.66144	gills (Imperial)
8	gills (U.S.)
9,46358	deciliters
57.75	cubic inches
94.6358	centiliters
946.358	milliliters
946.358	cubic centimeters
946,358	cubic millimeters
15,360	minims
32	ounces (fluid)
256	drams (fluid)
0.000930475	tons (long) water @ 62°F.
0.0009454	tons (metric) water @ 62°F.
0.00104213	tons (short) water @ 62°F.
0.9454	kilgrams water @ 62°F.
2.08425	pounds (Avoir.) water @ 62°F.
2.532943	pounds (Troy) water @ 62°F.
9.455	hektograms water @ 62°F.
94.55	dekagrams water @ 62°F.
30.39531	ounces (Troy) water @ 62°F.
33.348	ounces (Avoir.) water @ 62°F.
243.16248	drams (Troy) water @ 62°F.

QUART (U.S.—LIQUID): (cont'd)

533.568	drams (Avoir.) water @ 62°F.
607.9062	pennyweights water @ 62°F.
729.48744	scruples water @ 62°F.
945.5	grams water @ 62°F.
9,455	decigrams water @ 62°F.
14,589.75	grains water @ 62°F.
94,550	centigrams water @ 62°F.
945,500	milligrams water @ 62°F.

QUIRE: =

25 sheets

RADIAN: =

206,265	seconds or inches
3,437.75	minutes
57.29578	degrees
0.637	quadrants
0.159155	circumference or revolutions
0.5	pi (π)
57° 17′ 44.8″	(In degrees, minutes, and seconds)

RADIANS PER SECOND: =

57.29578	degrees per second
3,437.7468	degrees per minute
206,265	degrees per hour
4,950.355	degrees per day
0.637	quadrants per second
38.22	quadrants per minute
2,293.2	quadrants per hour
55,036.8	quadrants per day
0.159155	revolutions per second
9.5493	revolutions per minute
572.958	revolutions per hour
13,750.992	revolutions per day

RADIAN PER SECOND PER SECOND: =

57.29578	degrees per second per second
3,437.7468	degrees per minute per second
206,265	degrees per minute per minute
0.637	quadrants per second per second
38.22	quadrants per minute per second
2.293.2	quandrants per minute per minute
0.159155	revolutions per second per second
9.5493	revolutions per minute per second
572.958	revolutions per minute per minute

REAM: =

500 . . . . sheets
20 . . . . quires

REVOLUTION: =

1,296,000 . . . . seconds or inches
21,600 . . . . minutes
360 . . . . degrees
6.2832 . . . . radians
4 . . . . quadrants
2 . . . . Pi (π)
1 . . . . circumference

REVOLUTIONS PER SECOND PER SECOND: =

360 . . . . degrees per second per second
21,600 . . . . degrees per minute per second
1,296,000 . . . . degrees per minute per minute
6.2832 . . . . radians per second per second
376.9920 . . . . radians per minute per second
22,619.52 . . . . radians per minute per minute
4 . . . . quadrants per second per second
240 . . . . quadrants per minute per second
14,400 . . . . quadrants per minute per minute
1 . . . . revolutions per second per second
60 . . . . revolutions per minute per second
3,600 . . . . revolutions per minute per minute

REVOLUTIONS PER MINUTE PER MINUTE: =

0.1 . . . . degrees per second per second
6 . . . . degrees per minute per second
360 . . . . degrees per minute per minute
0.0017453 . . . . radians per second per second
0.104718 . . . . radians per minute per second
6.2832 . . . . radians per minute per minute
0.00111111 . . . . quadrants per second per second
0.06666667 . . . . quadrants per second per second
4 . . . . quadrants per minute per minute
0.000277778 . . . . revolutions per second per second
0.0166667 . . . . revolutions per minute per second
1 . . . . revolutions per minute per minute

REVOLUTIONS PER SECOND: =

111,974,400,000 . . . . seconds or inches per day
4,665,600,000 . . . . seconds or inches per hour

REVOLUTIONS PER SECOND: (cont'd)

77,760,000 . . . . seconds or inches per minute
1,296,000 . . . . seconds or inches per second
1,866,240,000 . . . . minutes per day
77,760,000 . . . . minutes per hour
1,296,000 . . . . minutes per minute
21,600 . . . . minutes per second
31,104,000 . . . . degrees per day
1,296,000 . . . . degrees per hour
21,600 . . . . degrees per minute
360 . . . . degrees per second
542,868 . . . . radians per day
22,619.52 . . . . radians per hour
376.9920 . . . . radians per minute
6.2832 . . . . radians per second
345,600 . . . . quadrants per day
14,400 . . . . quadrants per hour
240 . . . . quadrants per minute
4 . . . . quadrants per second
172,800 . . . . pi (π) per day
7,200 . . . . pi (π) per hour
120 . . . . pi (π) per minute
2 . . . . pi (π) per second
86,400 . . . . revolutions or circumferences per day
3,600 . . . . revolutions or circumferences per hour
60 . . . . revolutions or circumferences per minute
1 . . . . revolutions or circumferences per second

REVOLUTIONS PER MINUTE: =

1,866,240,000 . . . . seconds or inches per day
77,760,000 . . . . seconds or inches per hour
1,296,000 . . . . seconds or inches per minute
21,600 . . . . seconds or inches per second
31,104,00 . . . . minutes per day
1,296,000 . . . . minutes per hour
21,600 . . . . minutes per minute
360 . . . . minutes per second
518,400 . . . . degrees per day
21,600 . . . . degrees per hour
360 . . . . degrees per minute
6 . . . . degrees per second
9,047.808 . . . . radians per day
376.992 . . . . radians per hour
6.2832 . . . . radians per minute
0.10472 . . . . radians per second
5,760 . . . . quadrants per day
240 . . . . quadrants per hour
4 . . . . quadrants per minute
0.0666667 . . . . quadrants per second
2,880 . . . . pi (π) per day

REVOLUTIONS PER MINUTE: (cont'd)

120	pi (π) per hour
2	pi (π) per minute
0.0333333	pi (π) per second
1,440	revolutions or circumferences per day
60	revolutions or circumferences per hour
1	revolutions or circumferences per minute
0.016667	revolutions or circumferences per second

REVOLUTION PER HOUR: =

31,104,000	seconds or inches per day
1,296,000	seconds or inches per hour
21,600	seconds or inches per minute
360	seconds or inches per second
518,400	minutes per day
21,600	minutes per hour
360	minutes per minute
6	minutes per second
8,640	degrees per day
360	degrees per hour
6	degrees per minute
0.1	degrees per second
150.796512	radians per day
6.2832	radians per hour
0.104720	radians per minute
0.00174533	radians per second
96	quadrants per day
4	quadrants per hour
0.0666667	quadrants per minute
0.00111111	quadrants per second
48	pi (π) per day
2	pi (π) per hour
0.0333333	pi (π) per minute
0.000555556	pi (π) per second
24	revolutions or circumferences per day
1	revolutions or circumferences per hour
0.0166667	revolutions or circumferences per minute
0.000277778	revolutions or circumferences per second

REVOLUTIONS PER DAY: =

1,296,000	seconds or inches per day
54,000	seconds or inches per hour
900	seconds or inches per minute
15	seconds or inches per second
21,600	minutes per day
900	minutes per hour
15	minutes per minute
0.25	minutes per second
360	degrees per day

REVOLUTIONS PER DAY: (cont'd)

15 . . . . degrees per hour
0.25 . . . . degrees per minute
0.00416667 . . . . degrees per second
6.2618 . . . . radians per day
0.2618 . . . . radians per hour
0.00436333 . . . . radians per minute
0.0000727222 . . . . radians per second
4 . . . . quadrants per day
0.166667 . . . . quadrants per hour
0.00277778 . . . . quadrants per minute
0.0000462963 . . . . quadrants per second
2 . . . . pi (π) per day
0.0833333 . . . . pi (π) per hour
0.00138889 . . . . pi (π) per minute
0.0000231481 . . . . pi (π) per second
1 . . . . revolutions or circumferences per day
0.0416667 . . . . revolutions or circumferences per hour
0.000694446 . . . . revolutions or circumferences per minute
0.0000115741 . . . . revolutions or circumferences per second

ROD: =

0.00271363 . . . . miles (nautical)
0.003125 . . . . miles (statute)
0.00502922 . . . . kilometers
0.025 . . . . furlongs
0.0502922 . . . . hektometers
0.25 . . . . chains
0.502922 . . . . dekameters
1 . . . . rods
5.0292188 . . . . meters
5.5 . . . . yards
5.94 . . . . varas (Texas)
16.5 . . . . feet
22 . . . . spans
25 . . . . links
49.5 . . . . hands
50.292188 . . . . decimeters
502.921875 . . . . centimeters
198 . . . . inches
6,029.21875 . . . . millimeters
198,000 . . . . mils
5,029,219 . . . . microns
50,029,218,750 . . . . millimicrons
5,029,218,750 . . . . micromillimeters
7,811,165 . . . . wave lengths of red line of cadmium
50,292,187,500 . . . . Angstrom Units

SACK CEMENT: =

0.19592 . . . . barrels
94 . . . . pounds (Avoir.)

SACK CEMENT: (cont'd)

8.22857	gallons (U.S.—liquid)
1.1	cubic feet (set)
1,900.8	cubic inches
3.15	specific gravity
0.484	cubic feet (absolute volume)

SEAWATER GRAVITY: =

1.02 to 1.03

SECOND, FOOT (WATER): =

2,446.594024	kiloliters per day
101.941418	kiloliters per hour
1.699024	kiloliters per minute
0.0283171	kiloliters per second
2,446.594024	cubic meters per day
101.941418	cubic meters per hour
1.699024	cubic meters per minute
0.0283171	cubic meters per second
3,200.144983	cubic yards per day
133.339374	cubic yards per hour
2.222323	cubic yards per minute
0.0370378	cubic yards per second
15,388.959079	barrels per day
641.206628	barrels per hour
10.686777	barrels per minute
0.178113	barrels per second
24,465.940237	hektoliters per day
1,019.414177	hektoliters per hour
16.990236	hektoliters per minute
0.283171	hektoliters per second
69,429.445206	bushels (U.S.—dry) per day
2,892.893550	bushels (U.S.—dry) per hour
48.214893	bushels (U.S.—dry) per minute
0.803582	bushels (U.S.—dry) per second
67,273.995871	bushels (Imperial—dry) per day
2,803.0831613	bushels (Imperial—dry) per hour
46.718053	bushels (Imperial—dry) per minute
0.778634	bushels (Imperial—dry) per second
86,400	cubic feet per day
3,600	cubic feet per hour
60	cubic feet per minute
1	cubic feet per second
244,659	dekaliters per day
10,194.141766	dekaliters per hour
169.902363	dekaliters per minute
2.831706	dekaliters per second
277,718	pecks (U.S.—dry) per day
11,571.574201	pecks (U.S.—dry) per hour

SECOND, FOOT (WATER): (cont'd)

192.859570	pecks (U.S.—dry) per minute
3.214326	pecks (U.S.—dry) per second
646,336	gallons (U.S.—liquid) per day
26,930.679834	gallons (U.S.—liquid) per hour
448.844664	gallons (U.S.—liquid) per minute
7,480744	gallons (U.S.—liquid) per second
555,426	gallons (U.S.—dry) per day
23,142.740636	gallons (U.S.—dry) per hour
385.712344	gallons (U.S.—dry) per minute
6.428539	gallons (U.S.—dry) per second
538,202	gallons (Imperial) per day
22,425.0730560	gallons (Imperial) per hour
373.751218	gallons (Imperial) per minute
6.229187	gallons (Imperial) per second
2,585,340	quarts (liquid) per day
107,723	quarts (liquid) per hour
1,795.375258	quarts (liquid) per minute
29.922921	quarts (liquid) per second
2,221,757	quarts (dry) per day
92,573.205256	quarts (dry) per hour
1,542.886754	quarts (dry) per minute
25.713779	quarts (dry) per second
2,446,594	liters per day
101,941	liters per hour
1,699.0236276	liters per minute
28.317060	liters per second
2,446,594	cubic decimeters per day
101,941	cubic decimeters per hour
1,699.0236276	cubic decimeters per minute
28.317060	cubic decimeters per second
4,443,513	pints (U.S.—dry) per day
185,146	pints (U.S.—dry) per hour
3,085.766712	pints (U.S.—dry) per minute
51.429445	pints (U.S.—dry) per second
5,170,639	pints(U.S.—liquid) per day
215,443	pints (U.S.—liquid) per hour
3,590.716535	pints (U.S.—liquid) per minute
59.845276	pints (U.S.—liquid) per second
17,222,068	gills (Imperial) per day
717,586	gills (Imperial) per hour
11,959.767119	gills (Imperial) per minute
199.329452	gills (Imperial) per second
20,683,007	gills (U.S.) per day
861,792	gills (U.S.) per hour
14,363.205943	gills (U.S.) per minute
239.386766	gills (U.S.) per second
24,465,940	deciliters per day
1,019,414	deciliters per hour
16,990.236276	deciliters per minute
283.170605	deciliters per second
149,303,400	cubic inches per day

SECOND, FOOT (WATER): (cont'd)

6,220,975	cubic inches per hour
103,683	cubic inches per minute
1,728.048146	cubic inches per second
244,659,402	centiliters per day
10,194,142	centiliters per hour
169,902	centiliters per minute
2,831.706046	centiliters per second
2,446,594,024	milliliters per day
101,941,418	milliliters per hour
1,699,023	milliliters per minute
28,317.06046	milliliters per second
2,446,594,024	cubic centimeters per day
101,941,418	cubic centimeters per hour
1,699,023	cubic centimeters per minute
28,317.06046	cubic centimeters per second
2,446,594,024 x 10^3	cubic millimeters per day
101,941,417,656	cubic millimeters per hour
1,699,023,627	cubic millimeters per minute
28,317,060	cubic millimeters per second
82,730,841	ounces (U.S.) fluid per day
3,447,121	ounces (U.S.) fluid per hour
57,451.974260	ounces (U.S.) fluid per minute
957.532904	ounces (U.S.) fluid per second
86,110,312	ounces (Imperial—fluid) per day
3,587,941	ounces (Imperial—fluid) per hour
59,798.835597	ounces (Imperial—fluid) per minute
996.647260	ounces (Imperial—fluid) per second
661,847,490	drams (fluid) per day
27,576,966	drams (fluid) per hour
459,616	drams (fluid) per minute
7,660,271730	drams (fluid) per second
39,710,848,658	minims per day
1,654,618,682	minims per hour
27,576,966	minims per minute
459,616	minims per second
1.98347	acre feet
1	acre, inch per hour

SECOND, (ANGLE): =

1	seconds
0.0166667	minutes
0.000277778	degrees
0.0000048414	radians
0.00000308651	quadrants
0.00000154321	pi (π)
0.000000771607	circumference or revolutions

SQUARE CENTIMETER: =

0.00000000003831	square miles or sections
0.0000000001	square kilometers

SQUARE CENTIMETER: (cont'd)

0.00000000247104	square furlongs
0.0000000247104	acres
0.00000001	square hektometers or hectares
0.000000247104	square chains
0.000001	square dekameters or acres
0.00000395367	square rods
0.0001	square meters or centares
0.00011960	square yards
0.000139498	square varas (Texas)
0.00107639	square feet
0.00247104	square links
0.01	square decimeters
1	square centimeters
0.1550	square inches
100	square millimeters
155,000	square mils
197,350	circular mils
127.32	circular millimeters

SQUARE CHAIN: =

0.00015625	square miles or sections
0.000404687	square kilometers
0.01	square furlongs
0.1	acres
0.0404687	square hektometers or hectares
1	square chains
4.046873	square dekameters or acres
16	square rods
404.6873	square meters or centares
484	square yards
564.530690	square varas (Texas)
4,356	square feet
10,000	square links
40,468.73	square decimeters
4,046,873	square centimeters
627,265	square inches
404,687,300	square millimeters
627,265,000,000	square mils
798,650,386,550	circular mils
515,247,870	circular millimeters

SQUARE DECIMETER: =

0.000000003831	square miles or sections
0.00000001	square kilometers
0.000000247104	square furlongs
0.00000247104	acres
0.000001	square hektometers or hectares
0.0000247104	square chains

SQUARE DECIMETER: (cont'd)

0.0001	square dekameters or acres
0.000395367	square rods
0.01	square meters or centares
0.011960	square yards
0.0139498	square varas (Texas)
0.107639	square feet
0.247104	square links
1	square decimeters
100	square centimeters
15.5	square inches
10,000	square millimeters
15,500,000	square mils
19,735,000	circular mils
12,732	circular millimeters

SQUARE DEKAMETER: =

0.00003831	square miles or sections
0.0001	square kilometers
0.00247104	square furlongs
0.0247104	acres
0.01	square hektometers or hectares
0.247104	square chains
1	square dekameters or acres
3.95367	square rods
100	square meters or centares
119.60	square yards
139.498	square varas (Texas)
1,076.39	square feet
2,471.04	square links
10,000	square decimeters
1,000,000	square centimeters
155,000	square inches
100,000,000	square millimeters
155,000,000,000	square mils
197,350,000,000	circular mils
127,320,000	circular millimeters

SQUARE FOOT: =

0.0000000355913	square miles or sections
0.0000000929034	square kilometers
0.00000229568	square furlongs
0.0000229568	acres
0.00000929030	square hektometers or hectares
0.000229568	square chains
0.000929034	square dekameters or acres
0.00367309	square rods
0.0929034	square meters or centares
0.111111	square yards

SQUARE FOOT: (cont'd)

0.129598	square varas (Texas)
1	square feet
2.29568	square links
9.290341	square decimeters
929.0341	square centimeters
144	square inches
92,903.41152	square millimeters
144,000,000	square mils
183,346,560	circular mils
118,285	circular millimeters

SQUARE FURLONG: =

0.015625	square miles or sections
0.040687	square kilometers
1	square furlongs
10	acres
4.04687	square hektometers or hectares
100	square chains
404.6873	square dekameters or acres
1,600	square rods
40,468.73	square meters or centares
48,400	square yards
56,453.069	square varas (Texas)
435,600	square feet
1,000,000	square links
4,046,873	square decimeters
404,687,300	square centimeters
62,726,500	square inches
40,468,730,000	square millimeters
627,265 x 10^8	square mils
79,865,038,655 x 10^3	circular mils
51,524,787,000	circular millimeters

SQUARE HEKTOMETER: =

0.003831	square miles or sections
0.01	square kilometers
0.247104	square square furlongs
2.471044	acres
1	square hektometers or hectares
24.71044	square chains
100	square dekameters or acres
395.367	square rods
10,000	square meters or centares
11,959.8	square yards
13,949.8	square varas (Texas)
107,639	square feet
247,104	square links
1,000,000	square decimeters

SQUARE HEKTOMETER: (cont'd)

100,000,000	square centimeters
15,500,000	square inches
1×10^{10}	square millimeters
155×10^{11}	square mils
$19,735 \times 10^8$	circular mils
$12,732 \times 10^6$	circular millimeters

SQUARE INCH: =

0.00000000024908	square miles or sections
0.000000000645163	square kilometers
0.0000000159422	furlongs
0.000000159422	acres
0.0000000645163	square hektometers or hectares
0.00000159423	square chains
0.00000645163	square dekameters or acres
0.0000255076	square rods
0.000645163	square meters or centares
0.000771605	square yards
0.000899986	square varas (Texas)
0.00694444	square feet
0.0159423	square links
0.0645163	square decimeters
6.451626	square centimeters
1	square inches
645.16258	square millimeters
1,000,000	square mils
1,273,240	circular mils
821.423611	circular millimeters

SQUARE KILOMETER: =

0.383101	square miles or sections
1	square kilometers
24.71044	square furlongs
247.1044	acres
100	square hektometers or hectares
2,471.044	square chains
10,000	square dekameters or acres
39,536.7	square rods
1,000,000	square meters or centares
1,195,980	square yards
1,394,980	square varas (Texas)
10,764,000	square feet
24,710,440	square links
100,000,000	square decimeters
1×10^{10}	square centimeters
1,550,000,000	square inches
1×10^{12}	square millimeters
155×10^{13}	square mils

SQUARE KILOMETER: (cont'd)

19,735 10^{10} .... circular mils
12,732 x 10^8 .... circular millimeters

SQUARE LINK: =

0.000000015625 .... square miles or sections
0.0000000404687 .... square kilometers
0.000001 .... square furlongs
0.00001 .... acres
0.00000404687 .... square hektometers or hectares
0.0001 .... square chains
0.000404687 .... square dekameters or acres
0.0016 .... square rods
0.040469 .... square meters or centares
0.0484 .... square yards
0.0564531 .... square varas (Texas)
0.4356 .... square feet
1 .... square links
4.046873 .... square decimeters
404.6873 .... square centimeters
62.7265 .... square inches
40,468.73 .... square millimeters
62,726,500 .... square mils
79,863,380 .... circular mils
51,524.787 .... circular millimeters

SQUARE METER: =

0.0000003831 .... square miles or sections
0.000001 .... square kilometers
0.0000247104 .... square furlongs
0.000247104 .... acres
0.0001 .... square hektometers or hectares
0.00247104 .... square chains
0.01 .... square dekameters or acres
0.0395367 .... square rods
1 .... square meters or centares
1.19598 .... square yards
1.39498 .... square varas (Texas)
10.7639 .... square feet
19.13580 .... square spans
24.71044 .... square links
96.8750 .... square hands
100 .... square decimeters
10,000 .... square centimeters
1,550 .... square inches
1,000,000 .... square millimeters
1,550,000,000 .... square mils
197,350,000 .... circular mils
1,273,200 .... circular millimeters

SQUARE MIL: =

0.00000000000000024908 . . . . square miles or sections
0.00000000000000064516 . . . . square kilometers
0.0000000000000159422 . . . . square furlongs
0.000000000000159422 . . . . acres
0.000000000000064516 . . . . square hektometers or hectares
0.00000000000159423 . . . . square chains
0.00000000000645163 . . . . square dekameters or acres
0.0000000000255076 . . . . square rods
0.000000000645163 . . . . square meters or centares
0.000000000771605 . . . . square yards
0.000000000899986 . . . . square varas (Texas)
0.00000000694444 . . . . square feet
0.0000000159423 . . . . square links
0.0000000645163 . . . . square decimeters
0.00000645163 . . . . square centimeters
0.000001 . . . . square inches
0.000645163 . . . . square millimeters
1 . . . . square mils
1.273224 . . . . circular mils
0.000821424 . . . . circular millimeters

SQUARE MILE OR SECTION: =

1 . . . . square miles or sections
2.589998 . . . . square kilometers
64 . . . . square furlongs
640 . . . . acres
258.9998 . . . . square hektometers or hectares
6,400 . . . . square chains
25,899.98 . . . . square dekameters or acres
102,400 . . . . square rods
2,589,998 . . . . square meters or centares
3,097,600 . . . . square yards
3,612,995 . . . . square varas (Texas)
27,878,400 . . . . square feet
64,000,000 . . . . square links
258,999,800 . . . . square decimeters
25,899,980,000 . . . . square centimeters
4,014,489,600 . . . . square inches
2,589,998,000,000 . . . . square millimeters
40,144,969 x 10^8 . . . . square mils
51,114 x 10^{10} . . . . circular mils
32,976 x 10^8 . . . . circular millimeters

SQUARE MILLIMETER: =

0.0000000000003831 . . . . square miles or sections
0.000000000001 . . . . square kilometers

SQUARE MILLIMETER: (cont'd)

0.0000000000247104	square furlongs
0.000000000247104	acres
0.0000000001	square hektometers or hectares
0.0000000024104	square chains
0.00000001	square dekameters or acres
0.0000000395367	square rods
0.000001	square meters or centares
0.0000011960	square yards
0.00000139498	square varas (Texas)
0.0000107639	square feet
0.0000247104	square links
0.0001	square decimeters
0.01	square centimeters
0.00155	square inches
1	square millimeters
1,550	square mils
1,973.5	circular mils
1.2732	circular millimeters

SQUARE ROD: =

0.00000976563	square miles or sections
0.0000252929	square kilometers
0.000625	square furlongs
0.00625	acres
0.00252929	square hektometers or hectares
0.0625	square chains
0.252929	square dekameters or acres
1	square rods
25.2929	square meters or centares
30.25	square yards
35.283168	square varas (Texas)
272.25	square feet
625	square links
2,529.29	square decimeters
252,929	square centimeters
39,204	square inches
25,292,900	square millimeters
39,202,600,000	square mils
49,913,691,182	circular mils
32,202,992	circular millimeters

SQUARE VARA (TEXAS): =

0.000000274622	square miles or sections
0.000000716843	square kilometers
0.0000177135	square furlongs
0.000177135	acres
0.0000716843	square hektometers or hectares
0.00177135	square chains

UNIVERSAL CONVERSION FACTORS
SQUARE VARA (TEXAS): (cont'd)

0.00716843	square dekameters or acres
0.0283416	square rods
0.716843	square meters or centares
0.857332	square yards
1	square varas (Texas)
7.716	square feet
17.713467	square links
71.684263	square decimeters
7,168.426344	square centimeters
1,111.104	square inches
716,843	square millimeters
1,111,104,000	square mils
1,414,702,057	circular mils
912,687	circular millimeters

SQUARE YARD: =

0.000000322831	square miles or sections
0.000000836131	square kilometers
0.0000206612	square furlongs
0.000206612	acres
0.0000836131	square hektometers or hectares
0.00206612	square chains
0.00836131	square dekameters or acres
0.0330579	square rods
0.836131	square meters or centares
1	square yards
1.166382	square varas (Texas)
9	square feet
20.66112	square links
83.61306	square decimeters
8,361.306	square centimeters
1,296	square inches
836,131	square millimeters
1,296,000,000	square mils
1,650,119,040	circular mils
1.064,565	circular millimeters

TEMPERATURE, ABSOLUTE IN CENTIGRADE OR KELVIN: =

temperature in $C° + 273.18°$

TEMPERATURE, ABSOLUTE IN FAHRENHEIT OR RANKIN: =

temperature in $F° + 459.59°$

TEMPERATURE, DEGREES CENTIGRADE: =

5/9 (Temp. F° - 32°)
5/4 (Temp. Reaumur)

TEMPERATURE, DEGREES FAHRENHEIT: =

9/5 (Temp. C° + 32°)
9/4 (Temp. Reaumur + 32°)

TEMPERATURE, DEGREES REAUMUR: =

4/9 (Temp. F° - 32°)
4/5 (Temp. C°)

Degree Centigrade: =

0.8 or 4/5 degree Reaumur
1.00 degrees absolute, Kelvin
1.8 or 9/5 degrees Fahrenheit

Degree Fahrenheit: =

0.44444 or 4/9 degree Reaumur
0.55556 or 5/9 degree Centigrade

Degree Reaumur: =

1.25 or 5/4 degrees Centigrade
2.25 or 9/4 degrees Fahrenheit

TONS (LONG): =

1	tons (long)
1.0160470	tons (metric)
1.12	tons (net)
1,016.0470	kilograms
2,722.22	pounds (Troy)
2,240	pounds (Avoir.)
10,160.470	hektograms
101,605	dekagrams
32,667	ounces (Troy)
35,840	ounces (Avoir.)
1,016,047	grams
10,160,470	decigrams
101,604,700	centigrams
1,016,047,000	milligrams
15,680,000	grains
261,333	drams (Troy)
573,440	drams (Avoir.)
653,333	pennyweights
784,022	scruples
5,080,430	carats (metric)
6.19755	barrels of water @ 60°F.

TONS (LONG): (cont'd)

7.33627	barrels of oil @ 36° API
28.607	cubic feet
260.02971	gallons (U.S.—liquid)

TONS (METRIC): =

0.984206	tons (long)
1	tons (metric)
1.10231	tons (net)
1,000	kilograms
2,679.23	pounds (Troy)
2,204.622341	pounds (Avoir.)
10,000	hektograms
100,000	dekagrams
32,150.76	ounces (Troy)
35,273.96	ounces (Avoir.)
1,000,000	grams
10,000,000	decigrams
100,000,000	centigrams
1,000,000,000	milligrams
15,432,365	grains
257,206	drams (Troy)
564,384	drams (Avoir.)
643,015	pennyweights
771,618	scruples
5,000,086	carats (metric)
6.297	barrels of water @ 60°F.
7.454	barrels of oil @ 36° API
29.0662	cubic feet
264.474	gallons (U.S.—liquid)

TONS (NET): =

0.892858	tons (long)
0.907185	tons (metric)
1	tons (net)
907.184872	kilograms
2,430.56	pounds (Troy)
2,000	pounds (Avoir.)
9,071.84872	hektograms
90,718.4872	dekagrams
29,166.66	ounces (Troy)
32,000	ounces (Avoir.)
907,185	grams
9,701,849	decigrams
90,718,487	centigrams
907,184,872	milligrams
14,000,000	grains
233,333	drams (Troy)
512,000	drams (Avoir.)

TONS (NET): (cont'd)

583,333	pennyweights
700,020	scruples
4,536,000	carats (metric)
5.71255	barrels of water @ 60°F.
6.76216	barrels of oil @ 36° API
32.04	cubic feet
239.9271	gallons (U.S.—liquid)

TON OF REFRIGERATION: =

7,208,640,000	foot poundals per day
300,360,000	foot poundals per hour
5,006,000	foot poundals per minute
83,433.334	foot poundals per second
224,056,200	foot pounds per day
9,335,672	foot pounds per hour
15,594.53	foot pounds per minute
2,593.242	foot pounds per second
72,570.8	kilogram calories per day
3,023,784	kilogram calories per hour
50.396	kilogram calories per minute
0.83994	kilogram calories per second
2,559,836	ounce calories per day
107,060	ounce calories per hour
1,777.664	ounce calories per minute
29.62780	ounce calories per second
30,977,400	kilogram meters per day
1,290,722	kilogram meters per hour
21,512	kilogram meters per minute
358,534	kilogram meters per second
2,996,006	liter-atmospheres per day
124,834	liter-atmospheres per hour
2,080.554	liter-atmospheres per minute
34.676	liter-atmospheres per second
1,058,693,800	cubic foot atmospheres per day
44,112,200	cubic foot atmospheres per hour
735,200	cubic foot atmospheres per minute
12,253.4	cubic foot atmospheres per second
4.655	Cheval-vapeur hours
4.715	horsepowers
3.514	kilowatts
3.514	watts
303,609,600	joules per day
12,650,400	joules per hour
210,840	joules per minute
3,514	joules per second
19.7286	pounds of carbon oxidized with 100% efficiency per day
0.82202	pounds of carbon oxidized with 100% efficiency per hour

TONS OF REFRIGERATION: (cont'd)

0.0137	pounds of carbon oxidized with 100% efficiency per minute
0.00022834	pounds of carbon oxidized with 100% efficiency per second
296.64	pounds of water evaporated from and at 212°F. per day
12.36	pounds of water evaporated from and at 212°F. per hour
0.206	pounds of water evaporated from and at 212°F. per minute
0.0034334	pounds of water evaporated from and at 212°F. per second
288,000	BTU per day
12,000	BTU per hour
200	BTU per minute
3.3334	BTU per second

TONS (NET) OF WATER PER DAY: =

0.0907226	kiloliters per day
0.00378023	kiloliters per hour
0.0000630030	kiloliters per minute
0.00000105004	kiloliters per second
0.0907226	cubic meters per day
0.00378023	cubic meters per hour
0.0000630030	cubic meters per minute
0.00000105004	cubic meters per second
1.186579	cubic yards per day
0.0494404	cubic yards per hour
0.000824014	cubic yards per minute
0.0000137335	cubic yards per second
5.706411	barrels per day
0.237766	barrels per hour
0.00396285	barrels per minute
0.0000660467	barrels per second
0.907226	hektoliters per day
0.0378023	hektoliters per hour
0.00630030	hektoliters per minute
0.0000105004	hektoliters per second
32.0385859	cubic feet per day
1.334931	cubic feet per hour
0.0222490	cubic feet per minute
0.000370809	cubic feet per second
9.0722588	dekaliters per day
0.378023	dekaliters per hour
0.00630030	dekaliters per minute
0.000105004	dekaliters per second
239.664469	gallons (U.S.) per day
9.986099	gallons (U.S.) per hour
0.166433	gallons (U.S.) per minute

TONS (NET) OF WATER PER DAY: (cont'd)

0.00277388	gallons (U.S.) per second
199.563810	gallons (Imperial) per day
8.315159	gallons (Imperial) per hour
0.138586	gallons (Imperial) per minute
0.00230977	gallons (Imperial) per second
90.722588	liters per day
3.780228	liters per hour
0.0630030	liters per minute
0.00105004	liters per second
90.722588	cubic decimeters per day
3.780228	cubic decimeters per hour
0.0630030	cubic decimeters per minute
0.00105004	cubic decimeters per second
958.657876	quarts per day
39.944877	quarts per hour
0.665740	quarts per minute
0.011095	quarts per second
1,917.315752	pints per day
79.888076	pints per hour
1.331470	pints per minute
0.0221913	pints per second
7,669.263008	gills per day
319.551826	gills per hour
5.325824	gills per minute
0.0887645	gills per second
907.225881	deciliters per day
37.802277	deciliters per hour
0.630030	deciliters per minute
0.0105004	deciliters per second
55,362.492339	cubic inches per day
2,306.770514	cubic inches per hour
38.446974	cubic inches per minute
0.640767	cubic inches per second
9,072.258810	centiliters per day
378.0227670	centiliters per hour
6.300300	centiliters per minute
0.105004	centiliters per second
90,722.588095	milliliters per day
3,780.227670	milliliters per hour
63.00299561	milliliters per minute
1.0500419	milliliters per second
90,722.588095	cubic centimeters per day
3,780.227670	cubic centimeters per hour
63.00299561	cubic centimeters per minute
1.0500419	cubic centimeters per second
90,722,588	cubic millimeters per day
3,780,227	cubic millimeters per hour
63,002.995611	cubic millimeters per minute
1,050.0419380	cubic millimeters per second
2,000	pounds (Avoir.) per day
83.333333	pounds (Avoir.) per hour

TONS (NET) OF WATER PER DAY: (cont'd)

1.388888	pounds (Avoir.) per minute
0.0231481	pounds (Avoir.) per second
25.744757	bushels (U.S.—dry) per day
1.0726902	bushels (U.S.—dry) per hour
0.0178783	bushels (U.S.—dry) per minute
0.000297975	bushels (U.S.—dry) per second
24.958658	bushels (Imperial—dry) per day
1.0399521	bushels (Imperial—dry) per hour
0.0173323	bushels (Imperial—dry) per minute
0.000288868	bushels (Imperial—dry) per second

TONS (NET) OF WATER PER HOUR: =

2.177342	kiloliters per day
0.0907226	kiloliters per hour
0.00151204	kiloliters per minute
0.0000252007	kiloliters per second
2.177342	cubic meters per day
0.0907226	cubic meters per hour
0.00151204	cubic meters per minute
0.0000252007	cubic meters per second
28.476932	cubic yards per day
1.186579	cubic yards per hour
0.0197763	cubic yards per minute
0.000329611	cubic yards per second
136.953864	barrels per day
5.706411	barrels per hour
0.0951061	barrels per minute
0.00158512	barrels per second
21.773421	hektoliters per day
0.907226	hektoliters per hour
0.0151204	hektoliters per minute
0.000252007	hektoliters per second
768.925102	cubic feet per day
32.0385859	cubic feet per hour
0.533972	cubic feet per minute
0.00889970	cubic feet per second
217.734211	dekaliters per day
9.0722588	dekaliters per hour
0.151204	dekaliters per minute
0.00252007	dekaliters per second
5,751.947256	gallons (U.S.) per day
239.664469	gallons (U.S.) per hour
3.994488	gallons (U.S.) per minute
0.0665740	gallons (U.S.) per second
4,789.531441	gallons (Imperial) per day
199.56381	gallons (Imperial) per hour
3.326054	gallons (Imperial) per minute
0.0554344	gallons (Imperial) per second
2,177.342114	liters per day
90.722588	liters per hour

TONS (NET) OF WATER PER HOUR: (cont'd)

1.512043	liters per minute
0.0252007	liters per second
2,177.342114	cubic decimeters per day
907.225881	cubic decimeters per hour
1.512043	cubic decimeters per minute
0.0252007	cubic decimeters per second
23,007.789024	quarts (liquid) per day
958.657876	quarts (liquid) per hour
15.977711	quarts (liquid) per minute
0.266291	quarts (liquid) per second
46,015.578048	pints (liquid) per day
1,917.315752	pints (liquid) per hour
31.954464	pints (liquid) per minute
0.532282	pints (liquid) per second
184,062	gills per day
7669.263008	gills per hour
127.820251	gills per minute
2.130353	gills per second
21,773.421143	deciliters per day
9,072.258810	deciliters per hour
15.120431	deciliters per minute
0.252007	deciliters per second
1,328,700	cubic inches per day
55,362.492339	cubic inches per hour
922.708206	cubic inches per minute
15.378550	cubic inches per second
217,734	centiliters per day
90,722.588095	centiliters per hour
151.204313	centiliters per minute
2.520072	centiliters per second
2,177,342	milliliters per day
907,226	milliliters per hour
1,512.0431349	milliliters per minute
25.200719	milliliters per second
2,177,343	cubic centimeters per day
907,226	cubic centimeters per hour
1,512.0431349	cubic centimeters per minute
25.200719	cubic centimeters per second
2,177,342,114	cubic millimeters per day
907,225,881	cubic millimeters per hour
15,120.431349	cubic millimeters per minute
25,200.718915	cubic millimeters per second
48,000	pounds (Avoir.) per day
2,000	pounds (Avoir.) per hour
33.333333	pounds (Avoir.) per minute
0.555556	pounds (Avoir.) per second
617.874174	bushels (U.S.—dry) per day
25.744757	bushels (U.S.—dry) per hour
0.429071	bushels (U.S.—dry) per minute
0.00715135	bushels (U.S.—dry) per second
599.00778724	bushels (Imperial) per day

TONS (NET) OF WATER PER HOUR: (cont'd)

24.958658 . . . . bushels (Imperial) per hour
0.415986 . . . . bushels (Imperial) per minute
0.00693301 . . . . bushels (Imperial) per second

TONS (NET) OF WATER MINUTE: =

1,305.415814 . . . . kiloliters per day
54.434991 . . . . kiloliters per hour
0.907226 . . . . kiloliters per minute
0.0151207 . . . . kiloliters per second
1,306.415814 . . . . cubic meters per day
54.434991 . . . . cubic meters per hour
0.907226 . . . . cubic meters per minute
0.0151207 . . . . cubic meters per second
1,708.673452 . . . . cubic yards per day
71.194727 . . . . cubic yards per hour
1.186579 . . . . cubic yards per minute
0.0197764 . . . . cubic yards per second
8,217.231850 . . . . barrels per day
342.384660 . . . . barrels per hour
57.0641101 . . . . barrels per minute
0.0951061 . . . . barrels per second
13,064.158138 . . . . hektoliters per day
544.349908 . . . . hektoliters per hour
9.0722588 . . . . hektoliters per minute
0.151207 . . . . hektoliters per second
46,135.563668 . . . . cubic feet per day
1,922.315153 . . . . cubic feet per hour
32.0385859 . . . . cubic feet per minute
0.533972 . . . . cubic feet per second
130,642 . . . . dekaliters per day
5,443.499084 . . . . dekaliters per hour
90.722588 . . . . dekaliters per minute
1.512067 . . . . dekaliters per second
345,117 . . . . gallons (U.S.) per day
14,379.86814 . . . . gallons (U.S.) per hour
239.664469 . . . . gallons (U.S.) per minute
3.994488 . . . . gallons (U.S.) per second
287,372 . . . . gallons (Imperial) per day
11,973.828603 . . . . gallons (Imperial) per hour
199.563810 . . . . gallons (Imperial) per minute
3.326064 . . . . gallons (Imperial) per second
1,306,416 . . . . liters per day
54,434.990844 . . . . liters per hour
907.225881 . . . . liters per minute
15.120671 . . . . liters per second
1,306,416 . . . . cubic decimeters per day
54,434.990844 . . . . cubic decimeters per hour
907.225881 . . . . cubic decimeters per minute
151.206710 . . . . cubic decimeters per second
1,380,467 . . . . quarts per day

TONS (NET) OF WATER MINUTE: (cont'd)

57,519.472560 .... quarts per hour
958.657876 .... quarts per minute
15.977711 .... quarts per second
2,760,935 .... pints per day
115.039 .... pints per hour
1,917.315752 .... pints per minute
31.954464 .... pints per second
11,043,739 .... gills per day
460,156 .... gills per hour
7,669.263008 .... gills per minute
127.820970 .... gills per second
13,064,158 .... deciliters per day
544,350 .... deciliters per hour
9,072.258810 .... deciliters per minute
1,512.067104 .... deciliters per second
79,721,989 .... cubic inches per day
3,321,750 .... cubic inches per hour
55,362.492339 .... cubic inches per minute
922.708206 .... cubic inches per second
130,641,581 .... centiliters per day
5,443,499 .... centiliters per hour
90,722.588095 .... centiliters per minute
15,120.671014 .... centiliters per second
1,306,415,814 .... milliliters per day
54,434,991 .... milliliters per hour
907,226 .... milliliters per minute
151,207 .... milliliters per second
1,306,415,814 .... cubic centimeters per day
54,434,991 .... cubic centimeters per hour
907,226 .... cubic centimeters per minute
151,207 .... cubic centimeters per second
1,306,415,814 x 10³ .... cubic millimeters per day
54,434,991,000 .... cubic millimeters per hour
907,226,000 .... cubic millimeters per minute
151,207,000 .... cubic millimeters per second
2,880,000 .... pounds (Avoir.) per day
120,000 .... pounds (Avoir.) per hour
2,000 .... pounds (Avoir.) per minute
33.333333 .... pounds (Avoir.) per second
37,072.450454 .... bushels (U.S.) per day
1,544.685436 .... bushels (U.S.) per hour
25.744757 .... bushels (U.S.) per minute
0.429071 .... bushels (U.S.) per second
35,940.467234 .... bushels (Imperial) per day
1,497.519468 .... bushels (Imperial) per hour
24.958658 .... bushels (Imperial) per minute
0.415986 .... bushels (Imperial) per second

TONS (NET) OF WATER PER SECOND: =

783,849 .... kiloliters per day
3,266.0395345 .... kiloliters per hour

TONS (NET) OF WATER PER SECOND: (cont'd)

54.434991	kiloliters per minute
0.907226	kiloliters per second
783,849	cubic meters per day
3,266.0395345	cubic meters per hour
54.434991	cubic meters per minute
0.907226	cubic meters per second
102,520	cubic yards per day
4,271.683630	cubic yards per hour
71.194727	cubic yards per minute
1.186579	cubic yards per second
493,034	barrels per day
20,543.0796248	barrels per hour
342.384660	barrels per minute
5.706411	barrels per second
783,849	hektoliters per day
32,660.390552	hektoliters per hour
544,340322	hektoliters per minute
9.0722588	hektoliters per second
2,768,134	cubic feet per day
115,339	cubic feet per hour
1,922.315153	cubic feet per minute
32.0385859	cubic feet per second
7,838,494	dekaliters per day
326,604	dekaliters per hour
5,443.398425	dekaliters per minute
90.722588	dekaliters per second
20,707,010	gallons (U.S.) per day
862,792	gallons (U.S.) per hour
14.379868	gallons (U.S.) per minute
239.664469	gallons (U.S.) per second
17,242,313	gallons (Imperial) per day
718,430	gallons (Imperial) per hour
11,973.828603	gallons (Imperial) per minute
199.563810	gallons (Imperial) per second
78,384,937	liters per day
3,266,039	liters per hour
54,433.984253	liters per minute
907.225881	liters per second
78,384,937	cubic decimeters per day
3,266,039	cubic decimeters per hour
54,433.984253	cubic decimeters per minute
907.225881	cubic decimeters per second
82,828,040	quarts per day
3,451,168	quarts per hour
57,519.47256	quarts per minute
958.657876	quarts per second
165,656,081	pints per day
6,902,337	pints per hour
115,039	pints per minute
1,917.315752	pints per second
662,624,324	gills per day

TONS (NET) OF WATER PER SECOND: (cont'd)

27,609.347	gills per hour
460,156	gills per minute
7,669.263008	gills per second
783,849,373	deciliters per day
32,660,391	deciliters per hour
544,340	deciliters per minute
9,072.258810	deciliters per second
4,783,319,338	cubic inches per day
199,304,972	cubic inches per hour
3,321,750	cubic inches per minute
55,362.492339	cubic inches per second
7,838,493,732	centiliters per day
326,603,906	centiliters per hour
5,443,398	centiliters per minute
90,722.588095	centiliters per second
78,384,937,325	milliliters per day
3,266,039,055	milliliters per hour
54,433,984	milliliters per minute
907,226	milliliters per second
78,384,937,325	cubic centimeters per day
3,266,039,055	cubic centimeters per hour
54,433,984	cubic centimeters per minute
907,226	cubic centimeters per second
78,384,937,325 x 10³	cubic millimeters per day
3,266,039,055 x 10³	cubic millimeters per hour
54,433,984,000	cubic millimeters per minute
907,226,000	cubic millimeters per second
172,800,000	pounds (Avoir.) per day
7,200,000	pounds (Avoir.) per hour
120,000	pounds (Avoir.) per minute
2,000	pounds (Avoir.) per second
2,224,347	bushels (U.S.—dry) per day
92,681.126136	bushels (U.S.—dry) per hour
1,544.685436	bushels (U.S.—dry) per minute
25.744757	bushels (U.S.—dry) per second
2,156,428	bushels (Imperial) per day
89,851.168086	bushels (Imperial) per hour
150.605153	bushels (Imperial) per minute
24.958658	bushels (Imperial) per second
1,544.685436	bushels (U.S.—dry) per second
25.744757	bushels (U.S.—dry) per second
2,156,428	bushels (Imperial) per day
89,851.168086	bushels (Imperial) per hour
150.605153	bushels (Imperial) per minute
24.958658	bushels (Imperial) per second

VARA (TEXAS): =

0.000456840	miles (nautical)
0.000526094	miles (statute)
0.000846670	kilometers

VARA (TEXAS): (cont'd)

0.00420875	furlongs
0.00846670	hektometers
0.0420875	chains
0.0846670	dekameters
0.168350	rods
0.846670	meters
0.925926	yards
1	varas (Texas)
2.777778	feet
3.703704	spans
4.208754	links
8.333333	hands
8.466700	decimeters
84.667003	centimeters
33.333333	inches
846.670032	millimeters
33,333.333333	mils
846,670	microns
846,670,032	millimicrons
846,670,032	micromillimeters
1,315,011	wavelengths of red line of cadmium
8,466,700,319	Angstrom units

WATT: =

63,725.184	foot pounds per day
2,655.22	foot pounds per hour
44.2536	foot pounds per minute
0.73756	foot pounds per second
764.702	inch pounds per day
31,862.5920	inch pounds per hour
531.04320	inch pounds per minute
8.85072	inch pounds per second
20.640096	kilogram calories (mean) per day
0.860004	kilogram calories (mean) per hour
0.0143334	kilogram calories (mean) per minute
0.00023889	kilogram calories (mean) per second
45.503616	pound calories (mean) per day
1.895984	pound calories (mean) per hour
0.0315997	pound calories (mean) per minute
0.000526662	pound calories (mean) per second
728.0578547	ounce calories (mean) per day
30.335744	ounce calories (mean) per hour
0.505596	ounce calories (mean) per minute
0.00842659	ounce calories (mean) per second
20,640.096	gram calories (mean) per day
860.004	gram calories (mean) per hour
14.3334	gram calories (mean) per minute
0.23889	gram calories (mean) per second
81.930528	BTU (mean) per day
3.413772	BTU (mean) per hour

WATT: (cont'd)

0.056896	BTU (mean) per minute
0.00094827	BTU (mean) per second
0.001	kilowatts
1	watts
3,600	joules per hour
0.001341	horsepowers
0.0013597	horsepowers (metric)
0.0013597	Cheval-vapeur hours
0.000234	pounds carbon oxidized with 100% efficiency
0.00352	pounds water evaporated from and at 212°F.
0.852647	kiloliter-atmospheres per day
0.0355270	kiloliter-atmospheres per hour
0.000592116	kiloliter-atmospheres per minute
0.0000098686	kiloliter-atmospheres per second
852.647	liter-atmospheres per day
35.52695	liter-atmospheres per hour
0.592116	liter-atmospheres per minute
0.0098686	liter-atmospheres per second
8,808	kilogram meters per day
367.1	kilogram meters per hour
6.116667	kilogram meters per minute
0.101944	kilogram meters per second

WATT HOUR: =

2,655.22	foot pounds
31,982,64	inch pounds
0.860004	kilogram calories (mean)
1.895984	pound calories (mean)
30.385744	ounce calories (mean)
860.004	gram calories (mean)
3.413772	BTU (mean)
0.001	kilowatt hours
3,600	joules
0.001341	horsepower hours
0.0013597	horsepower hours (metric)
0.0355270	kiloliter-atmospheres
35.53695	liter-atmospheres
367.1	kilogram meters

WATT PER SQUARE INCH: =

8.1913	BTU per square foot per minute
6,372.6	foot pounds per square foot per minute
0.19310	horsepowers per square foot

YARD: =

0.000483387	miles (nautical)
0.000568182	miles (statute)
0.000914404	kilometers
0.00454545	furlongs
0.00914404	hektometers
0.0454545	chains
0.0914404	dekameters
0.181818	rods
0.914404	meters
1	yards
1.08	varas (Texas)
3	feet
4	spans
4.54545	links
9	hands
9.144036	centimeters
36	inches
914.40360	millimeters
36,000	mils
914,404	microns
914,403,600	millimicrons
914,403,600	micromillimeters
1,420,212	wavelengths of red line of cadmium
9,144.036345	Angstrom Units

PERSONAL CONVERSION FACTORS